# WELDING IN ENERGY-RELATED PROJECTS

PERGAMON PRESS

## Books of Related Interest

WELDING INSTITUTE OF CANADA: Pipeline & Energy Plant Piping: Design & Technology

MASUBUCHI: Analysis of Welded Structures

INTERNATIONAL INSTITUTE OF WELDING: The Physics of Welding

INTERNATIONAL INSTITUTE OF WELDING: Underwater Welding

## Journals of Related Interest

CANADIAN METALLURGICAL QUARTERLY

WELDING IN THE WORLD

## Electronic Databases

CHEMICAL ENGINEERING ABSTRACTS

ZINC, LEAD & CADMIUM ABSTRACTS

COMPENDEX

# WELDING
# IN
# ENERGY-RELATED PROJECTS

Pergamon Press
Toronto • Oxford • New York • Sydney • Paris • Frankfurt

| | |
|---|---|
| U.K. | Pergamon Press Ltd., Headington Hill Hall, Oxford OX3 0BW, England |
| U.S.A. | Pergamon Press Inc., Maxwell House, Fairview Park, Elmsford, New York 10523, U.S.A. |
| CANADA | Pergamon Press Canada Ltd., Suite 104 150 Consumers Rd., Willowdale, Ontario M2J 1P9, Canada |
| AUSTRALIA | Pergamon Press (Aust.) Pty. Ltd., P.O. Box 544, Potts Point, N.S.W. 2011, Australia |
| FRANCE | Pergamon Press SARL, 24 rue des Ecoles, 75240 Paris, Cedex 05, France |
| FEDERAL REPUBLIC OF GERMANY | Pergamon Press GmbH, 6242 Kronberg-Taunus, Hammerweg 6, Federal Republic of Germany |



**Canadian Cataloguing in Publication Data**

Main entry under title:
Welding in energy-related projects

Proceedings of an international conference, held in Toronto, Sept. 20-21, 1983.
Includes indexes.
ISBN 0-08-025412-8

1. Energy facilities - Welding - Congresses. I. Welding Institute of Canada.

TS227.2.W45 1984 671.5'2 C84-098681-5

*In order to make this volume available as economically and as rapidly as possible the authors' typescripts have been reproduced in their original forms. This method unfortunately has its typographical limitations but it is hoped that they in no way distract the reader.*

# CONTENTS

**CONTROL SYSTEMS**



**HIGH-ALLOY SYSTEMS**



**WELDING PROCEDURE OPTIMIZATION**



**QUALITY ASSURANCE AND INSPECTION**

# PREFACE

The Welding Institute of Canada is pleased to present these Proceedings of our Second International Conference which was held in Toronto, September 20-21, 1983, on the theme "Welding in Energy-Related Projects".

This conference follows the event hosted in Calgary in 1980 which addressed the topic of pipeline and piping technology in energy systems.

In the present conference we enlarged the topic presentations to cover the broader areas of design, materials and fabrication of total energy projects. Consequently, a much broader area of technology related to the construction of energy-related projects has been presented.

The contributions to the conference offer a unique overview of many areas of technology from research and development studies to construction and operation, and as such provide a comprehensive reference source.

This published volume of proceedings presents the contributed papers in full document form for the first time, with only the Abstracts being previously made available. In addition to the papers, since it is a post-conference document, we have included a summary of the discussions held at the technical sessions.

The Welding Institute of Canada is proud to make the documents available and trusts that the proceedings will be a valuable source of information on the technology of energy-related fabrication.

Dr. Norman F. Eaton
President
Welding Institute of Canada

February 28, 1984

# PROPERTIES OF WELDS IN THICK SECTION Nb-CONTAINING STEELS

R.B. Lazor*, R.D. McDonald**, and A.G. Glover*

**Welding Institute of Canada*

***CANMET, EMR*

## ABSTRACT

Three C/Mn steels microalloyed with Nb or Nb/Ti were subjected to thermal simulations using a Gleeble 1500 to produce structures representative of coarse-grained and fine-grained regions of weld heat-affected zones. Peak temperatures of 1350°C and 1000°C were used as were cooling times between 800°C and 500°C of 11, 16, 43, and 200 s. Microstructures were examined and Charpy specimens were tested at -40°C.

The microstructures of samples heated to 1350°C were predominantly mixtures of bainite and martensite, while some polygonal ferrite and proeutectoid ferrite were observed at slower cooling rates. Martensite content was directly related to the carbon content. The Ti alloyed steel developed the finest overall microstructural appearance in terms of prior austenite grain size, bainitic lath width, and ferrite grain size. For 1000°C peak temperature, the microstructures were similar to those of the original baseplates and there was negligible grain growth.

The impact values showed marked differences amongst the simulations. Poor toughness of the Nb steel suggests that dissolution of Nb precipitates can occur during welding and the grain size increases rapidly. The better properties obtained with the Nb/Ti steel, in which a relatively fine grain size was maintained, shows that TiN precipitates do not dissolve upon reheating and are available for ferrite nucleation during cooling. Overall, the Nb/Ti steel exhibited better toughness than the steel microalloyed with niobium.

## KEYWORDS

Welding, heat-affected zones, microalloyed steels, Gleeble simulations, toughness, precipitates.

## INTRODUCTION

The balance of strength and toughness of C/Mn microalloyed steels has led to their use in shipbuilding, offshore structures, and transportation. These applications quite often involve high heat input welding which has, on occasion, led to poor toughness of the weld zones. The embrittlement has been commonly attributed to the additions of Nb and V and their effects on microstructural transformations and precipitation.

The precipitation of niobium and/or vanadium carbonitrides during controlled rolling produces a fine grain size and good toughness. During the heating cycle of welding, some or all of the carbonitrides may dissolve. This results in austenitic grain growth adjacent to the fusion boundary and possible excessive re-precipitation in the ferrite phase during cooling. It has been shown that the coarse-grained zones of Nb/V steels are similar to those containing no grain refining elements at all (1). Toughness can be maintained by restricting grain growth at high temperatures and reducing the possibility of precipitation in the ferrite phase. The literature indicates that Nb microalloyed steels provide better toughness than V steels in the coarse-grained HAZ. The problem of excessive HAZ grain growth can be controlled also through Ti additions. The use of Ti also reduces the amount of free nitrogen which is also beneficial to toughness.

In view of the uncertainty with respect to HAZ behaviour of C/Mn microalloyed steels and the improved toughness reported with Ti additions, a limited study was undertaken using Gleeble simulations of heavy section plates. The results and discussion are concerned only with the properties of the heat-affected zone of steels alloyed with Nb and Nb/Ti combined.

## MATERIAL

The materials chosen for the tests were intended for application to offshore structures and marine applications. They include a steel supplied to Lloyd's grade EH36, British Standard BS4360 grade 50D, and a commercially produced low carbon steel, 272-2. The analyses of the plates are given in Table 1. These plates are supplied in the normalized condition and exhibit yield strengths above 340 MPa (50 ksi) and tensile strengths between 490 and 620 MPa (70 to 90 ksi).

## WELD SIMULATION

Different regions of the weld heat-affected zone were produced on a Gleeble 1500 thermal simulator using peak temperatures of 1350°C and 1000°C, and several cooling rates. The cooling rates were programmed as outlined in Table 2 to cover a range of conditions which could be expected in practice. The two conditions for the 50D steel, $t_{8-5}$ = 11 s and $t_{8-5}$ = 16 s, correspond to heat inputs of 2 kJ/mm and 3 kJ/mm, respectively. The other two steels were programmed for $t_{8-5}$ = 43 s and $t_{8-5}$ = 200 s. The 43 second tests correspond to the cooling of welds made using 3.4 kJ/mm and a preheat of 200°C. The 200 second cooling time would be obtained through very high heat inputs ($\sim$8 kJ/mm) such as for electroslag welding.

## RESULTS AND DISCUSSION

The microstructures of the Gleeble samples are shown in Figures 1, 2 and 3 for the three steels. In Fig. 1, steel EH36 heated to 1350°C and cooled between 800°C and 500°C for 43 s transformed to a mixture of bainite and martensite with delineated grain boundaries. At the slower cooling rate ($t_{8-5}$ = 200 s, the bainitic lath width increased and the prior austenite grain boundaries were less discernible. This sample also exhibited regions of polygonal ferrite and grain boundary ferrite which replaced most of the martensite.

The samples heated to 1000°C both exhibit microstructures of polygonal ferrite and pearlite, with a marked increase in grain size at the slower cooling rate.

The microstructures of steel 272-2 (Fig. 2) were similar to those observed for the EH36 for a 1350°C peak temperature, except that steel 272-2 developed a finer overall microstructural appearance in terms of prior austenite grain size, bainitic lath width, and ferrite grain size. For $t_{8-5}$ = 200 s, the structure is a mixture of ferrite and bainite with some polygonal ferrite. The lath dilineation and austenite grain boundaries at both cooling rates was less distinctive than for EH36.

As a comparison, Fig. 3 shows the microstructures of steel 50D heated to a 1350°C peak temperature and two cooling rates. The faster cooled test ($t_{8-5}$ = 11 s) is principally low carbon martensite (75%) with the remainder being bainite (25%). Prior austenite grain boundaries were not readily apparent. The sample cooled at $t_{8-5}$ = 16 s was almost the same mixture of low carbon martensite (70%) and bainite (29%), but there was some development of proeutectoid ferrite (1%). Carbide precipitation was also observed between bainitic laths for this sample as was some grain growth.

The samples for steel EH36 heated to 1000°C for both cooling times exhibit microstructures of polygonal ferrite and pearlite, with a marked increase in grain size at the slower cooling rate. For steel 272-2 at 1000°C, the final structures were composed entirely of polygonal ferrite at both cooling rates. The structures of EH36 and 272-2 compared for a cooling time of $t_{8-5}$ = 200 s and 1000°C peak temperature reflect the differences in the base metal microstructures.

Charpy specimens were prepared from steels EH36 and 272-2 with through-thickness notches and tested at -40°C. The energies reported in Table 2 reflect changes with peak temperatures and composition.

The peak temperatures of the Gleeble simulations produced marked differences in the microstructures which can be explained with reference to the chemical composition and the impact values.

For steel EH36 (no Ti), the niobium precipitates dissolve completely when heated to 1350°C and the austenite grain growth is essentially unrestricted. Impact values of 5 J at -40°C for both cooling rates show that although the transformation structures are more refined at the faster cooling rate, the cleavage resistance is very poor. The austenite grain size of steel 272-2 for a 1350°C peak temperature is slightly less than EH36. This, combined with a finer structure, results in improved toughness, although only slightly. The impact energy was better for the faster cooling rate which could be partly a result of the apparent mixture of structures and more numerous ferrite grains. The ferrite transformations suggest that TiN precipitates do not dissolve upon reheating and are therefore available for nucleation. TiN precipitate clusters were observed at both cooling rates for a peak temperature of 1350°C.

At 1000°C, the impact energy for EH36 shows an improvement but the slower cooling rate has a higher value than the fast rate. This is contrary to what one would expect since longer times are commonly associated with excessive precipitation and poor toughness. One can only guess that this improvement is due to a better precipitate distribution at the longer time. The microstructures of 272-2 at 1000°C were very similar as were their impact energies.

Overall, the Charpy values for steel 272-2 were considerably better than for EH36, with the better properties being the result of TiN precipitation and a finer grain size.

The relative hardness values were as expected in that the higher hardnesses were measured for faster cooling rates and higher carbon contents. The maximum hardness in a martensite-bainite mixture can be related to the chemical composition and the cooling time (2). The hardnesses over a range of cooling times can be used directly to ascertain the weldability of steels (3). This observation is confirmed by the present results in that the poorest impact values were obtained at the highest hardnesses and toughness improved as the hardness decreased.

The major concern when steels containing niobium are welded is not the weldability or resistance to cracking, but rather the properties of the heat-affected zone. The niobium readily forms carbides or carbonitrides which restrict austenite grain growth during controlled rolling and subsequently promote a fine grain size and toughness. When welded, some or all of the carbonitrides dissolve which then allows austenite grain growth and/or reprecipitation in the ferrite phase, both of which reduce toughness. Alloying with titanium is a practical solution to this problem since the precipitates are stable at high temperatures. While niobium carbonitrides will dissolve rapidly when heated above 1200°C (4,5), the dissolution of TiN is minimal up to 1400°C (6). The partial replacement of titanium for niobium will reduce excessive precipitation in the ferrite phase during welding. Precipitates are used mainly to restrict grain growth and to give precipitation hardening. The latter is desirable to reduce the total alloy content of the steel and to meet the strength requirement. Precipitates formed in the austenite phase are mainly responsible for retardation of recrystallization and grain growth (6), and these precipitates will grow to such a size ( $\sim$ 200 Å) (7) that their strengthening effect is minimal. Precipitates formed in the ferrite phase will be small (<50 Å) and are responsible for the strengthening effect (6). Since titanium nitride will form at a relatively high temperature, the chance of excessive precipitation in the ferrite phase is greatly reduced. Hence titanium nitride can restrict grain growth due to its low solubility at high temperatures and it can reduce the chance of excessive precipitation in the ferrite phase due to its minimal dissolution and the fact that it forms at a high temperature. Titanium also reduces the amount of free nitrogen which improves the toughness.

Previous attempts to explain the variations of fracture toughness with chemical composition have addressed both microstructural and mechanical aspects of the problem. In the former area, austenite grain size, microstructures, and precipitation effects are known to be significant. Within the mechanical area, an increase in hardness (or strength) is expected to lead to a deterioration in toughness, though it must be emphasized that this will only be generally true within the same microstructural category; for example, the replacement of a bainitic microstructure by martensite, though it is accompanied by an increase in hardness, generally improved toughness in low carbon steels.

The effects of composition measured using Charpy testing of Gleeble samples showed that the best transition temperatures were obtained with autotempered martensite and lower bainite, while coarse upper bainite or ferrite structures were detrimental (8). A steel essentially identical to the EH36 exhibited a rapid decrease in toughness as $t_{8-5}$ increased from 10 s to 20 s, with the latter thermal cycle promoting a coarse bainite lath.

It is usually a difficult task to identify the precise metallurgical factors which control toughness in niobium steels. When bainite is present, cracks readily propagate across ferrite-ferrite boundaries within a bainitic colony, with the critical failure event being the extension of a microcrack across a high angle boundary. Previously (9, 10) it has been noted that martensite-austenite (MA) particles tend to be trapped in linear arrays at such boundaries, but it is not suggested that cracks are likely to propagate preferentially along such arrays, and indeed this has never been observed in previous fractographic studies. However, the presence of a brittle microconstituent at high angle boundaries, accompanied by transformation stresses, may be expected to aid the propagation of a crack from one grain to another in much the same way that grain boundary carbides behave in polygonal ferrite structures.

The nature of the MA constituent itself has been shown to influence the toughness as well (11). They showed that a coarsening of either the gross microstructure or the MA led to poor toughness, with the MA exerting less influence at long compared to short cooling times. SEM studies of the fracture surfaces showed that the morphology of the bainite changed from elongated to massive shapes with increasing cooling time, and changed further to a decomposed ferrite and carbide aggregate. The MA was observed to include lath martensite, twinned martensite, and retained austenite. They concluded that toughness is adversely affected at low heat inputs primarily by the MA constituent, while at high heat inputs the coarsening of the microstructure overrides the loss of toughness due to the MA distribution.

## CONCLUSIONS

The microstructures of the Gleeble specimens examined in this study are generally associated with poor properties. The high percentages of bainite, with martensite, proeutectoid ferrite, and MA complexes have been shown, with reference to past results, as being the main cause of the poor toughness.

The problem of low impact values in the coarse-grained HAZ is more serious at higher heat inputs. Without Nb, slow cooling times improve toughness in the intercritical zone.

The higher toughness for the titanium steel showed that the TiN precipitates do not dissolve upon reheating, and that they are an effective means of maintaining a fine grain size. This in turn results in overall higher impact values for the Nb-Ti steel compared to the Nb alloyed steel.

## ACKNOWLEDGEMENTS

The authors wish to thank M. Letts, R.B. Narraway, and J. Ng-Yelim of CANMET for assistance in the Gleeble simulations, metallographic examination, and electron microscopy, respectively. The many valuable comments of Tom Lau are also greatly appreciated.

## REFERENCES

1. R. E. Dolby and G. G. Saunders, "Metallurgical Factors Controlling the Heat-Affected Zone Fracture Toughness of Carbon: Manganese and Low Alloy Steels", IIW Doc. IX-891-74.

2. K. Lorenz and C. Duren, "Evaluation of Large Diameter Pipe Steel Weldability by Means of the Carbon Equivalent", proc. Steels for Line Pipe and Pipeline Fittings, The Metals Society, London, 1983, p 322.

3. B.A.Graville, "Cold Cracking in Welds in HSLA Steels", proc. Welding of HSLA (Microalloyed) Structural Steels, Rome 1976, American Society for Metals, Metals Park, Ohio, 1978, p 85.

4. R.E. Dolby and G. C. Saunders, "Metallurgical Factors Controlling the Heat-Affected Zone Fracture Toughness of Carbon-Manganese and Low Alloy Steels", IIW Doc. IX-891-74.

5. I. Hrivnak, "Precipitation Processes in Welded Joints and Their Effect on Mechanical Properties of Steels", Zvanacske Spravy, 1974, p 1.

6, A. M. Sage, "A Review of the Physical Metallurgy of High Strength, Low Alloy Line Pipe and Pipe Fitting Steels", proc. Steels for Line Pipe and Pipeline Fittings, The Metals Society, London, 1983, p 39.

7. M. J. Crooks et al, "Precipitation and Recrystallization in Some Vanadium and Vanadium-Niobium Microalloyed Steels", Met Trans A, Dec. 1981, p 1999.

8. G. Bernard et al, "Properties of Welds in Submerged Arc Welding", proc. Welding of HSLA (Microalloyed) Structural Steels, Rome 1976, American Society for Metals, Metals Park, Ohio, 1976, p 187.

9. A. B. Rothwell et al, "Heat-Affected Zone Toughness in High Strength Pipeline Steels", Fifth Bolton Landing Conference, August 1978, General Electric.

10. J. G. Garland and P. Kirkwood, "Towards Improved Submerged Arc Weld Metal", Metal Construction, Vol 7(5), 1975, p 275.

11. H. Ikawa et al, "Effect of Martensite-Austenite Constituent on HAZ Toughness of a High Strength Steel", IIW Doc. IX-1156-80.

TABLE 1 Chemical Composition of Steels

| Steel | Chemical Composition, wt % | | | | | | | | | | |
|---|---|---|---|---|---|---|---|---|---|---|---|
| | C | Mn | Si | S | P | Al | Nb | Ti | Ni | Cr | Mo | Cu |
| EH36 | 0.135 | 1.89 | 0.27 | 0.015 | 0.02 | 0.06 | 0.03 | - | 0.01 | 0.17 | 0.06 | 0.015 |
| 50D | 0.20 | 1.43 | 0.33 | 0.006 | 0.022 | 0.05 | 0.05 | - | 0.04 | 0.10 | 0.02 | |
| 272-2 | 0.048 | 1.82 | 0.32 | 0.006 | 0.002 | 0.01 | 0.054 | 0.01 | - | - | 0.29 | 0.010 |

TABLE 2 Parameters and Results of Gleeble Testing

| Steel | Peak Temperature °C | Cooling Time $t_{8-5}$ | Charpy Energy at -40°C, J | Hardness DPH | Grain Size, μ m |
|---|---|---|---|---|---|
| EH36 | 1350 | 43 | 5 | 263 | 66 |
| | | 200 | 5 | 221 | 71 |
| | 1000 | 43 | 31 | 236 | 3.4 |
| | | 200 | 51 | 211 | 5.0 |
| 50D | 1350 | 11 | | 314 | 167 |
| | | 16 | | 271 | - |
| 272-2 | 1350 | 43 | 21 | 227 | 47 |
| | | 200 | 9 | 217 | 65 |
| | 1000 | 43 | 66 | 206 | 3.0 |
| | | 200 | 70 | 188 | 2.7 |

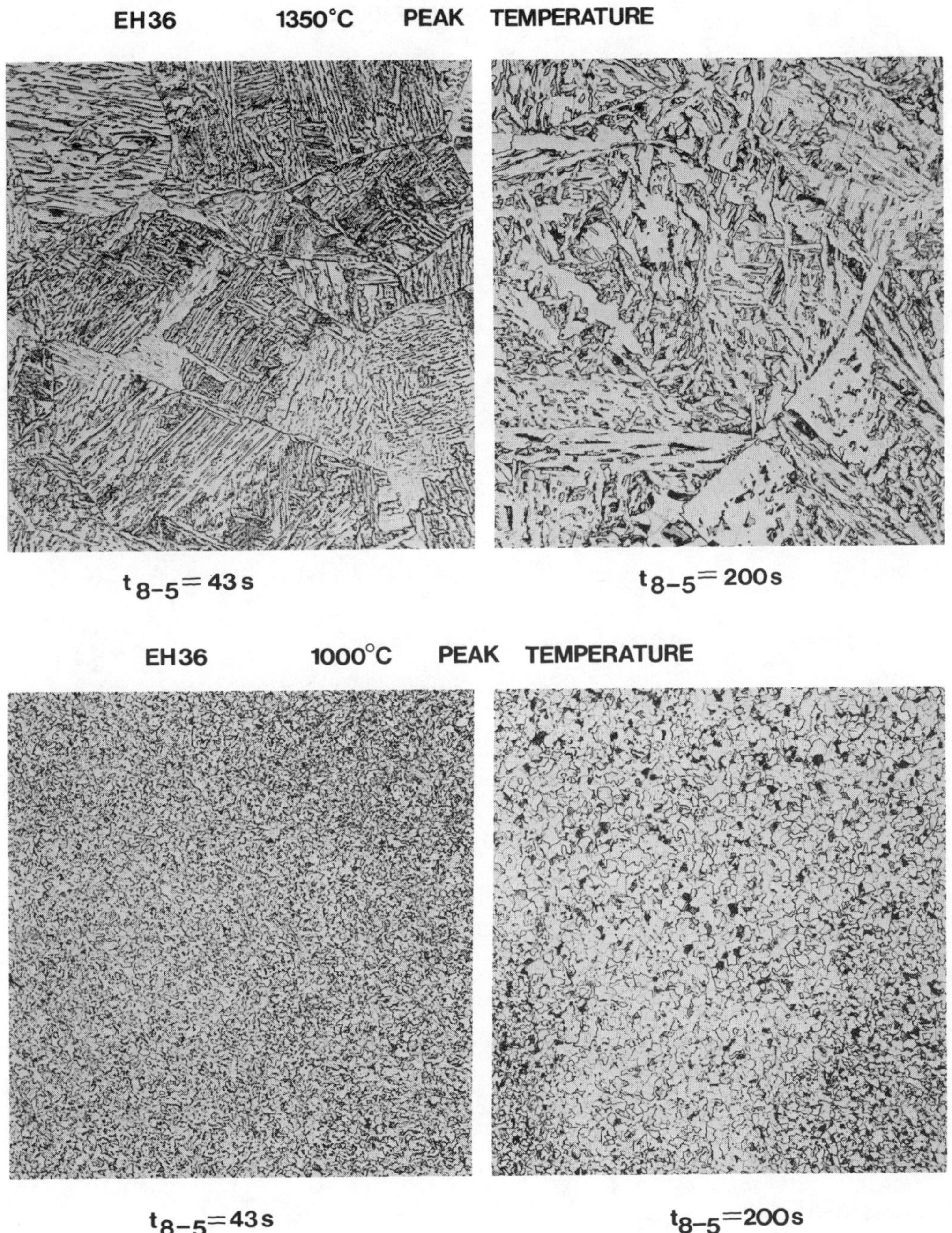


Fig. 1. Microstructures of Gleeble simulations of steel EH36 (500x).

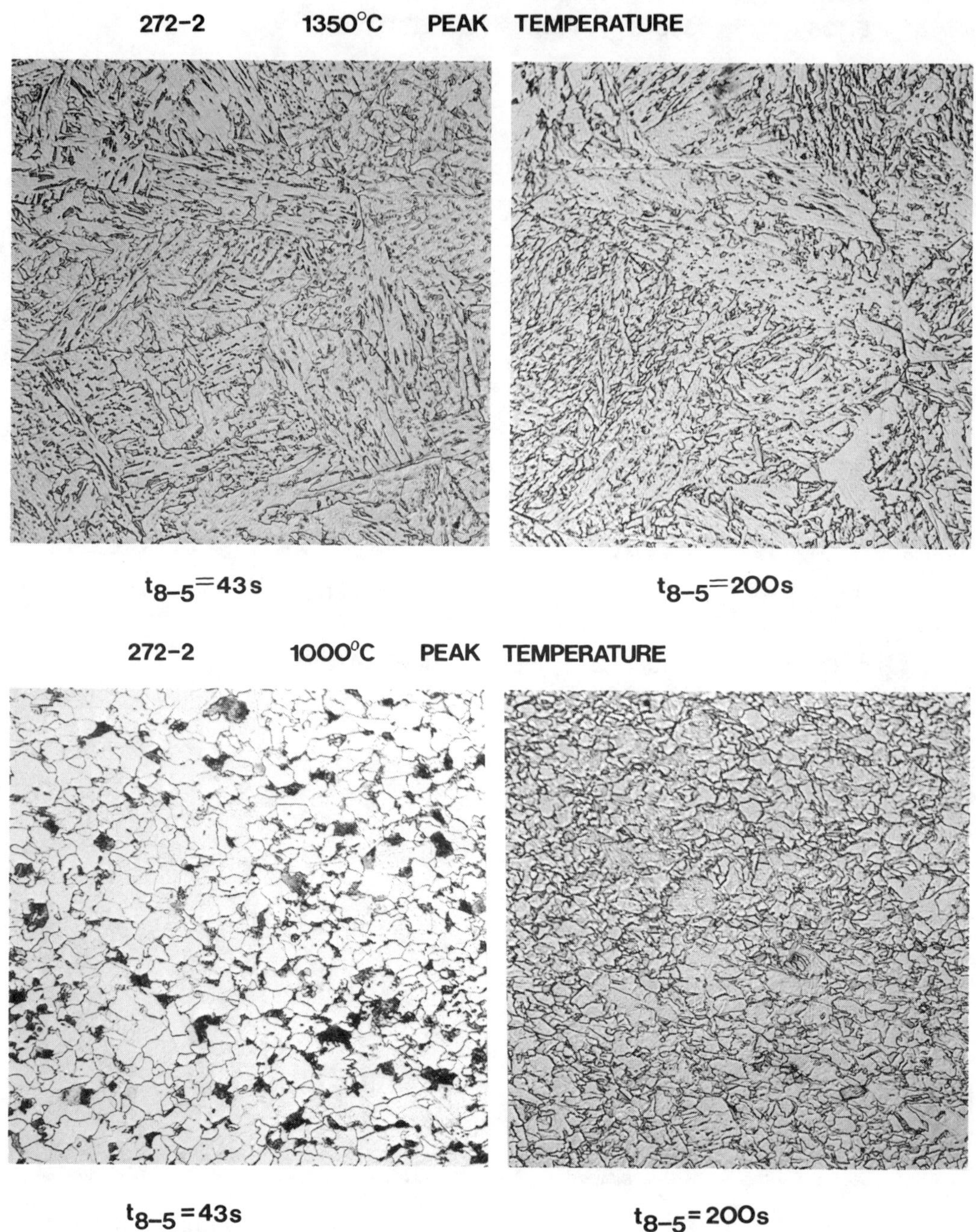


Fig. 2. Microstructures of Gleeble simulations of steel 272-2 (500x).

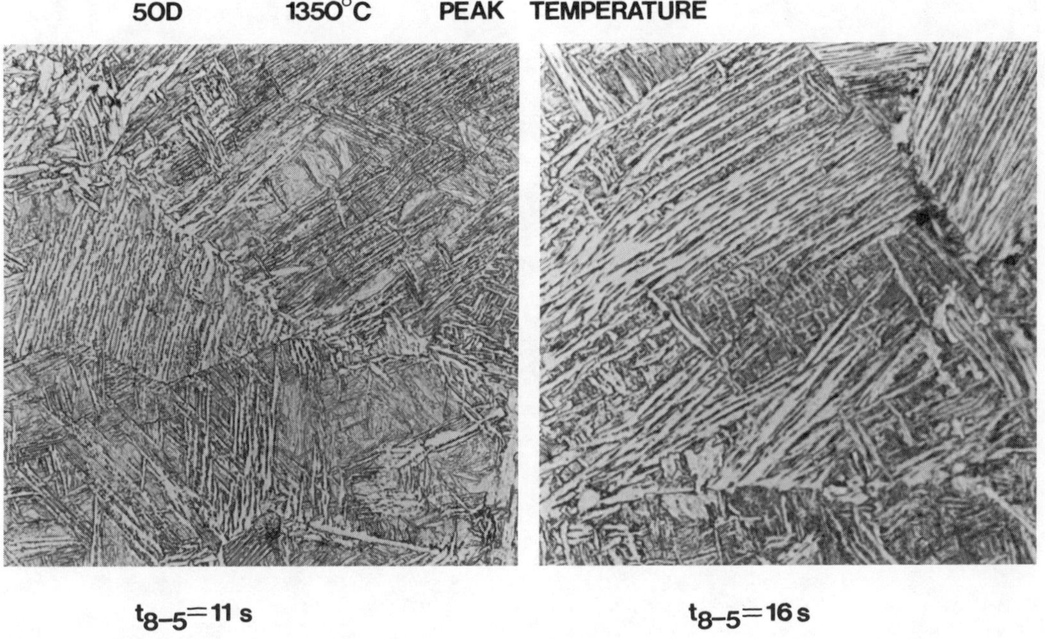


Fig. 3. Microstructures of Gleeble simulations of steel 50D (500x).

# DEVELOPMENT OF NORMALIZED AND AGED Mn-Mo-V-Nb STEEL FOR FITTINGS

M. Lafrance*, F.G. Bourdillon*, T. Wada**,
P. Boussel***, and W.E. Lauprecht****
*USINOR, Dunkerque, France
**Climax Molybdenum Company, Ann Arbor, Michigan, USA
***Climax Molybdenum S.A., now AMAX Tungsten Division, Paris, France
****Climax Molybdenum G.m.b.H., Düsseldorf, W. Germany

## ABSTRACT

Normalized and aged Mn-Mo-V-Nb steel has been developed for application including general construction and pipeline accessories. In the first phase of the development, four test steels were prepared and evaluated on the basis of continuous cooling transformation (CCT) characteristics and mechanical properties. Based on this evaluation, the composition of a trial heat was selected as 0.12% C, 1.7% Mn, 0.20% Mo, 0.07% V and 0.04% Nb. The trial heat was rolled to plates of four different thicknesses, 10, 30, 50, and 70 mm, and the mechanical properties, weldability, and strain-age hardenability of the plates were determined. The normalized and aged 30 mm thick plate exhibited approximately 480 MPa (70 ksi) yield strength and -40°C Charpy V-notch fracture appearance transition temperature (FATT). The plate showed good weldability and a modest susceptibility to strain-age hardening.

KEYWORDS

Pipeline fitting; normalized steel; Mn-Mo-V-Nb steel; implant test; CCT diagram; strain-aging; yield strength; toughness.

## INTRODUCTION

Normalized and aged high strength steels have several attractive features as compared with quenched and tempered steels. First, the properties are more thermally stable. If hot or warm forming is needed after heat treatment, normalized steels can tolerate higher forming temperatures than quenched and tempered steels. Second, if the product shape is not as simple as a flat plate or a tube, a normalizing treatment may be significantly more economical than quenching and tempering. For these reasons, the normalized steels have drawn attention in such energy systems applications as pressure vessels, pipe fittings and high strength steel castings. Mechanical properties of normalized and aged steels have been improved to the point that they are now comparable to some of the quenched and tempered steels. In particular, the normalized and aged Mn-Mo-V-Nb steels which have been recently developed (Lauprecht and others, 1978; Wada and others, 1979, 1981) provide excellent strength-toughness combinations.

This paper describes the latest research and development of normalized and aged Mn-Mo-V-Nb steels. A two-phase cooperative program was undertaken by Climax

Molybdenum Company and Usinor. In Phase I, four test steels with different combinations of molybdenum and vanadium were prepared by BOF melting, and mechanical properties and continuous cooling transformation (CCT) diagrams were determined to select an optimum composition. Then, in Phase II, a trial heat of the selected composition was finished to plates of four different thicknesses in the range from 10 to 70 mm and mechanical properties, including weldability and strain-aging response, were thoroughly investigated.

## PHASE I. SELECTION OF THE COMPOSITION

In order to evaluate the effect of molybdenum and vanadium in normalized and aged Mn-Mo-V-Nb steel plates, a trial BOF heat was produced and split into four compositions with two levels of molybdenum and two levels of vanadium. A 30 mm thick plate from each composition was prepared to determine whether the steels gave adequate toughness and which of the following minimum yield strength (0.2% offset) levels would be met:

(a) 358 MPa (52 ksi)
(b) 413 MPa (60 ksi)
(c) 448 MPa (65 ksi)

These yield strength levels correspond to Grades X52, X60 and X65 of the API specification, respectively. The investigation in Phase I included determination of continuous cooling transformation (CCT) diagrams at slow cooling rates to simulate the heat treatment and at rapid cooling rates to simulate conditions that occur during welding.

### Laboratory Procedures

Test steels and heat treatment. The compositions of the test steels are shown in Table 1. The test plates were heat treated to simulate normalizing of 50 mm thick

TABLE 1 Composition of Test Steels

| Steel | Element, % | | | | | | | | | |
|---|---|---|---|---|---|---|---|---|---|---|
| | C | Mn | Si | P | S | Mo | Nb | V | Al | N (ppm) |
| A | 0.113 | 1.703 | 0.295 | 0.014 | 0.006 | 0.201 | 0.036 | - | 0.058 | ND[a] |
| B | 0.114 | 1.697 | 0.296 | 0.015 | 0.006 | 0.201 | 0.037 | 0.065 | 0.059 | ND |
| C | 0.114 | 1.693 | 0.289 | 0.015 | 0.006 | 0.368 | 0.037 | - | 0.061 | ND |
| D | 0.115 | 1.692 | 0.293 | 0.015 | 0.006 | 0.372 | 0.038 | 0.068 | 0.057 | 126 |
| Test Steel Plates in Phase II | | | | | | | | | | |
| 50 mm Plate | 0.12 | 1.78 | 0.30 | 0.014 | 0.005 | 0.20 | 0.032 | 0.066 | 0.066 | 118 |
| 70 mm Plate | 0.12 | 1.78 | 0.30 | 0.015 | 0.005 | 0.21 | 0.033 | 0.066 | 0.063 | 126 |

[a]ND = Not Determined

plates by placing a 150 by 250 mm section of each plates in a steel box and air cooling the box from the austenitizing temperature. Two austenitizing temperatures, 910 and 940°C, were used for each steel; the average cooling rate was 12°C/min.* After normalizing, the plate sections were aged at selected temperatures between 500 and 700°C for one hour.

CCT diagrams. CCT diagrams were determined by dilatometry using a Formastor-F quench dilatometer combined with metallographic observations. The cooling rates simulated both the normalizing heat treatment and the thermal cycle expected during

*All cooling rates are averages between 800 and 500°C.

welding. The cooling rates for the normalizing heat treat cycle ranged from about 600 to 2.5°C/min* after holding the specimens for 20 minutes at a temperature 30°C above the $Ac_3$ temperature. The welding thermal cycle cooling rates ranged from 120 to 2.5°C/sec, cooling immediately after heating to 1300°C. The $Ac_1$ and $Ac_3$ temperatures were determined using the same Formastor-F dilatometer by heating the specimen at a constant rate of 2°C/min from 600 C to 960°C.

Mechanical tests. Charpy V-notch impact specimens were cut in the transverse and longitudinal directions from the center of each plate. Notches were cut perpendicular to the prior plate surface. Duplicate longitudinal tensile specimens were obtained from the center of each plate. The tensile test specimens had gauge sections 6.3 mm in diameter and 25 mm in length.

## Results of Phase I

CCT diagrams. The CCT diagrams for heat treatment and for welding, plotted as a function of cooling rate between 800 and 500°C, are shown in Fig. 1. The hardenability increased as the molybdenum content was increased from 0.20 to 0.37%,

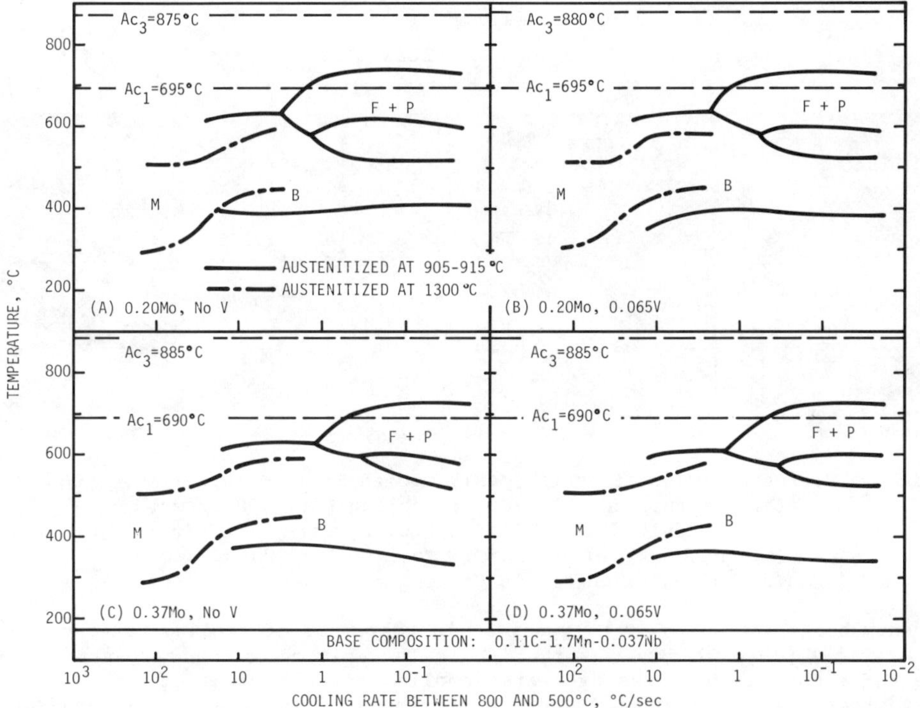


Fig. 1. CCT Diagrams of Mn-Mo-V-Nb Steels

primarily through retarded polygonal ferrite formation. Vanadium had a similar effect in the 0.20% Mo steels, but exhibited a lesser effect in the 0.37% Mo steels. The transformation behavior in the relatively rapid cooling range was little influenced by the variation of molybdenum or vanadium. When as-normalized hardnesses are compared as a function of cooling rate, Fig. 2, it is observed that the higher-

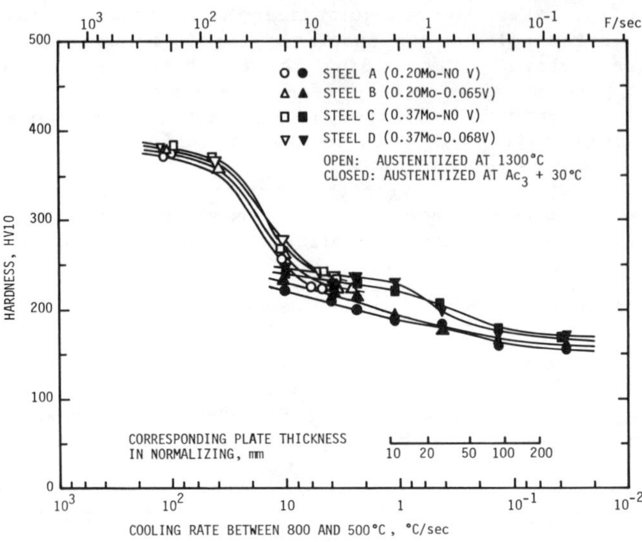


Fig. 2. Dependence of Hardness on Cooling Rate in the Test Steels

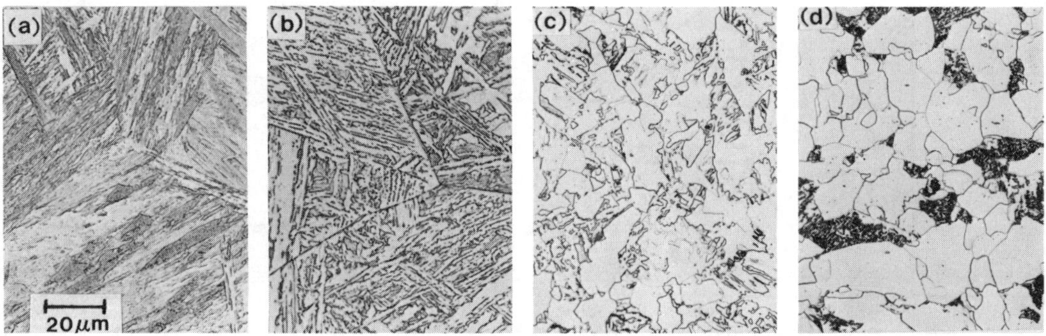


Fig. 3 Microstructures of Continuously Cooled Steel B (0.20% Mo, 0.065% V)
(a) Austenitized at 1300°C; Cooling Rate 100°C/sec.
(b) Austenitized at 1300°C; Cooling Rate 11°C/sec.
(c) Austenitized at 910°C; Cooling Rate 1.1°C/sec.
(d) Austenitized at 910°C; Cooling Rate 0.16°C/sec.

molybdenum heats are 30 to 50 HV10 harder in the range of cooling rates corresponding to typical plate thicknesses than their low molybdenum counterparts. There is little difference in hardness at faster cooling rates typical of welding. Fig. 3 shows micrographs of continuously cooled specimens of Steel B over a range of cooling rates; all four steels exhibited similar microstructures.

Mechanical properties. Fig. 4 summarizes the tensile properties of the test steels after various heat treatments. The yield/tensile ratio reaches a maximum at an aging temperature about 635 to 650°C. The two vanadium-containing steels (Heats B and D) normalized from 910°C showed a distinct maximum in 0.2% offset yield strength after aging at 635°C. These two steels also had a smaller decrease in tensile strength due to aging at temperatures higher than 600°C than the vanadium-free steels, the end result being that the change in yield/tensile ratio was similar in all the steels. The minimum yield strength level which each chemistry can meet, as represented by the test plates, is as follows:

Steel A - 358 MPa (52 ksi)
Steel B - 413 MPa (60 ksi)
Steels C and D - 448 MPa (65 ksi)

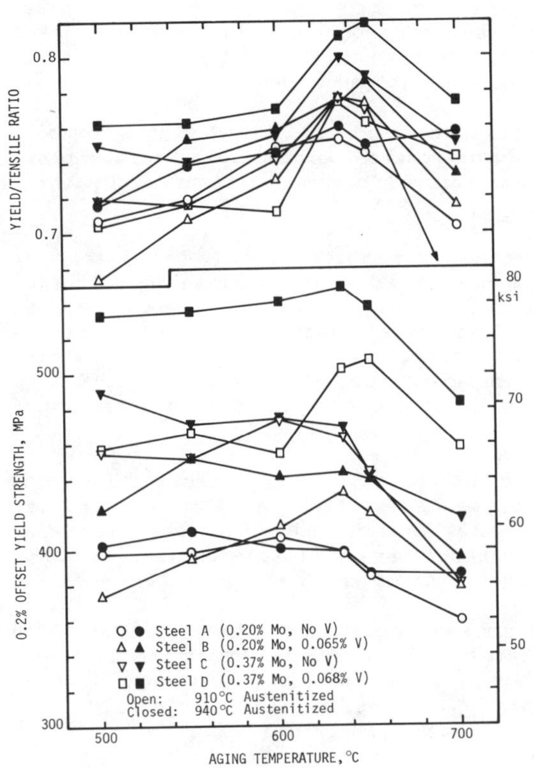


Fig. 4. Effect of Aging Temperature on Yield Strength with Yield/Tensile Ratio of Plates Normalized to Simulate 50 mm Thickness.

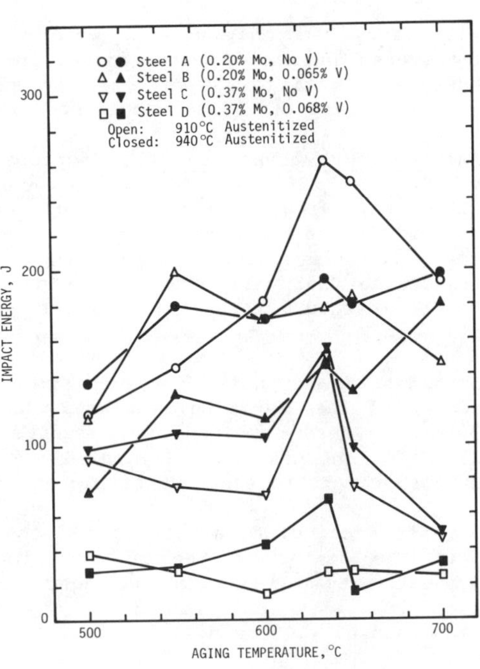


Fig. 5. Effect of Aging Temperature on Charpy Impact Energy at -32°C of Plates Normalized to Simulate 50 mm Thickness.

Charpy V-notch impact test results are shown in Fig. 5. The toughness of steels A, B and C is more than satisfactory.

Steel B with 0.20% Mo and 0.07% V exhibited the strength and toughness levels required for most of the potential applications. Its economical composition is also attractive. Thus, this composition was selected for further investigation.

## PHASE II. EVALUATION OF A TRIAL HEAT

Based on the results from the four test steels in Phase I, a composition which contained 0.20% Mo and 0.06% V was selected for further investigation. Plates of this composition were produced in four thicknesses, 10, 30, 50 and 70 mm. The plates were normalized, aged, and evaluated for mechanical properties, effects of strain aging and anisothermal stress relieving. Weldability was also investigated by the implant cracking test.

Laboratory Procedures

Steel and heat treatment. The compositions of two of the plates are shown in Table 1. Each plate was normalized from 940°C. Then, most plates were aged for 30 min- utes at 635° C; a portion of each plate was aged at 650°C for 30 or 60 minutes as noted.

Mechanical tests. Tensile test specimens were taken from the center of each plate in the transverse direction. The size of the test specimens for plates thicker than 30 mm was the same as in Phase I. The test specimens for plates 10 mm thick had reduced sections 4.76 mm in diameter and 25 mm in gauge length.

All Charpy V-notch specimens were taken from the centers of the plates in both transverse and longitudinal directions with the notches perpendicular to the plate surface. Testing was performed at six temperatures between 23 and -80°C; three specimens were tested at each temperature.

Anisothermal stress relief. Sections of the 30 mm thick plates aged either at 635°C or at 650°C for 30 min were further heat treated to simulate anisothermal stress relieving. The thermal cycle was about five hours including about one hour at 600°C and subsequent furnace cooling. After the simulated stress-relief treatment, two transverse tensile specimens were prepared from each plate and nine transverse Charpy V-notch specimens were prepared from the plate aged at 635°C. Room temperature tensile properties and low temperature impact values were determined on these specimens.

Strain-aging. Tensile specimens were taken in the transverse direction from the center of the 30 mm thick plates that had been aged either at 635 ot 650°C for 30 min. These specimens were subjected to 10% strain and subsequently aged at 250°C for one hour and air cooled. Tensile tests were performed on these strain-aged specimens to evaluate strain-age strengthening of the steel.

A plate section, about 12.5 by 50 by 400 mm was cut in the transverse direction from the center of the 30 mm thick plate aged at 635°C. This section was subjected to 10% strain and aged at 250°C for one hour. Six Charpy V-notch specimens were prepared from the strain-aged section, and tested at room temperature, -20°C and -60°C, two specimens at each temperature.

Implant tests. Implant tests were performed to evaluate the cold cracking susceptibility of the weld heat affected zone. Specimens were 8 mm in diameter with a circular notch having a root radius of 0.1±0.01 mm (Fig. 6), sectioned from a 30 mm thick plate normalized from 940°C and aged for one hour at 635°C. Each specimen was welded on a 20 mm thick flat plate. The implant test weld was

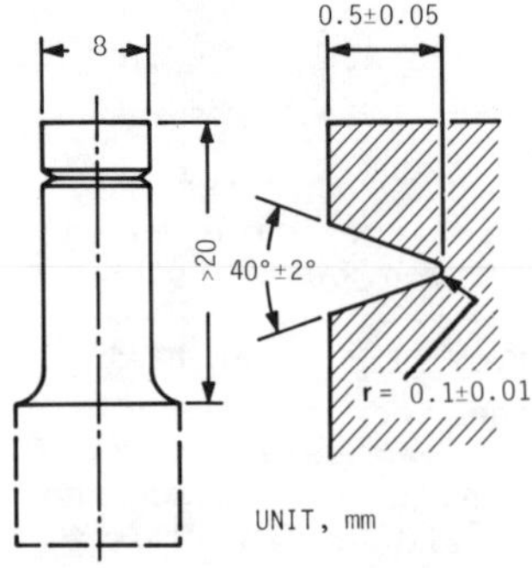


Fig. 6. Geometry of the Implant Specimen for Testing Weld HAZ Cracking

deposited by gravity with E8018G (low hydrogen) electrodes. Welding conditions were selected so as to produce several different cooling rates in the heat affected zone; the actual cooling rate of each test weld was measured with a thermocouple. After welding, the specimens were loaded when their temperature was in the range between 150 and 100°C, and were under stress for at least 16 hours or until failure occurred. Twelve specimens were tested with different loads and different cooling rates to determine the cold cracking susceptibility curve.

Experimental Results

Microstructure. The microstructures of 30 and 70 mm plates are shown in Fig. 7. The microstructure consists of polygonal ferrite, pearlite and bainite. The amount of polygonal ferrite increases, while that of upper bainite decreases, with in- creasing plate thickness.

Tensile properties. Room temperature tensile properties are shown in Table 2. Yield and tensile strengths were quite high in 10 and 30 mm thick plates, whereas the strength were somewhat lower in the heavier section sizes. The effect of aging time on the yield and tensile strengths was not significant.

Fig. 8 shows elevated temperature tensile properties of the plates. The yield strength decreases monotonically with temperature, whereas the tensile strength exhibits a maximum at about 300°C.

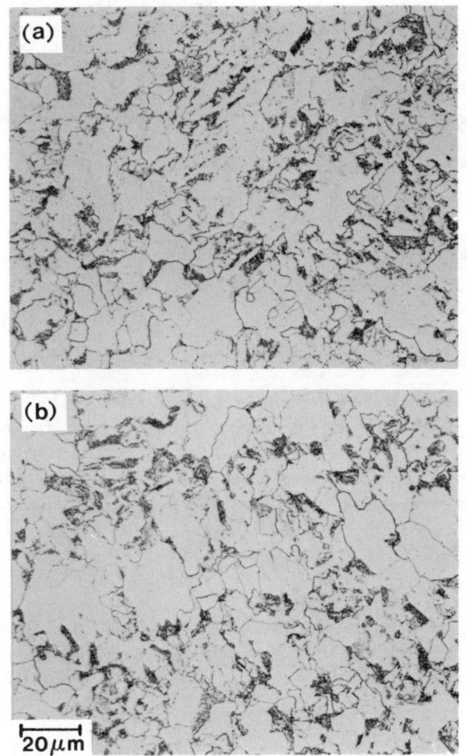


Fig. 7. Microstructures of Normalized and Aged Mn-Mo-V-Nb Steel Plates. (a) 30 mm Thick, (b) 70 mm Thick

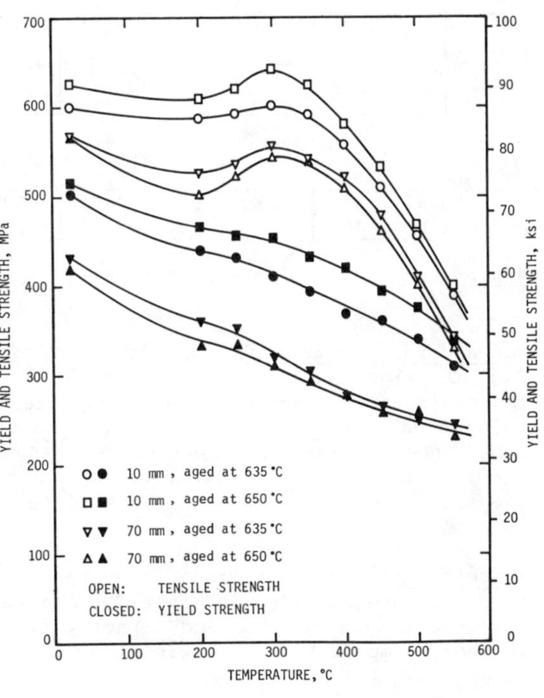


Fig. 8. 0.2% Offset Yield and Tensile Strength of Test Steel at Elevated Temperatures.

TABLE 2 Room Temperature Tensile Properties and the 50% Shear Fracture Appearance Transition Temperature (FATT) in the Charpy Test of the Normalized and Aged Mn-Mo-V-Nb Steel Trial Heat (Phase II)

| Plate Thick- ness mm | Aging Temp. and Time °C, min | Direc- tion | Room Temperature Tensile Properties | | | | | FATT °C |
|---|---|---|---|---|---|---|---|---|
| | | | Yield Point MPa (ksi) | 0.2% Offset Yield Str. MPa (ksi) | Tensile Strength MPa (ksi) | Elonga- tion % | Red. of Area % | |
| 10 | 635, 30 | T | 512 (74.3) | 503 (73.0) | 600 (87.0) | 22.5[a] | 71.0 | -40 |
| | | L | - | - | - | - | - | -70 |
| | 650, 30 | T | 537 (77.9) | 517 (75.0) | 627 (91.0) | 23.0[a] | 73.5 | -25 |
| | | L | - | - | - | - | - | -50 |
| 30 | 635, 30 | T | 516 (74.8) | 496 (71.9) | 627 (90.9) | 28.0 | 70.0 | -40 |
| | | L | - | - | - | - | - | -45 |
| | 650, 30 | T | 495 (71.8) | 472 (68.5) | 596 (86.5) | 30.5 | 69.0 | -40 |
| | | L | - | - | - | - | - | -40 |
| | 650, 60 | T | 490 (71.1) | 470 (68.1) | 591 (85.7) | 30.5 | 74.0 | -30 |
| | | L | - | - | - | - | - | -45 |
| | 650, 90 | T | 510 (73.9) | 483 (70.1) | 603 (87.4) | 28.5 | 69.5 | -15 |
| | | L | - | - | - | - | - | -40 |
| 50 | 635, 30 | T | 431 (62.3) | 410 (59.5) | 551 (79.9) | 33.5 | 72.0 | -60 |
| | | L | - | - | - | - | - | -60 |
| | 650, 30 | T | 438 (63.5) | 424 (61.5) | 560 (81.2) | 33.0 | 73.0 | -50 |
| | | L | - | - | - | - | - | -60 |
| 70 | 635, 30 | T | 447 (64.8) | 433 (62.8) | 569 (82.5) | 30.5 | 70.0 | -20 |
| | | L | - | - | - | - | - | -40 |
| | 650, 30 | T | 441 (64.0) | 420 (62.1) | 569 (82.5) | 31.0 | 71.5 | -35 |
| | | L | - | - | - | - | - | -45 |
| ANISOTHERMALLY STRESS-RELIEVED | | | | | | | | |
| 30 | 635, 30 | T | 512 (74.2) | 487 (70.7) | 626 (90.8) | 28.5 | 70.0 | -35 |
| | 650, 30 | T | 487 (70.7) | 465 (67.4) | 598 (86.7) | 33.0 | 74.0 | - |

[a]Specimens had 4.75 mm diameter and 25 mm gauge length.

Notch impact toughness. The Charpy V-notch impact test results are summarized in Fig. 9. Each value in this figure is the average of three tests. The impact toughness varied somewhat with the plate thickness, whereas the effect of different aging conditions was not as significant. Table 2 contains the 50% shear fracture appearance transition temperatures (FATT) of the test plates. The thick plates, i.e., 50 and 70 mm in thickness, did not necessarily exhibit lower toughness than the thin plates, though the thick plates tended to have a lower strength than the thin plates.

Anisothermal stress relief. The test results of anisothermally stress relieved specimens are included in Table 2. Virtually no effect of the stress relieving was observed on the tensile and impact properties. As shown in Fig. 10, the decrease in Charpy impact toughness was also small.

## Strain-Age Hardening

Strain-aging of the Mn-Mo-V-Nb steel results in a modest hardening as shown in Fig. 11 and Table 3. At 10% elongation and before aging, the test steels exhibited

TABLE 3 Test Results of Strain-Age Hardening 30 mm Plate

| Specimen No. | Aging Condition of Plate, °C, min. | Specimen Condition | Yield Point, MPa (ksi) | 0.2% Offset Yield Strength, MPa (ksi) | Tensile Strength, MPa (ksi) | Elongation, % | Red. of Area, % |
|---|---|---|---|---|---|---|---|
| 31-1 | 635, 30 | Before SA[a] | 519 (75.2) | 490 (71.0) | 611 (88.6)[b] | (10) | - |
| | | After SA | 678 (98.4) | 672 (97.5) | 678 (98.4) | 25.0[c] | 67.5 |
| 31-2 | 635, 30 | Before SA | 508 (73.7) | 496 (71.9) | 623 (90.3)[b] | (10) | - |
| | | After SA | - | 650 (94.3) | 650 (94.3) | 27.5[c] | 65.0 |
| 32-1 | 650, 30 | Before SA | 496 (71.9) | 472 (68.4) | 588 (85.3)[b] | (10) | - |
| | | After SA | 649 (94.1) | 630 (91.9) | 649 (94.1) | 23.0[c] | 69.5 |
| 32-2 | 650, 30 | Before SA | 493 (71.5) | 466 (67.6) | 587 (85.1)[b] | (10) | - |
| | | After SA | 646 (93.8) | 637 (92.4) | 646 (93.8) | 26.5[c] | 71.0 |

[a]SA = Strain-aging at 250°C after 10% strain.
[b]Tensile stress at 10% elongation.
[c]Total elongation including 10% prestrain.

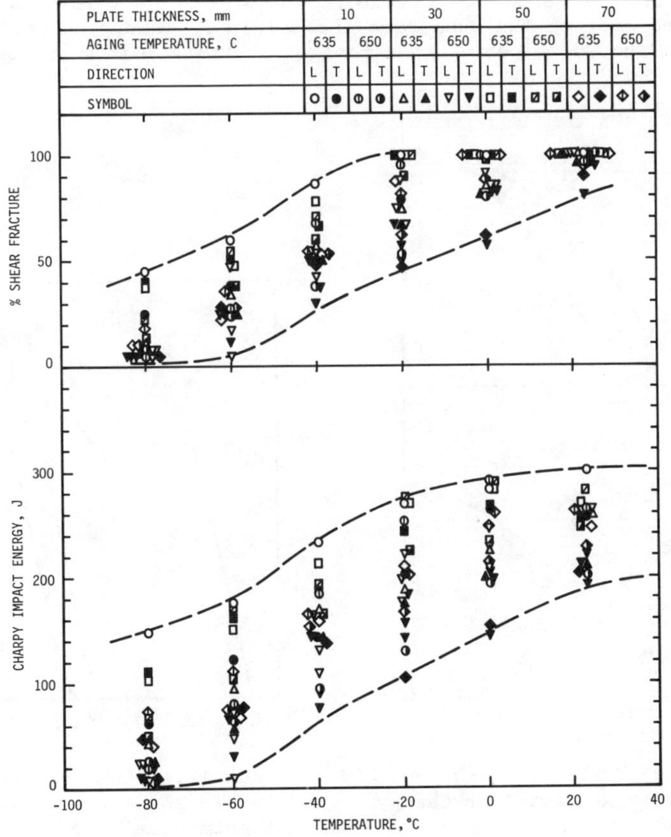


Fig. 9. Charpy Test Results of Normalized and Aged Mn-Mo-V-Nb Steel

587 to 623 MPa tensile flow stress. After aging at 250°C, most specimens exhibited 60 to 70 MPa increase in strength, showing a stress peak within a few tenths of a percent of plastic deformation. This peak was taken as the yield point after strain aging and listed in Table 3. One specimen did not show this type of maximum, the maximum stress being observed at about 0.2% offset. After passing through the initial maximum, the flow stress for all the samples gradually and continuously decreased until the specimens fractured. The tensile strength was therefore taken as the maximum stress reached during the test.

The strain-age hardening may be measured as the difference between the stresses at 10% strain in the first stress-strain curve and the yield stress in the second curve. This difference was in the range of 5 to 11% of the total yield strength, and three out of four specimens were in the range of 10 to 11%. The elongation was also decreased by 3 to 5% by strain-aging. Charpy FATT increased about 30°C by the strain-aging treatment (Fig. 10). It is concluded that the strain-aging effect in Mn-Mo-V-Nb steel is moderate.

## Weld HAZ Cracking

The implant test results, Figs. 12 and 13, show that Mn-Mo-V-Nb steel exhibits good weldability. Compared to a 0.19% C-V-Nb steel, the Mn-Mo-V-Nb steel showed lower HAZ hardness over the range of cooling rates examined, and higher critical stresses at cooling times greater than 9 sec between 800 and 500°C. This cooling rate range covers the manual butt welding of plates in thicknesses up to about 25 mm (1 in.) and most of the submerged arc welding. Thus, the weldability of the Mn-Mo-V-Nb steel should be better than that of Nb-V steel in most applications.

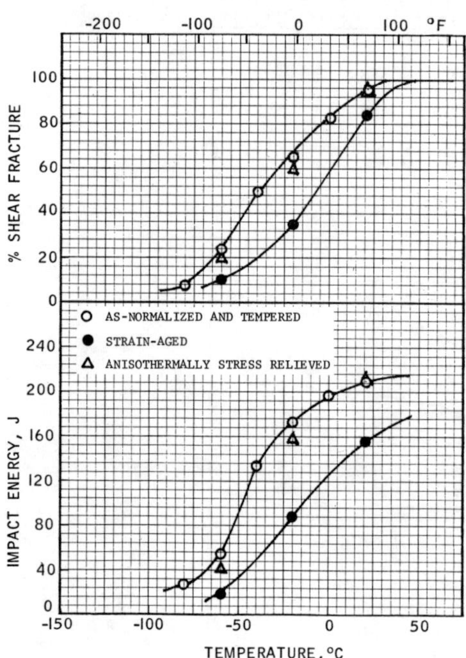


Fig. 10. Effects of Strain-Aging and Anisothermal Stress Relieving on Charpy V-Notch Toughness
Plate; 30 mm Thick, aged at 635°C
Direction; Transverse

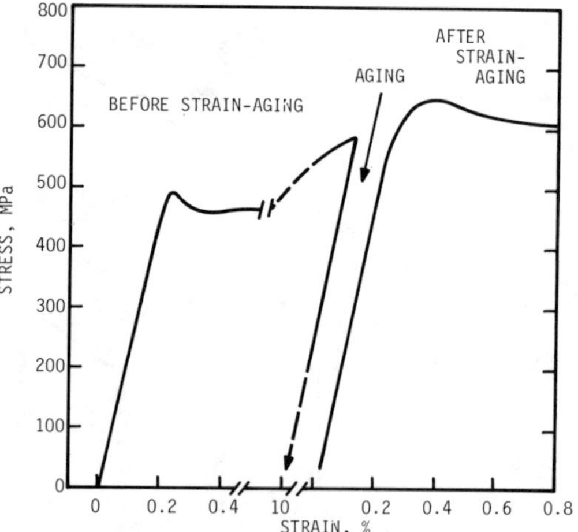


Fig. 11. Stress-Strain Curves Before and After Strain-Aging Treatment.

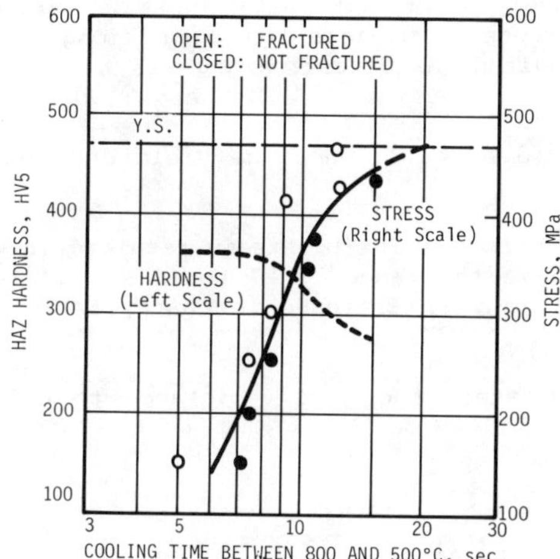


Fig. 12. Implant Test Results on Weld Cracking Susceptibility

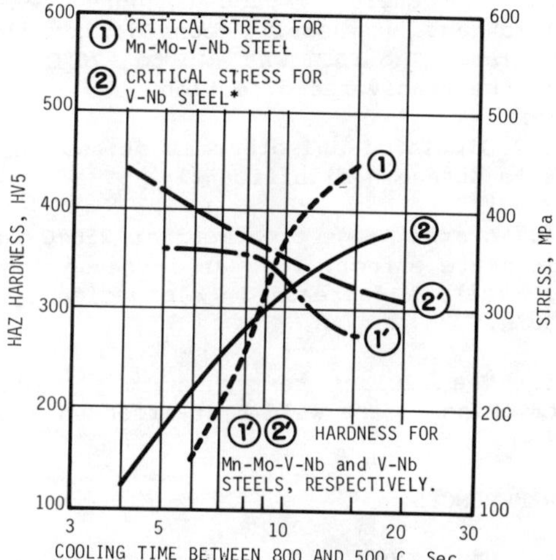


Fig. 13. Cold HAZ Cracking Susceptibility of Mn-Mo-V-Nb Steel as Compared with V-Nb Steel*

(*Composition: 0.19C-1.4Mn-0.085V-0.03Nb)

## DISCUSSION

Normalized and aged Mn-Mo-V-Nb steel exhibits well balanced strength and toughness, and may be produced without specialized heat treat equipment. Molybdenum contributes not only to strength and hardenability, but also increases resistance to softening during warm forming. This steel is particularly suitable to applications such as pipeline fittings.

The yield strength levels which Mn-Mo-V-Nb steel could cover are 400 to 480 MPa (58 to 70 ksi) in plates up to 70 mm thick. The 450 MPa (65 ksi) yield strength is the highest level among the existing normalized constructional steels. It should be noted that this strength level can be achieved with a relatively low carbon content (<0.12%). There is an interrelationship between strength and carbon and manganese contents; a regression analysis of existing data on Mn-Mo-V-Nb steel showed that the carbon equivalent of manganese for yield strength is %Mn/10 (Lauprecht and others, 1978). For an aim yield strength of 450 MPa (65 ksi), it is recommended that %C + %Mn/10 be in the range of 0.27 to 0.29 (Wada and others, 1981). Impact toughness of the normalized and aged Mn-Mo-V-Nb steel is satisfactory for pipelines with design temperatures down to -46 C (-50 F).

## CONCLUSION

Normalized and aged Mn-Mo-V-Nb steel plates exhibit excellent properties in the thickness range from 10 to 70 mm. Test results may be summarized as follows:

1. The yield strength at room temperature depended on the thickness of the plates. It was about 500 MPa (73 ksi) in the 10 and 30 mm thick plates, and about 415 MPa (60 ksi) in the 50 and 70 mm thick plates. The effect of aging condition (635 or 650°C) was rather small.

2. The Charpy impact toughness was relatively high and variation with plate thickness was small, considering the differences in yield strength among the plates. The FATT was -40 to -70°C in the longitudinal direction and -15 to -60°C in the transverse direction.

3. Simulated anisothermal stress relieving did not influence the yield and tensile strength significantly.

4. A strain-age treatment at 250°C for 1 hour after a 10% prestrain resulted in a moderate effect, i.e. an increase in strength in the range of 27 to 68 MPa (4 to 10 ksi), a decrease in elongation by 3 to 5%, and an increase in Charpy FATT by 30°C.

5. The implant test results showed that the steel had good resistance to cold cracking in the weld heat-affected zone.

## REFERENCES

1. Lauprecht, W.E., Wada, T. and Boussel, P., "Normalized High Strength Steels with 450 N/mm$^2$ Yield Strength and Improved Weldability for General Construction." Proceeding of International Conference on "Trends in Steels and Consumables for Welding," The Welding Institute, (1978), p. 555.

2. Wada, T., Smith, Y.E., and Lauprecht, W.E., "Mn-Mo-V-Cb Steel for Pressure Vessels." A paper in ASME publication MPC-10, "Applications of Materials for Pressure Vessels and Piping," (1979), p. 31.

3. Wada, T., Diesburg, D.E., Boussel, P., and Lauprecht, W.E., "High Strength Steels for Pipeline Fittings," Proceedings of the Conference, "Steels for Line Pipe and Pipeline Fittings," The Metals Society, London, 21-23 October 1981.

# DEVELOPMENT OF HEAVY WALL HIGH STRENGTH BENT PIPE FOR ARCTIC USAGE

T. Yamura*, T. Hashimoto**, Y. Komizo**, T. Sawamura***
I. Hoshi**** and H. Nakate****

**Osaka Head Office, Sumitomo Metal Industries, Ltd.*
*5-15 Kitahama, Higashiku, Osaka City, JAPAN*
***Central Research Laboratory, Sumitomo Metal Industries, Ltd.*
****Tokyo Head Office, Sumitomo Metal Industries, Ltd.*
*****Wakayama Steel Works, Sumitomo Metal Industries, Ltd.*

ABSTRACT

Heavy wall bent pipe with high strength and toughness at low temperature have been demanded in recent pipeline construction. Ordinarily, the line pipes can obtain strength and toughness by means of controlled rolling process using low carbon equivalent steel. However, the design of base and welding material has not been considered for the bent pipes subjected to induction heating for bending.
The mechanical properties, especially toughness, of welding metal should be improved for the development of heavy wall high strength bent pipe for arctic usage.
This paper presents the properties after heat treatment with bending for various steels of low carbon equivalent, i.e. low Pcm, steels and various welding materials.
As the results of examination, the products with excellent properties were developed demanded for high grade bent pipe.

KEYWORDS

Induction bending; Low carbon equivalent steel; High grade bent pipe;
Heavy wall bent pipe; Arctic usage;

## INTRODUCTION

Recently, the demand of bent pipes has been increased intended for riserbend and complex expansion loops along offshore pipeline. Consequently, heavy wall line pipes with high strength and toughness at low temperature are demanded.
Generally, controlled-rolling plate of low carbon equivalent steel (i.e. low Pcm steel) for line pipes is applied to obtain high strength, toughness at low temperature and good girth-weldability. However, the bent pipe is produced by induction heating process over pipe full body.
On the above-mentioned production of line pipes, the design of base material together with welding material is not satisfactory on the stand point of the properties after bending.
Therfore, the base material and heat treatment condition for the fabrication of bent pipes have been examined, and the most preferable design of the mother pipe has been obtained.

The aiming properties of the bent pipes are as follows;

Wall thickness: up to 35mm
Grade : up to API 5LX X-70
Toughness : 70J at -30°C
(Charpy impact test 2mm V notch)

## BASIC CONCEPT OF BENT PIPE

The bent pipe is manufactured traveling through induction heating coil for bending followed by water cooling to obtain high strength for high grades. Both heating temperature and cooling speed of bending are examined in actual bending test and also in the simulated heat treatment thermotester. The results obtained are as follows;

(1) Temperature difference between outside and inside surface is about 100°C for heavy wall pipe up to 35mm.

(2) Water cooling is ordinarily done only from outside surface and cooling speed is ranging from 5 to 15°C/sec. through pipe wall as shown in Fig. 1.

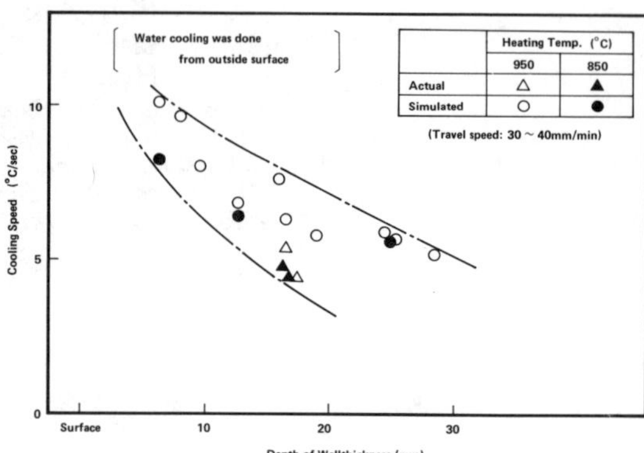


Fig. 1 Change of cooling speed

Therefore, range of heating temperature and cooling speed through pipe wall is ranging very widely, and its microstructure is heterogeneous which leads to variation of properties of the bent pipe. Fig. 2 shows the test results of base material for various testing conditions of heating temperature and cooling speed by means of simulated thermotester together with the simulated heating conditions of bent pipe.

When the heating temperature is high and cooling speed is rapid, the strength of bent pipe is high. Structure transforms austenitic phase at 950°C and large amount of structure is change in bainite with further rapid cooling, so that strength is high.

In case of heating at 850°C, the structure is ferrite-austenite and is changed bainite partially during cooling, so the strength is lower than that in case of heating at 950°C. While, in case of as-water-cooled condition after bending, yield ratio in tensile test is very low, and by further tempering, tensile strength is slightly decreasing although yield strength

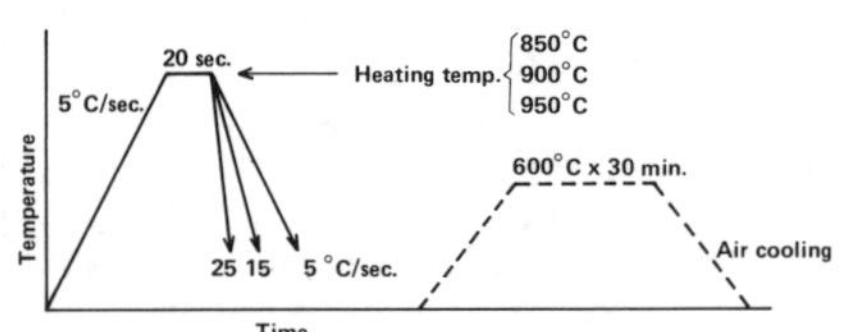


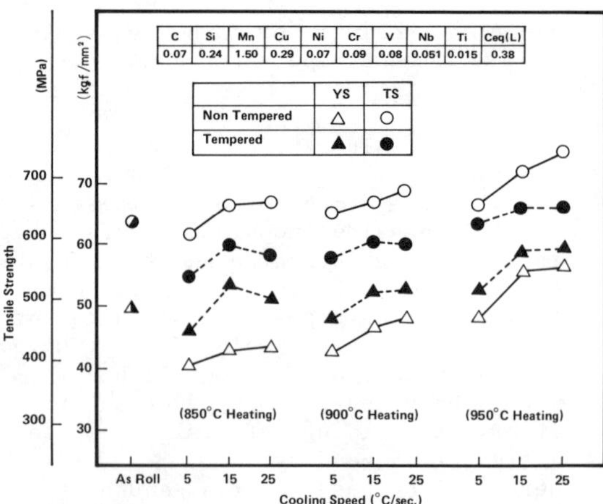


Fig. 2 Influence of heat treatment

increases.
The relationship between strength and toughness by induction heating is shown in Fig. 3 according to the former data. In case of line pipe, as base material is controlled rolling plate, good mechanical properties are obtained by ferrite-pearlite structure.
As bent pipe is heated up to higher temperature than that of austenite transformation, and then cooled, behavior of transformation differs from that of controlled rolling plate. Especially, in as bent condition, decrease of yield ratio is remarkable so the strength of higher grade is not obtained even if the material of high carbon equivalent is applied. By tempering after bending, yield strength recovers and the strength equivalent to that of controlled rolling is reached. On the other hand on low carbon equivalent material, as the structure of as bent condition is ferrite, its toughness is good. However, with increase of carbon equivalent, the amount of bainite structure increases. Although toughness of as-bent pipe is lower, toughness recovers by tempering after bending and the higher value equivalent to that of controlled rolling is obtained.
Therefore, on high grade bent pipe, high strength is obtained through transformation into bainite structure, while tempering is necessary to maintain the stability of strength and toughness.
So, we mainly examined conditions of tempering after bending in the following examinations.

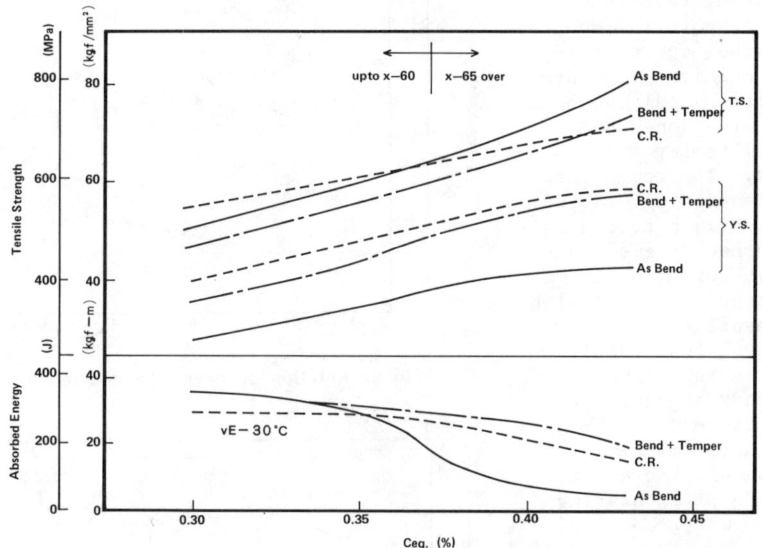


**Fig. 3 Relation between Ceq. and mechanical properties**

## DESIGN OF BASE MATERIAL

Ordinarily, for high grade line pipe the material having carbon equivalent of over 0.35% is used. We selected six kinds of steels including various elements, and examined variation of mechanical properties after heat treatment by means of simulated heat-cycle tester. Table 1 shows chemical compositions of tested materials. As shown in Fig. 4, the material having carbon equivalent of over 0.38% satisfies 5LX X-65 in case of tempering after heating followed by water cooling, and good toughness is obtained by tempering as shown in Fig. 3.
To examine the influence of heating temperature, we tested niobium-vanadium steel and chromium-molybdenum steel in the heating range of 800 to 1050°C. As shown in Fig. 5, the strength and toughness of chromium-molybdenum steel are stable in the

TABLE 1 Chemical Composition

(wt %)

| Steel | C | Si | Mn | Cu | Ni | Cr | Mo | V | Nb | Ti | Ceq. | Pcm | Remark |
|---|---|---|---|---|---|---|---|---|---|---|---|---|---|
| A | 0.03 | 0.22 | 1.52 | 0.18 | 0.17 | - | 0.17 | 0.06 | 0.048 | 0.017 | 0.35 | 0.143 | Mo-V |
| B | 0.05 | 0.17 | 1.74 | - | - | - | - | - | 0.103 | 0.013 | 0.35 | 0.145 | High Nb |
| C | 0.07 | 0.24 | 1.50 | 0.29 | 0.07 | 0.09 | - | 0.08 | 0.051 | 0.015 | 0.38 | 0.181 | Nb-V |
| D | 0.06 | 0.22 | 1.48 | - | 0.15 | 0.12 | 0.20 | - | 0.033 | 0.014 | 0.39 | 0.163 | Cr-Mo |
| E | 0.04 | 0.22 | 1.54 | - | 0.46 | 0.20 | 0.19 | - | 0.047 | 0.011 | 0.41 | 0.156 | Cr-Mo |
| F | 0.10 | 0.26 | 1.43 | - | - | 0.24 | 0.20 | - | 0.028 | 0.016 | 0.43 | 0.207 | Cr-Mo |

Ceq. = C+Mn/6+(Cr+Mo+V)/5+(Ni+Cu)/15

Pcm = C+Si/30+(Mn+Cu+Cr)/20+Ni/60+Mo/15+V/10+5B

wider range of heating temperature. While, the toughness of niobium-vanadium steel is lower at the higher heating temperature than that of chromium-molybdenum steel due to formation of bainite structure. However, this steel can be more economical one when the heating temperature is controlled. The toughness of steel containing 0.1% of high niobium is better than those of other steels and the strength of this ferritic steel is lower with low hardenability.
Through these test results the pipes having high strength and toughness at low temperature were obtained with chromium-molybdenum steel. Moreover, by tempering after heating followed by water cooling, the bent pipes having stable strength and toughness were obtained.

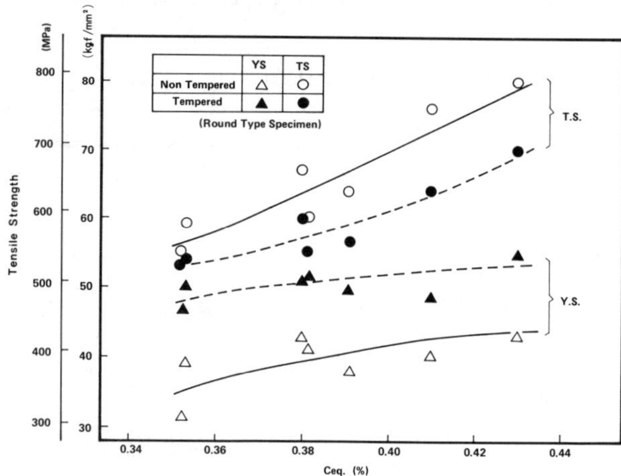


Fig. 4 Relation between Ceq. and tensile strength

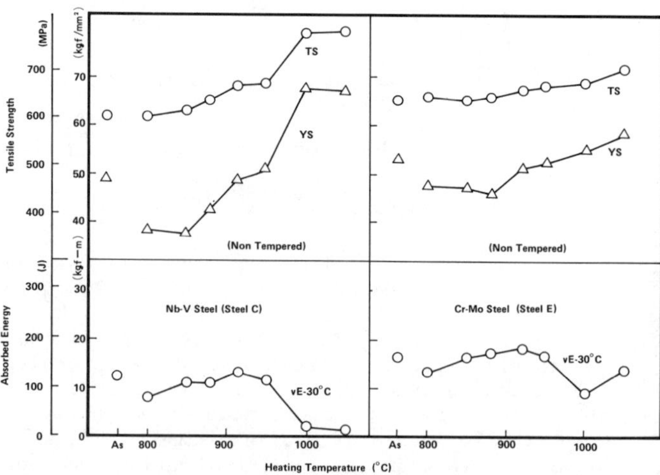


Fig. 5 Influence of heattreatment

## DESIGN OF WELD METAL

On the line pipes, properties of submerged-arc welded portion are considered to be important as well as that of base metal. Improvement of welded portion is done by decreases of oxygen contents and addition of alloy elements in welded metal, but design of weld material heat treated during bending has not yet been examined.

In this test, we examined the influence of mechanical properties depending element of welding metal and heat treatment of bending. Fig. 6 shows test results of strength and toughness after quenching and tempering. Toughness of titanium-added steel is better than that of non-titanium-added steel, because of grain refinement and disappearance of primary precipitated ferrite by addition of titanium. While, the effect of molybdenum as additional element is detrimental, i.e. toughness of high molybdenum is lower than that of low molybdenum because of high hardenability.

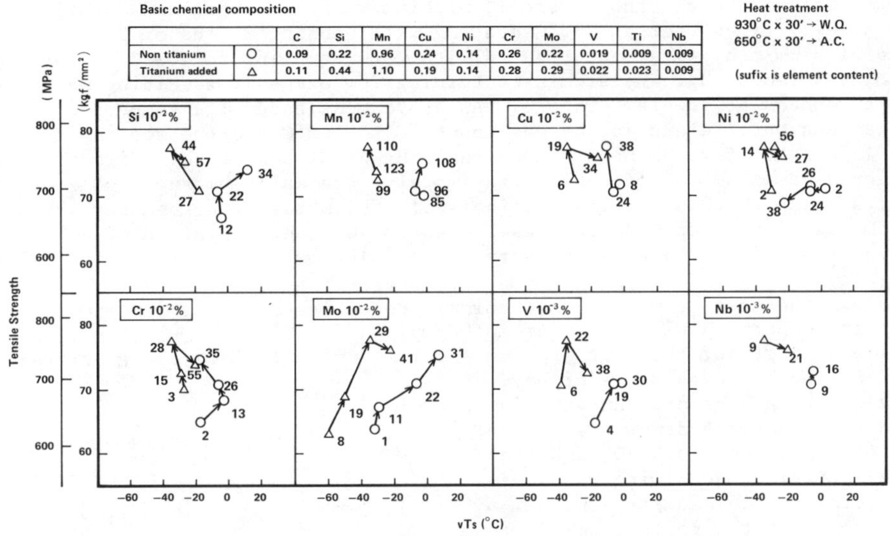
Basic chemical composition

| | | C | Si | Mn | Cu | Ni | Cr | Mo | V | Ti | Nb |
|---|---|---|---|---|---|---|---|---|---|---|---|
| Non titanium | ○ | 0.09 | 0.22 | 0.96 | 0.24 | 0.14 | 0.26 | 0.22 | 0.019 | 0.009 | 0.009 |
| Titanium added | △ | 0.11 | 0.44 | 1.10 | 0.19 | 0.14 | 0.28 | 0.29 | 0.022 | 0.023 | 0.009 |



Fig. 6 Influence of element in weld metal

Fig. 7 shows the influence of oxygen contents in welding metal, toughness of low oxygen content is higher. In case of high oxygen content, the nose of CCT curve shifts to short time zone and upper bainite generates during cooling. While in low oxygen content, lower bainite generates and toughness is improved with decrease of hardenability.

Toughness is improved with decrease of 250-350 ppm from 500-900 ppm in ordinary materials. Low oxygen content can be achieved using highly basic flux developed by Sumitomo Metal Industries, Ltd..

Therefore, improvements of toughness in welding metal are achieved as a result of low hardenability by decrease of

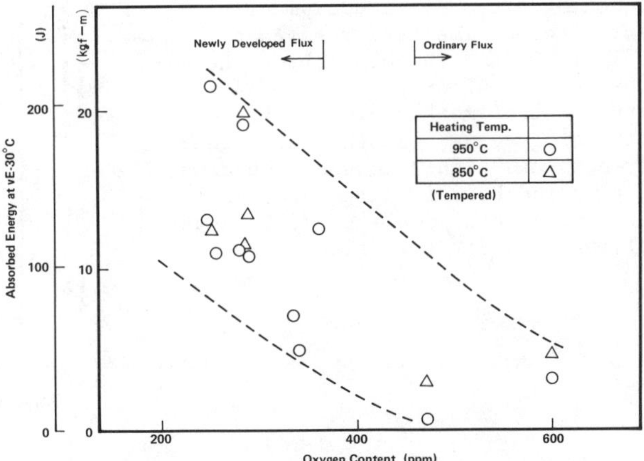


Fig. 7 Influence of oxygen content in weld metal

oxygen contents and high hardenable elements, but more stable toughness can be obtained by tempering after bending.

## ACTUAL PRODUCTION OF BENT PIPE

Through simulation tests we selected the chemical composition, welding material and heat treatment condition for suitable bent pipe. We produced mother pipes using the selected steels, and then produced bent pipe by varing heat treatment after bending. Consequently, some properties of the bent pipes have been examined. Tested materials are high niobium steel, niobium-vanadium steel and chromium-molybdenum steel. Pipes bent at 950°C heating with bending radius of 3DR and traveling speed of 35mm/min. were heat treated with water cooling followed by tempering or quenching & tempering. Materials containing low oxygen for welding were used to maintain toughness at low temperature. As above-mentioned, the temperature difference between outside and inside was about 50 to 100°C and the cooling speed was was 5 to 15°C/sec. Fig. 8 shows dimensional change of bent pipe. Circumferential length decreases at amount of 0.6% approximately, the wall thickness of compression side increases up to 20% over its original wall thickness after bending while the wall thickness of tension side decreases down to 10%, and variation of circumferential length and wall thickness is very small. The ovality of bent pipe is up to 2%.

Table 2 shows the test results of mechanical properties after bending and heat treatment in comparison with those of mother pipes.

The tensile strength of the high niobium steel is equivalent to those of grade 5LX X-52 whereas those toughness is higher.

The tensile strength of niobium-vanadium steel and chromium-molybdenum steel equal to those of grade X-65 to X-70, and those toughness satisfies the aiming value of the base metal and welded portion. Microstructure of bent pipe is finer grain size in comparison with that of furnace heat treatment as shown in Photo. 1. Microstructure of welded portion is fine grain as shown in Photo. 2, while that of heat affected zone is as same as that of base metal and toughness is higher. Additional tests using the induction quenching & tempering equipment has been carried out. The test results are shown in Table 3, and Photo. 3 shows the microstructure after test in comparison with that of mother pipe. This equipment gave us pipes having higher strength and higher toughness because of tempered-banitic structure due to very rapid water cooling and very short holding time for tempering.

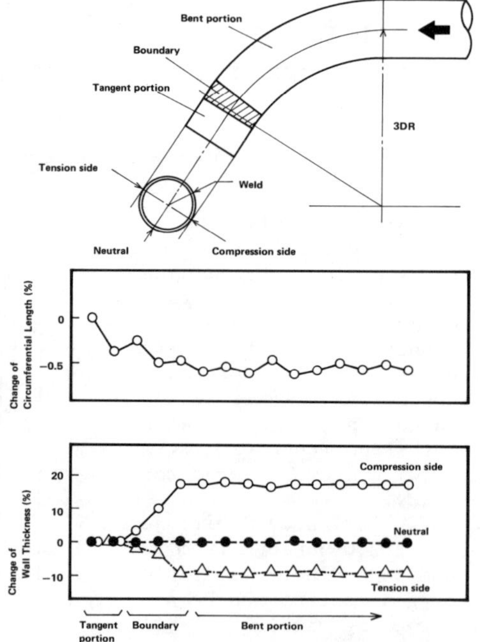


Fig. 8 Dimensional change of bent pipe (28"φ x 1,250"t, 950°C heating)

TABLE 2 Mechanical Properties of Bent Pipe

| Steel Pipe Size | Heat Treatment | Tensile Test Base Metal YS (MPa) | TS (MPa) | El (%) | Weld TS (MPa) | Charpy Test vE-30°C (J) Base Metal | Weld Metal | HAZ |
|---|---|---|---|---|---|---|---|---|
| Steel B | Mother Pipe | 445 | 538 | 58.5 | 607 | 309 | 76 | 257 |
| 711.2mm O.D. | Bend+Temper | 401 | 523 | 62.0 | 513 | 322 | 110 | 328 |
| x31.75mm W.T. | Bend+Q & T | 392 | 522 | 62.2 | 541 | 328 | 152 | 352 |
| Steel C | Mother Pipe | 521 | 620 | 52.7 | 667 | 230 | 72 | 173 |
| 762.0mm O.D. | Bend+Temper | 432 | 578 | 56.9 | 574 | 261 | 97 | 264 |
| x33.35mm W.T. | Bend+Q & T | 510 | 640 | 46.1 | 631 | 93 | 89 | 83 |
| Steel F | Mother Pipe | 509 | 612 | 49.4 | 670 | 161 | 95 | 81 |
| 711.2mm O.D. | Bend+Temper | 494 | 655 | 44.9 | 609 | 84 | 184 | 154 |
| x31.75mm W.T. | Bend+ Q & T | 490 | 652 | 43.1 | 634 | 112 | 76 | 176 |
| Steel F | Mother Pipe | 483 | 585 | 51.4 | 650 | 83 | 96 | 83 |
| 711.2mm O.D. | Bend+Temper | 456 | 602 | 48.8 | 601 | 85 | 87 | 252 |
| x35.00mm W.T. | Bend+Q & T | 466 | 627 | 48.6 | 613 | 89 | 176 | 252 |

Bending condition: Bending radius 3DR
Bending temp. 950°C
Traveling speed 35mm/min.

Heat treatment : Temper 650°C x 1hr. → Air cooling
Q & T 950°C x 1hr. → W.Q. → 650°C x 1hr.→ A.C.

Tensile test : API specimen, Transverse direction

Charpy test : 10x10 size, 2mm V notch

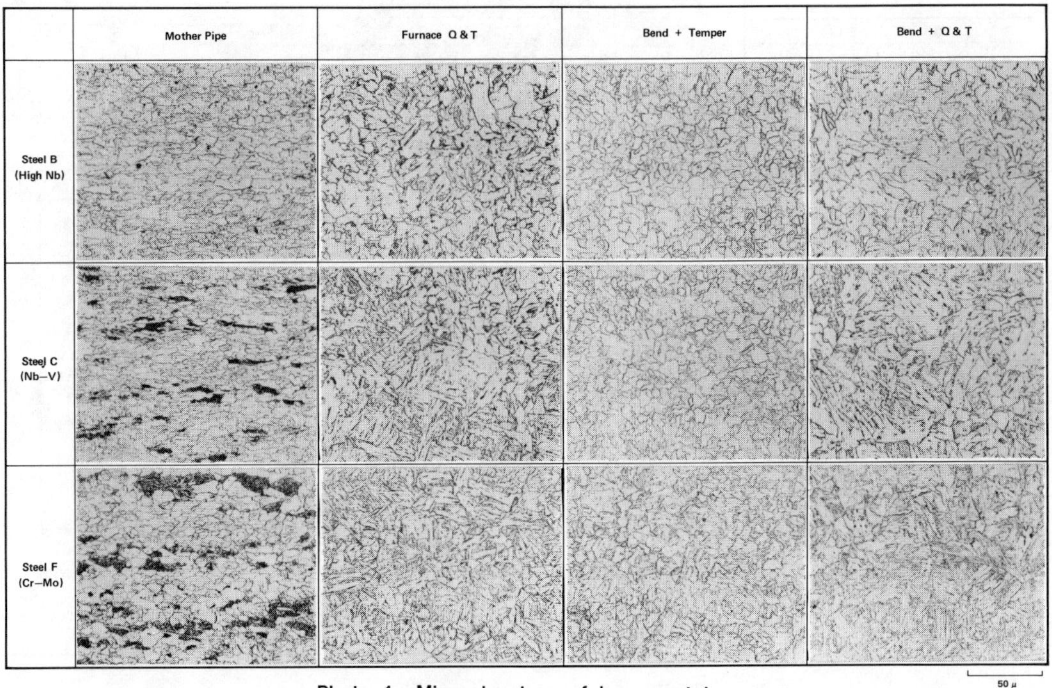

| | Mother Pipe | Furnace Q & T | Bend + Temper | Bend + Q & T |
|---|---|---|---|---|
| Steel B (High Nb) | | | | |
| Steel C (Nb–V) | | | | |
| Steel F (Cr–Mo) | | | | |

50 μ

Photo 1 Microstructure of base metal

Mother Pipe
Bend + Temper
Weld
HAZ
50 μ

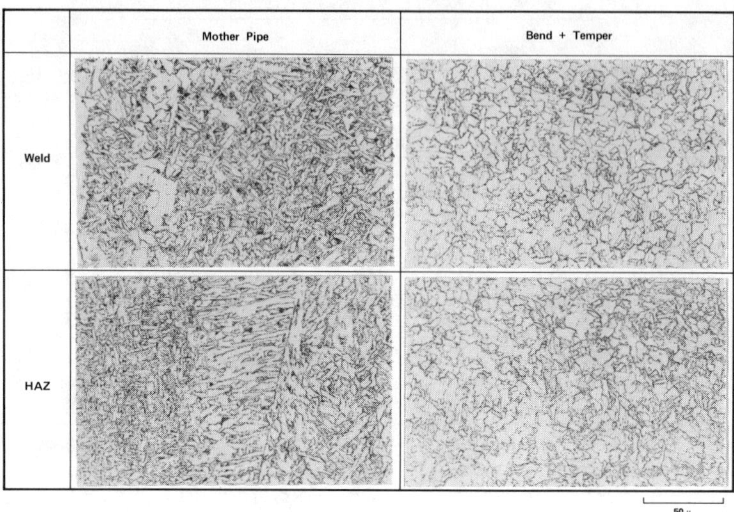

Photo 2 Microstructure of weld portion (Steel F)

TABLE 3
Mechanical Properties of Induction Q & T
(Steel F)

| Tensile Test | | | |
|---|---|---|---|
| Base Metal | | | Weld |
| YS (MPa) | TS (MPa) | El (%) | TS (MPa) |
| 527 | 685 | 49.1 | 671 |

| Charpy Test vE-30°C (J) | | |
|---|---|---|
| Base Metal | Weld Metal | HAZ |
| 96 | 165 | 233 |

Heat treatment condition:
950°C→W.Q.→650°C→A.C.
Tensile test specimen: API specimen, T-direction
Charpy test specimen: 10x10 2mm V notch

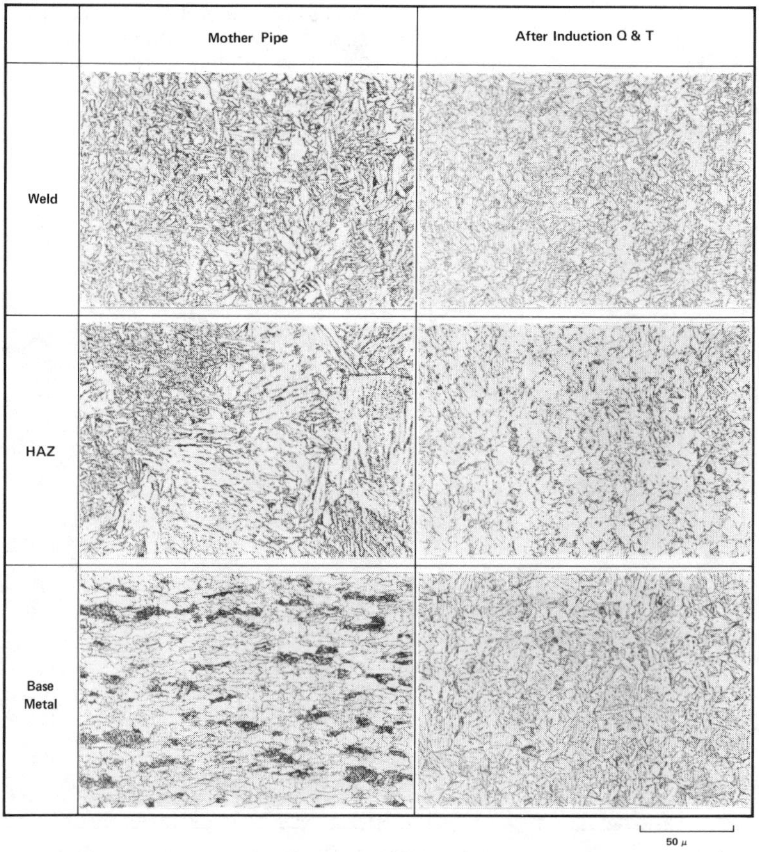

Mother Pipe
After Induction Q & T
Weld
HAZ
Base Metal
50 μ

Photo 3 Microstructure after induction Q & T (Steel F)

## CONCLUSION

We carried out simulated and actual bending tests on heavy wall line pipes using various steels.

The test results obtained are as follows;

1. The bent pipes of grade API 5LX X-60 to X-70 with wall thickness up to 35mm have been produced by using steels having low carbon equivalent value with range of 0.38 to 0.43% (Pcm is 0.18 to 0.22%).
2. Heat treatment of tempering or quenching & tempering after bending at 950°C followed by water cooling yields stable strength and toughness at low temperature.
3. Chromium-molybdenum steel is suitable as base material because its toughness is stable for wider range of heating temperature.
4. The toughness of the welding portion is better with low oxygen content and low hardenable elements.

Recently, for purpose of cost minimization, bending process of line pipe with good combination heating condition and traveling speed have been discussed.

We are now examining this process and the presentation will be made in near future.

# STEELS FOR PIPES OVER GRADE X 70 AND THEIR WELDABILITY

J.P. Jansen and M. LaFrance (USINOR)
H. Pero and Y. Provou (VALLOUREC)

ABSTRACT

With approximately 10% of the international market, VALLOUREC is considered to be among the world's most important pipe producers. Its manufactured products are essentially for export. The department of VALLOUREC manufacturing large diameter welded pipe produces approximately 500,000 metric tons per year, of which the greatest portion is fabricated from steel delivered by USINOR.

The development of high pressure pipelines has rendered mandatory the use of steels with the following properties:
- high yield strength
- high values of ductility, especially in the transverse direction
- good weldability, usually characterized by a low value of carbon equivalent.

After a survey of the evolution in specifications for large diameter linepipe this report will demonstrate how VALLOUREC and USINOR have adapted their production facilties to satisfy the most stringent demands.

It will also show how these new production procedures have resulted in steel grades superior to X.70 which are still weldable at temperatures of -30°C.

## 1 - EVOLUTION OF SPECIFICATIONS FOR LARGE DIAMETER WELDED PIPES

The specifications for large diameter welded pipes have changed very quickly, which is evident to us when we look back at the important orders we have executed over the past 15 years.

Figure 1 illustrates this fact and shows how the steel grades have progressed for pipes manufactured by USINOR and VALLOUREC between 1966 and 1982.

Fifteen years ago, the production of large diameter longitudinal seam welded pipes was limited to steel grades inferior to X.60.

The wall thicknesses were held to between 6.35 and 12.7 mm (1/4" to 1/2") and the diameters did not exceed 40", but were predominantly 30" to 36".

The aim to increase the working pressures for the transport of gas or oil has resulted in the development of steels with better mechanical properties.

The result was that we booked our first important order from Coastal for a steel grade of X.65 at the end of the sixties, (X.65 - 0 36" - t = 10.3 mm). This job was followed in 1972 by an order from LONE STAR (X.65 - 0 36" - t = 13.5 mm) and an order from GREAT LAKES (X.65 - 0 36" - t = 9.52 mm).

The order from GREAT LAKES also indicates another requirement: here Charpy V notch impact values of 66 J at -4°C were specified for the first time. Previously the level that was asked for did not exceed 28 J at + 10°C or 0°C. It was then that we began tests to reduce the sulphur content of steel and to control the sulphide shape. The order for GREAT LAKES was the first to be completed using steel treated with Cerium for sulphide shape control.

In 1973-1974, these more demanding specifications extended to pipes with heavy walls and high steel grades. Here we have the orders from FRIGG (X.65 - 0 32" - t = 20 mm) and DISTRIGAZ (X.70 - 0 36" - t = 11.1 mm).

These two jobs also highlight a second point: due to these revised wall thicknesses and high grades with low carbon equivalent, we had to employ a low carbon steel containing Molybdenum which also gives an advantage in pipe forming and field weldability. We also used for the first time, on the FRIGG order, the MULTIPHY three phase system for controlled rolling of the steel plates. This system was set up with help of IRSID and DILLINGER.

During the following years, the X.65 grade steel of average to high thickness became more generally available and at the same time the pipeline diameters were increased to boost their transport capacity.

This development reflected in the orders for the URSS (0 48" - t = 16 mm) with ys > 460 N/mm$^2$ and UTS > 590 N/mm$^2$.

During this time the demands for higher ductility compelled us to rely on calcium, which is more efficient then cerium for sulphide shape control.

The transition temperatures also had to come down as Figure 2 shows the evolution of the requirements based on the BDWTT between 1975 and 1982.

This evolution towards lower transition temperatures dictates that mill rolling schedules have to be more severe.

The high price increases of Molybdenum in the years 1978-1979 lead to the abandonment of the Nb-Mo formulae and to the replacement of these with Nb-V formulae. During manufacture these steels are calcium treated and in subsequent pipe forming they are subjected to severe cold rolling and form-rolling schedules in the biphase regions. To accommodate these Nb-V steels we have so far introduced several changes in our production procedures.

## II - THE EVOLUTION IN TECHNOLOGY AND PRODUCTION OF X.75 AND X.80 GRADES OF PIPE

### 1 - Actions on the Steel Structure

The characteristics of the plates depend on the chemical composition and the metallurgical structure of the steel. Since the chemical composition of the steel has been increasingly limited, it is the steel microstructure which becomes the main factor and primarily the size of the ferritic grains. It is this factor which is responsible for the rapid development of the controlled rolling techniques.

Controlled rolling is a particular case of thermo-mechanical treatment at high temperatures with the intention of obtaining small ferritic grains which permits an increase in the yield strength and a decrease in the transition temperature.

The basic principle of controlled rolling consists of fixing the temperature as required by the different rolling passes and to control in this way the deformation and recrystallization. The following factors should be borne in mind:

- the passes executed at high temperature are not metallurgically efficient since the recrystallization and the growth of the austenite grains as fast
- the passes executed at intermediate temperatures can lead to smaller austenite grain if there is not grain growth after recrystallization
- the passes executed at low temperatures are very efficient since the austenite grains can not recrystallize between passes which lead to the formation of fine ferritic grains.

Heavy reductions included in the zone of non-recrystallization of the austenite are a must to obtain small ferritic grains. The addition of elements like niobium or vanadium is very interesting since these elements delay the recrystallization.

The grains obtained are the finest when the rolling is ended close to the Ar3 temperature. On top of this, reductions in the two phase regin (temperature < Ar3) are now used to increase the yield strength.

Figure 4 shows the evolution of the finish-rolling temperatures between 1970 and 1982.

These temperatures decreased to a level close to Ar3 somewhere in 1972. Rolling in the two phase region was introduced to industry in 1979.

The low finish-rolling temperatures produce an increase yield strength and a reduction in transition temperature, but this is not always sufficient, in particular when very low transition temperatures are imposed for BDWTT. In this case there was to be a combination of important reductions and reheating in the low temperature region (temperature < Ar3) where recrystallization of the austenite does not occur. Figure 5 shows the evolution of the reheating temperatures of slabs corresponding to the new requirements of DBWTT.

2 - Action if sulphide inclusions and their morphology

The Charpy V impact value at a given temperature depends on the shape of the transition curve of which the basic two elements are the transition temperature and the shelf energy.

We have seen how it is possible to reduce the transition temperature by adopting steels with a low carbon content and severe rolling schedules.

The shelf energy depends essentially on the shape and the distribution of sulphide inclusions. These inclusions, which are highly anisotropic when deformed, create concentrations of local stresses which can increase the risk of ductile failure.

This phenomenon increases as the plane of fracture coincides with greater inclusion length which is the case where samples are taken in a transverse direction.

It is of course relevant that in longitudinally welded pipe the greatest strains occur in the transverse direction and it is the transverse Charpy V Notch values which interest the pipe producer and which have to be increased if possible.

The ever higher level of toughness which is demanded by specifications has led to the production of steel with a lower sulphur content.

Figure 6 shows how this evolution occurrred. The sulphur content was first reduced and the shape of the sulphide inclusions has been modified by treatment with cerium or calcium.

The calcium treatment is actually the best way to reduce sulphur content and achieve inclusion shape control.

### 3 - Application: X.75 and X.80 grades

The examination of past developments leads us to feel that in years to come steel grades superior to X.70 will be produced.

Different observations of market conditions support such a conclusion and we look forward to a progression of steel grades through to X.75, X.80, X.85 and even X.100.

One already hears about X.75 and X.80 and several companies have studied the possibilities of producing such grades. We have here in Dunkirk produced heats in two chemistries:

- Nb-V steel for X.75 grade
- Nb-V-Mo steel for X.80 grade

The plates have been produced with very severe controlled rolling schedules and have been used in the production of 48" pipes with a wall thickness of 15.4 mm (0.606"). The analyses and mechanical properties of the pipes are listed in Table 7.

## III - WELDABILITY AND WELDING OF STEEL GRADES X.75 AND X.80

### 1 - Weldability parameters

For the conduct of the tests we have been particularly interested in the properties in the heat affected zone (HAZ), which are dependant on the cooling rates of the welds. The properties have been checked as a function of the standard cooling parameter 800-500 established by the International Institute of Welding.

Figure 8 gives the underbead hardness curve obtained by simulation for an austenization temperature of 1300°C. We can see that the low carbon content of the steels led to a maximum Vickers hardness of 380. The lower carbon content of the formula containing Niobium, Vanadium and Molybdenum gives this steel a lower maximum hardness.

Since there is a direct relationship between the implant cold cracking test curve and the underbead hardness curve, the Nb-V-Mo steel is seen to demonstrate lower cold cracking susceptibility (Figure 9).

This information has permitted us to set up a welding program to test these materials in the fixed conditions.

2 - Longitudinal welding in pipe-mill

The pipes produced by VALLOUREC by the U.O.E. process have been welded with the submerged-arc system employing the conditions described in Figure 10.

For this production operation all welding has been done with a molybdenum-titanium alloyed wire and a flux containing boron. The Charpy V Notch values obtained in welded joints are slightly more favourable with the Nb-V-Mo steel (Figure 11).

3 - Field welding at -30°C

These tests have attempted to obtain useful information on the performance and productivity expected from recommended basic welding electrodes on the market, and on the weldability of the steel in exceptional climatic conditions.

3.1 - Welding program

The demonstration of the weldability of our steel grades superior to X.70 is based on the following tests executed with coated electrodes:

- in the laboratory primary tests have been performed on bevelled plates (to simulate pipe joints welded in the vertical position with cellulosic or basic electrodes. The aim was to see if an industrial solution was available in which the weld metal could be matched to the base metal.
- Tests of circumferential butt welds in field conditions at -309°C have been performed using such electrodes. This was made possible thanks to the "Etablissements Techniques de Bourges" who do possess a low temperature facility for this purpose.

The API 1104 Standard has been applied to check the quality of the welds.

3.2 - Selection of the filler products: laboratory tests

The aim was here to check the different solutions for filler products with basic or cellulosic coating.

The weld was executed in the vertical position on plates 1000 mm long and 200 mm wide. The two plates were unrestrained and the weld was executed in the transverse direction.

The test conditions were left to the producers of the welding electrodes who also used their own welders.

All the welds were in agreement with the API 1104 Standard.

The Charpy V Notch specimen with notch perpendicular to the weld surface gave a curve for each type of welding electrode. A clear difference between basic and cellulosic electrodes was observable (Figure 12).

Welding electrodes with cellulosic coatings seem to have reached their technical limits in terms of the toughness/yield strength relationship of the weld metal (Figure 13).

3.3 - Tests at -30°C

Tests in the low temperature room at -30°C have been conducted on pipes of 48", wall thickness 15.4 mm, length 3 m.

A wind force of 10 km/h was simulated parallel to the axis of the pipe.

(a) Welding sequence and conditions

The welds were made with the welding electrodes which gave the best results in primary tests executed on plates under the following conditions: 2 welds with basic coated electrodes, vertical-down position for the all weld, one rod containing nickel, the other molybdenum, a third weld with basic coated electrode in first layer and a fourth weld with a cellulosic coated electrode for the first layer and a nickel allied basic coated electrode down for the other layers.

The preparation of the joint was of API type. Each weld was executed simultaneously by four welders.

The first weld with the cellulosic coated electrodes was executed when the pipe was at 100°C preheated to 140°C (starting temperature - 30°C). For the basic coated electrodes only a warm up and drying of the pipe were undertaken.

The heat inputs expressed in kJ/cm are given for each type of welds, each electrode and each quarter of the circumference (in other words, for each welder). We have observed large variations of heat input compared to the tests on the plates probably due to the effect of the cold on the welders and in each quarter of the circumference on higher intermediate heat input (4th or 5th weld) (Figure 4).

Note:
In the climatic conditions of the tests (cooling rate of the pipes approximately 15% faster at -30°C than at the ambient 20°C) we have observed that initial heating needed to be 250°C at the level of the burners to obtain a temperature of 140°C on the bevel at the commencement of welding.

After this initial preheat approximately 10 minutes were available to put the pipes into position and secure them before the temperature dropped to the desired preheat level.

Figure 15 shows also that the interpass temperature could be maintained at an average value of 100°C (Position a = 3 cm from bevel).

Heat flow conditions in the weld area have permitted us to maintain the plate temperature (Position b = 23 cm from bevel) at 70°C over the ambient temperature (here -30°C).

Under such conditions no cracks were observed.

b) Control

All welds have been radiographed. The examinations of films showed no defects according to API 1104. The first weld can be executed with a basic or cellulosic electrode but the fill and capping passes have to be completed with basic electrodes.

c) Inspection of the welded joints

The results of the transverse tensile tests, the bending tests and the texture of the weld are in agreement with the API 1104 Standard.

A Charpy V impact test in the HAZ has been executed in position 0 and 2 mm from the fusion line at -20°C.

For the weld metal the impact specimens have been taken in the weld axis at temperatures -60°C, -40°C, -20°C and 0°C. The notches were perpendicular to the weld surface. The results are given for two quarters of the circumference (Figure 16).

One can consider that the HAZ width is about 2 mm.

The obtained Charpy values are greater than 80 J. One can also see the significant differences between electrodes, due to variations in the heat input.

The impact values of the weld metal are generally similar to but somewhat lower than the values obtained in the preliminary tests on plates.

Figure 17 gives the hardnesses obtained following the desired pattern.

The highest values of hardness in the HAZ can be related to low welding heat input. They are predicted from the last pass due to the absence of any subsequent heat treatment.

## CONCLUSIONS

The tests executed on plates and pipes and the welding test undertaken at a temperature of -30°C with basic electrodes do show that pipes in grades superior to X.70 produced by Usinor and Vallourec do have excellent weldability and mechanical properties.

## REFERENCES

(1) LAFRANCE, M. - PERO, H - PROUVOU, Y. - BOURDILLON, F. "Evolution de la qualite des gros tubes soudes". Seminaire ONU TURIN (Juin 1982).

(2) LIEGEOIS, J. - LECLERC, J. - BOURDILLON, F. - LAFRANCE, M. "Welding X.70 Arctic Grade Line Pipe at frigid temperatures" Pipeline and gas journal - septembre 1978.

(3) LIEGEOIS, J. - "Une experience de soudage par - 30°C de tubes de 48 pounces en acier X.70 au Nb-Mo-Ce" Congres sur les aciers HLE - VERSAILLES (Janvier 1979).

(4) PROVOU, Y. - PERO, H. "Soudabilite des aciers X.75" Symposium de Moscou (juin et novembre 1981).

(5) DENELE, B. - JANSEN, J.P. - LAFRANCE, M. - PERO, H. "Essais de soudage par -30°C de tubes de 48 pources de diametre en aciers de grade superieur a l'X.70". Revue des aceris speciaux 1982.

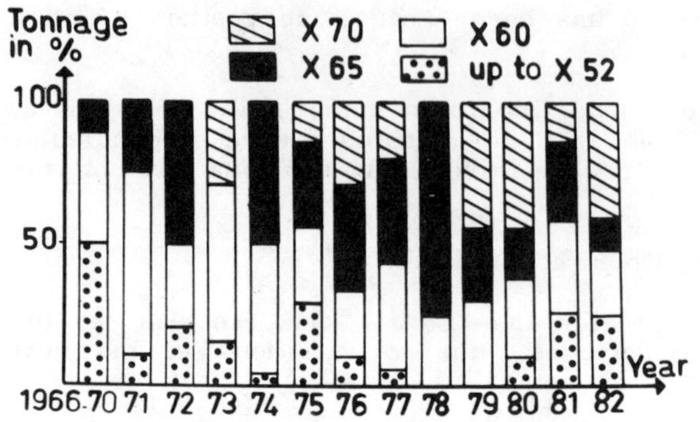


FIG.1: Evolution of the pipelines steel grades from 1966 to 1982

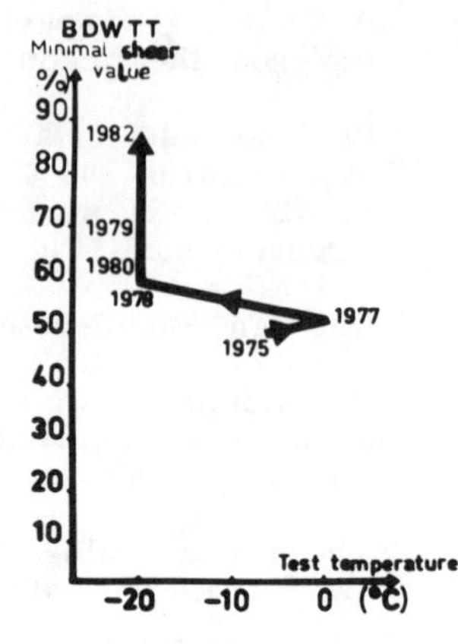


FIG.2: Evolution of BDWTT requirements from 1975 to 1982

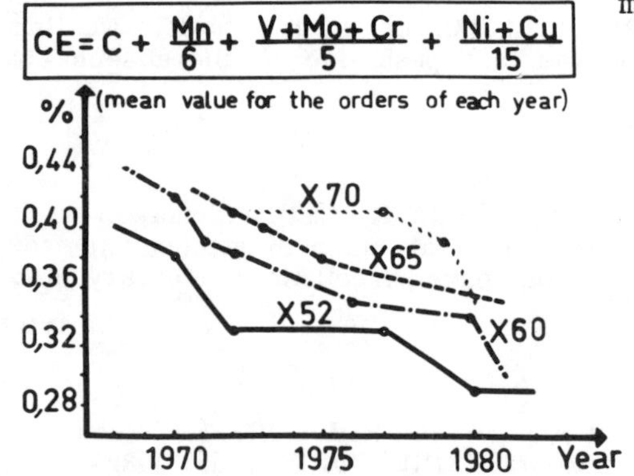


FIG.3: Evolution of carbon equivalent from 1970 to 1982

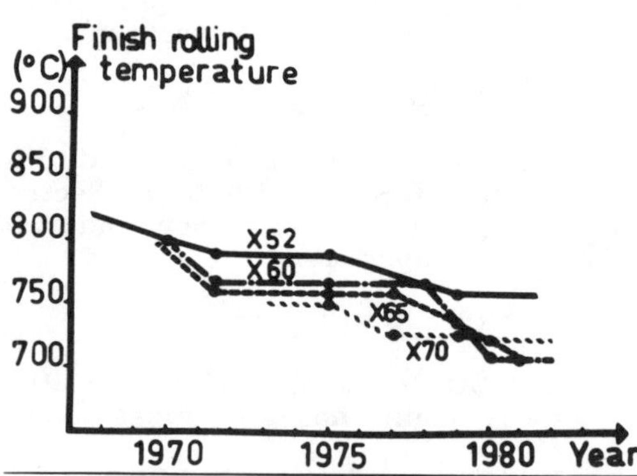


FIG.4: Evolution of finishing-rolling temperatures from 1970 to 1982

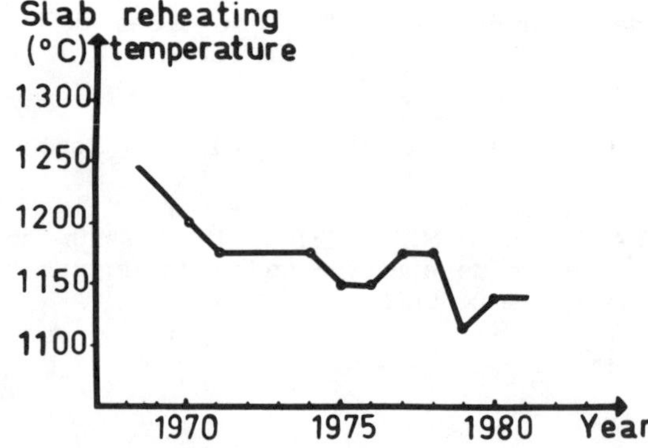


FIG.5: Evolution of the slab reheating temperatures from 1970 to 1982

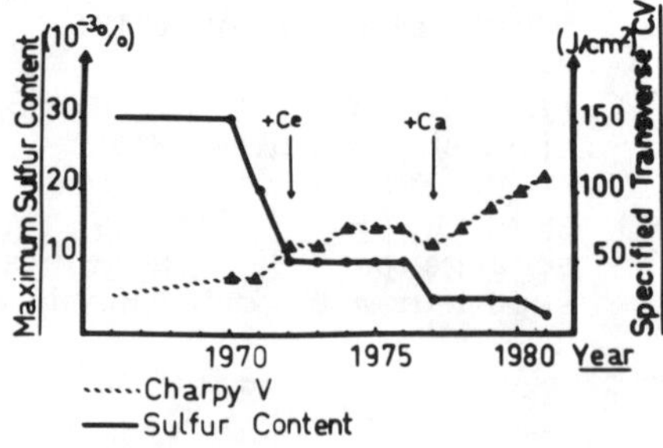


FIG.6: Evolution of the maximal sulfur content according to the specified transverse charpy V value

**Nb.V FORMULA**

| Ladle Analysis(%) |
|---|
| C = 0.091 |
| Mn = 1.690 |
| P = 0.020 |
| S = 0.002 |
| Si = 0.377 |
| Al = 0.277 |
| Nb = 0.058 |
| V = 0.100 |

| | |
|---|---|
| Yield Strength | 538 N/mm² |
| Tensile Strength | 656 N/mm² |
| YS/TS | 0.82 |
| Elongation(5.65√S) | 22 % |

| Transverse Charpy.V notch (10×10×55mm³) | | | |
|---|---|---|---|
| CV − 40°C | CV − 20°C | CV + 20°C | TT 28J |
| 99 J | 108 J | 124 J | − 90°C |

**Nb.V.Mo FORMULA**

| Ladle Analysis (%) |
|---|
| C = 0.071 |
| Mn = 1.692 |
| P = 0.020 |
| S = 0.002 |
| Si = 0.364 |
| Al = 0.035 |
| Nb = 0.048 |
| V = 0.075 |
| Mo = 0.208 |

| | |
|---|---|
| Yield Strength | 572 N/mm² |
| Tensile Strength | 717 N/mm² |
| YS/TS | 0.79 |
| Elongation(5.65√S) | 20 % |

| Transverse Charpy.V notch (10×10×55mm³) | | | |
|---|---|---|---|
| CV − 40°C | CV − 20°C | CV + 20°C | TT 28J |
| 77 J | 82 J | 101 J | − 95°C |

Table 7: Pipe properties

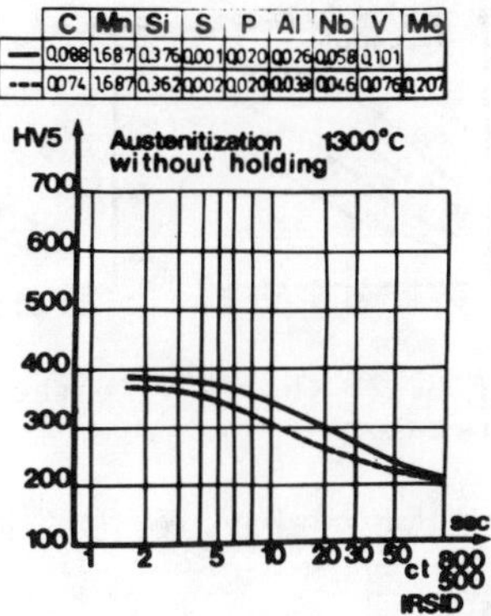

| | C | Mn | Si | S | P | Al | Nb | V | Mo |
|---|---|---|---|---|---|---|---|---|---|
| —— | 0,088 | 1,687 | 0,376 | 0,001 | 0,020 | 0,026 | 0,058 | 0,101 | |
| ---- | 0,074 | 1,687 | 0,362 | 0,002 | 0,020 | 0,038 | 0,046 | 0,076 | 0,207 |



FIG.8: Underbead hardness curves

| | MAG | | | SAW | | | | | |
|---|---|---|---|---|---|---|---|---|---|
| | tack welding | | | inside welding | | | outside welding | | |
| | Ø inch | VOLTAGE Volts | CURRENT Amps | Ø inch | VOLTAGE Volts | CURRENT Amps | Ø inch | VOLTAGE Volts | CURRENT Amps |
| 1st wire | 1/8 | 23 | 700 | 1/8 | 32 | 900 | 1/8 | 32 | 800 |
| 2nd wire | | | | 5/32 | 36 | 1000 | 5/32 | 40 | 950 |
| 3rd wire | | | | 5/32 | 36 | 800 | 1/8 | 42 | 800 |
| Speed | 4m/minute | | | 1.65 m/minute | | | 1.65 m/minute | | |

FIG.10: Welding parameters

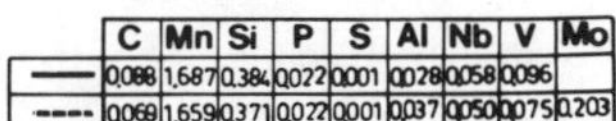

| | C | Mn | Si | P | S | Al | Nb | V | Mo |
|---|---|---|---|---|---|---|---|---|---|
| —— | 0,088 | 1,687 | 0,384 | 0,022 | 0,001 | 0,028 | 0,058 | 0,096 | |
| ---- | 0,069 | 1,659 | 0,371 | 0,022 | 0,001 | 0,037 | 0,050 | 0,075 | 0,203 |

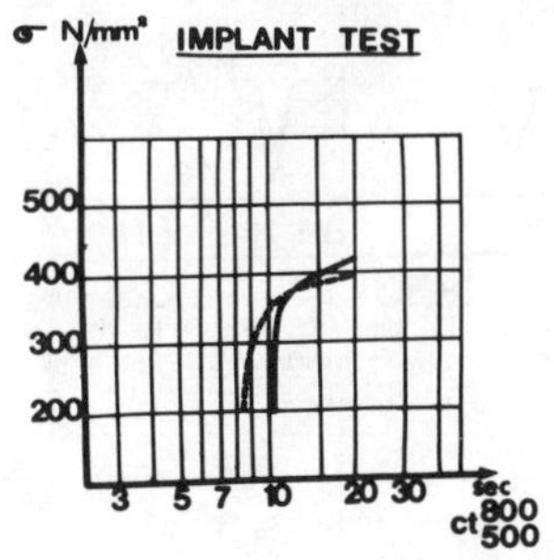



FIG.9: Implant cold cracking test

| GRADE | Position \ T°C | -60°C | -40°C | -20°C | 0°C |
|---|---|---|---|---|---|
| X 75 | Weld Metal | 14 | 26 | 58 | 69 |
| Nb.V | Fusion Line | 20 | 25 | 55 | 70 |
| X 80 | Weld Metal | 51 | 58 | 65 | 80 |
| Nb.V.Mo | Fusion Line | 18 | 70 | 81 | 103 |

(JOULES)

FIG.11: Impact values on welded joints

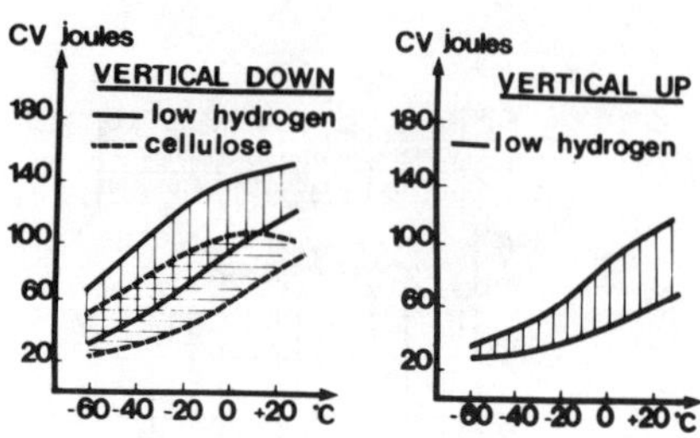


FIG.12: Charpy V in the weld metal

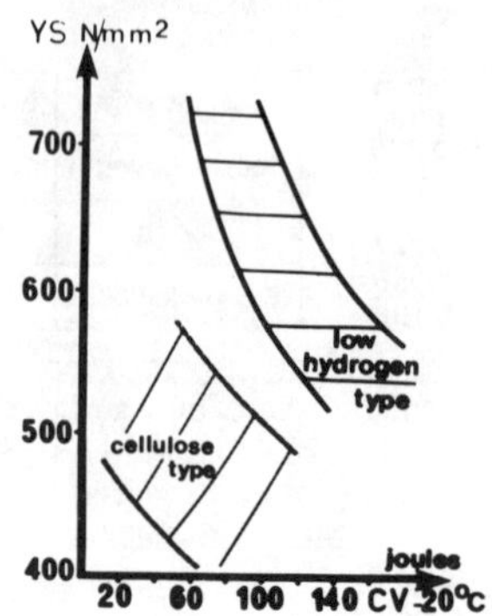


FIG.13: YS-CV relation in the weld metal

Low Hydrogen Type – Vertical Down

| Ni 4 | Ni 3 | mm | Pass | mm | Mo 4 | Mo 3 |
|---|---|---|---|---|---|---|
| 16,6 | 19 | Ø4 | 8 | Ø4 | | |
| 11,8 | 12,3 | | 7 | | 5,2 | 5,5 |
| 13,4 | 10,5 | | 6 | | 17,3 | 15,5 |
| 16,3 | 16,9 | | 5 | | 27,7 | 21,2 |
| 19,1 | 25,3 | | 4 | | 21,8 | 23,1 |
| 17,3 | 16,6 | | 3 | | 15,5 | 13,7 |
| 16,56 | 10,8 | Ø4 | 2 | Ø4 | 9,8 | 18 |
| 13,2 | 12,5 | Ø3,2 | 1 | Ø3,2 | 6,8 | 11,3 |
| kJ/cm | | mm | | mm | kJ/cm | |

Low Hydrogen Type Vertical Up (Ni) | Cellulose Type Vertical Down (Ni)

Low Hydrogen Type Vertical Down (Mn-Ni) | Low Hydrogen Type Vertical Down (Mn-Ni)

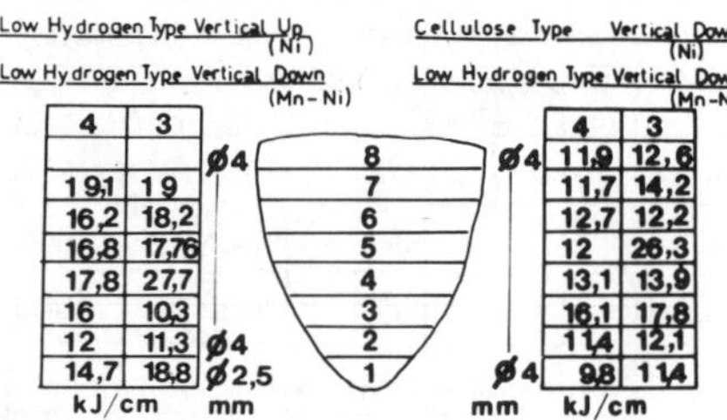

| 4 | 3 | mm | Pass | mm | 4 | 3 |
|---|---|---|---|---|---|---|
| | | Ø4 | 8 | Ø4 | 11,9 | 12,6 |
| 19,1 | 19 | | 7 | | 11,7 | 14,2 |
| 16,2 | 18,2 | | 6 | | 12,7 | 12,2 |
| 16,8 | 17,76 | | 5 | | 12 | 26,3 |
| 17,8 | 27,7 | | 4 | | 13,1 | 13,9 |
| 16 | 10,3 | | 3 | | 16,1 | 17,8 |
| 12 | 11,3 | Ø4 | 2 | | 11,4 | 12,1 |
| 14,7 | 18,8 | Ø2,5 | 1 | Ø4 | 9,8 | 11,4 |
| kJ/cm | | mm | | mm | kJ/cm | |

FIG.14: Heat inputs and electrodes used

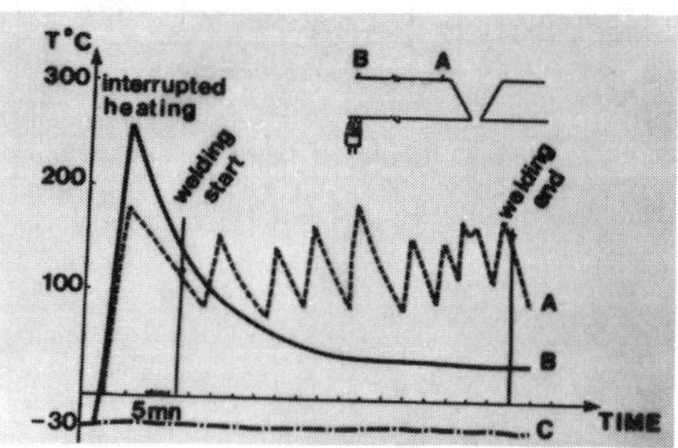


FIG.15: Temperatures during welding (cellulosic electrodes with 150° C preheating)

CHARPY.V TESTS (sample 10×10×55mm)

(Joules)

| electrode | position | FUSED METAL -60°C | FUSED METAL -40°C | FUSED METAL -20°C | FUSED METAL 0°C | HAZ -20°C longitudinal FL | HAZ -20°C longitudinal FL+2 | HAZ -20°C longitudinal BM |
|---|---|---|---|---|---|---|---|---|
| LHVD (Ni) | 4 | 42 | 63 | 88 | 109 | 80 | 152 | 152 |
| | 3 | 23 | 58 | 62 | 97 | 102 | 160 | 145 |
| LHVU (Mn-Ni) + | 4 | 37 | 55 | 85 | 111 | 175 | 114 | 135 |
| LHVD (Mn-Ni) | 3 | 26 | 37 | 68 | 95 | 96 | 165 | 152 |
| CVD + | 4 | 28 | 43 | 60 | 95 | 111 | 206 | 141 |
| LHVD (Mn Ni) | 3 | 21 | 36 | 59 | 106 | 150 | 154 | 150 |
| LHVD (Mo) | 4 | 19 | 33 | 70 | 80 | 97 | 116 | 142 |
| | 3 | 20 | 39 | 52 | 80 | 102 | 177 | 148 |

LHVD = Low Hydrogen Vertical Down CVD = Cellulose Vertical Down
LHVU = Low Hydrogen Vertical Up FL = Fusion Line BM = Base Metal

FIG.16: Impact values in weld metal

HARDNESS 2mm

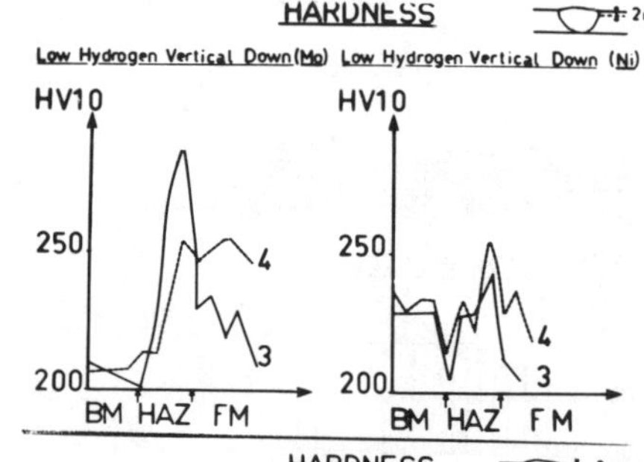


HARDNESS 2mm

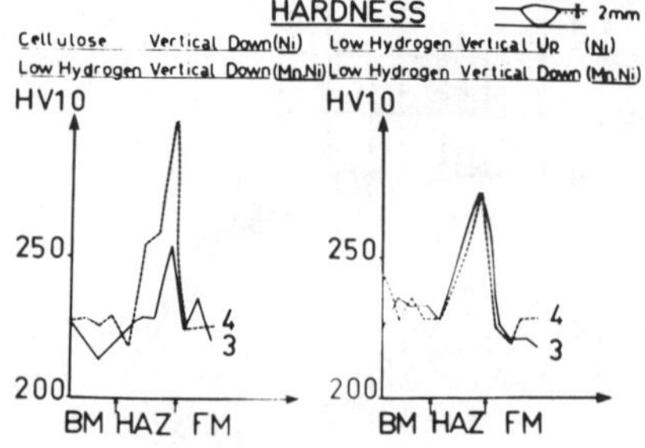


FIG.17: Weld hardness

# THE WELDABILITY OF RARE-EARTH TREATED LINEPIPE STEEL

E.F. Nippes*, J. Mathew*, and G.A. Ratz**
*Rensselaer Polytechnic Institute, Troy, New York
**Molycorp, Inc., Pittsburgh, Pennsylvania

ABSTRACT

A study was made to determine the arc and fusion-zone characteristics of rare earth metal (REM)-treated linepipe steel in the gas tungsten arc, gas metal arc, and flux-cored arc welding modes. Impulse decanting was used to study the fusion-zone characteristics for the welding processes studied. Varestraint tests were also made to study the effect of REM on hot-cracking susceptibility. The data obtained showed that at the usual levels of REM present in commercial steels, about 0.03 to 0.05%, one can expect no significant adverse changes in the arc and fusion-zone characteristics of the steel. Moreover, the susceptibility to hot-cracking is actually reduced by small REM additions. Hence, REM additions to linepipe or high strength low alloy steels should not present any significant adverse problems in welding.

KEYWORDS

Rare earth metals; cerium; linepipe steel; weldability tests; GTA welding; GMA welding; flux cored arc welding; varestraint testing; impulse decanting.

## INTRODUCTION

During construction of a large natural gas pipeline in Australia in 1975-76, some manual metal arc welding difficulties were experienced during the laying of 508 and 864 mm (20-and 34-inch) diameter linepipe. The purported welding difficulties included cracking in the girth welds, undercutting and erractic arc behavior. Both rare earth metal (REM)-treated and non REM-treated steels were used for the pipeline. The purported welding difficulties were initially attributed only to the REM-treated steels. Subsequently, it was determined that the welding difficulties were not solely restricted to the REM-treated steels.

Because welding is the major fabricating process employed in the construction of pipelines, it is imperative that thorough understanding of the role of REM and other alloying elements in the weldability of HSLA steels be developed. Submerged arc welding is now the most often used welding process for seam welding linepipe. Although manual shielded metal arc welding is widely used, automatic gas metal arc welding (GMAW) is finding wider application in the girth welding of linepipe. This process offers improved productivity and economy through a higher welding speed, reduced manpower requirements, and smooth operation at sub-zero temperatures.

Accordingly, Molycorp, Inc. sponsored a series of research projects at Rensselaer Polytechnic Institute (RPI) to understand the nature and extent of the problems in weldability associated with some of the commonly used microalloying elements in linepipe steels. This paper reports on the effects of REM additions on the weldability of Mn-Mo-Cb (Nb) linepipe steels. The investigation was carried out in four phases:

(1) Gas tungsten arc (GTA) spot welds to study the arcing and fusion-zone characteristics (FZC).
(2) Gas metal arc (GMA) bead-on-plate welds in the short-circuiting mode to study short-circuiting and FZC.
(3) Flux-cored arc (FCA) bead-on-plate welds to determine the combined effect of REM, in both the electrode wire and in the base plate, on FZC.
(4) Varestraint tests to study the effect of REM additions on the hot-cracking susceptibility of linepipe steel using REM-treated base plates and REM-containing electrode wires.

## MATERIALS AND EXPERIMENTAL WORK

Laboratory-melted, high strength low alloy linepipe steels (API X-60 grade) were used for the study. REM as 96% mischmetal or 50% RE silicide (with 3% Ca) was added to the steels. The steels contained REM within the range 0 to 0.098%. For the FCAW experiments, lots of commercially prepared flux-cored electrodes were used. These electrodes contained varying amounts of REM (as 30% RESi) within the range 0 to 0.092%.

REM are usually added to steels as mischmetal (containing about 96% REM), or rare earth silicide (containing 30% REM). The various elements usually comprising the REM are cerium (48/50%), lanthanum (32/34%), neodymium (13/14%), praesodymium (4/5%), and other rare earths (1.5%) regardless of product form. Because cerium comprises 50% of the REM, analyses are usually made for Ce only and the total REM content is considered to be twice the Ce content. Therefore, in subsequent discussions in the paper, when a Ce value is given, the total REM content will be twice that value.

### Gas Tungsten Arc Welding

GTA spot welds were made with a constant-current welding system with a function-controlled welding supply. Direct-current straight-polarity (DCSP) welds (work was the anode) were made in a constant-arc-length (CAL) mode and a constant-arc-voltage (CAV) mode as listed in Table 1.

Table 1 GTA Spot Weld Conditions

Electrode: 3.18 mm 2% thoriated tungsten with 90° tip angle
Shielding Gas: Argon at 18.9 l/min. Arc Time: 20 sec.

Constant arc length mode
Current: 300 and 350 A
Initial arc length: 1.6 mm (at 300 A) and 3.3 mm (at 350 A)

Constant arc voltage mode (14.0 ± 0.2 V)
Current: 250 A Initial arc length: 3.3 mm

Gas Metal Arc Welding

The GMA bead-on-plate welds were made using a variable-inductor, rectifier-type power supply in the direct current, reverse-polarity (DCRP) mode. The welding conditions are listed in Table 2.

Table 2 GMA Bead-On-Plate Weld Conditions

Electrode: 1.14 mm steel wire E70S-1B, Cu coated
Shielding Gas: 75% Ar, 25% $CO_2$ at 18.9 l/min.
Electrode Feedrate and Travel Speed: 4300 mm/min and 380 mm/min.
Arc Voltage and Current: 20 V and 180 A
Contact Tube-Work Distance: 13 mm

Flux-Cored Arc Welding

FCA bead-on-plate welds were made on the REM steels using the same welding system as used for the GMAW welds. The welding conditions are listed in Table 3.

Table 3 Flux-Cored Arc Weld Conditions

Electrodes: 1.6 mm dia. medium and high REM
Shielding Gas: 75% Ar, 25% $CO_2$ at 18.9 l/min.
Travel Speed: 380 mm/min.
Arc Voltage and Current: 28 V and 300 A

Varestraint Tests

Varestraint tests were conducted at a constant weld heat input of about 16.5 kJ/mm. The tests were conducted at one, two and four % augmented strain. The test conditions are listed in Table 4.

Table 4 Varestraint Test Conditions

Electrode: 3.18 mm 2% thoriated tungsten
Shielding Gas: Argon at 18.9 l/min.
Travel Speed: 114 mm/min.
Arc Voltage and Current: 13 V and 240 A

Measurement of the Fusion-Zone Dimensions

The weld fusion-zone measurements were made using the impulse decanting and replicating techniques developed by Savage, Nippes and Zanner. The following fusion-zone dimensions were measured (or calculated) for each of the decanted GTA samples:

(1) Volume of the weld crater.
(2) Inner surface area (ISA) of the weld crater.
(3) Weld spot area (SA).
(4) Average depth of the weld crater.
(5) Average width and depth-to-width ratio of the weld crater.

## RESULTS

Gas Tungsten Arc Welds

Figure 1 shows the behavior of the arc voltage at current values of 350 and 300 amperes (A) in the constant arc length mode.

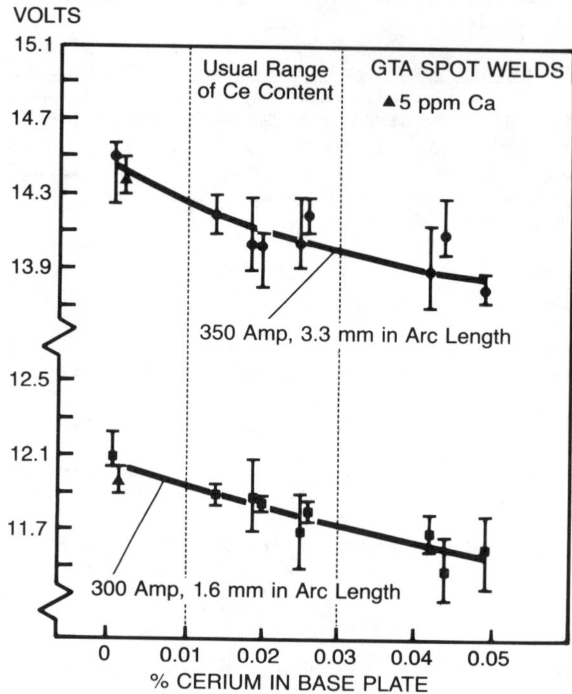


Fig. 1. Arc voltage versus % cerium

There was a tendency for arc voltage to decrease with increasing Ce content of the base plate. An addition of 0.014% Ce decreased the arc voltage (at 350 A) approximately two % (from 14.5 to 14.2 V). A further increase in Ce content to 0.049% (well beyond the normal residual amount of Ce) produced an additional decrease of only about two %. Similar trends were observed for the weld-puddle volume, spot area, inner surface area, average depth and depth-to-width ratio measurements. In fact, the depth-to-width ratio of specimens at a high Ce content of 0.049% was only slightly less than that of the base composition, Fig. 2. The heats within the usual levels of Ce in commercial steels, of about 0.010 to 0.030%, were comparable to the base composition without cerium.

The variation of arc length, when welding in the constant-arc-voltage mode, is shown in Fig. 3. An addition of about 0.015% Ce caused an increase in the arc length. However, further additions of Ce (to 0.049%) did not appear to cause a significant change. This behavior was consistent with that observed for the constant-arc-voltage mode.

## Gas Metal Arc Welds

The results of the GMA bead-on-plate weld tests are shown in Fig. 4 and indicate that the weld-spot area and weld-puddle volume decreased with increasing Ce content of the plate. A similar trend was also observed for short-circuiting frequency, Fig. 5. This effect for GMAW was in contrast to the relatively shallow slope of the curves obtained in GTAW.

However, maximum penetration measurements displayed an increase from 0% Ce to about 0.03% Ce before decreasing, Fig. 5. It should also be noted that the amount of melting observed in GMAW was substantially lower than that obtained in GTA spot welding.

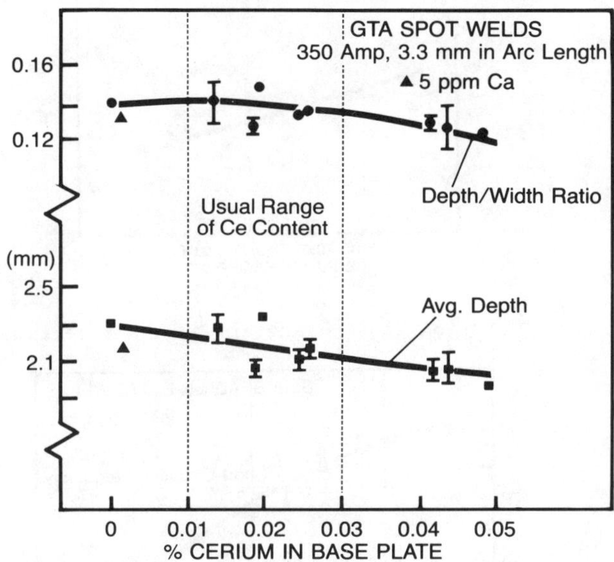


Fig. 2. Fusion zone parameters

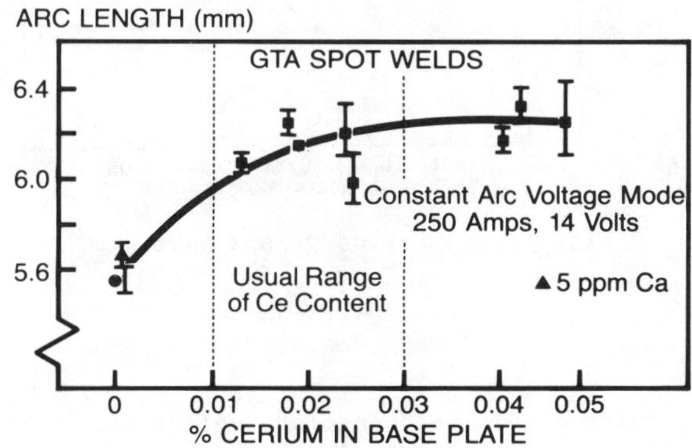


Fig. 3. Arc length versus % cerium

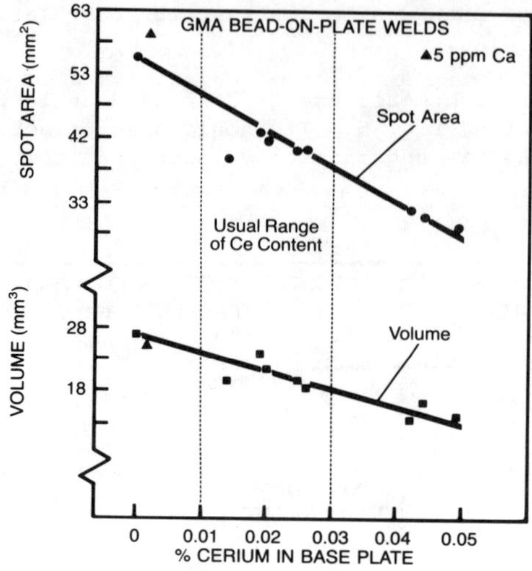


Fig. 4. Fusion zone parameters versus % cerium

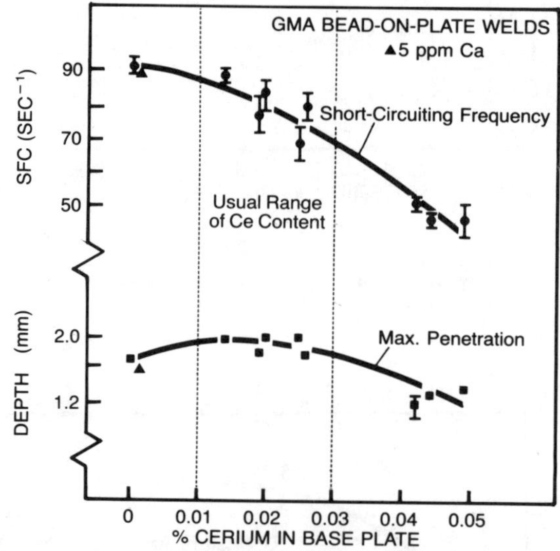


Fig. 5 Penetration versus % cerium

## Flux-Cored Arc Welds

Figure 6 shows the results of FCA bead-on-plate weld tests made with electrode wires containing medium and high levels of REM in the flux-core electrodes. The following points can be noted from Fig. 6.

(a) Both weld-puddle volume and weld penetration showed a slight initial increase with increase in the Ce content of the base plate before decreasing. When welding with an electrode containing a medium Ce content, the puddle volume increased by about 18% as the Ce content of the base plate increased from 0 to about 0.025%. For the plate containing 0.049% Ce, the puddle volume decreased to nearly the same level as that of the base composition.

(b) The effect of increasing Ce content in the flux-cored electrodes tends to flatten the curves, thus reducing the extent of variation obtained with change of Ce content in the base-plate composition. It can be construed that relatively high amounts of Ce in a flux-cored electrode tends to "swamp" or override the effects of Ce in the base plate.

(c) The weld-spot area behaved in an opposite manner observed for puddle volume and the penetration. That is, the spot area initially decreased, then increased, as the Ce content of the base plate was increased.

Thus, it appears that additions of Ce in the electrode wire may improve the fusion-zone characteristics of base-plate steels containing low levels of REM. In addition the variation of the fusion-zone dimensions obtained with increasing Ce content of the base plate is reduced by adding Ce to the electrode wire.

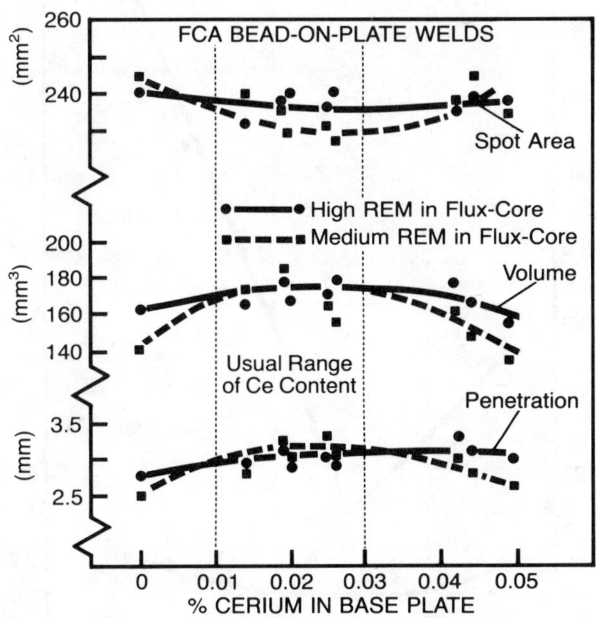


Fig. 6. Fusion zone parameters versus % cerium

## Varestraint Tests

Figure 7 shows the variation of the total crack length versus Ce content of the base plate at two levels of augmented strain. The addition of small amounts of Ce was beneficial to the hot-cracking resistance of the steel. The optimum amount of Ce was about 0.02% or 0.04% REM. The addition of 0.02% Ce decreased the total crack length by almost 70% at four % strain, and by about 40% at two % strain. However at four % strain, as Ce content was increased beyond 0.02%, the extent of cracking increased sharply.

The results of Varestraint tests conducted on weld pads prepared from the REM-containing flux-cored wires showed similar trends and substantiated the results shown in Fig. 7 regarding the beneficial effects of small additions on the hot-cracking resistance of steel.

It must be noted that the amount of Ce usually present in most commercial steels with present steel-making practices is about 0.015% (0.030% REM) and often below about 0.025% (0.050% REM). Therefore, the data obtained in this investigation showed, that at the usual levels of Ce present in commercial steel one can expect no significant adverse changes in the fusion-zone characteristics of the steel. Moreover, the susceptibility to hot-cracking is actually reduced by small REM additions. Hence, REM additions to linepipe and HSLA steels, should not present any significant adverse problems in welding.

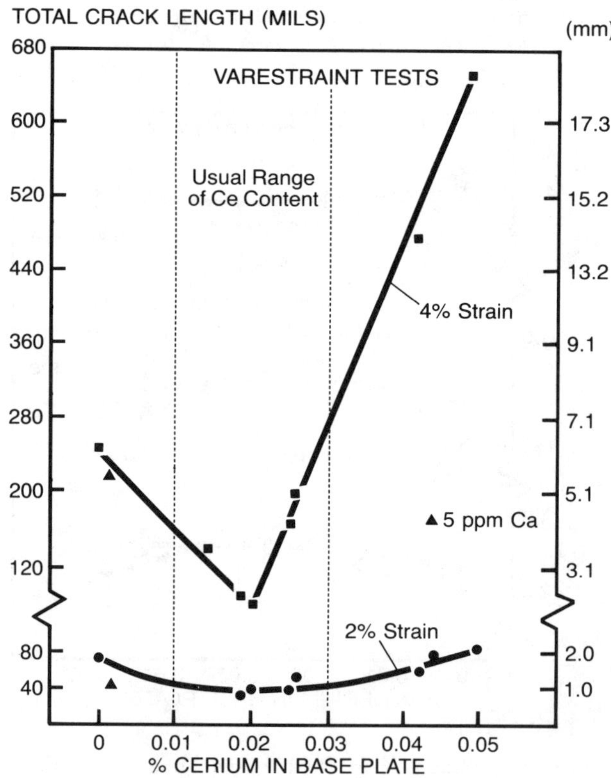


Fig. 7. Total crack length versus % cerium

These results are substantiated by the findings of Banks and Gunn, Cotton, Croll, Fletcher and Morrison, Kelville, Maume and Sasaki regarding the weldability of REM-treated linepipe steels when welded by manual metal arc (MMA) or automatic GMA welding. That is, even though REM can cause a decrease in SCF and penetration or increase in weld spatter, the problems appear to be significant only when the REM content is well above about 0.030%. Furthermore, the problems appear to be exacerbated at high heat inputs and travel speeds.

Patchett and Tien have suggested the following steps to alleviate any possible welding problems associated with relatively high REM additions in steels. Reduction of (1) oxidation potential of the shielding gas, (2) travel speed, (3) heat input, and (4) the use of DCSP and a pulsed-power supply in GMAW.

The improvement in hot-cracking resistance with small additions of REM, and the improvement in weldability with additions of REM to the welding electrode, are substantiated by the findings of Agusa, Blackwood, Pechennikov, and Slutskaya and their co-workers.

## CONCLUSIONS

### Gas Tungsten Arc Welding

Direct current, straight-polarity GTA spot welds, made in the constant-current (CC) and constant-arc-length modes, showed a decrease in arc voltage, weld-puddle volume, and inner-surface area of the fusion-zone with increase in REM content of the plate. In the CC, constant-arc-voltage mode, arc length increased with increase in REM content.

### Gas Metal Arc Welding

DCRP, GMA bead-on-plate welds, made in the short-circuiting mode, showed a decrease in short circuiting frequency, puddle volume, and weld-spot area with increase in REM content. Weld penetration showed an initial increase, and then a gradual decrease with increase in REM content.

### Flux Cored Arc Welding

The results of DCRP, FCA bead-on-plate weld tests indicated that the addition of REM to the electrode had little effect on the welding parameters measured in the base plates with REM in the usual addition range. However, at relatively low and high REM in the base plate, melting and penetration was increased. Increasing REM in the electrode wire suppressed the REM effects in the base plate.

### Varestraint Tests

Varestraint test results indicated that susceptibility to hot-cracking initially decreased and then increased with increasing REM content in the base plate. For the lowest level of hot-cracking, the optimum REM content of the base plate appeared to be around 0.04%. Varestraint tests conducted on weld pads from REM-containing flux-cored wires corroborated these findings.

## REFERENCES

Agusa, K, Nishiyama, N. and Tsuboi, J. (1981). MIG welding with pure argon shielding arc stabilization by RE additions to electrode wires. Metal Construction, 12,

No. 9, 570-574.

Banks, E. E. and Gunn, K. W. (1978). Australian experience in the welding of cerium-treated C/Mn/Cb steels for structural and pipeline usage. In A. B. Rothwell and J. M. Gray (Ed.), International Conference on Welding of HSLA (Microalloyed) Structural Steel, ASM, Metals Park, OH, pp. 467-492.

Blackwood, E. and Squires, I. F. (1980). Modern development in construc. steels for use in weld. fab., Welding in The 80's, Australian Weld. Inst. Meeting, Melbourne, Australia.

Cotton, H. C. (1981). The current status of linepipe and its girth welding. Symposium-Pipeline Weld. in the 80's, Melbourne, Aust., Australian Weld. Research Assoc., 1-25.

Croll, J. E. and co-workers (1979). Pipeline steels in Australia-past experience, present and future applications. Aust. Inst. of Metals, Metals Congress, Perth, Australia, pp. 1-47.

Fletcher, A. L. and Morrison, R. J. (1980). The effect of diff. pipe steels on the cellulosic MMA stovepipe weld. process. Proced. Intl. Conf. Weld Pool Chemistry and Met., Weld. Inst., London, England, pp. 55-64.

Kelville, B. R. (1976). The effect of REM additions on the weld. of controlled rolled pipe plate. British Steel Corp. Report GS/Y/PROD/176/27/76/C, Sheffield, England.

Maume, D. (1978). Disc. In A. B. Rothwell and J. M. Gray (Ed.), International Conf. on Weld. of HSLA (Microalloyed) Struc. Steel, ASM, Metals Park, OH, pp. 495-6.

Patchett, B. M. and Tien, L. S. (1983). Weld. RE-treated pipeline steels with GMAW-S process. 64th AWS Convention, Philadelphia, PA, April 24-29.

Pechennikov, V. I. and co-workers (1977). Influence of RE on properties of a weld in heat-treated low temperature resistant steels. Welding Research Abroad, 23,6.

Sasaki, H. and co-workers (1977). $CO_2$ short arc weld. of RE-treated pipeline steel in circum. welds. Welding Research Abroad, 23, 11.

Savage, W. S., Nippes, E. F. and Zanner, F. J. (1978). Det. of GTA weld-puddle config. by impulse decanting. Welding Journal, 57, 7, 201-s - 210-s.

Slutskaya, T. M. and co-workers (1979). The prop. of wire alloyed with cerium and yttrium used for $CO_2$ weld. Welding Research Abroad, 25, 9.

# SOLIDIFICATION CRACKING PREVENTION IN PIPELINE GIRTH WELDING OF LOW CARBON STEEL

N. Yurioka, S. Ohshita, S. Saito and H. Sakurai
*R & D Laboratories II, NIPPON STEEL CORPORATION*

## ABSTRACT

Solidification cracking in weld metal was found to become more likely as a result of reducing the carbon content of the weld metal. In order to prevent the solidification cracking in the low carbon range, an increase in carbon and nickel in weld metal was effective. It was also beneficial to control the bead shape in girth welding for crack prevention.

## KEYWORDS

Solidification cracking, Girth welding, Line-pipe steel, Carbon, Nickel, Weld metal, SMAW, GMAW

## INTRODUCTION

There is an increasing demand for line-pipe steels having a higher toughness at low temperatures and excellent field weldability. In order to be weldable line-pipe steels must have high resistance to cold cracking and this requires limitation of hardness at their heat-affected zone (HAZ). Furthermore, line-pipe steels are required to resist the hydrogen induced cracking (step-wize cracking) in sour gas environment service.

In order to satisfy the above mentioned demands concurrently, the reduction of carbon in steel is essential. However, the authors recently found that low-carbon steels are somewhat susceptible to solidification cracking when they are welded with low-carbon welding materials and with high travel speed (Ohshita, 1983).

The objectives of the present experiment were to furnish some information to welding engineers for the prevention of solidification cracking at lower carbon contents.

## EXPERIMENTAL PROCEDURE

### Materials

The weld cracking tests employed pipes with large diameter and flat steel plates.

The plate thickness and wall thickness of the pipes ranged between 15.9 mm (0.625 in.) and 25.4 mm (1 in.). Outer diameters of pipes were 609 mm (24 in.) and 1,219 mm (48 in.). The chemical compositions of the steel plates and pipes are shown in Table 1. These are mainly API-5LX-X60 and X-70 grade of line-pipe steels.

For investigating the effect of alloying elements such as Si and Ni, steels melted in a laboratory vacuum furnace with 500 kg capacity were also used. The chemical compositions of these steels are shown in Table 2. As shown in Tables 1 and 2, the contents of C ranged between 0.005 and 0.129%, Si between 0.15 and 0.81%, and Ni between nil and 3.52%.

TABLE 1 Chemical Compositions of Steels

| Symbol | Steel | Shape | Thick. (mm) | Chemical Compositions | | | | | | | | | | | | |
|---|---|---|---|---|---|---|---|---|---|---|---|---|---|---|---|---|
| | | | | C | Si | Mn | P | S | Cu | Ni | Cr | Mo | V | Nb | Ti | B |
| PA | BNT (1) | Pipe | 15.9 | .020 | .16 | 1.83 | .023 | .003 | — | — | — | — | — | .043 | .017 | .0013 |
| PB | BNT (2) | do. | 15.9 | .034 | .16 | 1.61 | .016 | .003 | .29 | .17 | — | — | — | .046 | .016 | .0010 |
| PC | Cr-V-Nb (1) | do. | 15.9 | .039 | .21 | 1.06 | .016 | .001 | — | — | .48 | — | .069 | .037 | .013 | — |
| PD | Cr-V-Nb (2) | do. | 15.9 | .065 | .25 | 1.08 | .009 | .001 | — | — | .42 | — | .069 | .037 | .017 | — |
| PE | V-Nb | do. | 15.9 | .072 | .24 | 1.57 | .018 | .003 | — | .22 | — | — | .071 | .044 | .018 | — |
| PF | Ti | do. | 15.9 | .090 | .29 | 1.57 | .015 | .005 | — | .17 | — | — | — | — | .077 | — |
| PG | Mn-V-Nb | do. | 19.1 | .192 | .27 | 1.31 | .016 | .013 | — | — | — | — | .045 | .041 | — | — |
| FA | BNT (1) | Plate | 22.0 | .011 | .15 | 1.87 | .022 | .007 | — | — | — | — | — | .042 | .020 | .0010 |
| FB | BNT (2) | do. | 15.9 | .034 | .16 | 1.61 | .016 | .003 | .29 | .17 | — | — | — | .046 | .016 | .0010 |
| FC | AF (1) | do. | 18.3 | .021 | .30 | 1.61 | .014 | .002 | — | .20 | — | .20 | .072 | .038 | .013 | — |
| FD | AF (2) | do. | 18.3 | .048 | .28 | 1.57 | .017 | .001 | — | .18 | — | .18 | .073 | .040 | .011 | — |
| FE | V-Nb (1) | do. | 15.2 | .072 | .24 | 1.57 | .018 | .003 | — | .22 | — | — | .071 | .044 | .018 | — |
| FF | V-Nb (2) | do. | 16.5 | .098 | .33 | 1.49 | .017 | .004 | — | .27 | — | — | .074 | .048 | — | — |
| FG | Ti | do. | 15.7 | .090 | .29 | 1.57 | .015 | .005 | — | .17 | — | — | — | — | .077 | — |
| FH | Si-Mn-V | do. | 20.0 | .166 | .33 | 1.39 | .025 | .011 | — | — | — | — | .028 | — | — | — |
| FI | Mn-V-Nb | do. | 19.1 | .192 | .27 | 1.31 | .016 | .013 | — | — | — | — | .045 | .041 | — | — |
| FJ | 3.5Ni | do. | 25.0 | .030 | .60 | .50 | .008 | .004 | — | 3.52 | — | .10 | — | — | — | — |

TABLE 2 Chemical Compositions of Laboratory Melt Steels

| Symbol | Shape | Thick (mm) | Chemical Compositions | | | | | | | | | | | | |
|---|---|---|---|---|---|---|---|---|---|---|---|---|---|---|---|
| | | | C | Si | Mn | P | S | Cu | Ni | Cr | Mo | V | Nb | Ti | B |
| SA | plate | 20.0 | .020 | .41 | 1.42 | .004 | .004 | – | – | – | – | – | – | – | – |
| SB | do. | do. | .021 | .63 | 1.40 | .004 | .004 | – | – | – | – | – | – | – | – |
| SC | do. | do. | .020 | .81 | 1.41 | .004 | .004 | – | – | – | – | – | – | – | – |
| NA | do. | do. | .006 | .36 | 1.42 | .010 | .010 | – | .54 | .01 | – | – | – | – | .0002 |
| NB | do. | do. | .005 | .38 | 1.43 | .011 | .011 | – | 1.11 | – | – | – | – | – | .0002 |
| NC | do. | do. | .006 | .37 | 1.44 | .012 | .010 | .03 | 2.15 | – | – | – | – | – | .0002 |
| ND | do. | do. | .007 | .36 | 1.44 | .011 | .011 | – | 3.24 | – | – | – | – | – | .0001 |
| NE | do. | do. | .041 | .37 | 1.41 | .011 | .011 | – | .06 | .01 | – | – | – | – | .0002 |
| NF | do. | do. | .047 | .38 | 1.43 | .010 | .010 | – | 1.10 | – | – | – | – | – | – |
| NG | do. | do. | .047 | .38 | 1.48 | .010 | .012 | .03 | 2.20 | – | – | – | – | – | – |
| NH | do. | do. | .048 | .38 | 1.43 | .012 | .011 | .04 | 3.20 | – | – | – | – | – | – |

The weld cracking tests employed two welding methods - shielded metal arc welding (SMAW) and gas metal arc welding (GMAW). In SMAW, cellulosic electrodes ranging from AWS E6010 to E9010 grades with 4 mm (0.16 in.) diameters were used. In GMAW, both 100% $CO_2$ and 50% $CO_2$-50% Ar mixture gas were used as shielding gas, together with electrodes of 0.9 mm (0.035 in.) diameter. The chemical compositions of the welding materials examined by the all-weld metal tests are shown in Table 3. The cellulosic electrodes and GMAW electrodes in Table 3 were commercially available except those from GE to GH, which were produced from laboratory-melted ingots.

TABLE 3 All-Weld Metal Chemical Compositions of Welding Materials Used in Tests

| Welding Method | Symbol | Dia. (mm) | Chemical Compositions (%) | | | | | | | |
|---|---|---|---|---|---|---|---|---|---|---|
| | | | C | Si | Mn | P | S | Ni | Cr | Mo |
| SMAW (Cellulose Type) | HA | 4.0 | .12 | .14 | .35 | .011 | .016 | .04 | .03 | .57 |
| | HB | do. | .18 | .24 | .98 | .014 | .008 | .02 | .02 | — |
| | HC | do. | .24 | .20 | 1.01 | .014 | .011 | .24 | .25 | — |
| | HD | do. | .11 | .21 | .62 | .019 | .012 | .02 | .04 | — |
| | HE | do. | .14 | .12 | .88 | .021 | .012 | .23 | .02 | — |
| | HF | do. | .14 | .11 | .75 | .014 | .007 | .21 | .02 | .22 |
| | HG | do. | .09 | .14 | .40 | .015 | .011 | .02 | .02 | — |
| | HH | do. | .12 | .20 | .49 | .013 | .012 | 2.40 | — | .18 |
| GMAW (100% $CO_2$) | GA | 0.9 | .04 | .53 | 1.01 | .005 | .012 | — | — | — |
| | GB | do. | .09 | .94 | 1.53 | .006 | .004 | — | — | — |
| | GC | do. | .24 | .50 | 1.40 | .008 | .007 | — | — | — |
| | GD | do. | .14 | .50 | 1.40 | .009 | .006 | — | — | — |
| | GE | do. | .17 | .85 | 1.37 | .004 | .005 | — | — | — |
| | GF | do. | .23 | .86 | 1.38 | .001 | .005 | — | — | — |
| | GG | do. | .29 | .87 | 1.38 | .002 | .004 | — | — | — |
| | GH | do. | .33 | .87 | 1.41 | .002 | .005 | — | — | — |

## Groove Shape and Restraint

In most of the tests, single root-pass welds were made, because solidification cracking is more likely to occur in root pass rather than in filler-pass welds. In SMAW, a 60 deg. V-groove as shown in Fig. 1 was used in both pipes and flat plates welding. When girth-weld testing pipes, two pipes of pipe length of 1,000 mm were tack-welded, followed by downhill welding by three welders for pipes with 1,219 mm (48 in.) diameter and by two welderes in the case of 609 mm (24 in.) diameter.

SMAW flat plate welding tests employed restraint specimens shown in Fig. 2. The groove shapes in these tests were the same as that shown in Fig. 1 and welding was conducted by vertical down techniques. The restraint against joint shrinkage in the flat plate tests was considered higher than that of the pipe tests.

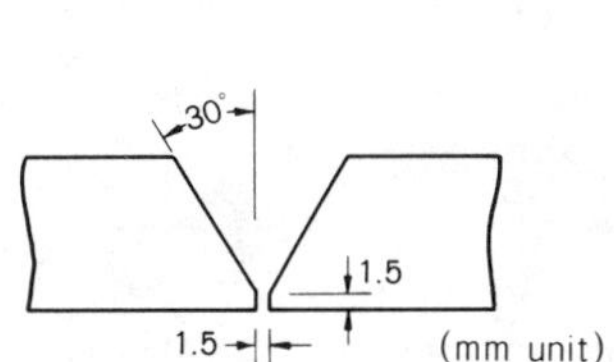


Fig. 1. Groove shape for SMAW.

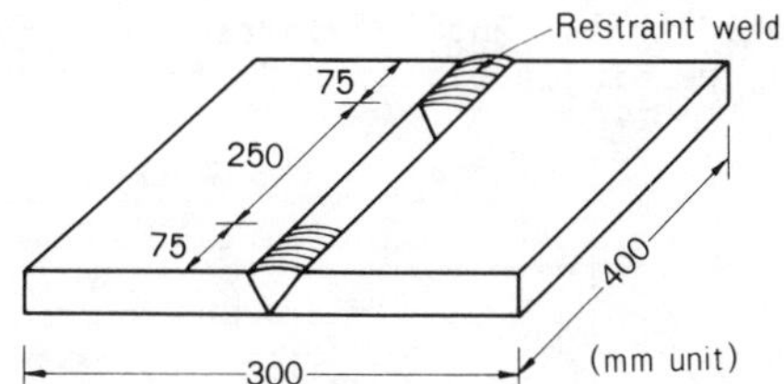


Fig. 2. Test specimen for root weld.

A fully automatic process was employed for GMAW pipe tests. Fig. 3 shows the groove shape with copper backing in GMAW tests. In flat plate GMAW tests, a test piece shown in Fig. 4 was used. The groove shape was the same as that in Fig. 3 in most tests, but the groove shape was changed in the tests for investigating the effect of the groove shape on solidification cracking as shown in Fig. 5.

In order to compare the joint restraint in the tests, the shrinkage of joints in the direction perpendicular to the welding line was measured by a contact-type strain gauge. The contact balls were inserted in the back side of the plates and the inner side of pipes. The gauge length was 20 mm (0.8 in.), and the amount of shrinkage was obtained by measuring the change in gauge length after welding.

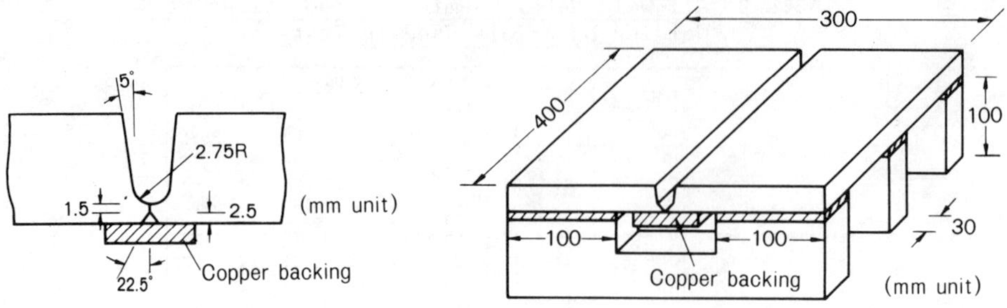


Fig. 3. Groove shape for GMAW.

Fig. 4. Test specimen for root weld.

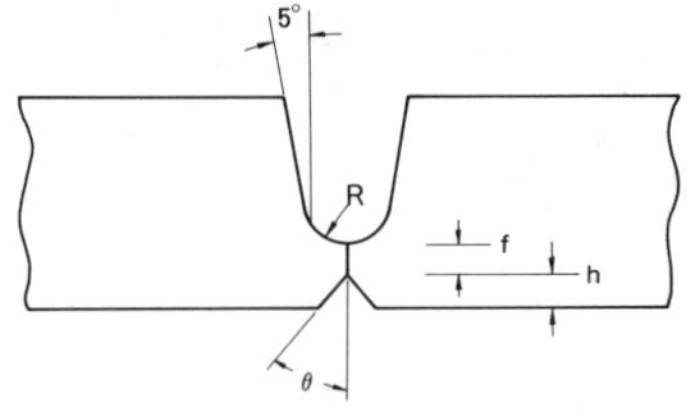


| Symbol | h (mm) | f (mm) | θ (deg) | R (mm) |
|---|---|---|---|---|
| XA | 3.0 | 1.0 | 22.5 | 2.75 |
| XB | 2.0 | 1.0 | 22.5 | 2.75 |
| XC | 1.0 | 1.0 | 45.0 | 2.75 |
| XD | 1.0 | 1.0 | 15.0 | 2.75 |
| XE | 1.0 | 1.0 | 22.5 | 2.75 |
| XF | 1.0 | 0.5 | 15.0 | 2.75 |
| XG | 1.0 | 0.5 | 22.5 | 2.75 |
| XH | 1.0 | 0.5 | 45.0 | 2.75 |

Fig. 5. Groove shape with varying root face.

## Welding Conditions

The welding conditions for each test method are summarized in Table 4. The SMAW employed cellulosic electrodes with 4 mm (0.16 in.) diameter, and the travel speed ranged from 240 to 590 mm/min (9.4 to 23.2 ipm). The GMAW employed 0.9 mm (0.04 in.) electrode with travel speed ranging from 300 to 1,600 mm/min (12 to 63 ipm). The welding current and voltage employed in the tests varied depending on welding speed, welding position and groove shape.

TABLE 4 Testing Conditions

| Welding Method | Test Method | Welding Condition | Material: Base Metal | Material: Weld Material | Test Purpose |
|---|---|---|---|---|---|
| SMAW (4mmφ) | Pipe [Fig.1] (All position) | 140～200 A, 240～590 mm/min | PA～PG | HA～HG | Carbon Effect |
| | Plate (Vertical Down) Fig.1 Fig.2 | 140～190 A, 290～560 mm/min | FA～FI | HA～HG | Carbon Effect |
| | | | FA, FG NA～NH | HA, HH | Nickel Effect |
| GMAW (0.9mmφ) | Pipe [Fig.3] (All position) | 200～300A, 300～1,500 mm/min 24～31V (100%$CO_2$) | PA～PG | GA～GH | Carbon Effect |
| | Plate (Flat position) Fig.3 Fig.4 | 250～300A, 400～1,600 mm/min 26～31V (100%$CO_2$) | FA～FI | GA～GH | Carbon Effect |
| | | 250～300A, 500～1,500 mm/min 27～31V (100%$CO_2$) | SA～SC | GA, GB, GE | Silicon Effect |
| | | | FA, FJ NA～NH | GA, GB, GC GE | Nickel Effect |
| | Plate (Flat position) Fig.4 Fig.5 | 260～280A, 600～1,000 mm/min 26～28V (50%Ar + 50%$CO_2$) Oscillate(5Hz 2mm Width) | FA | GA, GB, GE | Groove Effect |

## RESULTS AND DISCUSSION

### Effect of Travel Speed and Weld Metal Carbon Content

The occurrence of solidification cracking was determined not only radiographically, but macroscopically by observing sections cut from welds. The reason was that the macroscopic observation made it possible to distinguish between solidification cracks and cold cracks. Fig. 6 shows a macrograph of a solidification crack in a root-weld made using a cellulosic electrode. Solidification cracking mostly occurred at the position where horizontally growing dendrites finally meet each other (a head-on collision type of dendritic growth) as seen in Fig. 6.

The surfaces of some solidification cracks were observed by a scanning electron microscope (SEM). Fig. 7 shows an example of SEM observation which clearly shows a typical type of weld solidification cracking. It should be noted that root solidification cracks, for all practical purposes, disappeared in most cases by subsequent second pass deposition as shown in Fig. 8. Fig. 9 shows an example of cracking initiated at a root bead by a GMAW process with copper backing.

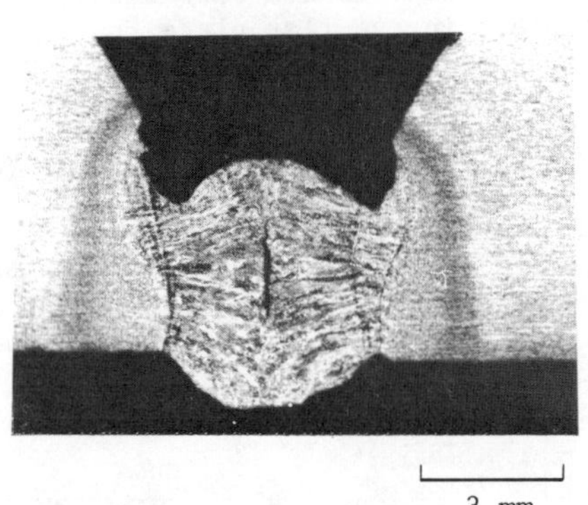


Fig. 6. Macrograph of GMAW root-pass bead with a solidification crack.

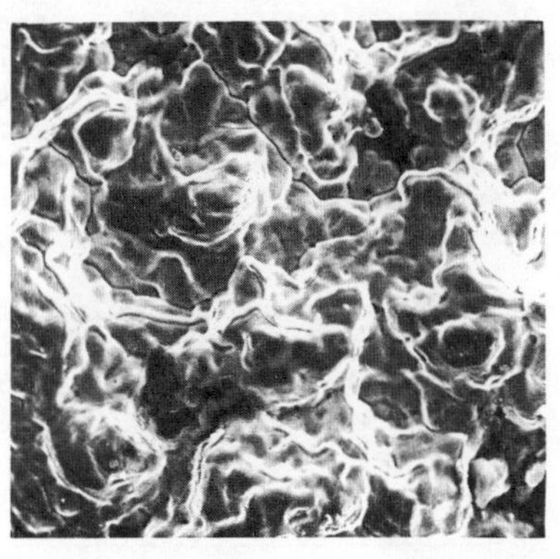


Fig. 7. Scanning electron micrograph of surface of a solidification crack.

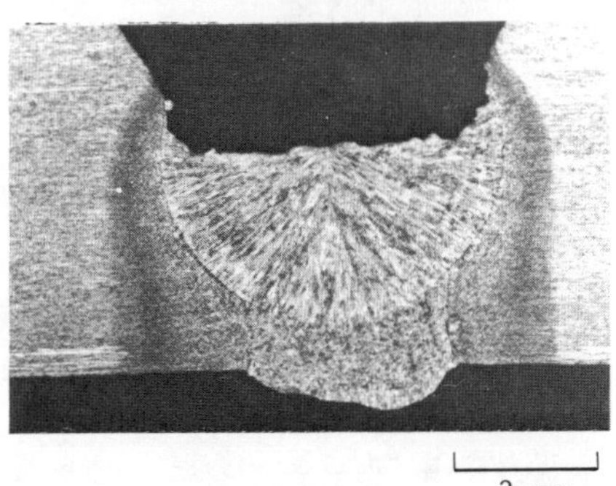


Fig. 8. Macrograph of root-and-hot pass bead in cellulosic electrode welding.

Fig. 9. Macrograph of root-pass bead with a solidification crack in GMAW pipe girth welding.

The results of SMAW pipe girth weld tests were plotted with respect to the weld metal carbon content and the travel speed in Fig. 10. The weld metal carbon content ($[C]_{weld}$) was determined from that of base metal ($[C]_{base}$), that of all-weld metal ($[C]_{all\ depo}$) and the base metal dilution rate (DR) as (Ohshita, 1983):

$$[C]_{weld} = DR\cdot[C]_{base} + (1 - DR)\cdot[C]_{all\ depo} \tag{1}$$

where DR is 0.56 or thereabout in the present experiments.

Fig. 10 reveals that the solidification cracking occured as the carbon content in weld metal became less than 0.06% and the travel speed exceeded 330 mm/min (13 ipm). $\delta$ shown in Fig. 10 denotes the amount of joint shrinkage measured by the contact-type strain gauge. These values ranged between 0.35 and 0.50 mm (0.014 and 0.020 in.) in pipe girth-welding.

Fig. 11 is the result of flat plate SMAW cracking tests. The critical line for the occurrence of solidification cracking at Ni = 0% in Fig. 11 is similar to that in Fig. 10, since the weld metal in the tests shown in Fig. 10 did not contain a substantial amount of Ni. The critical carbon content in the flat plate restraint cracking tests was 0.065%. The critical carbon content at which susceptibility to cracking increases is slightly greater in the case of flat plates than girth welds. This can be attributed to the higher restraint present when flat plate welding, that is suggested by the lesser shrinkage which was between 0.30 and 0.35 mm (0.010 and 0.014 in.).

As seen in Figs. 10 and 11, solidification cracking is not initiated when travel speed is under than 330 mm/min (13 ipm). It follows that the cracking is not expected in the welding with low-hydrogen electrodes since they are generally used with travel speeds less than 250 mm/min (10 ipm).

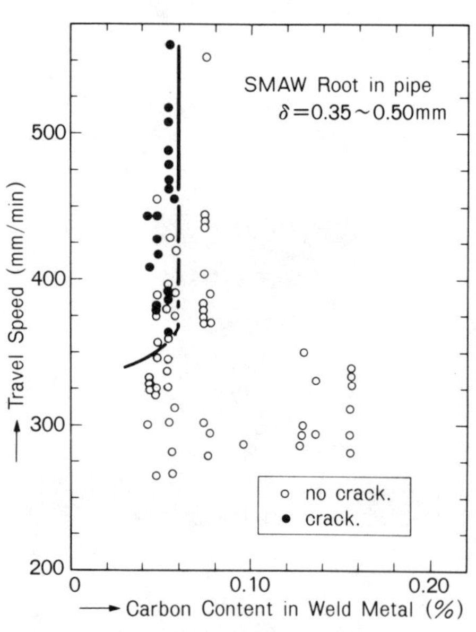


Fig. 10. Critical conditions for solidification cracking in SMAW root welds in pipes.

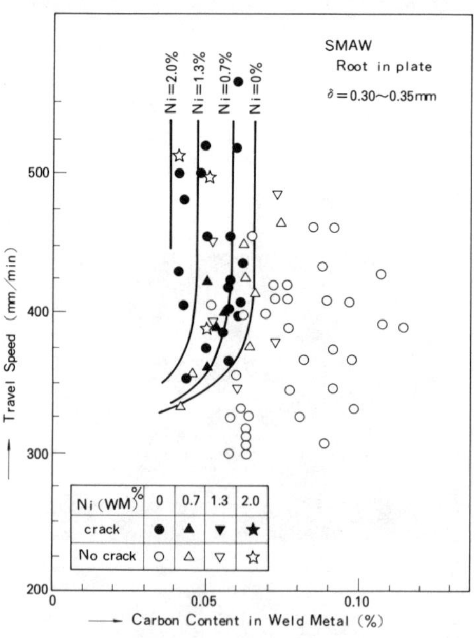


Fig. 11. Critical conditions for solidification cracking in SMAW root welds in flat plates.

GMAW results are shown in Fig. 12 for the test of root-weld pass in girth welding and in Fig. 13 for root-weld pass in flat plates welding. The carbon content of root-weld in GMAW was roughly given by the following relation (Ohshita, 1983):

$$[C]_{weld} = C_o + 0.89 \{DR \cdot [C]_{base} + (1 - DR) \cdot [C]_{wire}\} \tag{2}$$

where $C_o$: carbon pick-up from shielding gas, which is 0.02% for 100% $CO_2$ and 0.01% for 50% Ar-50% $CO_2$

DR: base metal dilution rate, which is 0.58 or thereabout in the groove shown in Fig. 3

$[C]_{wire}$: carbon content of GMAW electrode

As in the case of SMAW, solidification cracking became more likely to occur as the carbon content decreased and travel speed increased. The critical carbon content was 0.08% for the pipe girth-welding and 0.10% for the flat plated tests. This difference can be also attributed to the difference in the amount of joint shrinkage.

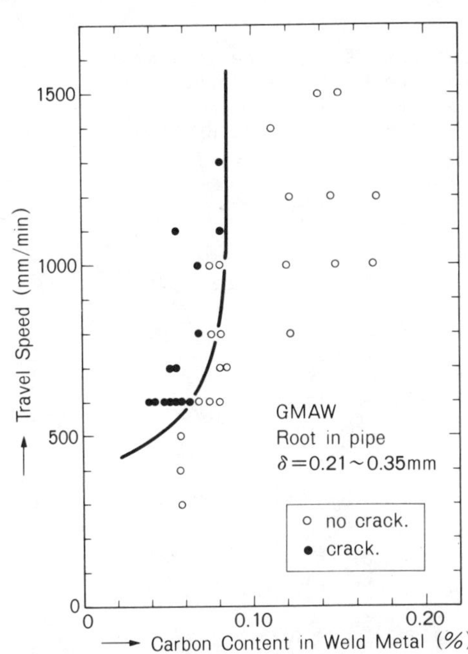


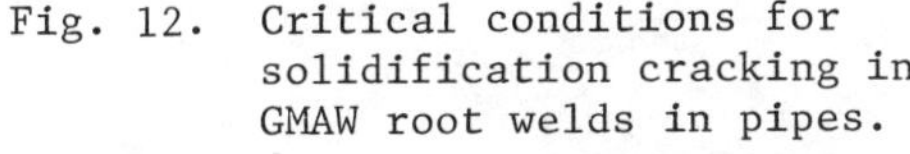

Fig. 12. Critical conditions for solidification cracking in GMAW root welds in pipes.

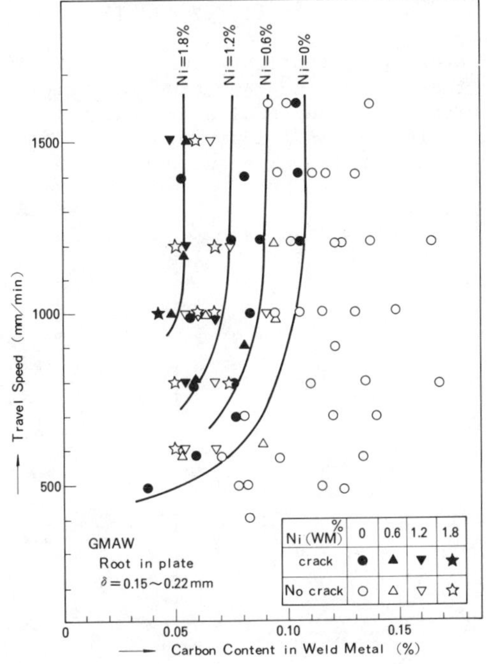


Fig. 13. Critical conditions for solidification cracking in GMAW root welds in flat plates.

## Effect of Weld Metal Silicon and Nickel Content

It had been common knowledge that steel welds with higher carbon contents were more susceptible to the solidification cracking. In contrast the present experiment showed that the solidification cracking becomes more likely to occur when carbon content is decreased. The solidification cracking did not occur in weld metals with carbon content as high as 0.20%.

In the equilibrium state, steels solidify in a δ-phase state when they contain carbon contents less than 0.09%. When the steels transform from δ (ferrite) to γ (austenite) phase during cooling, transformation shrinkage occurs. The lateral shrinkage at the transformation from δ to γ is 0.0011 (Briggs, 1946). Although the transformation raises the shrinkage of solidifing metals by only a small amount, it is believed to be responsible for the increase in susceptibility accompanying reduction in carbon content.

Alloying elements such as Al, Cr, Si, Ti, Mo, V, W and Zr are known as δ-phase stabilizing elements in steels. C, Ni, Mn, and Cu are γ-phase stabilizing elements. There is a carbon equivalent expressing the element's contribution to γ-phase solidification of steel welds as (Wada, 1967):

$$CE_{\gamma} = C + \frac{Ni}{28} + \frac{Mn}{110} + \frac{Cu}{83} - \frac{Si}{15} - \frac{Mo}{21} - \frac{Cr}{76} \quad (3)$$

Equation (3) predicts that C and Ni are beneficial and Si is not beneficial to prevent solidification cracking enhanced by δ-phase solidification. This prediction was partly verified by the experimental result that solidification cracking was prevented by maintaining the weld metal carbon over the critical amount as shown in Figs. 10, 11, 12 and 13.

In order to investigate an effect of Si on cracking, flat plate root-weld tests were conducted in GMAW using laboratory melted steels with varing Si contents shown in Table 2. The results are shown in Fig. 14. It is seen that Si had no effect on the critical conditions for solidification cracking when the weld metal carbon content and the travel speed were varied over ranges.

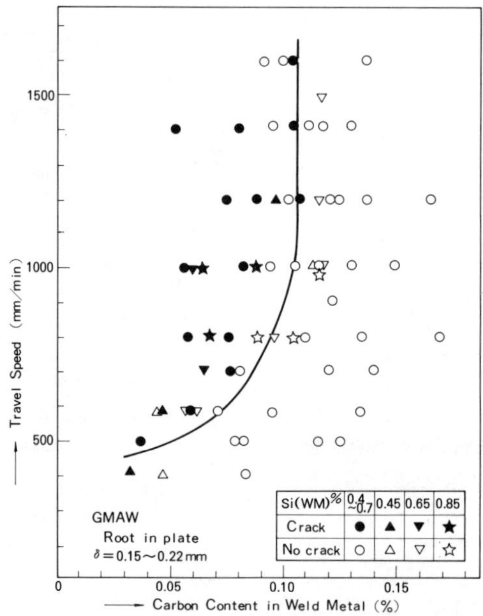


Fig. 14. Effect of weld metal silicon on solidification cracking.

The nickel effect was examined in flat plate root-weld tests with both SMAW and GMAW processes using Ni bearing commercial product steels shown in Table 1 and laboratory melted steels shown in Table 2. The experimental results for SMAW and GMAW are shown in Fig. 11 and Fig. 13, respectively. It is seen that an increase in the Ni content in weld metal was very beneficial for the prevention of solidification cracking.

Effect of Shape of Root-Pass Bead

In order to examine the effect of the bead shape, flat plate root-weld tests with GMAW were conducted with varying groove shape as shown in Fig. 5. These tests employed 50% Ar-50% $CO_2$ as shielding gas and oscillation techniques with 5 Hz and 2 mm width in order to change the bead shape more effectively. Fig. 15 shows the macrophotographs of root-welds by steel FA and GMAW electrode GA with varing weld groove under the condition of 800 mm/min of welding speed. The height-width ratio of a root-weld was found to decrease as the total root face in the groove (f + h in Fig. 5) decreased.

The result of macrographic observation of sectioned pieces from welds was shown in Fig. 16, where the ordinate denotes the percentage of numbers of sections with solidification crack among 7 sections taken from 400 mm long root-weld. The results revealed that the probability of solidification cracking became less when the bead height-width ratio was controlled below 1.0. The susceptibility to solidification cracking in the test whose result is shown in Fig. 16 was slightly less than that in 100% $CO_2$ GMAW test. An employment of 50% Ar-50% $CO_2$ shielding gas and oscillation techniques made the shape of a root bead concave, which seem to reduce the cracking susceptibility.

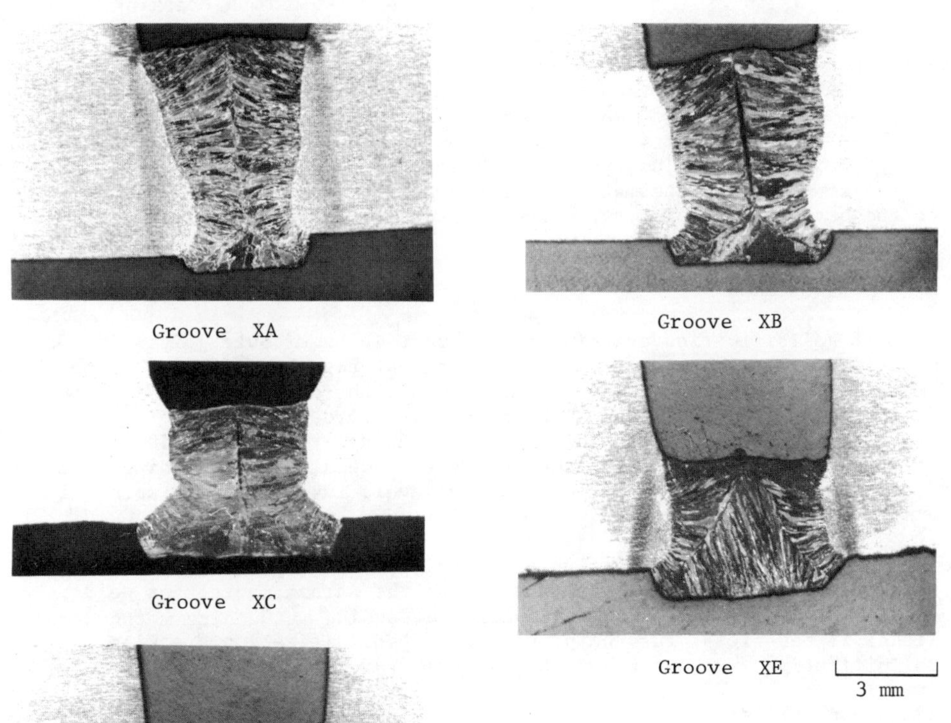


Fig. 15. Change in root-bead shape depending on groove shape in GMAW with travel speed of 800 mm/min.

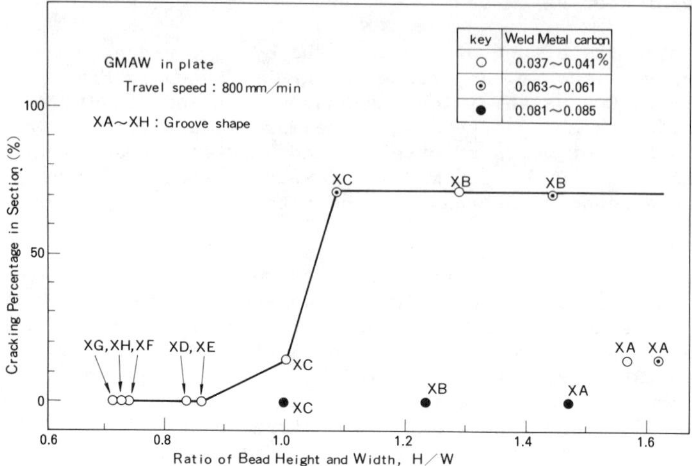


Fig. 16. Effect of bead shape on solidification cracking.

Since the present GMAW test employed Cu backing, heat flowed in two directions - to the base metal (horizontal) and to the Cu backing (vertical). Thus, dendritic structure of weld metal grew in two directions. As the ratio of bead height to its width decreased, the portion of weld metal which grew vertically increases relatively and head-on collision type of dendritic structure eventually disappeared. One of the important countermeasures in preventing solidification cracking is to reduce the head-on collision type of dendritic structures in weld metal.

## CONCLUSION

The risk of solidification cracking in the root-welds of butt joints is increased by the use of filler metals of low carbon content together with low carbon steel base metals. Welding tests were conducted under the combination of various steels and filler metals with SMAW and GMAW processes in order to find suitable conditions for preventing this type of solidification cracking. The solidification cracking may be affected by many factors such as state of welding arc, welding position, fluidity of molten weld pool and others. However, in the present work only some of these many factors were studied.

Findings within the range of variables of the present experiments are as follows:

1. Solidification cracking is prevented when the carbon content of weld metal exceed the critical amount which varies depending on welding method employed.
2. Nickel is beneficial for the prevention of solidification cracking.
3. Solidification cracking never occurs when travel speed is less than the critical level.
4. A head-on collision type of dendritic structure is susceptible to cracking. This structure can be reduced by limiting the level of the ratio of the weld bead's height to its width.

## REFERENCES

Ohshita, S., and N. Yurioka (1983). Welding Journal 62, 129s-136s

Briggs, C.W. (1946). The Metallurgy of Steel Casting, McGraw Hill, New York.

Wada T. (1967) J. Japan Welding Society 36, 319

# DEVELOPMENT OF LOW PCM X70 GRADE LINE PIPE FOR PREVENTION OF CRACKING AT GIRTH WELDING

T. Hashimoto*, Y. Komizo*, K. Bessyo**, Y. Yamaguchi*** and T. Yamura****

*Central Research Labs., Sumitomo Metal Industries, Ltd.*
*1-3 Nishinagasu Hondori, Amagasaki City, JAPAN 660*
***Kashima Steel Works, ***Wakayama Steel Works,*
*****Osaka Head Office, Sumitomo Metal Industries, Ltd.*

## ABSTRACT

Based on implant test, it is possible to predict and determine the welding condition without cracking in girth welds. The maximum $P_{CM}$ value to be performed by the stovepipe welding technique without preheating is 0.20% for line pipe of 25 mm and 0.13% for 35 mm in wall thickness. It is emphasized that low carbon X70 grade line pipe showing good weldability are developed by combination of alloy design using $P_{CM}$ formula and thermomechanical rolling technique of plate rolling.

According to these basic requirement, the production method combined with controlled rolling and dynamic accelerated cooling(DAC process) was newly developed to maintain extremely low $P_{CM}$ level for X70 grade steel. For example, ultra low carbon -Nb-Ti-B steel ($C \leq 0.03$, $P_{CM}$ 0.12%), low carbon -Nb-Ti-B steel ($C \leq 0.05\%$, $P_{CM}$ 0.14%), and low carbon -high Nb-Ti steel ($C \leq 0.05\%$, $P_{CM} \leq 0.15\%$).

Newly developed X70 grade line pipes have (a) excellent girth weldability and HAZ toughness (b) high absorbed energy and (c) high resistance to hydrogen induced cracking.

## KEYWORDS

X70 line pipe; controlled rolling; accelerated cooling; high niobium steel; low carbon boron steel; $P_{CM}$ formula; field weldability; girth welding

## INTRODUCTION

High grade line pipes used for the nothern district and or deep sea tend to demand higher strength and heavier wall thickness. In the ordinary process of plate making and alloy designing, alloy elements must be increased for satisfying such requirements, and it results in lowering toughness of mother metal and welded portion and also field weldability at girth welding. But recent technique of special controlled rolling and or accelerated cooling make possible to produce high grade line pipe with extremely low carbon steel.

The $P_{CM}$ formula was introduced by Ito and Bessyo (1968, 1969) as an effective criterion on cold cracking susceptibility after welding. It makes possible to predict the condition of crack free welding followed by Tanaka and co-workers (1980). The lower the $P_{CM}$ value calculated using chemical composition of steel becomes, the lower the hardenability of weld portion is. The low $P_{CM}$ steel relaxes the some restrictions on welding condition and can be welded without preheating at the critical $P_{CM}$ value.

The aim of this work is to develope the X70 grade line pipe showing good field weldability without preheating a girth welding. According to limitation of maximum $P_{CM}$ value, the production method combined with alloy designing and thermomechanical rolling process was researched to obtain the X70 grade line pipe steel. Selected steels are Nb-Ti-B steel of 0.02 to 0.04% carbon and 0.10% high Nb steel. Investigation has been achieved to produce the line pipes in good qualities, showing low cracking susceptibility, high toughness and high resistance to hydrogen induced cracking.

## CRITERION FOR ASSESSMENT OF WELDABILITY

Ito and Bessyo (1968, 1969) have established the $P_C$ evaluation for the prevention of the weld cold cracking. The $P_C$ is the indicates of cold cracking susceptibility and is expressed as follows.

$$P_C = P_{CM} + H_D/60 + t/600 \quad (1)$$

$$P_{CM} = C + Si/30 + Mn/20 + Cu/20 + Ni/60 + Cr/20 + Mo/15 + V/10 + 5B \quad (2)$$

Where $H_D$ is the diffusible hydrogen content measured by JIS glycerin method (mℓ/100 gr), t is the plate thickness (mm).

By extension of this formula, the critical stress of tested material ($\sigma_{cr}$) could be obtained by implant test. And it has been known that the selection of suitable material and welding conditions. Ito and co-workers (1979) presented the following equation.

$$\sigma_{cr} = -0.167Hv - 22.5 \log Hc + 98 \quad (3)$$

Hv is the maximum hardness of the inner side of the welded joint, and Hc is the residual hydrogen content when it is cooled down to 100°C. When $\sigma_{cr}$ in equation (3) is larger than the restraint stress ($\sigma_T$), cold cracking on girth welding will be prevented. Then, the relation between Hv and Hc is shown as follow by Ito and co-workers (1982).

$$[Hv]_{cr} = 587 - 9t - 135 \log Hc' \quad (4)$$

Relation between $P_{CM}$ value and Hv is obtained experimentally. Here the correlation of $P_{CM}$, Max. Hv and Hc can be presented. These results were shown graphically by Tanaka, Ito and co-workers (1980, 1982).

Table 1 shows the maximum $P_{CM}$ value for girth welding without preheating on 25 mm wall thickness of X70 grade line pipe. The other important factor is the sulfide stress corrosion cracking (SSCC) for line pipes applied to the environment of wet hydrogen sulfide. It is generally understood that prevention of SSCC needs lower hardness than Rc 22 or Hv 248. When the MIG welding technique is applied, the limitation of $P_{CM}$ is necessary from the view point of SSCC more than cold cracking. The both limitations are shown in Table 1. According this table, it is considered that almost of the girth welding methods do not require the preheating if developed the X70 grade line pipes with low $P_{CM}$ value is equal or less than 0.13% or 0.15%.

Table 1 Maximum $P_{CM}$ for Crack-free Welding without Preheat
— X-70, 25mm$^t$ —

| Welding Method and Material | | Welding Condition | Max. $P_{CM}$, % | |
|---|---|---|---|---|
| | | | Prevention of Cold Cracking | Prevention of SSCC by $H_2S$ |
| SMAW | Cellulose Type | 1 pass = 8 kJ/cm | 0.16 | 0.16 |
| | | Stringer + Hot pass | 0.20 | 0.18 |
| | Low-H Type | 1 pass = 8 kJ.cm | 0.22 | 0.16 |
| GMA | Wire | 1 pass = 4 kJ/cm | 0.21 | 0.13 |
| | | 1 pass = 8 kJ/cm | 0.28 | 0.16 |

Ambient Temp. = 10°C

## DEVELOPMENT OF MANUFACTURING PROCESS

### Sellection of Alloying System

The decreasing of carbon content in alloying elements is the most effective method to lower the $P_{CM}$ value of line pipe steel. Then, the low carbon steels of 0.02% and 0.04% were studied in this research. As shown in Fig. 1, the strength decreases with lowering the carbon content in ordinary Nb-V steel. It is emphasized that the research should be put on the strengthening of low carbon steel. Based on consideration of simple alloying system without raising the $P_{CM}$ value, two kinds of steels of Nb-Ti-B and 0.10% high Nb were sellected in our test data. The effect of strengthening with boron addition to low carbon steel is shown in Fig. 1. The addition of small amount of Nb, Ti and B strengthens the steel extremely with changing the micro-structure from ferrite and pearlite to micro duplex structure of ferrite and a small amount of fine martensite (Nakasugi and workers, 1980; Ohtani and co-workers, 1983). Besides, the high Nb steel was picked up as B free steel. Figure 2 shows the effect of Nb content on strength and toughness in 0.03% and 0.08% carbon steel. The 0.08%C steel shows higher strength than 0.03%C up to 0.05% Nb addition but lower at over range of 0.05% Nb. This is because that lowering the carbon content enhances the precipitation hardening with NbC. The 0.03%C steel obtained high toughness, also (Heisterkamp and co-workers, 1979).

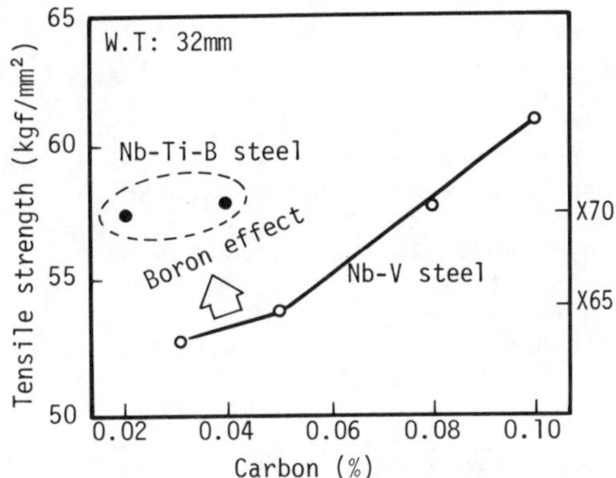


Fig. 1 Change of strength with decreasing carbon content in Nb-V steel and strengthening with Nb-Ti-B addition.

## Dynamic Accelerated Cooling Process

In recent years, the accelerated cooling technique has been adopted practically. Dynamic accelerated cooling equipments have been installed on production line as shown in Fig. 3. No. 1 cooling equipment is laminar flow type and practices to interrupted cooling with slow cooling rate. No. 2 cooling equipment is jet spray type and practices to cool down to room temperature rapidly. DAC process combined with controlled rolling can be applicable to strengthen the steel without loss of toughness.

Figure 4 shows the effect of DAC process on strengthening of Nb-Ti-B steels with varied Mn content. DAC process makes it possible to obtain strength of X70 with less than 1.6% Mn specified API specification. And, X80 grade can be obtained by increaseing Mn content. In the high Nb steel of 0.03%C-0.10%Nb, the thick wall plate more than 25 mm does not satisfy X70 grade in the as rolled condition but in the DAC process.

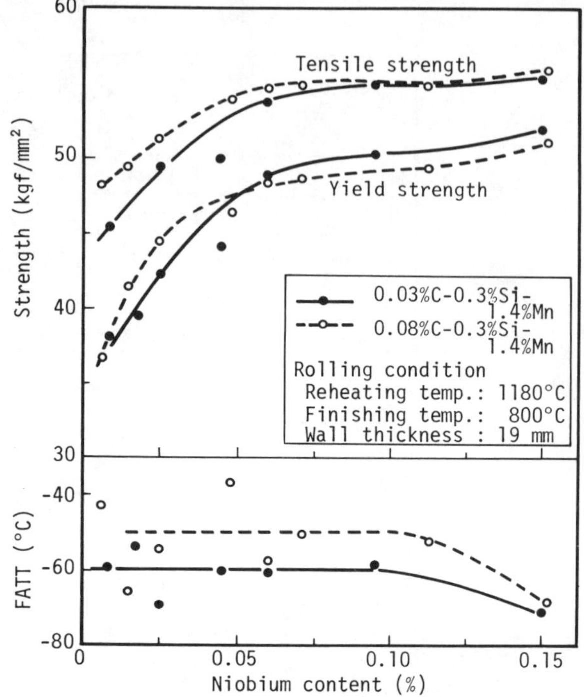


Fig. 2 Change of strength and Charpy FATT with niobium content in 0.03% and 0.08% carbon steel.

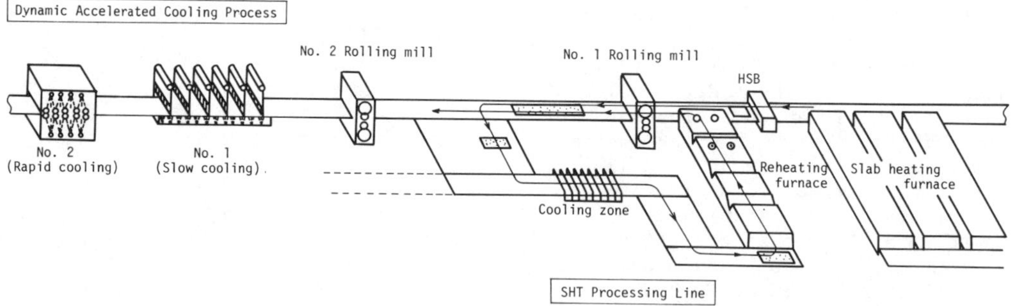


Fig. 3 Layout for thermomechanical processing in plate mill of Kashima Steel Works

## PROPERTIES OF X70 GRADE LINE PIPE

### Manufacturing Process

The chemical compositions of tested heats are shown in Table 2. Steels A and B are Nb-Ti-B steels of 0.02% and 0.04% C. Steel A aimed at 0.02%C with less than 0.13% $P_{CM}$ and steel B for 0.04% C of less than 0.15% $P_{CM}$. Steels C and D are high Nb steels, and they contained 1.7% Mn for X70 grade. Both heats of high Nb were practiced Ca treatment and slab soaking process for resistance to hydrogen induced cracking. Figure 5 shows production process of these 4 heats. After steel making of 250 or 160 ton converters and vacuum degassing, continuous casting slabs were produced. On the plate rolling, steels A and B were reheated to 1050°C, finished rolled at 700°C and then followed DAC process of cooling down to room temperature. The other plates of finished rolled at 650°C were cooled without DAC. These plates were formed to UO pipe of OD 762 mm with thickness of 19, 25 and 32 mm. Steels C and D were reheated to 1180°C and finished rolled at 800°C to interrupted cooling down to 600°C. They also were formed to pipes of OD 711 mm with 20 and 35 mm thickness.

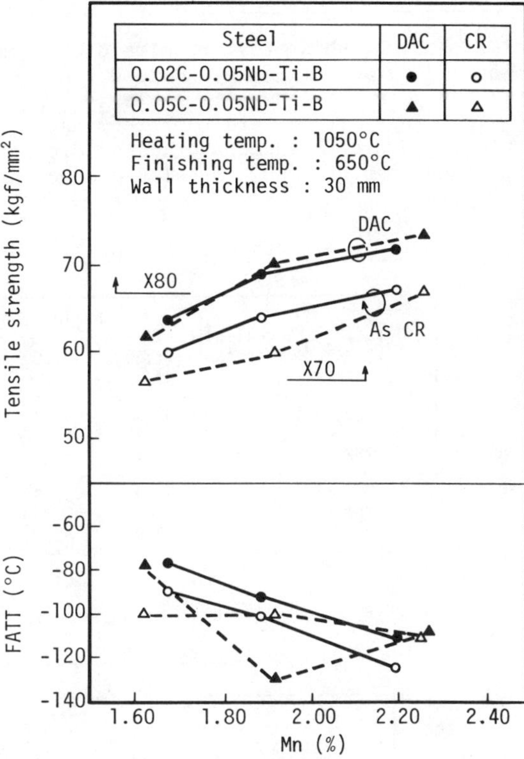


Fig. 4 Effect of Mn content on tensile strength Charpy FATT in controlled rolled and DAC processed steel.

Table 2 Chemical Composition (%)

| Mark | Steel | C | Si | Mn | P | S | Nb | Ti | B | N | Others | sol.Al | Ceq | $P_{CM}$ |
|---|---|---|---|---|---|---|---|---|---|---|---|---|---|---|
| A | 0.02C-Nb-Ti-B | 0.02 | 0.12 | 1.55 | 0.006 | 0.001 | 0.033 | 0.012 | 0.0012 | 0.0043 | Cu | 0.024 | 0.31 | 0.12 |
| B | 0.04C-Nb-Ti-B | 0.04 | 0.15 | 1.58 | 0.021 | 0.003 | 0.047 | 0.014 | 0.0014 | 0.0044 | Cu | 0.031 | 0.33 | 0.14 |
| C | 0.03C-0.1Nb | 0.03 | 0.20 | 1.71 | 0.016 | 0.001 | 0.103 | 0.013 | — | 0.0064 | Ca | 0.030 | 0.32 | 0.13 |
| D | 0.04C-0.1Nb | 0.04 | 0.22 | 1.73 | 0.021 | 0.002 | 0.115 | 0.016 | — | 0.0069 | Ca 0.08 Mo | 0.037 | 0.35 | 0.14 |

$$Ceq = C + \frac{Mn}{6} + \frac{Cu + Ni}{15} + \frac{Cr + Mo + V}{5} \qquad P_{CM} = C + \frac{Si}{30} + \frac{Mn}{20} + \frac{Cu}{20} + \frac{Ni}{60} + \frac{Cr}{20} + \frac{Mo}{15} + \frac{V}{10} + 5B$$

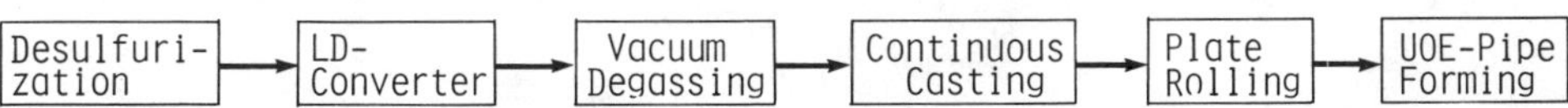


Fig. 5 Production process

## Properties of X70 Pipes

Table 3 shows test results of base metal in steels A and B of Nb-Ti-B. They show X70 grade strength and good toughness of Charpy and DWTT. The test results of SAW joint are shown in Table 4. The high Charpy energy obtained at HAZ and weld metal. It is said that they have good quality for arctic usage lower than -20°C. Hydrogen induced cracking did not occur in steel A of 0.002% C at the low pH condition of 3 pH. Steel B did not have cracking only at solution of pH 5.

Table 3 Mechanical Properties of Base Metal in Nb-Ti-B Steel Pipe

| Steel | Pipe size O.D x W.T (mm) | Rolling process | Tensile test | | | | 2V-Charpy test | | | DWTT (19mm) | |
|---|---|---|---|---|---|---|---|---|---|---|---|
| | | | Y.S (kgf/mm²) | T.S (kgf/mm²) | Y.R (%) | El. (%) | FATT (°C) | $vE_{-20}$ (kgf·m) | S.I ($mm^{-1}$) | 85% FATT (°C) | $E_{-20}$ (kgf·m) |
| A | 762 x 19 | DAC | 52.4 | 59.7 | 88 | 42 | -104 | 47.1 | 0.02 | -67 | 1461 |
| | 762 x 25 | DAC | 52.1 | 59.1 | 88 | 48 | -102 | 38.5 | 0.03 | -52 | 1330 |
| B | 762 x 32 | DAC | 51.9 | 59.3 | 88 | 45 | -98 | 30.3 | 0.06 | -33 | 863 |
| | | As CR | 52.5 | 58.8 | 89 | 46 | -72 | 21.3 | 0.14 | -24 | 813 |

Table 4 Mechanical Properties of SAW Joint in Nb-Ti-B Steel Pipe

| Steel | Pipe size O.D x W.T (mm) | Rolling process | Transverse weld Tensile test | 2V-Charpy test | | | | Hardness test | | |
|---|---|---|---|---|---|---|---|---|---|---|
| | | | | Weld metal | | HAZ (2mm) | | Max. Hv, Load : 10 kg | | |
| | | | T.S (kgf/mm²) | FATT (°C) | $vE_{-20}$ (kgf·m) | FATT (°C) | $vE_{-20}$ (kgf·m) | Weld | Bond | HAZ |
| A | 762 x 19 | DAC | 62.1 | -33 | 9.2 | -32 | 15.2 | 216 | 209 | 206 |
| | 762 x 25 | DAC | 63.1 | -41 | 20.9 | -34 | 27.8 | 207 | 205 | 198 |
| B | 762 x 32 | DAC | 63.3 | -60 | 22.6 | -60 | 37.5 | 220 | 213 | 213 |
| | | As CR | 62.8 | -50 | 21.3 | -47 | 24.7 | 213 | 212 | 213 |

Table 5 shows test results of mother metal and SAW joint in high Nb steels of C and D. They show good results as well as steel A and B. The typical microstructure of base metal in these 4 heats are shown in Photo. 1. Steels A and B are micro duplex structure of fine ferrite and a small amount of martensite and steels C and D are fine ferrite with pearlite reduced.

Table 5 Test Results of High Nb Steel Pipe with 711 mm O.D

| Steel | W.T | Tensile test | | | | 2V-Charpy test | | | | Pre-Crack DWTT 85% FATT | HIC test | Max. Hv on Girth weld* |
|---|---|---|---|---|---|---|---|---|---|---|---|---|
| | | | | | | Base metal | | Weld | HAZ | | | |
| | (mm) | Y.S (kgf/mm$^2$) | T.S (kgf/mm$^2$) | Y.R (%) | El. (%) | $vE_{-30}$ (kgf·m) | FATT (°C) | $vE_{-30}$ (kgf·m) | $vE_{-30}$ (kgf·m) | (°C) | (NACE) | |
| C | 20 | 52.0 | 61.4 | 85 | 49 | 31.6 | -87 | 12.4 | 22.3 | -42 | No crack | 201 |
| | 35 | 51.8 | 58.5 | 88 | 58 | 31.4 | -71 | 8.0 | 16.5 | -30 | No crack | 217 |
| D | 20 | 54.7 | 65.2 | 84 | 46 | 28.5 | -81 | 9.8 | 18.3 | -32 | No crack | 213 |
| | 35 | 52.9 | 62.1 | 85 | 53 | 28.0 | -70 | 13.7 | 11.4 | -23 | No crack | 229 |

* SMAW : Welding rod (E9016G), Heat input (19 to 22 kJ/cm)

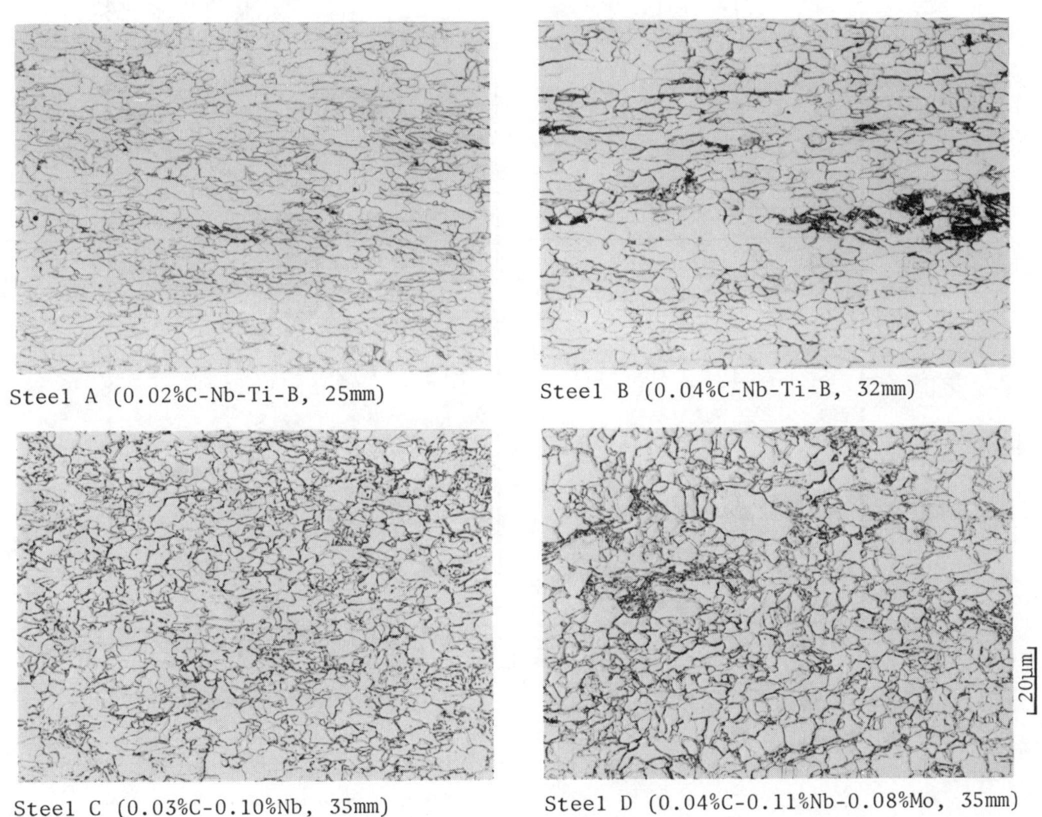


Steel A (0.02%C-Nb-Ti-B, 25mm)

Steel B (0.04%C-Nb-Ti-B, 32mm)

Steel C (0.03%C-0.10%Nb, 35mm)

Steel D (0.04%C-0.11%Nb-0.08%Mo, 35mm)

Photo. 1 Optical microstructure of DAC processed steel

Field Weldability

As simulated girth weldability test, shielded manual arc welding (SMAW) and MIG welding were practiced using the pipes. Welding condition is shown in notes of Fig. 6 and Table 6. Figure 6 shows hardness test result in steels A and B of Nb-Ti-B. Steel A of 0.02% C has low hardness but steel B of 0.04% slightly beyond Hv 250 at SMAW. Steel B may require the post heat treatment for prevention SSCC in some case

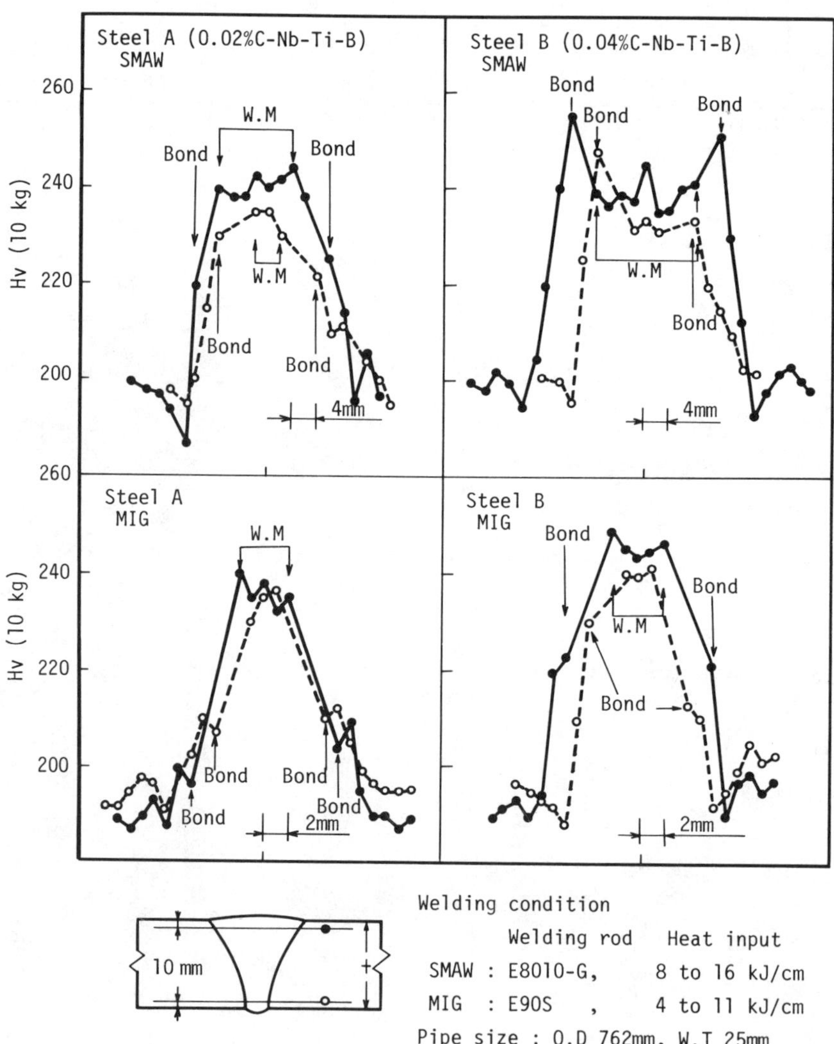


Fig. 6 Distribution of hardness across SMAW or MIG weld joint in steel A and B.

The maximum hardness of high Nb steels C and D deposited by SMAW are shown in Table 5. They have low hardness. In Table 6, mechanical properties of simulated girth welded portion are shown. All of the steels show good results in strength, Charpy value and COD value.

Table 6 Mechanical Properties of Simulated Girth Welded Joint

| Mark | Steel (Pipe size) | Welding method | T.S ($kgf/mm^2$) | Charpy test Temp (°C) | Charpy test Weld (kgf·m) | Charpy test HAZ (kgf·m) | COD test Temp (°C) | COD test Weld (mm) | COD test HAZ (mm) |
|---|---|---|---|---|---|---|---|---|---|
| A | 0.02%C-Nb-Ti-B | SMAW1) | 60.1 | -20 | 4.4 | 13.3 | 0 | — | — |
| | (762 x25mm) | MIG2) | 61.3 | | 3.5 | 12.6 | | 0.17 | 0.69 |
| B | 0.04%C-Nb-Ti-B | SMAW1) | 64.4 | -20 | 4.8 | 9.8 | 0 | — | — |
| | (762 x 25mm) | MIG2) | 63.8 | | 3.5 | 11.2 | | 0.18 | 0.74 |
| C | 0.03%C-0.11%Nb (711 x 35mm) | SMAW3) | 60.7 | -30 | 5.4 | 20.3 | — | — | — |

| Welding condition | Welding rod type | Heat input |
|---|---|---|
| 1) | E8010-G | 8 to 16 kJ/cm |
| 2) | E90S | 4 to 11 kJ/cm |
| 3) | E9016-G | 19 to 22 kJ/cm |

## CONCLUSION

1. X70 grade line pipe showing low $P_{CM}$ value of less than 0.13 or 0.15% can be developed by special controlled rolling and or Dynamic Accelerated Cooling Process.

2. Selected chemical compositions are extremely low carbon high niobium or boron treated steel.

3. These pipes show so low crack susceptibility of weld cold cracking that it can be welded without preheating at girth welding.

## REFERENCES


Ito, Y., and K. Bessyo (1968). IIW Doc. IX-576-68

Ito, Y., and K. Bessyo (1969). IIW Doc. IX-631-69

Ito, Y., and co-workers (1982). Sumitomo Search, No. 27, pp 119

Tanaka, T., and co-workers (1980). Proceedings of an International Conference on "Pipeline and Energy Plant Piping", Welding Institute of Canada, Calgory, pp 77 - 84

Ito, Y., and co-workers (1979). Sumitomo Search, No. 22, pp 156

Nakasugi, H., and co-workers (1980). Alloys for the Eighties, Climax Molybdenum Company, pp 213 - 224

Ohtani, H., and co-workers (1983). To be presented at International Conference on "Technology and Applications of HSLA Steels", (Philadelphia).

Heisterkamp, F., and co-workers (1979). International Conference on "Pipewelding" (London).

# FIELD WELDABILITY OF THICK SECTION MATERIAL

by A.G. Glover* and A.B. Rothwell**
*Welding Institute of Canada
**NOVA, An Alberta Corporation

## ABSTRACT

The field welding operations of a line pipe can exert considerable stresses on the root pass of the girth weld. Previous finite element studies have shown the importance of lift height and wall thickness on the stresses generated in the root region. This present work has simulated these field variables using full scale and small scale tests. The results are analyzed in terms of a cracking model involving preheat, stress, and composition.

## KEYWORDS

Weldability, line pipe, weld metal, cracking, stresses, preheat, thickness.

## INTRODUCTION

Field welding operations comprise the rate limiting step during pipeline construction. Acceptable girth welds must be made in a wide range of weather conditions and consequently the field weldability of line pipe compositions directly affect pipeline integrity and construction economics. During the sequence immediately following the deposition of the root bead considerable stresses can be imposed on the weld region. The likelihood of cracking in these circumstances depends upon the chemical composition of the weldment, the hydrogen content of the deposited weld metal, and how well the joint handles the external stresses caused by the skidding operations. In terms of the factors influencing the stress level, ongoing studies (1,2,3,4) have shown that the critical variables in pipeline construction are the lift height, misalignment, and the pipe wall thickness. This present paper analyzes the full scale and small scale test behaviour of line pipes of different thicknesses.

## ANALYTICAL BACKGROUND

The earlier studies (1,2) with the University of Waterloo considered the stresses in the root pass region that develop due to (a) differences in ovality, (b) misalignment of the joint faces, (c) changes in pipe diameter and wall thickness, and (d) lifting of the pipe after deposition of the root pass. This theoretical analysis, carried out using finite element techniques, identified misalignment, the lifting (or lowering) operation and wall thickness as the main variables which contribute to the stresses in the root pass.

The original finite element analysis (all based on elastic behaviour) determined the effect of an applied stress in terms of the stress ratio in the root pass. It was determined that there is little variation in the stress ratios between a 1067 mm and 1220 mm diameter line pipe. Figure 1 summarizes the variation in stress ratios with diameter for several locations in the weld area. Although the differences in stress ratio are small when considering large diameter pipes, Figure 1 shows that it would be wrong to use a much smaller pipe (eg 300 mm) to model the lifting stresses for large pipes.

Considering the effect of wall thickness with a fixed root pass thickness,the finite element analysis shows that the stress ratio depends much more on the pipe wall thickness than on the diameter (Figure 2). For these results it will be noticed that these lines converge at the co-ordinate (3.2:1) where the wall thickness equals the root pass thickness and the stress ratio is unity.

The pipe lifting operation can be modelled as a beam on a solid foundation and the detailed analysis of the stresses generated by this operation are given by Higdon et al (3). Generally, the stress is a function of the lift height, and the critically stressed joint also depends on the lift height. This lift height will also control whether the toe or the root controls the initiation of cracking (Figure 3).

Previous results (3) on 13.72 mm wall pipe showed that steels with low $P_{cm}$ values can be preheated to moderate levels to prevent cracking. However, at higher wall thicknesses (Figure 2) preheat control on its own may not be enough and stress control may be required. The objective of this present study is to evaluate these factors and quantify their relative significance.

## EXPERIMENTAL

The experimental program used 1067 mm diameter line pipe, Grade 483, with wall thicknesses of 12 mm and 19 mm. In order to minimize the number of variables, the composition of the two pipes was carefully selected. The compositions are given below:

| Pipe | C | Mn | Si | S | P | Cr | Mo | Nb | V | C.E. | $P_{cm}$ |
|---|---|---|---|---|---|---|---|---|---|---|---|
| 12 mm | 0.06 | 1.71 | 0.25 | - | - | 0.22 | 0.13 | 0.049 | 0.051 | 0.43 | 0.184 |
| 19 mm | 0.07 | 1.76 | 0.25 | - | - | 0.20 | 0.22 | 0.052 | 0.053 | 0.45 | 0.20 |

A series of small scale restraint tests were carried out on the 19 mm material using the WIC restraint cracking test (5). These were performed over a range of temperatures using E8010 and E6010 electrodes.

The six field weldability tests used the test arrangement shown in Figure 4 and the following format:

| Test Number | Wall Thickness mm | Misalignment mm | Lift Height mm | Preheat °C |
|---|---|---|---|---|
| 1 | 19 | 2.4 | 294 | 20 |
| 2 | 19 | 2.4 | 294 | 100 |
| 3 | 19 | 2.4 | 294 | 50 |
| 4 | 19 | 2.4 | 294 | 75 |
| 5 | 12 | 2.4 | 312 | 75 |
| 6 | 12 | 2.4 | 312 | 20 |

Each test was fully instrumented with strain gauges, thermocouples and pressure transducers and the same procedures were followed as would be used in the field. A five minute delay time between the end of welding and loading was utilized for each test. All welds were made using a stovepipe technique with 4 mm diameter E8010 electrodes by four pipeline welders with the following parameters, 160/175A, 22/24 volts, and 250-300 mm/min giving a nominal heat input of 0.83 kJ/mm.

## RESULTS

The results of the small scale tests are given in Figure 5a, which compares the results for the 8010 and 6010 electrodes on the 19 mm material. The predicted critical preheat is lowered by about 20°C for a decrease in electrode strength.

For the full scale tests all of the 19 mm tests gave complete failure except for the one preheated at 100°C. Neither of the 12 mm tests gave full failure and only some minor cracking was found in thin wall test that was not preheated. The results of percentage cracking versus preheat for both thicknesses are given in Figure 5b.

## DISCUSSION

The results of the full scale tests show that as the wall thickness increases the likelihood of cracking increases considerably. The critical preheat for the 12 mm material is close to 30°C whereas for the 19 mm pipe it is greater than 80°C, and this is for only a moderate lift height of 294 mm. Very similar results were obtained for the small scale restraint tests where a critical preheat of 80°C for the 19 mm material was obtained (6). Previous tests on thinner wall material have shown a shift in critical preheats between the full scale and small scale tests. Typically for 14 mm material a slight increase in critical preheat of 25 to 30°C has been shown from the full scale to the small scale tests. This had been attributed to the much higher restraint levels present in the small scale test. In the particular case of the 19 mm wall pipe the restraint levels between the small and full scale tests appear to be similar. This may arise in part because of the geometry effects. The relatively small root bead compared to the thickness generates very high stresses only locally in the root bead. This effect can be seen in Figure 6 which plots the strain along the pipe remote from the weld due only to the application of the load. The hoop strain local to the weld increases dramatically for the 19 mm pipe and this will enhance the likelihood to crack. The local effect of contraction around the weld causing the pipe to bend can be seen in Figure 6 which illustrates the geometical effect between the 12 and 19 mm pipes. The results of a separate experiment on the local straindistribution at the weld due to loading are given in Figure 7, showing high tensile axial stresses at the misalignment.

In all the tests the nature of the cracking was similar to what is generally considered to be of a hydrogen-induced nature. The three main factors which are considered to determine the susceptibility of the completed joint to hydrogen cracking are:

- the susceptibility of either the weld metal or the heat affected zone is related to the the overall microstructure and the orientation of the inclusions, both of which are determined by the chemical composition.
- the amount of diffusible hydrogen remaining in the weld area after it has cooled to near ambient temperatures.
- the stress or strain levels in the area of the joint at the point of fracture initiation.

In the present tests, the chemical composition remains constant although the microstructure may change slightly. The microstructure will be influenced by cooling rate, which in turn is controlled by heat input, preheat, and thickness. Although the thickness does vary, the cooling rate of the root is sufficiently rapid that a relatively hard transformation product is produced in the weld metal, with a moderate hardness in the HAZ for these low $P_{cm}$ materials. Hence this aspect can be considered as a constant. The variables are therefore the hydrogen and stress levels.

In establishing a model for the effects of preheat, hydrogen content and chemical composition on cracking it is assumed that the only effect of preheat is to allow the hydrogen to diffuse out of the weld before it cools to ambient. The effect of preheat on the removal of hydrogen from a bead-on-plate weld is shown in Figure 8. A similar effect exists for root pass welds.

The most critical factor is therefore the stress level applied to the root pass. The general stress level in the pipe due to lifting can be calculated from reference 3, however, the geometry of the weld has been shown to increase the stress level considerably. Apart from geometry (i.e. alignment) and residual stress, the overall stress level also depends on the eccentricity of the weld. Japanese work by NKK used the approach of multiplying by $t/t_w$, where t is the wall thickness and $t_w$ is the thickness of the partially completed weld. Figure 2 is the result of finite element analysis and gives stress ratios of nearly 9 for 12 mm and 16 for 19 mm (compared to 4 and 6 for NKK's work). Even for a modest lift height at these levels of stress ratios, local yielding is expected to occur and regions of the weld will become plastic. This aspect remains to be evaluated. The results of the full scale tests also show that time-temperature mechanisms are applicable and that local stresses in the root pass should be included in the analysis.

The majority of the cracking was in the weld metal just adjacent to the fusion boundary (Figure 3). The crack susceptibility of the weldment has previously been related to composition by considering the diffusion of hydrogen (7). This has been further modified by Lazor (8 to take into account cracking in the weldment giving rise to the following equation.

$$\beta\tau c = 5.896\ (P_{cm}) + \log_{10} H_f$$

where $\tau c$ = thermal factor corresponding to the preheat to give zero cracking

$H_f$ = original hydrogen content, and

$P_{cm}$ = composition factor

This relationship is plotted in Figure 9 for 14 mm thick small scale high restraint tests and the present 19 mm and 12 mm test results are included on the figure. This figure illustrates that the crack susceptibility of the joint is controlled by the diffusion of hydrogen away from the weld area, the chemical composition of the weld joint and the stresses in the area of high crack susceptibility.

## CONCLUSIONS

It has been shown from the present results that the likelihood of cracking is very stress dependent and that the generalized stress relationships for increasing wall thickness are true. However considerable variation in local strain distribution occurs during welding and subsequent loading which gives rise to local plasticity. Under these circumstances the restraint levels for the thick wall material are similar for both full scale and small scale tests. For the condition of high stress and hydrogen levels, preheat remains a very effective mean of preventing cracking in the root pass.

ACKNOWLEDGEMENTS

The authors wish to acknowledge financial support and permission to publish this paper from NOVA, An Alberta Corporation

REFERENCES

1. Higdon, H.I., "Root Pass Stresses in Pipeline Girth Welds" M.A.Sc. Thesis University of Waterloo, 1978.

2. Weickert, C.A., "Analysis of Detailed Weld Geometry in Root Pass Welds of Pipelines," M.A.Sc. Thesis, University of Waterloo 1980.

3. Higdon, H.I., Weickert, C.A., Pick, R.J. and Burns, D.J., "Root Pass Stresses in Pipeline Girth Welds Due to Lifting" Conference "Pipeline and Energy Plant Piping", WIC Calgary November 1980, Pergamon Press.

4. North, T.H., Rothwell, A.B., Glover, A.G. and Pick, R.J., "Weldability of High Strength Line Pipe Steels," Welding Journal Vol 61(8), 1982, p.243s.

5. Lazor, R., Glover, A.G. and Graville, B.A., "Properties and Problems in Structural Steel Welds," 64th AWS Annual Meeting, Philadelphia, April 1983.

6. Ko, K., "Hydrogen Cracking in Weld Metal", M.A.Sc. Thesis, University of Toronto, 1983.

7. Graville, B.A., "Cold Cracking in Welds in HSLA Steels" Conference, Welding of HSLA Structural Steels ASM, Rome 1976, American Society for Metals, Metals Park, Ohio.

8. Lazor, R.B., "Prediction of Weld Cracking Susceptibility," WIC Project RC76, November 1981.

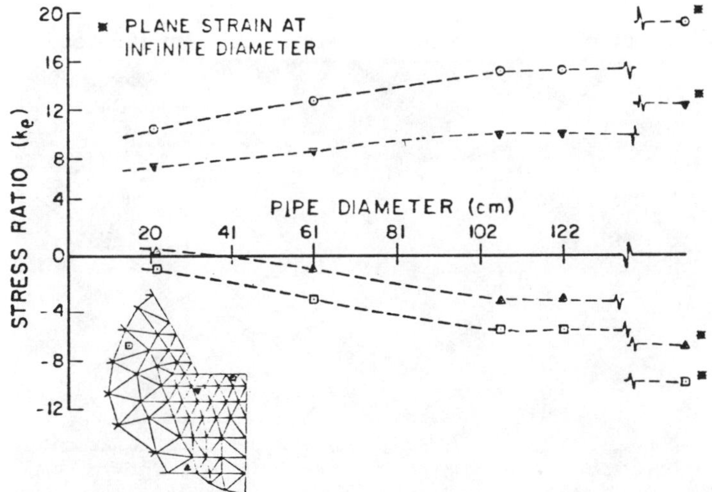


Figure 1. Effect of pipe diameter on stress ratio in the root pass (1,2).

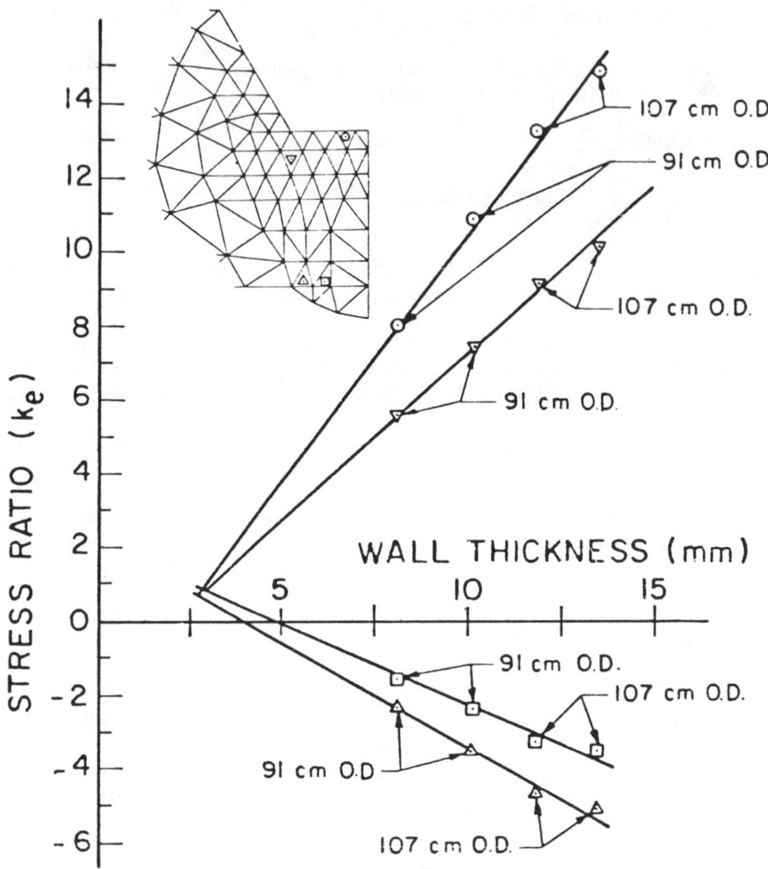


Figure 2. Effect of wall thickness on stress ratio in the root bead (1).

Figure 3a. Cracking from the root; low lift height

Figure 3b. Cracking from the toe; high lift height

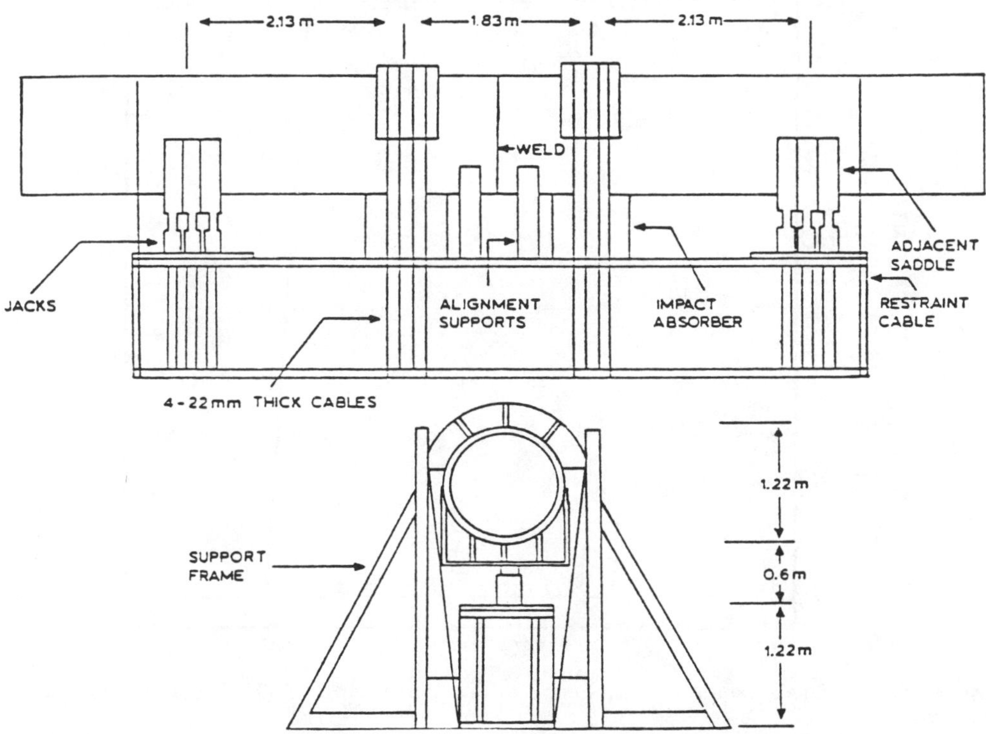


Figure 4. Schematic of field weldability rig.

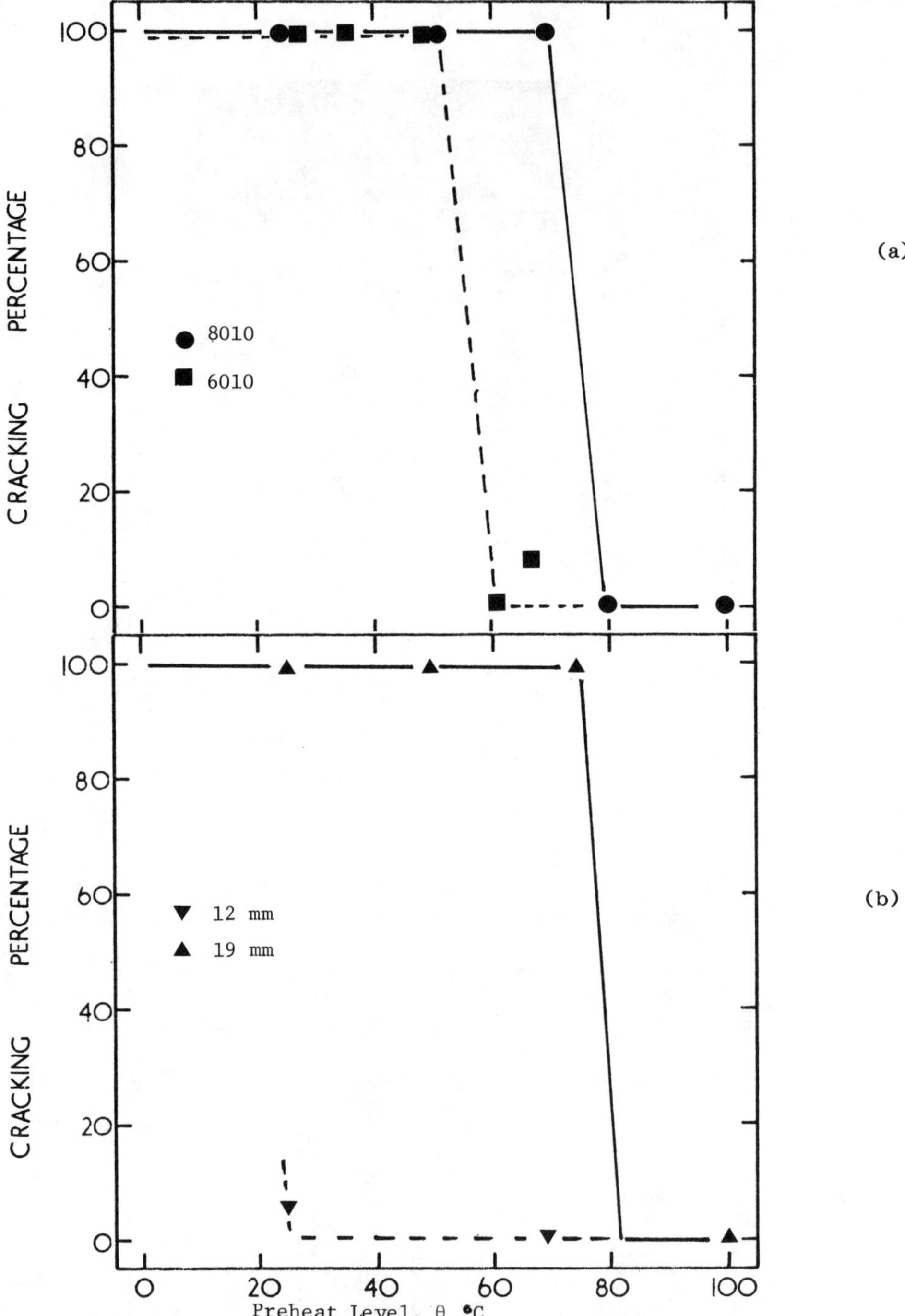


Figure 5a. Small scale results for the E8010 and E6010 electrodes, 2.4 mm misalignment.
Figure 5b. Full scale results for E8010 electrodes, 12 and 19 mm pipe, 2.4 mm misalignment.

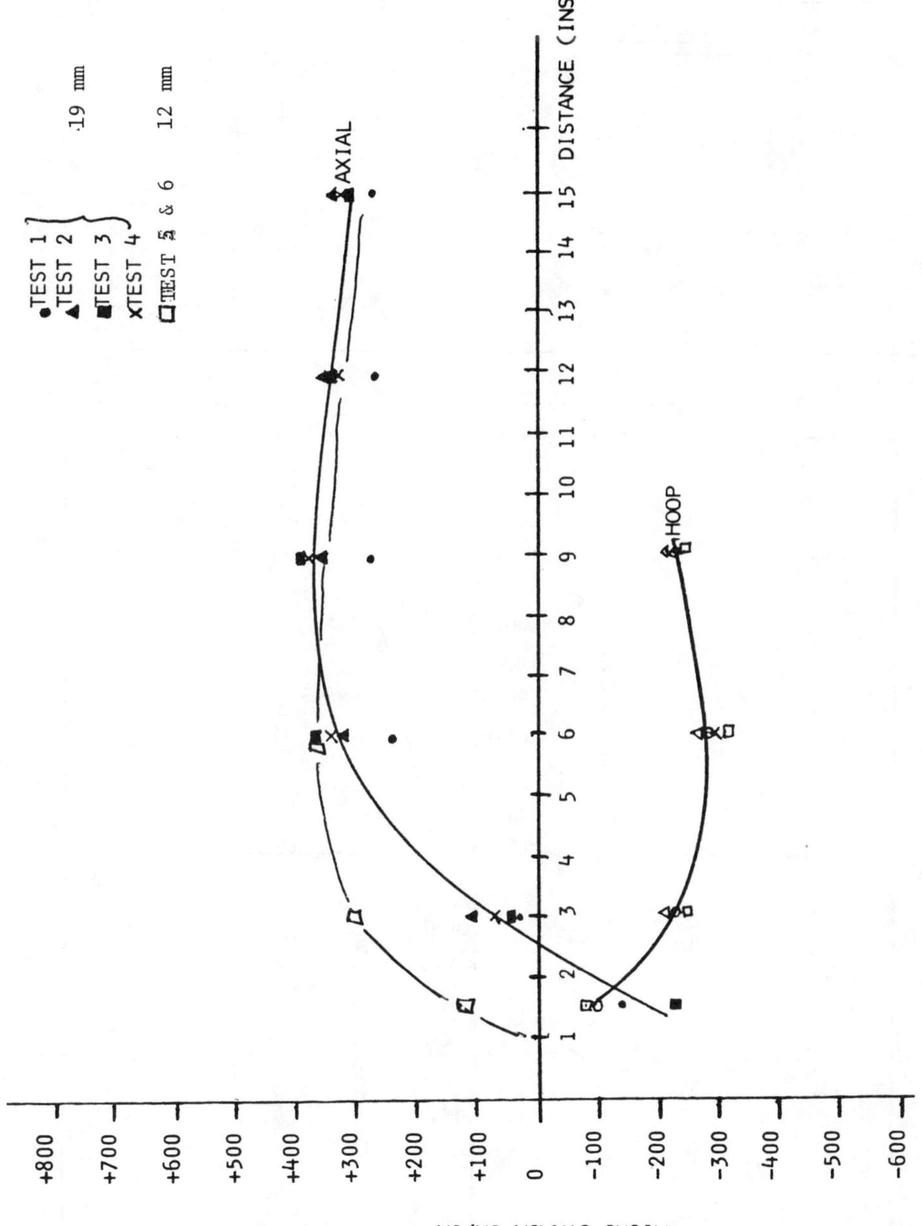


Figure 6. Strain distribution along pipe due only to loading.

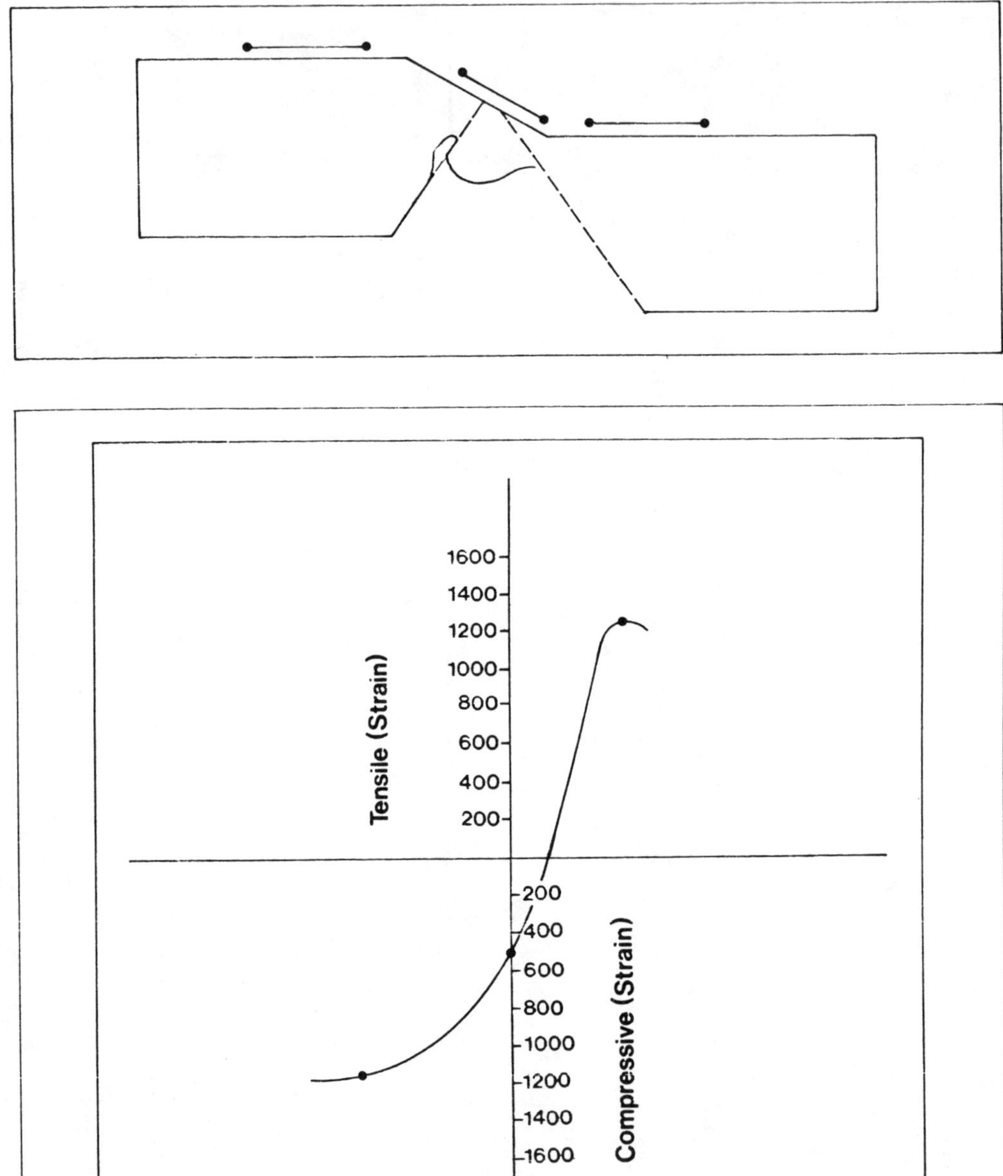


Figure 7. Local strain around weld due to loading, for 12 mm pipe.

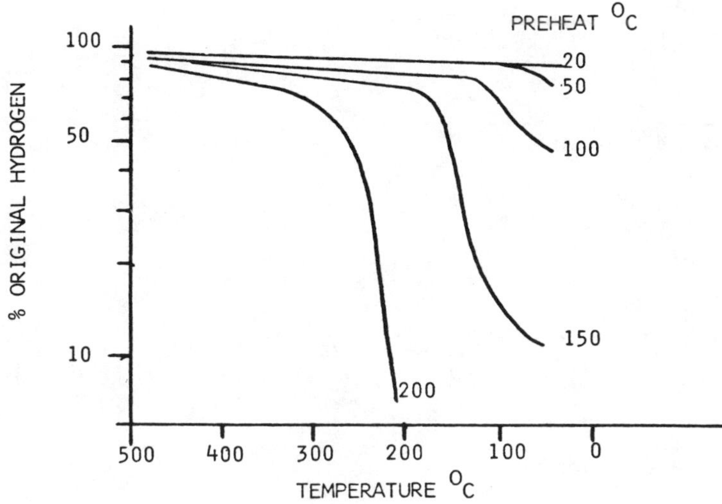


Figure 8. Effect of preheat on the amount of hydrogen remaining in the weld (reference 7).

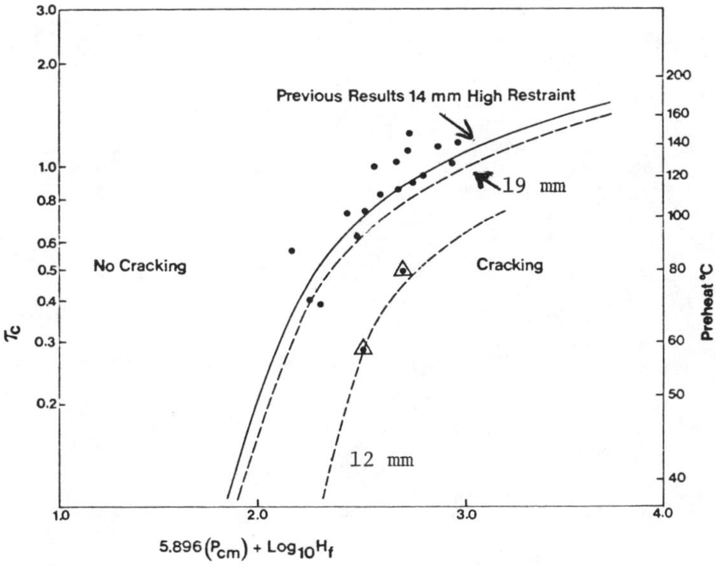


Figure 9. Present results in terms of a cracking criteria for weld metals.

# PIPELINE GIRTH WELDING USING THE FLUX-CORED ARC WELDING PROCESS

R.M. Huntley, D.V. Dorling and A.B. Rothwell
*NOVA, AN ALBERTA CORPORATION*

## ABSTRACT

Flux-cored arc welding (FCAW) is a more economical alternative to shielded metal arc welding (SMAW), since it is a continuous-wire process and good, effective deposition rates can be achieved as a result of the inherently high duty cycle. The process also lends itself readily to mechanization. Recently-developed self-shielded and gas-shielded consumables offering all-position capability have been evaluated, and welding procedures have been developed and qualified for both the vertical downhill welding of large-diameter, cross-country pipelines and the vertical uphill welding of compression and metering facilities.

Construction scheduling has not yet permitted the field use of FCAW for mainline welding by NOVA or its affiliates. However, experience with semi-automatic, self-shielded FCAW and mechanized, gas-shielded FCAW applied to station facility construction on the Eastern Leg of the Alaska Highway Gas Pipeline Project (AHGPP) is discussed in the present paper.

## KEYWORDS

Pipeline construction, flux-cored arc welding, welding procedure qualification, semi-automatic welding, mechanized welding.

## INTRODUCTION

The increasing physical scale and cost of pipeline construction activities, together with greater demands for positive assurance of the technical integrity of pipeline systems, has led to considerable interest in the development of welding processes which offer both better productivity and improved reproducibility and control, relative to the traditional, SMAW techniques. While a great deal of progress has been made in recent years in the application of mechanized GMAW to overland pipelines, the occasional incidence of high levels of fusion defects associated with this process constitutes a potential problem. In addition, a more productive means of making fixed position welds between heavy-wall pipe and components in station construction has become a pressing need. With the dimensions of components now being reached, it may take well over a day to complete a weld using the SMAW process, and the large number of passes necessitated by heavy wall thicknesses offers great potential for slag defects.

In principle, the FCAW process provides solutions to most of these problems. A number of consumables offering all-position capability have become available. Because of the mode of metal-transfer, the process is not inherently susceptible to fusion defects; deposition rates are high and, like all continuous-wire processes, it lends itself readily to mechanization.

In order to assess the applicability of different variants of the FCAW process, which has not been widely used to date in the pipeline industry, it was necessary to establish the handling characteristics of the available consumables and to determine whether they were capable of depositing weld metal which would meet mechanical property requirements for both buried and above-ground service. This paper describes initial investigations of both self-shielded and gas-shielded techniques; in the latter case, both semi-automatic and mechanized approaches were examined. For semi-automatic welding, both uphill and downhill progressions were adopted; the former is customarily required for station construction in Alberta, while the latter is more appropriate to cross-country pipeline construction. The development and qualification of procedures for specific applications is described; field experience during the construction of compressor station facilities for the AHGPP Eastern Leg is reviewed, and some considerations regarding quality and productivity are discussed.

## EXPERIMENTAL DETAILS

### Equipment

For the gas-shielded, semi-automatic welding trials, a Miller Trailblazer 55D DC generator was used, in conjunction with a Miller Portamig wire-feeder and a Bernard E-Z-Feed 400 A gun. For the mechanized trials, the same, constant-potential power source was used, coupled with a CRC Model M200 welding bug.

For the trials on self-shielded consumables, a Lincoln SA200 constant current DC generator was used, coupled with a K808 CV converter to provide constant potential characteristics. A Lincoln LN 27P wire-feeder and a K345-10 350 A welding gun with a 90° nozzle were used for this phase of the program.

Table 1 Chemical Composition - Parent Materials

| Material | Diameter (mm) | Thickness (mm) | C | Mn | Nb | V | Mo | Cr | Ni | Cu |
|---|---|---|---|---|---|---|---|---|---|---|
| CSA Z245.2 Gr.483 MPa | 1067 | 12.0,16.0 | 0.080 | 1.86 | 0.023 | 0.020 | 0.21 | 0.23 | 0.02 | 0.03 |
| CSA Z245.2 Gr.414 MPa(A) | 1067 | 24.0 | 0.044 | 0.35 | 0.022 | - | 0.23 | - | 0.85 | 0.50 |
| CSA Z245.2 Gr.414 MPa(B) | 1067 | 24.0 | 0.09 | 1.74 | 0.016 | - | 0.13 | - | 0.028 | - |
| CSA Z245.2 Gr.414 MPa | 1067 | 28.8 | 0.09 | 0.52 | 0.01 | 0.005 | 0.19 | 0.026 | 1.02 | 0.70 |
| ASTM A633 MOD.E | 1067 | 28.8 | 0.18 | 1.20 | 0.005 | 0.02 | 0.19 | 0.12 | 0.085 | 0.08 |

## Materials

Welding was carried out on a range of pipe and fitting materials, including ASTM A633E and CSA Z245 Gr. 483 and Gr. 414. Chemical compositions are shown in Table 1.

Though a considerable number of wires were evaluated during the course of the program, most of the work was concentrated on a limited number which were found to be acceptable from the point of view of handling and mechanical properties. The nominal chemical compositions of these wires are summarized in Table 2.

Table 2 Typical Wire Chemical Compositions*

| Consumable Classification | Process | C | Mn | Si | Ni | Cr | Mo | Al | V | Ti |
|---|---|---|---|---|---|---|---|---|---|---|
| 1 E81T1-G | Gas-Shielded Vertical Down Semi-Automatic | 0.035 | 1.25 | 0.29 | 0.90 | - | 0.20 | - | - | - |
| 2 E81T1-Ni1 | Gas-Shielded Vertical Up Mechanized | 0.05 | 1.46 | 0.40 | 0.95 | - | - | - | - | - |
| 3 E71T-GS | Self-Shielded Vertical Down Semi-Automatic (Root) | 0.134 | 0.54 | 0.23 | - | - | - | 0.35 | - | 0.48 |
| 4 E71T8-K2 | Self-Shielded Vertical Down Semi-Automatic (Fill & Cap) | 0.085 | 0.73 | 0.04 | 1.41 | 0.02 | 0.02 | 1.00 | <0.005 | - |
| 5 E71T8-K6 | Self-Shielded Vertical Up Semi-Automatic | 0.07 | 0.88 | 0.02 | .83 | 0.03 | 0.02 | 0.76 | <0.005 | - |

*Manufacturers' quoted typical figures for 'undiluted' weld-metal

Electrodes with various diameters were assessed, including 1.73 mm and 1.98 mm self-shielded and 1.14, 1.37 and 1.60 mm gas-shielded. Shielding gas, when used, was generally 75% Ar/25% $CO_2$.

## Evaluation Program

Consumables were evaluated on the basis of operating characteristics, weld-metal soundness and mechanical properties. Productivity aspects connected with the operating parameters were also considered. After initial screening, complete girth welds were made in the fixed, 5G position (pipe axis horizontal), using a standard CSA bevel with a $60^\circ$ included angle. In most cases, the root pass was deposited with an E8010G stick electrode, though one self-shielded wire specially developed for root pass welding was evaluated.

A consideration of the operator appeal of the various consumables, though necessarily subjective, formed an important part of the assessment. Consumables were evaluated on the basis of ease of manipulation, range of operable parameters, amount of smoke and spatter and anticipated training requirements.

Consumables were required to deposit uniform beads with no visible surface defects, and to meet the radiographic requirements specified in ASME Section VIII, Clause UW-51 or CSA Z184-M1983 Clause 5.2.11, as appropriate to the projected service. Particular care was necessary concerning the radiography of joints made using self-shielded consumables forming barium-rich slags, which may lead to unusual contrast effects.

A range of consumables which had been judged suitable on the basis of the above criteria was subjected to full procedure qualification testing, according to ASME Section IX or CSA Z184, as appropriate. In addition, company Charpy impact requirements had to be met; generally, requirements were a minimum of 40 J at -5°C for buried, cross-country pipelines, and 27 J at -45°C for station piping in above-ground service.

In order to assess the productivity of the different FCAW procedures, calculated welding times and actual production times were compared with typical values for the SMAW process. This comparison was carried out for three diameters and three wall-thicknesses. Two of the qualified procedures were used in the construction of compressor stations on the Eastern Leg of the AHGP system, which presented the opportunity for an assessment of their productivity and quality performance under field conditions.

## OBSERVATIONS AND RESULTS

### Mechanical Properties

Gas-shielded FCAW. Five consumables were initially evaluated, one being run in the semi-automatic mode, vertical down, and the others in the mechanized mode, vertical up. While all met basic procedure test requirements, only one of the consumables used for uphill welding met the appropriate Charpy toughness requirements (27 J at -45°C) for above ground service. Some details of typical welding parameters and mechanical properties for three acceptable gas-shielded procedures are shown in Table 3.

Self-shielded FCAW. Three self-shielded consumables were evaluated in the semi-automatic mode, but of these, one had been developed for root bead deposition only. Table 4 shows typical welding parameters and mechanical properties for a downhill procedure using FCAW only, and for an uphill procedure applied to two different base materials. Adequate Charpy impact properties for the projected applications were attained in all cases.

### Weld Metal Soundness

For all processes and consumables, initial parameters were selected to produce an acceptable bead profile at all positions around the circumference, with no visible defects. Parameters were then systematically adjusted to optimize the deposition characteristics of each individual wire. The resulting deposits showed good penetration and profile characteristics (Fig. 1). All procedure qualification welds met the radiographic requirements of the appropriate code (CSA Z184-M1983 for cross-country pipelines, ASME Section VIII for stations).

### Operating Characteristics

Gas-shielded FCAW. One wire (wire 1, Table 2) was evaluated for semi-automatic, downhill welding on cross-country pipelines. This reverse-polarity wire, which was specifically designed for multi-pass, vertical down applications, exhibited a soft arc in the spray mode. Slag cover was thin and easily removed by power brushing: smoke was not a problem. Minimal spatter and easy cleaning resulted in this wire being rated high in operator appeal. Though various gas mixtures were tried, 75% Ar/25% $CO_2$ provided the best operating characteristics.

The training of stovepipe welders to use this consumable system could be expected to be minimal, since the techniques are essentially the same as for conventional SMAW. It is felt that most qualified welders would be able to produce quality weldments consistently after at most one week.

The gas-shielded vertical-up trials used a mechanized welding system incorporating a bug which travels circumferentially around the pipe on a steel band (Fig. 2). Individual controls for dwelling at the bevel edges and for adjusting travel speed and wire-feed speed ease the welding operation. This allowed for procedures to be developed which optimize the deposition characteristics of the process and minimize operator manipulation.

Table 3 Welding Parameters & Mechanical Properties - Gas Shielded

1) Semi-Automatic Vertical Down - 1067 mm O.D. x 16.0 mm W.T. Gr. 483

| Pass # | Size (mm) | Filler Metal | Amps | Volts | Travel Speed (mm/min) | Wire Feed (mm/min) | Shielding Gas | Polarity |
|---|---|---|---|---|---|---|---|---|
| 1 | 0.90 | E70S-2 | 200 | 22 | 203 | 9650 | 100% $CO_2$ | Reverse |
| 2-5 | 1.14 | E81-T1G | 200 | 25 | 146-246 | 9650 | 75Ar/25$CO_2$ | Reverse |
| 6 | 1.14 | E81-T1G | 195 | 24 | 103 | 9650 | 75Ar/25$CO_2$ | Reverse |

| Yield Strength (MPa) | Ultimate Strength (MPa) | Weld Centerline Cv(J) -5°C | Weld Centerline Cv(J) -45°C |
|---|---|---|---|
| 596, 596, 574, 595 | 674, 676, 669, 677 | 52, 69, 49 | 24, 25, 14 |

2) Mechanized Vertical Up - 1067 mm O.D. x 24.0 mm W.T. Gr. 414 (A)

| Pass # | Size (mm) | Filler Metal | Amps | Volts | Travel Speed (mm/min) | Wire Feed (mm/min) | Shielding Gas | Polarity |
|---|---|---|---|---|---|---|---|---|
| 1-2 | 3.2 | E8010G | 85 | 24 | 76 | N/A | N/A | Reverse |
| 3-7 | 1.4 | E81T1-Nil | 225 | 21 | 178-203 | 7620 | 75Ar/25$CO_2$ | Reverse |
| 8-11 | 1.4 | E81T1-Nil | 215 | 21 | 178-203 | 6350 | 75Ar/25$CO_2$ | Reverse |

| Yield Strength (MPa) | Ultimate Strength (MPa) | Weld Centerline Cv(J) -5°C | Weld Centerline Cv(J) -45°C |
|---|---|---|---|
| 465, 488 | 535, 551 | 95, 93, 93 | 86, 80, 71 |

3) Mechanized Vertical Up - 1067 mm O.D. x 28.83 mm W.T. Gr. 414 to ASTM A633

| Pass # | Size (mm) | Filler Metal | Amps | Volts | Travel Speed (mm/min) | Wire Feed (mm/min) | Shielding Gas | Polarity |
|---|---|---|---|---|---|---|---|---|
| 1 | 3.2 | E8010G | 85 | 24 | 76 | N/A | N/A | Reverse |
| 2 | 1.4 | E81T1-Nil | 210 | 23.5 | 152 | 6096 | 75Ar/25$CO_2$ | Reverse |
| 3-8 | 1.4 | E81T1-Nil | 210 | 24.5 | 152 | 6096 | 75Ar/25$CO_2$ | Reverse |
| 9-11 | 1.4 | E81T1-Nil | 210 | 24 | 127 | 6096 | 75Ar/25$CO_2$ | Reverse |

| Yield Strength (MPa) | Ultimate Strength (MPa) | Weld Centerline Cv(J) -5°C | Weld Centerline Cv(J) -45°C |
|---|---|---|---|
| 477, 480 | 545, 544 | 69, 80, 75 | 63, 61, 67 |

Table 4 Welding Parameters & Mechanical Properties - Self-Shielded

1) Semi-Automatic Vertical Down - 1067 mm O.D. x 12.0 mm W.T. Gr. 483

| Pass # | Size (mm) | Filler Metal | Amps | Volts | Travel Speed (mm/min) | Wire Feed (mm/min) | Polarity |
|---|---|---|---|---|---|---|---|
| 1 | 1.73 | E71T8-GS | 200 | 13.5 | 311 | 2540 | Straight |
| 2-5 | 1.98 | E71T8-K2 | 250 | 18.5-19.5 | 107-259 | 2032-2286 | Straight |

| Yield Strength (MPa) | Ultimate Strength (MPa) | Weld Centerline Cv(J) -5°C | Weld Centerline Cv(J) -45°C |
|---|---|---|---|
| 555, 581, 581, 574 | 749, 708, 727, 742 | 41, 67, 89 | 19, 41, 19 |

2) Semi-Automatic Vertical Up - 1067 mm O.D. x 24.0 mm W.T. Gr. 414 (A)

| Pass # | Size (mm) | Filler Metal | Amps | Volts | Travel Speed (mm/min) | Wire Feed (mm/min) | Polarity |
|---|---|---|---|---|---|---|---|
| 1 | 3.2 | E8010G | 85 | 25 | 76 | N/A | Reverse |
| 2 | 1.98 | E71T8-K6 | 180 | 18 | 102 | 1778 | Straight |
| 3-6 | 1.98 | E71T8-K6 | 180 | 18 | 51 | 1778 | Straight |

| Yield Strength (MPa) | Ultimate Strength (MPa) | Weld Centerline Cv(J) -5°C | Weld Centerline Cv(J) -45°C |
|---|---|---|---|
| 485, 483 | 552, 553 | 60, 172, 122 | 28, 32, 25 |

3) Semi-Automatic Vertical Up - 1067 mm O.D. x 24.0 mm W.T. Gr. 414 (B)

Welding Parameters as Per 2) Above
Weld Centerline Cv(J)

| -5°C | -45°C |
|---|---|
| 110, 108, 96 | 35, 43, 49 |

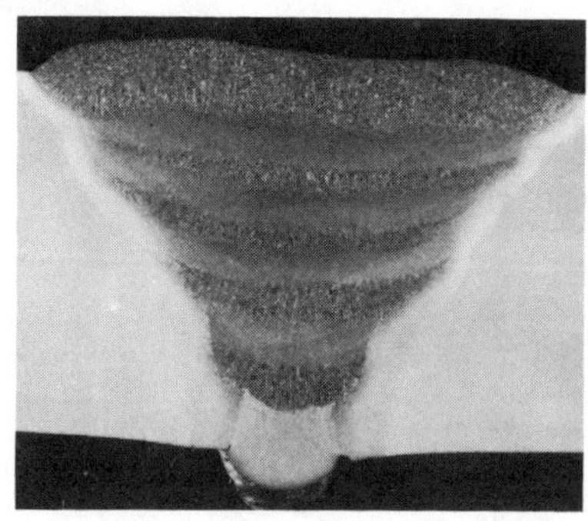

Semi-automatic, self-shielded, uphill

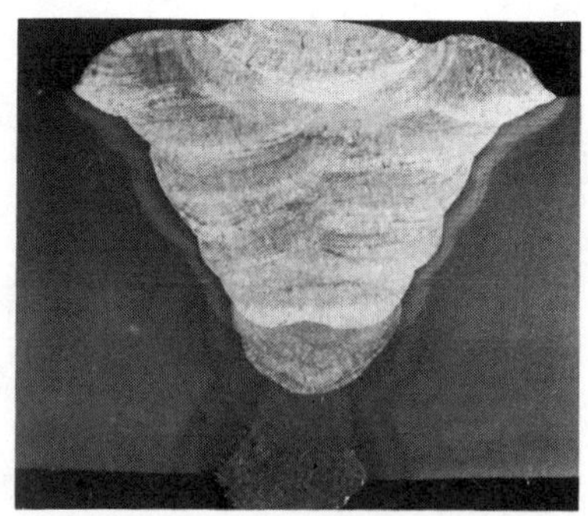

Mechanized, gas-shielded, uphill

Fig. 1 Macrographs of typical FCAW deposits in heavy-wall pipe

Fig. 2 CRC M200 welding bug

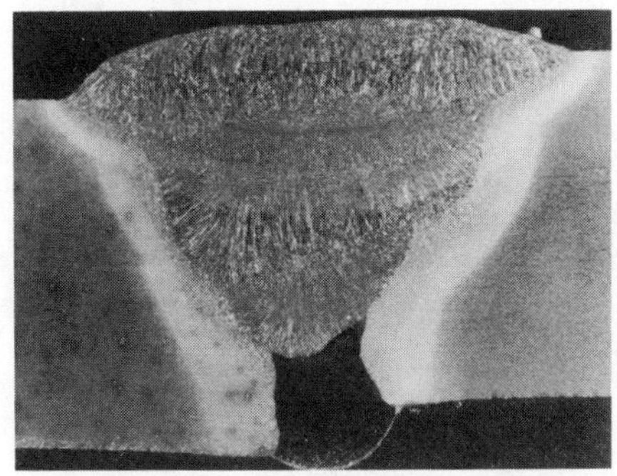

Fig. 3 Self-shielded, all-FCAW weld with ~2 mm misalignment

In developing parameters and techniques, it was found that the critical point was in the 45° overhead position. Since the fluidity of the weld pool and slag is relatively high, there is a tendency for the molten pool to "fall out" at this point. Another critical aspect was bead placement. Because of the rather shallow penetration characteristics of these wires, each pass must maintain a reasonably flat profile: if a previously-deposited pass is convex and is not ground properly, lack of inter-pass fusion can occur.

## Semi-Automatic, Self-Shielded FCAW

A straight-polarity, 1.7 mm wire was investigated for downhill root-bead applications (see Tables 2 and 4). Using a low arc voltage for optimum control, the root is deposited with a stringer technique, a leading torch angle being adopted at the top (flat position) and a lagging torch angle at the bottom (overhead).

Travel speeds in the order of 300 mm/min are required to ensure that the arc is always deep in the groove and in the land portion only. If this is not maintained, lack of penetration can result. A minimum root gap of 2.4 mm was required for this wire, in order to maintain good penetration; larger gaps (up to 4 mm) could be handled. Provided the gap was adequate, it was found that high-low up to 2.4 mm could be accommodated (Fig. 3).

Fill and cap passes were deposited downhill with a 1.98 mm wire. Techniques were very similar to those used in SMA "stovepipe" welding. All passes could be cleaned easily by power brushing, and overall handling characteristics were good. Both wires expelled a great deal of smoke, which would probably necessitate the use of local exhaust systems in a shop environment. Because the wires do not employ auxiliary shielding, they are ideally suited to field applications: moderate wind conditions do not destroy the shielding integrity.

As regards training requirements, it is estimated that a minimum of two weeks would be required by qualified welders to achieve consistency with the root pass. Conversely, the fill and cap passes present few difficulties.

Uphill welding (fill and cap passes only) was carried out with an alloyed, 1.98 mm wire, again on straight polarity (see Tables 2 and 4). Determining factors in achieving sound welds were found to be electrical stick-out (ESO) and electrode angle; a simple side-to-side motion with a leading torch angle of 75-90° and an ESO of 12.7 mm gave consistently satisfactory results.

An initial difficulty was encountered in maintaining the required minimum ESO in the region of the 1 o'clock position. At this point there is a tendency to shorten the stick-out, giving an increase in arc voltage; this increase results in porosity. The final solution to this problem was to install gas nozzles which maintain a minimum ESO of 12.7 mm (Fig. 4).

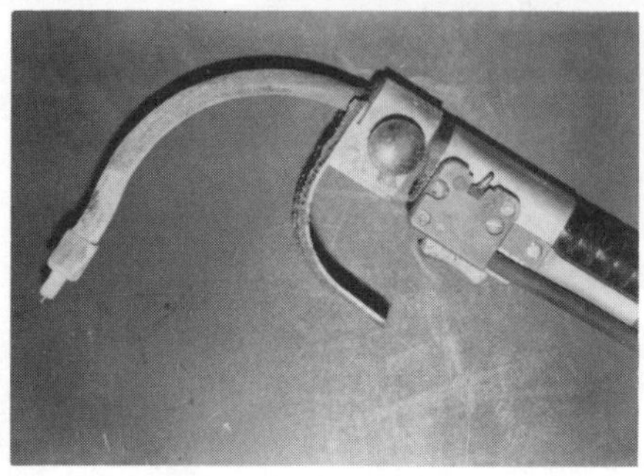

Without insulated guide

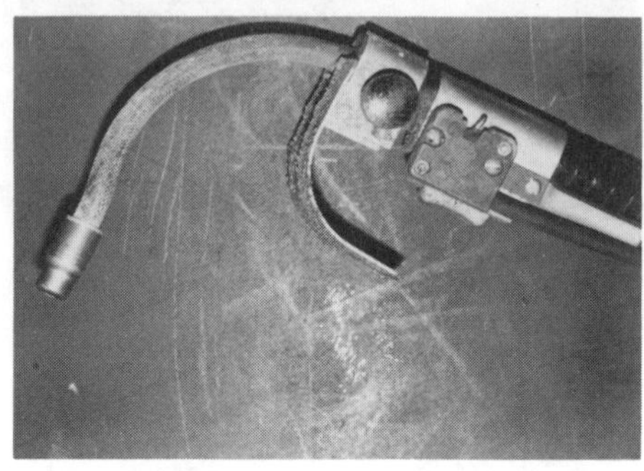

With insulated guide - min. ESO 12.7 mm

Fig. 4 Semi-automatic gun for self-shielded consumables

In general, this wire was found to be very sensitive to changes in arc parameters; the 1.98 mm wire was found to operate effectively only within a narrow range of parameters (wire-feed speed 2000-2500 mm/min, 18-20 V).

Although some training would be necessary to allow the operator to become familiar with the process and equipment, it is not believed that a prolonged period would be needed for most qualified welders.

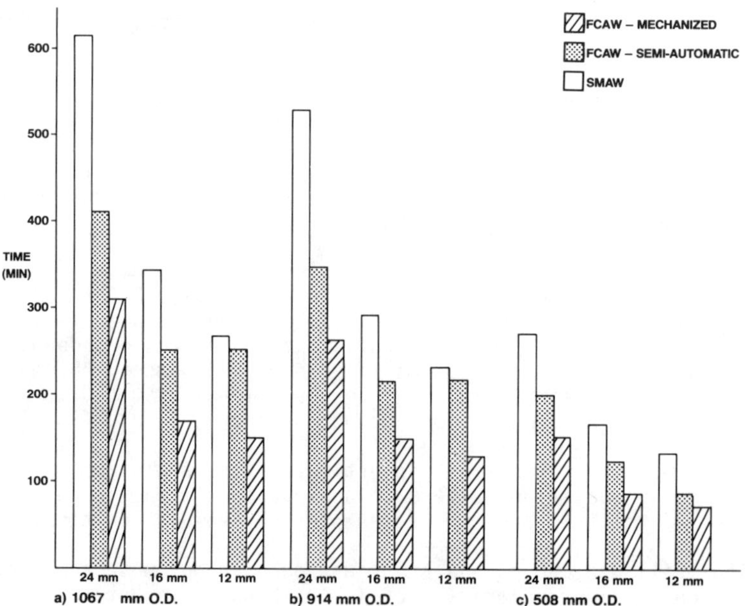


Fig. 5 Productivity Data for SMAW and FCAW Processes

### Productivity

In comparing weld-out times for various welding techniques, the continuous-wire FCAW processes were found to be much faster than conventional SMAW methods. Fig. 5 shows comparisons between self-shielded, semi-automatic FCAW, mechanized, gas-shielded FCAW and conventional SMAW, for uphill procedures; three diameters and three wall thicknesses are considered. The mechanized system is consistently 80-90% faster than SMAW, regardless of pipe size. It is also faster than the semi-automatic techniques, but this advantage decreases as the wall thickness increases. This is attributable to the fact that the gas-shielded wires tested were not capable of accommodating a weave wider than ~20 mm, whereas the self-shielded consumable can deposit beads 30 mm and wider. For the former case, this necessitates the use of a split weave on final fill and cap passes at wall thicknesses of 24 mm and above. (See Fig. 1).

## FIELD EXPERIENCE

The construction schedule of the Eastern Leg of the AHGPP did not permit the introduction of FCAW for mainline or section welding. It did prove possible, however, to gain experience with both the semi-automatic self-shielded (uphill) process and the mechanized gas-shielded process through their application to the construction of compression facilities.

### Mechanized, Gas-Shielded FCAW

The mechanized, gas-shielded process was used to make fixed-position joints in piping and components at Monchy compressor station, 45 km south-east of Val Marie, Saskatchewan, on the Montana border. Typical joints were 1067 mm O.D. x 28.8 mm wall thickness; in total fifty-six welds were made.

It proved possible to complete one weld per day from the outset with the mechanized system, as compared with one weld every two days with the SMAW technique. The best time achieved for a 1067 mm O.D. x 28.8 mm W.T. joint, exclusive of fit-up and root pass, was three hours.

Of the fifty-six mechanized welds made, it proved necessary to make repairs (generally for minor slag defects in the 12 o'clock region) to twenty-seven. In general, these repairs were not extensive, and the ratio of total length of repair to total length of weld was 3.3%. This figure assumes an even more positive aspect when the heavy wall-thickness is taken into account; before the adoption of the mechanized FCAW process, no SMAW heavy-wall joints had been accepted without multiple repairs. It should also be noted that the percentage of welds requiring repair decreased continuously as experience was gained, (Fig. 6) and the repair frequency had reached a very low level by the later stages of the job.

### Semi-Automatic, Self-Shielded FCAW

Only limited experience was gained with this process, at the Richmound compressor station, 21 km east of the Alberta-Saskatchewan border. Again, productivity was good from the outset, and the quality of the few welds which were completed was excellent, with a 1.1% repair length ratio. However, problems were experienced as a result of the inexperience of the personnel with this process, specifically with control of the ESO; as described above, the process is particularly susceptible to variations of ESO below the specified minimum. It was not possible to correct this problem and restore the confidence of the welders in the process before welding was completed at this site. The limited trial which did occur suggested that no major problems exist, but that in practice, the productivity and quality benefits (when compared with a particularly competent SMAW operation) may be marginal, at least until extremely heavy wall-thicknesses are reached.

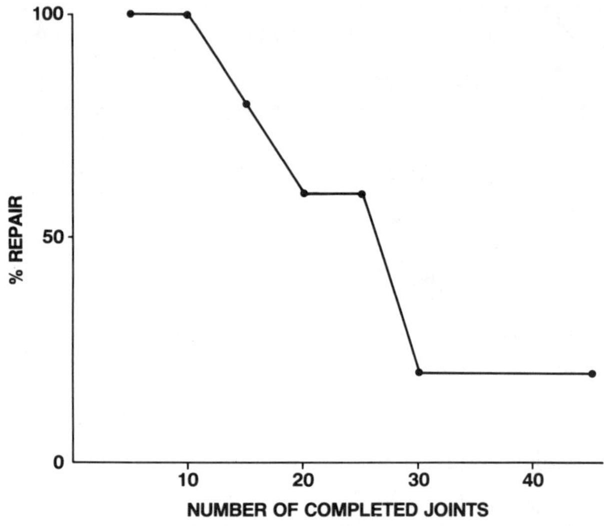


Fig. 6 Weld Quality Data - Mechanized FCA Welds, 1067 mm O.D. Station Piping

## CONCLUSIONS

Consumables have been evaluated for the vertical downhill welding of cross-country pipelines, using semi-automatic, gas-shielded and self-shielded FCAW and for the uphill welding of station facilities, using semi-automatic, self-shielded FCAW and mechanized, gas-shielded FCAW processes. Wires showing acceptable handling characteristics and weld-metal mechanical properties were identified for all four methods. Sensititivy to changes in welding conditions and requirements for special manipulation techniques varied from process to process, so that projected training needs would depend on the particular method adopted.

All the techniques examined offered significant productivity advantages over conventional, SMA welding, with the mechanized, gas-shielded approach being particularly attractive for joining heavy-wall station-piping and components. This technique was applied to the construction of the Monchy compressor station on the Eastern Leg of the AHGPP, and very encouraging results were achieved as regards quality and productivity.

With increasing reliance on fitness-for-purpose concepts in pipeline construction, the consistent attainment of specific weldment properties in production takes on a much greater importance. The semi-automatic and, particularly, mechanized processes lend themselves to much-improved standardization of welding variables, relative to SMAW techniques. For this reason, allied to the considerable improvements in productivity which they offer, continuous-wire processes are expected to assume an increasingly-important role in the construction of major pipeline systems.

# A NEW HIGH-SPEED CIRCUMFERENTIAL ROOT PASS WELDING FOR PIPELINE CONSTRUCTION

T. Fujimoto*, K. Akahide*, M. Hara*

**Structure Research Lab., Kawasaki Steel Corp., Naganuma-cho 351, Chiba City, Japan*

## ABSTRACT

Root pass speed can be raised to about 1,300 mm/min by applying the pulsed-MAG process to the pipe's outer surface using the one-side downhill technique. This is attributable to the use of a rare earths-bearing wire in the DC electrode negative polarity. The flow rate of shielding gas mixture consisting of 90 % Ar and 10 % $CO_2$ is 20 l/min. For the good alignment of pipes, an oil pressure internal clamper which has been newly developed is employed, and, around the outer surface of the clamper, a copper backing band is mounted for the formation of internal bead. During welding, the copper band makes direct contact with the internal bead, thereby promoting cooling and solidifying of molten metal. The result is a sound root pass bead free from hot cracks and copper chip pickup.

## KEYWORDS

Pipeline; circumferential welding; copper backing; hot crack.

## INTRODUCTION

The promotion of circumferential welding efficiency at pipeline construction sites which are usually under severe climatic conditions is one of the ardent desires of field engineers. In the case of the one-side all-position welding, the joining of fixed pipe is at present made usually by using the cellulose type stick electrodes. The root pass speed limited to only 500 mm/min-man, so requires a plural number of welders depending upon the pipe diameter if a higher welding efficiency is desired. Since the line pipes are characterized by their large diameter, thick wall and high tensile strength recently, the gas shielded welding is more desirable than the manual welding for higher welding efficiency and hydrogen crack prevention; hence, it is applied increasingly widely. Even in this case, as the root pass speed per head is approximately equal to that of the manual welding, the multi-head automatic welding is used for the sake of increasing the efficiency. The present paper will discuss a new high-speed one-side downhill technique using the GMA welding which ensures a root pass speed per head of about 1,300 mm/min for the pipe groove with a root gap.

## EXPERIMENTAL APPARATUS

### Welding Power Supply

Two types of power supply were prepared for the experiment; the conventional constant potential DC power supply (output voltage: max 32 volt, output current: max 300 amp), and the model PA-3M pulsed-arc power supply of AIRCO., the United States.

### Welding Equipment

The equipment shown in Fig. 1 was used for welding flat plates, and a modified version of the model APW-400 (Fig. 2) of Denyo Co., Japan, for circumferential welding.

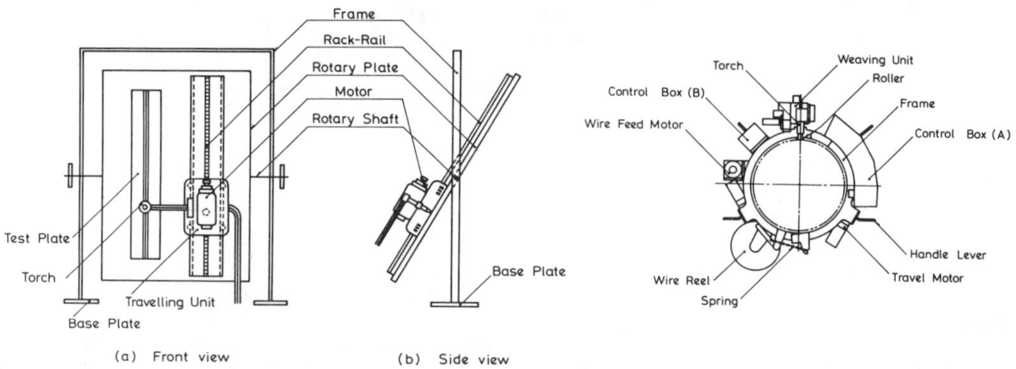


Fig. 1. Equipment used for welding flat plates.

Fig. 2. Side view of APW-400 used for circumferential welding.

### Internal Clamper

By employing the oil-pressure internal clamper shown in Fig. 3, the groove is constrained and released as the clamp-pieces shift radially in and out through the moving of a cone linked to a piston along the axial direction of the pipe. A copper backing band is mounted for an internal bead is wound around the outer periphery of the clamp-pieces. A circumferential welding equipment and an internal clamper are shown in Fig. 4.

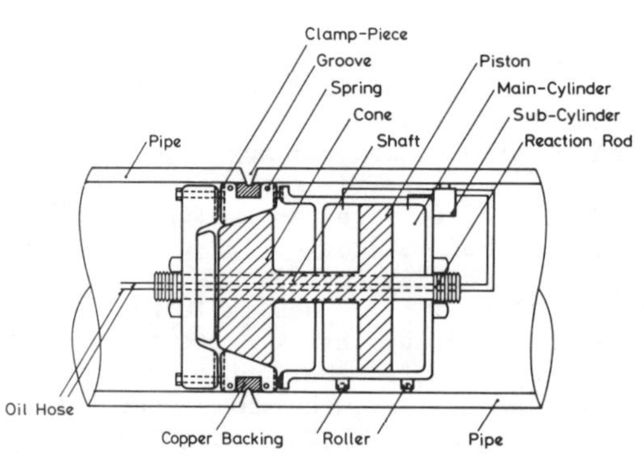


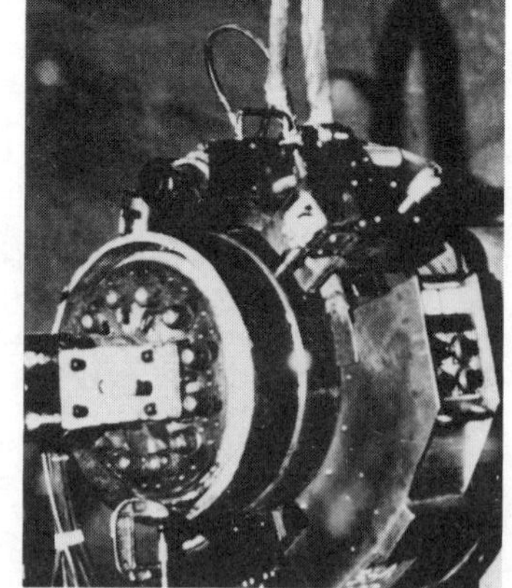

Fig. 3. Oil-pressure internal clamper.

Fig. 4. Circumferential welding equipment and internal clamper.

## WELDING PRACTICES

### DC Electrode Positive Welding

Materials and method. A mild steel flat plate 12 mm in thickness and a 1.2 mm dia. KC-50 wire were used as test materials. The chemical compositions of these materials are given in Table 1. Using the conventional constant potential DC power supply, 1-to-1 mixture of carbon dioxide with argon was run at a flow rate of 20 l/min as shielding gas mixtures. The groove geometry is of V-shape with 60° angle. Welding was done at speeds up to 1,000 mm/min for various positions.

TABLE 1 Chemical Compositions of Plate and Filler Wire

(wt %)

| | | C | Si | Mn | P | S |
|---|---|---|---|---|---|---|
| Plate | SS 41 | 0.15 | 0.20 | 0.65 | 0.020 | 0.013 |
| Filler wire | KC 50 | 0.09 | 0.45 | 0.98 | 0.015 | 0.013 |

Results and discussions. When welding is done without backing materials with welding current of 160-230 amp and arc voltage of 16-20 volt, burn-through occurs at a 1.3 mm or wider root gap in the flat position. At a 1.0 mm or narrower, the penetration is poor. Similar defects result from a deviation of ± 10 amp and/or ± 1 volt from optimum current and voltage, respectively. During welding, no keyhole is noticed. In the vertical down position, the internal bead is concave, failing to give a good stringer bead at a welding speed of 1,000 mm/min. In the overhead position, penetration bead produces undercut. When welding is performed using backing material under the conditions shown in Table 2, the results are as follows.

TABLE 2 Welding Conditions for Root Pass Using Backing Materials (DC Electrode Positive)

| Test No. | Backing material | Welding current (A) | Arc voltage (V) | Root face (mm) | Root gap (mm) | Welding speed (mm/min) |
|---|---|---|---|---|---|---|
| 1 | 8-ply glass tape | 240 — 260 | 22 | 1.4 — 1.6 | 1.0 — 1.2 | 1000 |
| 2 | 2-ply glass tape | 240 — 250 | 22 — 23 | 1.2 — 1.4 | 0.9 — 1.1 | 1000 |
| 3 | Scotch tape | 250 — 270 | 22 — 24 | 1.3 — 1.6 | 1.0 — 1.3 | 1000 |
| 4 | Copper plate | 240 — 260 | 22 — 23 | 1.2 — 1.5 | 0.9 — 1.2 | 1000 |

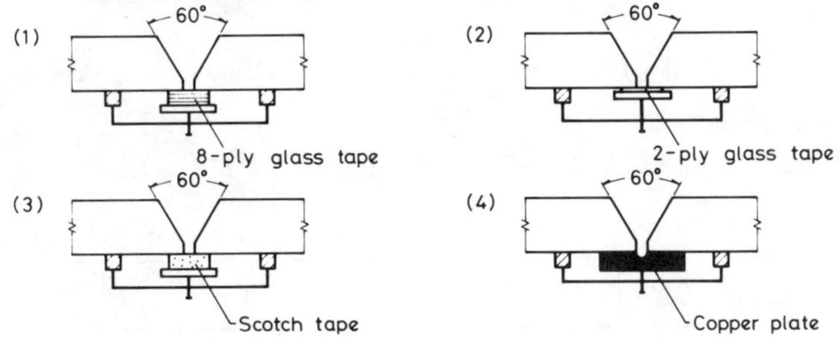


8-ply glass tape (4 mm thick). In the vertical down position, burn-through occurs frequently, and the internal bead tends to be concave-shaped. While good bead is obtained in the flat and overhead positions, hot cracks take place at the center of bead cross-section.

2-ply glass tape (1 mm thick). In the flat, vertical down and overhead positions, good bead is obtained, but hot cracks develop at the upper center part of cross-section through the bead (Fig. 5).

Scotch backing tape. In the vertical down position, burn-through occurs, and in the flat and overhead positions arc becomes unstable to a point that no welding can be performed.

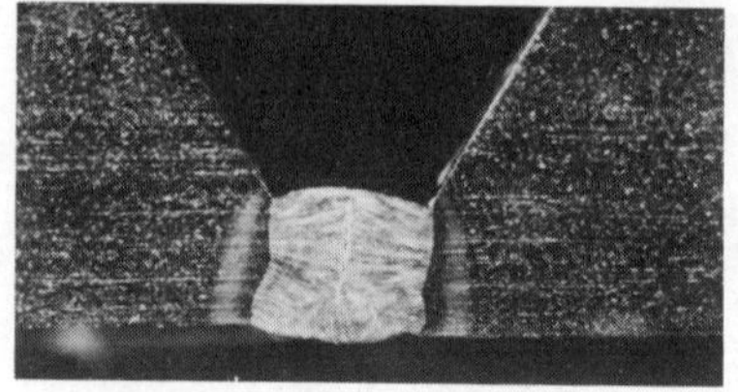

Fig. 5. Cross sectional view of root pass bead with hot crack.

Copper backing alone. In the three positions previously mentioned, satisfactory bead is obtained, but copper chips pickup on the surface of the internal beads. No hot cracks result.

Figure 6 shows dendrite structures in the bead cross-section with and without a hot crack. When copper backing with a high heat conductivity is used, heat dissipation is accelerated, dendrites grow rapidly toward the outer surface of root pass bead, and do not intersect with each other in the beads. Consequently, no hot cracks generate.

On the other hand, when glass tape or Scotch tape with heat retaining property is used, dendrites intersect in the cross-section because of a low heat conductivity, causing cracks to induce more readily. The foregoing makes it apparent that the use of copper backing is desirable for avoiding hot cracks and establishing good-shaped internal beads in all positions. Whereas, it is necessary to prevent copper chip pickup on the surface of the bead. To cope with this problem, copper backing with alumina, titania, tungsten carbide or zirconium powder coated was prepared, but no effective results were obtained.

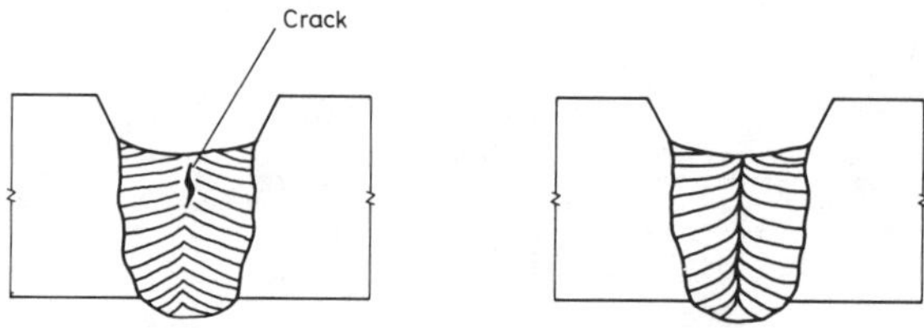


Fig. 6. Dendrite patterns with and without hot crack.

TABLE 3 Chemical Composition of Filler Wire Used

(wt %)

| C | Si | Mn | P | S | Ce | Total REM |
|---|---|---|---|---|---|---|
| 0.05 | 0.62 | 1.31 | 0.003 | 0.003 | 0.041 | 0.083 |

DC Electrode Negative Welding

Materials and method. Steel plate used is the same as that described in the above, and wire used is of 1.2 mm dia. containing REM. The chemical composition of this wire is shown in Table 3. If the cerium content is lower than 0.025 %, the effect of stabilizing arc of straight polarity is insufficient. In the present experiment, the total REM content of the trial wire is within a range between 0.05 and 0.08 %. The welding power supply is the same as previously mentioned except for a pulsed-arc power supply additionally used.
The flow rate is the same as that for DC electrode positive welding, but mixing ratio of shielding gas was altered. The groove geometry is V-shaped with 60° angle.

Results and discussions. When welding is conducted under the conditions given in Table 4, using the constant potential DC power supply at a speed of 1,000 mm/min, the arc spreads at an arc length of 2-3 mm at any position. The wire does not stretch long enough to reach the groove bottom; therefore, good penetration bead is scarcely formed in the vertical down position. These characteristics are common to all types of backing materials tried out. On the other hand, when welding is done under the conditions shown in Table 5, using the pulsed-arc power supply at a speed of 1,100-1,300 mm/min, satisfactory beads without copper pickup are obtained stably (Fig. 7).

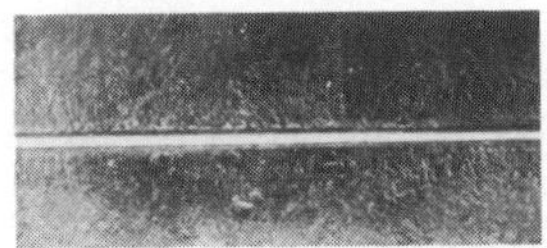

(a) Internal appearance

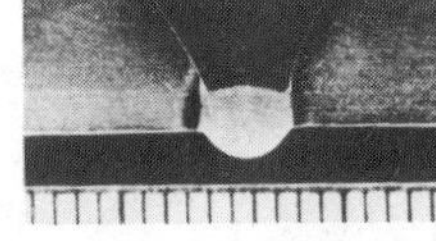

(b) Cross sectional view

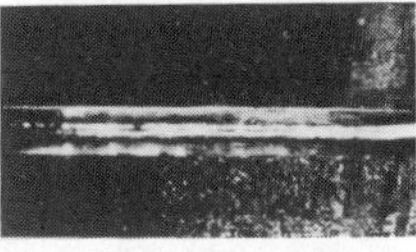

(c) External appearance

Fig. 7. Examples of satisfactory stringer bead.

However, even in this case if glass tape is applied as backing, hot cracks are present as in the case of DC electrode positive welding. It is required, therefore, to use copper band for backing. The current waveform during welding is in Fig. 8, showing peak current 830 amp and base current 140 amp.
The above are possible because in the case of "DC electrode negative" in which the copper band side becomes the anode, opportunities will be decreased for generating copper pickups which were caused in the case of "DC electrode positive" due to mainly formation of the cathode spot. Further, the following facts overlap one another: 1) metal transfer is performed in spray instead of short-circuit, 2) penetration is properly deep, 3) arc is comparatively longer with a sufficient spread and 4) adequate root faces and the narrow root gap suppress arc penetration. As a result, it has become possible to form penetration beads which are successful and free of copper pickups, and a favorable prospect has been obtained in establishing the high-speed circumferential root pass welding method. Figure 9 illustrates an example of EPMA, giving the result of line analysis for copper component from the weld metal to the outer surface of internal bead.

TABLE 4 Welding Conditions for Root Pass Using Constant Potential DC Power Supply (DC Electrode Negative)

| Test No. | Position* | Welding current (A) | Arc voltage (V) | Root face (mm) | Root gap (mm) |
|---|---|---|---|---|---|
| 1 | F | 270 — 280 | 22 | 1.0 — 1.2 | 1.0 — 1.1 |
| 2 | F | 270 — 280 | 23 | 1.0 — 1.2 | 0.9 — 1.0 |
| 3 | F | 270 — 290 | 23 — 24 | 1.0 — 1.2 | 0.8 — 1.0 |
| 4 | VD | 250 | 22 | 1.0 | 0.9 — 1.2 |
| 5 | VD | 270 — 280 | 22 | 1.0 | 0.9 |
| 6 | OH | 270 — 280 | 22 | 1.0 — 1.2 | 0.9 — 1.0 |
| 7 | OH | 270 — 290 | 22 | 1.0 — 1.2 | 1.0 |
| 8 | OH | 270 — 290 | 22 | 1.0 — 1.2 | 0.9 — 1.2 |

* "F" designates flat ; "VD" vertical down ; "OH" overhead

TABLE 5 Welding Conditions for Root Pass Using Pulsed-Arc Power Supply (DC Electrode Negative)

| Test No. | Position* | Welding current (A) | Arc voltage (V) | Root face (mm) | Root gap (mm) | Welding speed (mm/min) |
|---|---|---|---|---|---|---|
| 1 | F | 240 — 250 | 19 — 20 | 1.8 — 2.0 | 1.8 — 2.0 | 1100 |
| 2 | F | 250 — 260 | 19 — 20 | 1.8 — 2.0 | 1.6 — 1.9 | 1200 |
| 3 | F | 260 — 270 | 20 | 1.4 — 1.6 | 1.5 — 1.8 | 1200 — 1300 |
| 4 | VD | 260 — 270 | 20 | 1.4 — 1.6 | 1.5 — 1.8 | 1200 — 1300 |
| 5 | OH | 260 — 270 | 20 — 21 | 1.5 — 1.7 | 1.5 — 1.8 | 1200 — 1300 |

* "F" designates flat ; "VD" vertical down ; "OH" overhead

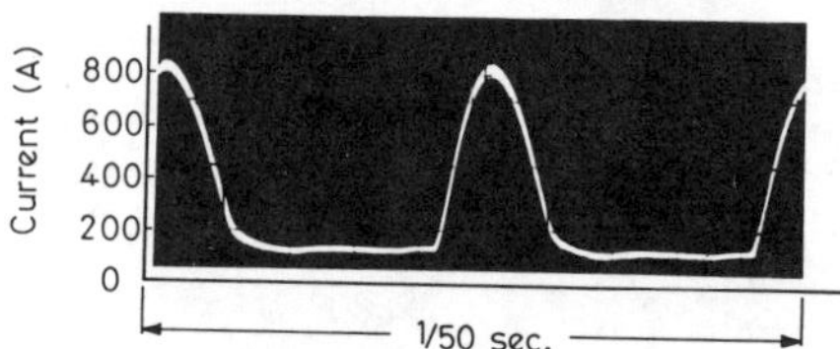


Fig. 8. Current waveform during pulsed-arc welding.

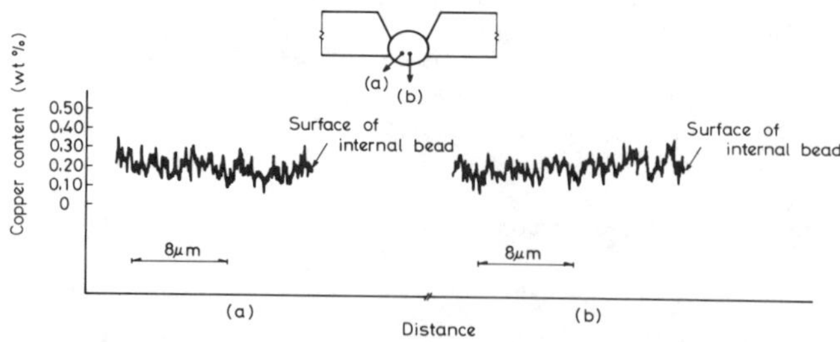


Fig. 9. Copper distribution along line (a) and (b). (EPMA)

## MECHANICAL PROPERTIES OF WELDED JOINT

Steel pipes used are API 5LX X52 and X60, with 16 inch outer dia. The wall thickness is 0.5 inch for X52 and 0.25 inch for X60. Welding wire used is of 1.2 mm dia., containing REM. Chemical compositions of pipes and the filler wire are listed in Table 6.

TABLE 6 Chemical Compositions of Pipes and Filler Wire Used for Welded Joint Tests

(wt %)

| | | C | Si | Mn | P | S | Nb | Ce | Total REM |
|---|---|---|---|---|---|---|---|---|---|
| Pipe | ×52 | 0.13 | 0.27 | 1.45 | 0.016 | 0.002 | — | — | — |
| | ×60 | 0.07 | 0.21 | 1.19 | 0.016 | 0.004 | 0.023 | — | — |
| Filler wire | | 0.05 | 0.60 | 1.23 | 0.006 | 0.004 | — | 0.026 | 0.052 |

The circumferential welding equipment and the internal clamper used are shown in Fig. 4. For X52 pipe, the groove geometry is shown in Fig. 10 and welding conditions listed in Table 7. The root and hot passes were made with machine welding, and the filler and cap passes with SMA welding.
Figure 11 represents Vickers hardness values of root pass bead after the hot pass is completed. No abnormal hardening is observed as the maximum value is 200 or so. Figure 12 shows the Charpy impact test results. For X60 pipe, the hardness value of stringer bead itself after the root pass welding is shown in Fig. 13 and these results do not give any practical difficulty because of the tempering effect expected of the subsequent passes. Table 8 gives the tensile and bend test results, and Fig. 14 and 15 show these test specimens tested for X52 and X60, respectively, suggesting satisfactory strength and ductility. Figure 16 shows the micro-structures of welded joint for X52.

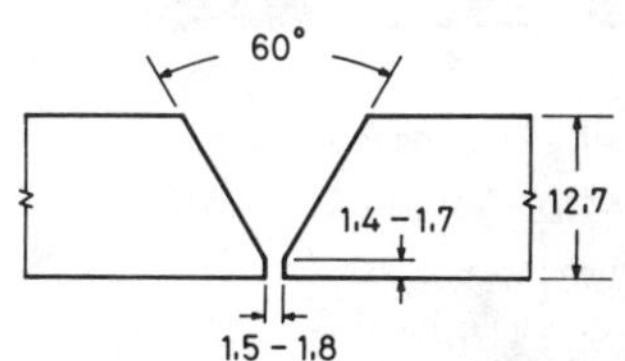


Fig. 10. Groove geometry.

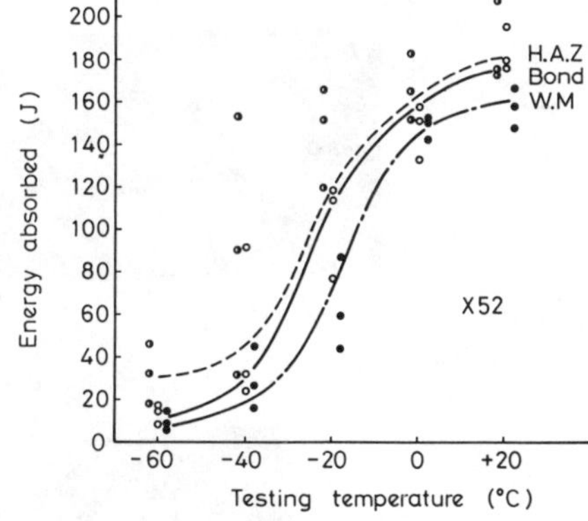


Fig. 12. Charpy impact test results of X52 welded joint.

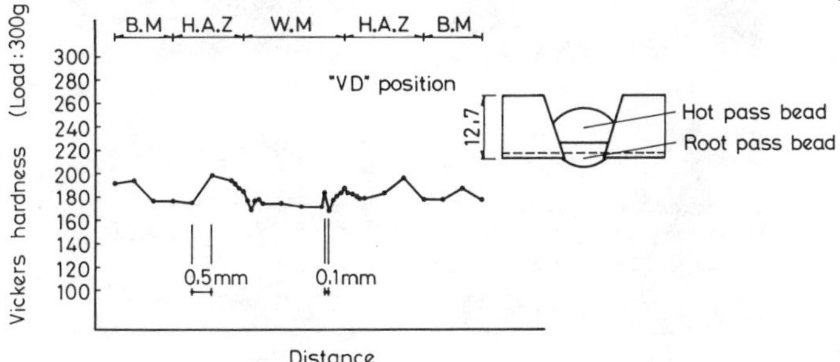


Fig. 11. Hardness values of root pass bead after hot pass is completed.

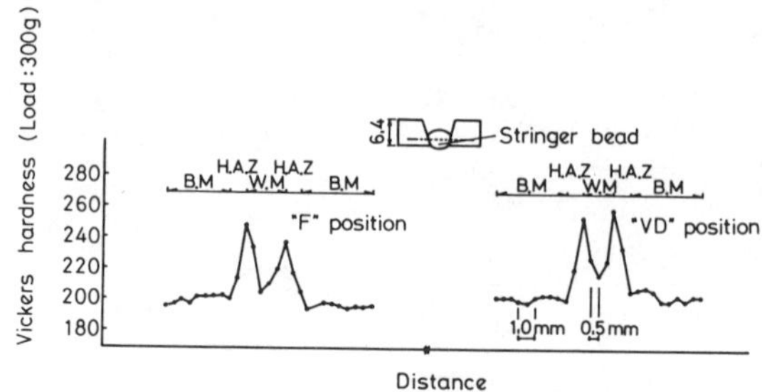


Fig. 13. Hardness values of stringer bead.

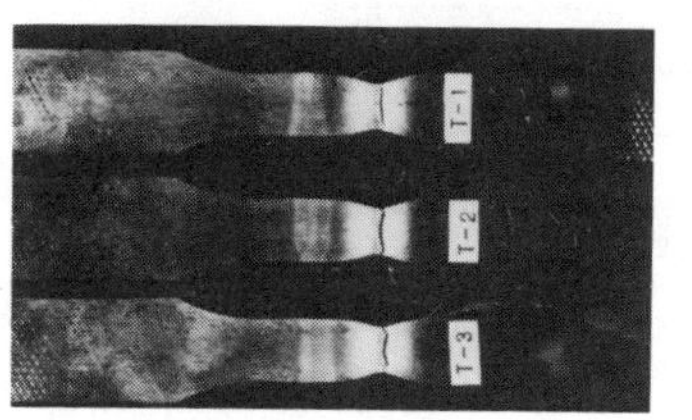


(a) Tensile test

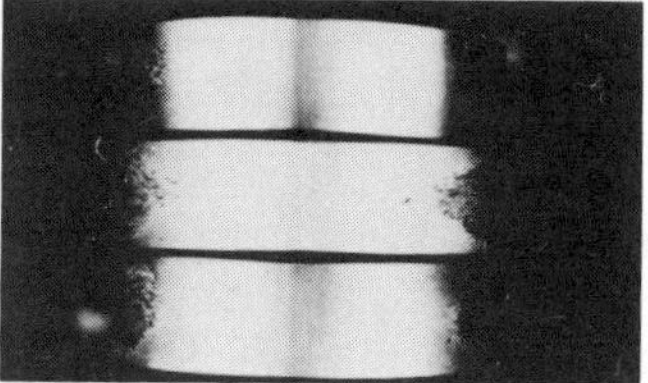

(b) Root bend test

Fig. 14. Tensile and bend test specimens using X52 welded joints.

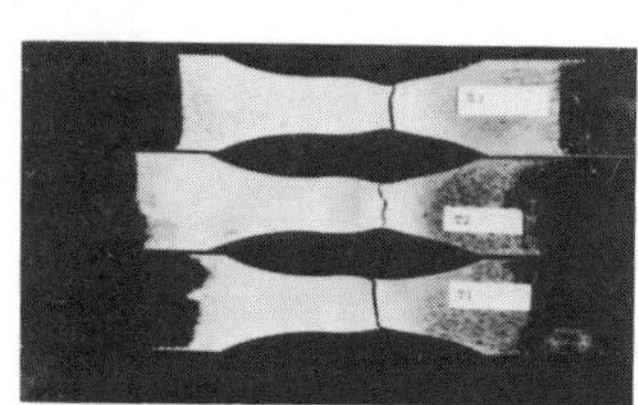

(a) Tensile test

(b) Root bend test

Fig. 15. Tensile and bend test specimens using X60 welded joints.

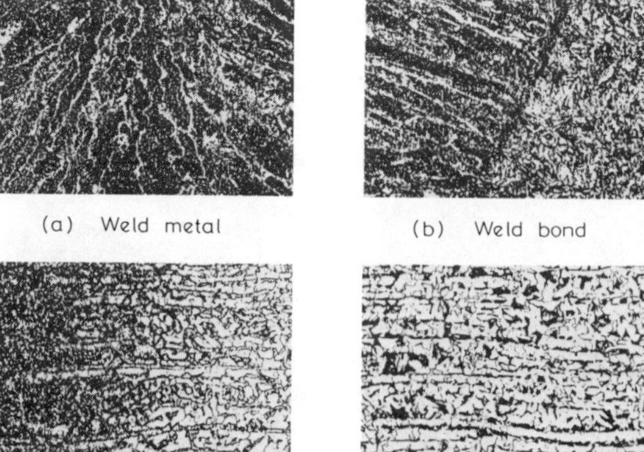

(a) Weld metal

(b) Weld bond

(c) Heat affected zone

(d) Base metal

Fig. 16. Micro-structures for X52 welded joint. (x100)

TABLE 7 Welding Conditions and Pass Sequence

| Pass | Welding current (A) | Arc voltage (V) | Welding speed (mm/min) | Pass sequence |
|---|---|---|---|---|
| Root * | 250 — 270 | 19 — 20 | 1300 | 4, 3, 2, 1 — Wall thickness |
| Hot * | 250 — 270 | 20 — 21 | 780 | |
| Filler ** | 130 — 140 | 25 | 70 | |
| Cap ** | 120 — 130 | 25 | 55 | |

* GMA welding (Machine) ** SMA welding

TABLE 8 Tensile and Bend Test Results of Welded Joints

| Pipe | Tensile test | | Bend test |
|---|---|---|---|
| | T. S. ($kg/mm^2$) | Location of rupture | Root bend |
| ×52 | 53.0 | B.M. | No defect |
| | 52.0 | B.M. | No defect |
| | 54.0 | B.M. | No defect |
| ×60 | 61.0 | B.M. | No defect |
| | 60.0 | B.M. | No defect |
| | 60.0 | B.M. | No defect |

## CONCLUSION

A high-speed circumferential root pass welding method for the large diameter pipeline application has been studied. The main conclusions are as follows;

(1) In the case of the constant potential power supply, and copper backing using a solid wire containing REM in order to realize DC electrode negative welding, satisfactory penetration bead is scarcely formed in the vertical down position. Whereas, in the case of the pulsed-arc power supply, it is possible to obtain sound beads stably without copper pickup by setting the welding speed at 1,100-1,300 mm/min.

(2) The mechanical properties of X52 and X60 welded joints including the root pass welds produced by the newly developed process are satisfactory.

(3) In the case of DC electrode positive welding using the conventional constant potential DC power supply, it is difficult to achieve the one-side downhill welding at a speed of 1,000 mm/min without backing material.

(4) In the DC electrode positive welding using backing materials such as glass tape or Scotch tape, many hot cracks occur in the bead and arc tends to be unstable. On the other hand, when using copper backing, good internal beads are realized although copper chips are frequently picked up on the surface of the internal bead.

## REFERENCES

E. Tsunetomi, T. Bada, K. Saito, H. Sakurai and T. Saito (1982). Study on high-speed automatic welding of piping. Journal of the J.W.S., 51, 9, 30-36.

K. Agusa, N. Nishiyama and J. Tsuboi (1980). Stabilization of pure-argon-shielded metal-arc by rare earth additions to electrode wires. IIW Doc., XII, B-287-80, E-56-80.

Slutskaya, T.M., Asuits, A.E. and Tyurin, A. Ya. (1978). The properties of wire alloyed with cerium and yttrium used for $CO_2$ welding. Automatic Welding, 31, 2, 44-46.

# POSITION BUTT WELDING OF TECHNOLOGICAL PIPELINES OF POWER ENGINEERING STRUCTURES

I.K. Pokhodnya, V.N. Shlepakov, Yu.A. Gavrilyuk, L.N. Orlov

*E.O. Paton Welding Institute of the Ukrainian Academy of Sciences, USSR*

## ABSTRACT

The E.O.Paton Welding Institute has developed the technology and equipment for the automatic position butt welding of 325-630 mm dia. technological pipelines. The technology and materials are specified for welding the technological pipelines with operating temperature up to +400°C.

KEYWORDS

Forced weld formation; low temperature brittleness; resistance to cyclic loadings.

The improvement of technology of fabrication of power engineering welded structures is being developed in the direction of automation and mechanization of all kinds of welding jobs. A solution of the problem of safety and quality of joints becomes important with an increase of the capacity of power systems.

The position welding of the technological pipelines is one of the most complex operations in site construction of nuclear and heat power stations. In the majority of cases the position welding of technological pipelines is performed by a manual argon-arc non-consumable electrode filler welding of the root weld, while the rest of the weld is welded by rod electrodes. In this case the main scope of jobs is taken up by welding the filling welds, this seriously affecting the efficiency and quality, especially with large pipe wall thicknesses.

The E.O.Paton Electric Welding Institute of the Ukrainian Academy of Sciences has developed the technology, equipment and consumables for flux-cored wire position welding of pipe butts with a forced weld formation. The principle of the method is described in the paper[1] and consists in the formation of the welding pool in the melting space between the pipe edges and the forming cooled device.

According to the technology developed upward position welding of a butt is performed in turn by two heads, placed on both sides of the butt welded. The place of welding start for both heads is 6 o'clock position, and the welding ends in the 12 o'clock position.

The present paper contains the results of studying the characteristics of mechanical properties of pipe welds in the testing temperature interval from -60° up to 450°C, for pipelines of power engineering structures with the following limit operating parameters: wall temperature is from -30° up to 400°C, atmospheric pressure is up to 5MPa. Welded joints of 15ГС, 16ГС and 20 steel pipes of 325mm up to 630 diameter and 25 mm wall thickness were tested.

Tables 1 and 2 give pipe metal composition and the mechanical properties in the as-delivered state. The weld root with a 2 mm technological gap was argon-arc welded manually for 6 mm of the thickness. The remainder of the groove was $CO_2$ welded in three runs with 1.6 mm dia. ПП-АН31 flux-cored wire. Reversed polarity direct current welding was performed at the following conditions: 300-320 A welding current; 29-30 V arc voltage; 0.4 $s^{-1}$ electrode oscillation frequency.

TABLE 1 Pipe steel composition

| Steel grade | Content,% | | | | |
|---|---|---|---|---|---|
| | C | Si | Mn | S | P |
| | | | | not more than | not more than |
| 15ГС | 0.12-0.18 | 0.70-1.00 | 0.90-1.30 | 0.025 | 0.035 |
| 16ГС | 0.09-0.15 | 0.50-0.80 | 0.80-1.20 | 0.040 | 0.040 |
| 20 | 0.17-0.24 | 0.17-0.37 | 0.35-0.65 | 0.025 | 0.030 |

TABLE 2 Mechanical properties of pipe steels

| Steel grade | T.S. MPa | Y.S. MPa | El. % | R.A. % | J/cm² |
|---|---|---|---|---|---|
| 15ГС | 500 | 300 | 18 | 45 | 60 |
| 16ГС | 490 | 290 | 21 | - | 60 |
| 20 | 410 | 220 | 24 | 45 | 50 |

The chemical composition of the weld metal on various steels is given in Table 3. Tensile specimens were made of the weld metal for all the studied steel grades to test the short-time mechanical properties. During tensile tests the characteristics of the mechanical properties for the weld metal were determined in the temperature interval from 20 up to 450°C after each 50°C. Fig.1 and 2 show the dependence of the mechanical properties on test temperature.

The analysis of the specimen testing for static tension shows, that a certain increase of the strength properties and a drop in the ductile properties is observed in welded joints made on 15ГС, 16ГС, and 20 steels. This effect is usually associated with the strain ageing

TABLE 3. Weld metal composition

| Steel grade | Element content, % | | | | |
|---|---|---|---|---|---|
| | C | Si | Mn | S | P |
| 15ГС | 0.09-0.11 | 0.14-0.17 | 1.17-1.23 | 0.018-0.022 | 0.02-0.024 |
| 16ГС | 0.09-0.12 | 0.17-0.21 | 1.75-1.22 | 0.018-0.022 | 0.02-0.025 |
| 20 | 0.12-0.15 | 0.15-0.20 | 0.92-1.05 | 0.018-0.024 | 0.018-0.025 |

[2]. The effect of the strain ageing on the weld metal is displayed at the temperatures of 250-300°C. The ultimate rupture strength grows by 50-80 MPa, depending on the steel grade. With the temperature rise the effect of the yield point increase is displayed less prominently than in case of the ultimate rupture strength. The increase of the weld metal yield point for all the studied steel grades was about 30 MPa (at 275°C temperature). The elongation and reduction in area are decreased by 5% and 12%, respectively. With the further increase of the testing temperature the strength characteristics are lowered, and those of the ductile properties grow. For 400°C testing temperature the mechanical properties of welds, made on 15Г С, 16 Г С and 20 steels are equal to those properties at room temperature.

The impact bending tests in the temperature interval from 20 down to -60°C were performed to determine the critical transition temperature. The notch was oriented to the weld center and along the HAZ. Fig.3 and 4 give the curves of the impact toughness dependence on the testing temperature. The critical transition temperature was determined from the impact toughness level of 50 J/cm$^2$. The welded joint made on 20 steel has the lowest transition temperature -43°C (notched along the weld); -40°C (notched along the HAZ).

The specimens were soaked for 1000 hours at the pipeline working temperature of 350°C to determine the temperature ageing effect on the mechanical properties of the welded joints.

Table 4 gives the mechanical properties of the weld metal, obtained on 15ГС, 16ГС and 20 steels. The shifting of the critical tansition temperature ($\Delta tcm$) due to the temperature ageing was determined as a difference of the critical transition temperature of specimens, subjected to temperature ageing and specimens in as-welded state.

TABLE 4 Mechanical properties of the weld metal (ageing at 350°C for 100 hours).

| Steel grade | T.S. MPa | Y.S. MPa | El. % | R.A. % | Critical brittleness temperature, °C | |
|---|---|---|---|---|---|---|
| | | | | | Weld | HAZ |
| 15Г С | 510 | 400 | 24 | 62 | -37 | -32 |
| 16 Г С | 520 | 410 | 22 | 65 | -37 | -32 |
| 20 | 480 | 375 | 24 | 59 | -43 | -40 |

The shifting of the critical transition temperature is zero for the weld metal and 2°C for the HAZ. Thus, the temperature ageing does not noticeably affect the weld metal mechanical properties.

Low-cyclic tests of pipe welded joints were carried out to determine the resistance to the fatigue fracture. Series of 15 samples were tested on five levels of strain for each zone of the welded joint (weld metal, HAZ and base metal). Testing was performed on the nick-break test specimens of Koffin type with recording of the transverse strain in the given section of the specimen, and with further recalculation of it to obtain the longitudinal strain, using the Roisson's ratio for plastic (0.475) and elastic (0.25) deformation [3]. The type of loading was tension-compression, the loading conditions being severe and the strain cycle was symmetric. Testing was carried out at the 20°C temperature and the 350°C working temperature of the pipelines.

When determining the strain ageing effect on the low-cyclic fatigue resistance tests were performed at 275°C temperature. The low-cyclic fatigue curves were plotted from the experimental data in the coordinates: amplitude of a complete elastic-plastic longitudinal deformation versus number of cycles before fracture (Fig.5,6,7).

The coefficients of cyclic strength drop were determined from the comparison curves of low-cyclic fatigue, as a ratio of weld metal (HAZ) strain amplitude and the base metal strain amplitude (Table 5)

TABLE 5 Coefficient of cyclic strength decrease in the pipe welded joint zones.

| Base metal | Cycle number | Test temperature, °C | | |
|---|---|---|---|---|
| | | 20 | 350 | 275 |
| 15ГС | $10^2$ | $\frac{1.093}{0.906}$ | $\frac{1.111}{0.888}$ | $\frac{1.074}{0.926}$ |
| | $10^3$ | $\frac{1.111}{0.888}$ | $\frac{1.125}{0.875}$ | $\frac{1.125}{0.975}$ |
| | $10^4$ | $\frac{1.280}{0.880}$ | $\frac{1.080}{0.880}$ | $\frac{1.277}{0.944}$ |
| 16ГС | $10^2$ | $\frac{1.142}{0.928}$ | $\frac{1.076}{1.000}$ | $\frac{1.200}{1.080}$ |
| | $10^3$ | $\frac{1.158}{0.951}$ | $\frac{1.120}{1.000}$ | $\frac{1.230}{1.092}$ |
| | $10^4$ | $\frac{1.129}{0.871}$ | $\frac{1.093}{0.875}$ | $\frac{1.352}{1.176}$ |
| 20 | $10^2$ | $\frac{1.178}{0.892}$ | $\frac{1.130}{0.956}$ | $\frac{1.077}{1.040}$ |
| | $10^3$ | $\frac{1.125}{0.887}$ | $\frac{1.092}{0.923}$ | $\frac{1.148}{1.081}$ |
| | $10^4$ | $\frac{1.480}{1.280}$ | $\frac{1.174}{0.869}$ | $\frac{1.000}{1.000}$ |

Note: The numerator gives the weld metal coefficients, and the denominator - those of the HAZ.

The testing results showed that the fatigue fracture resistance at 350°C temperature is lower than that at 20°C temperature. The strain amplitudes, leading to the fracture after a given number of loading cycles are approximately 1.1-1.3 times lower at 350°C temperature than at the normal temperature. This decrease is typical for both the base metal and the welded joint. The weld metal possesses the highest resistance to low-cyclic fatigue, and the HAZ has the lowest resistance at the 20°C test temperature. At the test temperature of 350°C the HAZ metal on 16ГC steel has the same level of the resistance as the base metal. At 275°C test temperature the base metal has the lowest resistance to low-cyclic fatigue for all steel grades. At 275°C temperature a decrease of the general level of resistance to low-cyclic loads $\varepsilon^{a275}/\varepsilon^{a20}$ = 0.88 + 0.92 is observed for the weld metal, it being $\varepsilon^{a275}/\varepsilon^{a20}$ = 0.82 + 0.85 for the pipe steel. Low-cyclic testing was performed on cylindrical specimens with a slot according to a standard procedure to determine the endurance limit [4]. The loading type is tension-compression, the loading cycle being symmetric. The testing of the welded joint zones was accomplished at 20,350 and 275°C temperatures. The testing base was $10^6$ cycles. In Table 6 the values of the welded joint endurance limit are compared to those for the base metal.

TABLE 6 Endurance limit $\sigma^{-1}$ of metal of pipe welded joint zones, MPa.

| Zone tested | Test temperature °C | Base metal 15 C | 16 C | 20 |
|---|---|---|---|---|
| Weld | 20 | 326 | 325 | 335 |
| | 350 | 323 | 320 | 330 |
| | 275 | 317 | 315 | 320 |
| HAZ | 20 | 305 | 300 | 310 |
| | 350 | 304 | 300 | 308 |
| | 275 | 300 | 295 | 300 |
| Base metal | 20 | 290 | 287 | 295 |
| | 350 | 290 | 290 | 295 |
| | 275 | 285 | 282 | 290 |

The testing results show that the endurance limits of the weld metal and the HAZ are higher for the steel grades studied, than of the base metal. The endurance limits of the weld metal and the HAZ of welded joints, made on 15Г C, 16ГC steels do not differ. The weld metal has the highest endurance limit, and the base metal has the lowest one.

The work done allowed to study the dependences of mechanical properties and cyclic strength within the -60° to 400°C temperature interval for weld metal and welded joints of pipes of 15ГC, 16ГC and 20 steels, using the technology of flux-cored wire welding with a forced weld formation.

The analysis of results of the investigations performed showed that the level of welded joint metal properties is not lower than that of the base metal of pipes, and corresponds to the requirements for welding the technological pipelines of nuclear and heat power stationes.

The experience of application of the technology and equipment for position welding of technological pipeline butts showed, that the efficiency grows 2-3 times and the joint quality is high, as compared to the manual electric arc welding.

## REFERENCES

Paton B.E., Pokhodnya I.K., et al. Automatic position butt welding of large diameter pipes with the self-shielding flux-cored wire.- "Stroitel'stvo truboprovodov", 1981, No.2.

Karzov G.P., Leonov V.P., Timofeev B.T. High-pressure welded vessels: strength and service life - L. Machinostrojenije,Leningradskoe otdelenije, 1982.

Strength at low-cyclic loading: Basis for the calculation and testing procedures. (Ed. S.V.Serensen), M. Nauka, 1975.

Monze V., Kh. Fatigue strength of steel welded structures, M., "Mashinostrojenije", 1968.

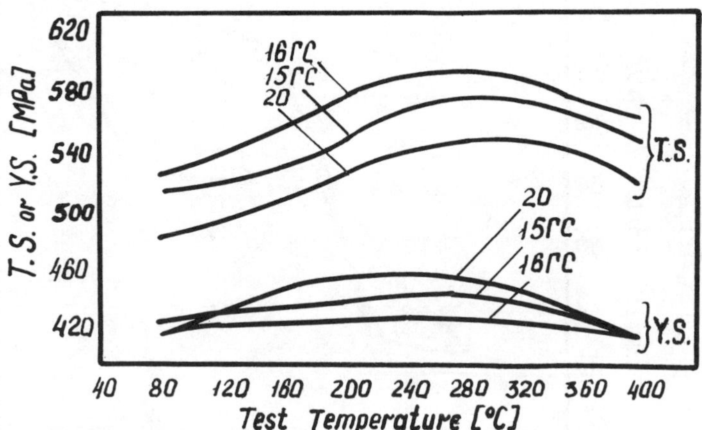


Fig.1. Dependence of strength properties of weld metal made with ПП-АН31 wire on test temperature. 1-15ГС steel; 2-16ГС steel; 3-20 steel.

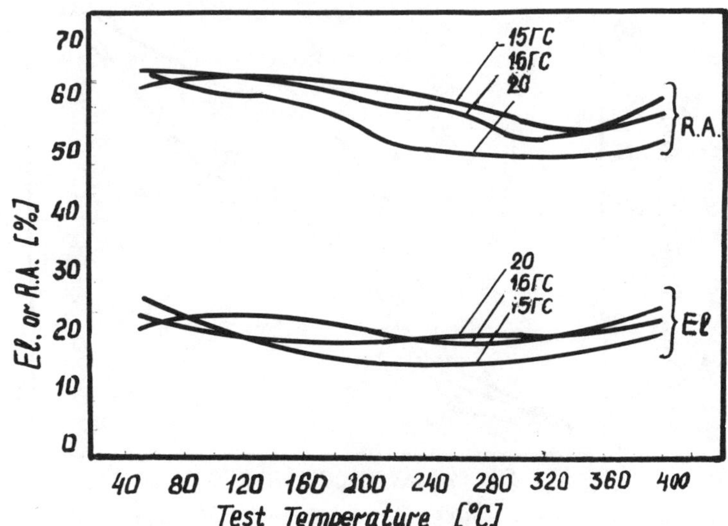


Fig.2. Dependence of ductile properties of weld metal made with ПП -АН31 wire on test temperature: 1-15ГС steel; 2-16ГС steel; 3-20 steel.

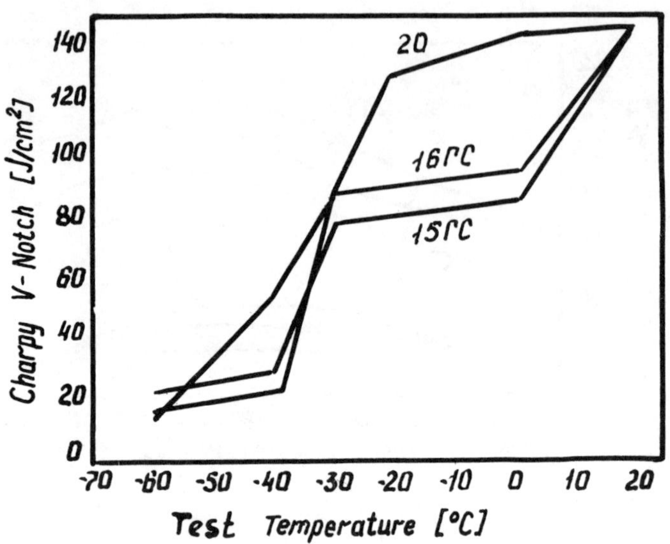


Fig.3. Dependence of impact toughness of the welded joint on test temperature (notch along the weld): 1-15ГС steel; 2-16ГС steel; 3-20 steel.

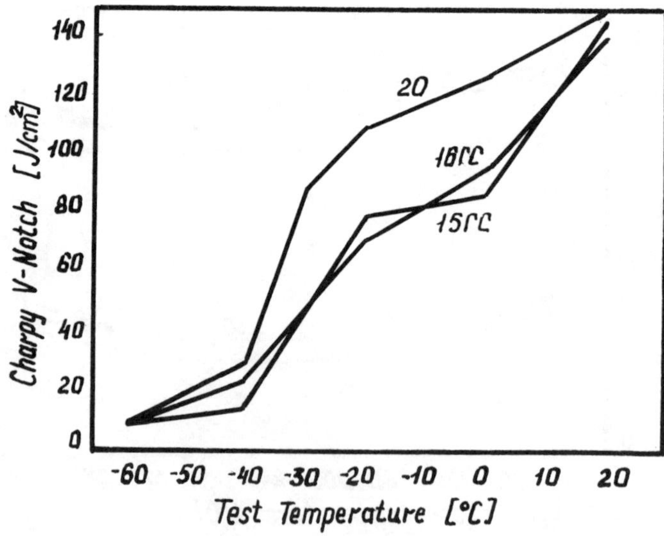


Fig.4. Dependence of impact toughness of a welded joint on test temperature (notch in the HAZ). 1-15ГС steel; 2- 16ГС steel; 3-20 steel.

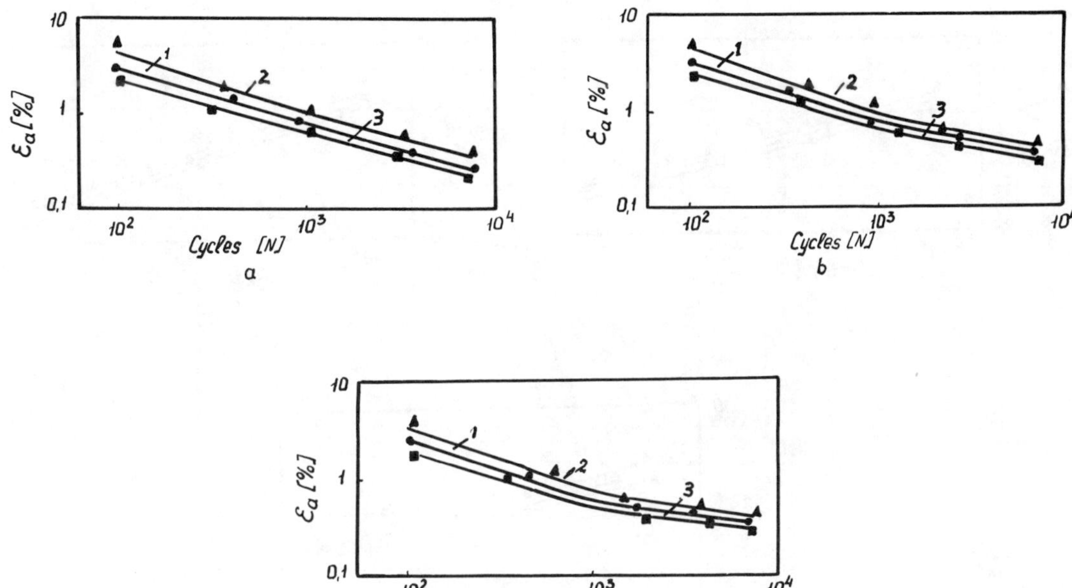


Fig.5. Comparison of low-cyclic fatigue curves of welded joints of 15ГС(a); 16ГС(b); and 20 (c) steels at 20°C test temperature: 1-base metal; 2-weld metal; 3-HAZ.

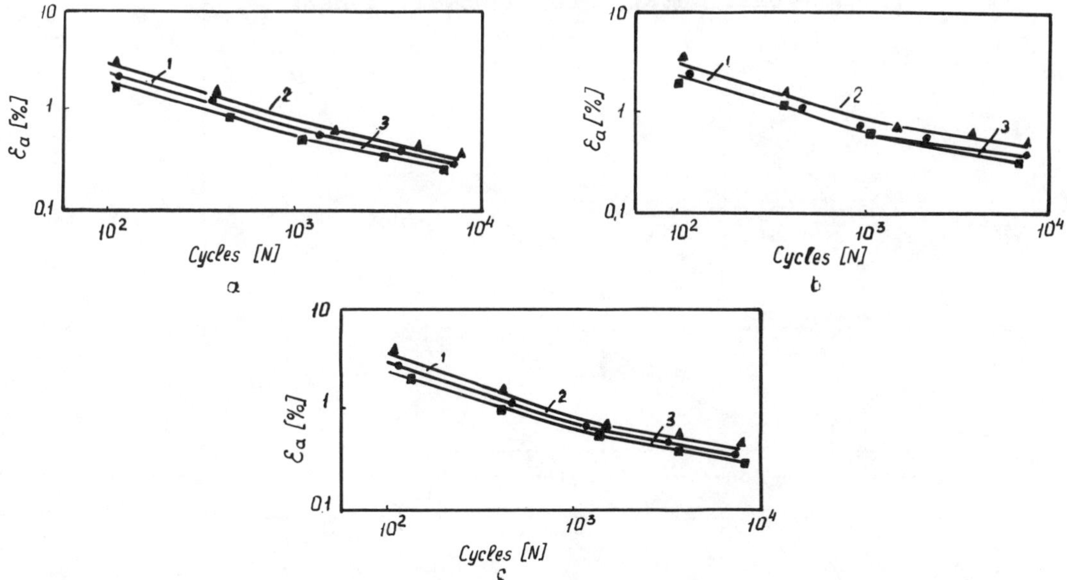


Fig.6. Comparison of low-cyclic fatigue curves of welded joints of 15ГС(a); 16ГС(b); 20(c) steels at 350°C test temperature: 1-base metal; 2-weld metal; 3-HAZ.

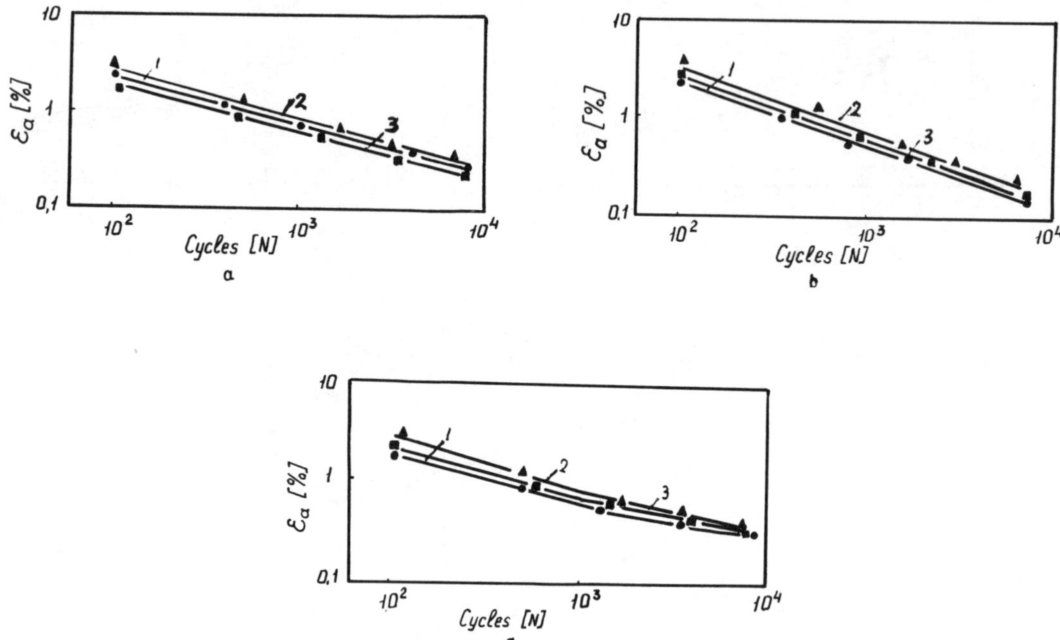


Fig.7. Comparison of low-cyclic fatigue curves of welded joints of 15ГС(a), 16ГС(b), 20(c) steels at the test temperature 275°C, 1-base metal, 2-weld metal, 3-HAZ.

# AUTOMATIC WELDING OF PIPELINES WITH THE 'SATURNE' PROCESS ON A LAYBARGE

G. Dreyfuss* and A.A. Smith**
*Serimer-Dasa/ETPM, France*
***Consultant, United Kingdom*

## ABSTRACT

A description is given of the use of the 'Saturne' automatic MIG welding system for welding pipelines from a laybarge; theoretical production rates are given.

KEYWORDS

Automatic MIG welding; 'Saturne' welding system; pipelines; welding parameters.

## INTRODUCTION

Of all the manual welding tasks which are capable of mechanisation, the circular joints involved in the butt welding of pipes must be surely the most suitable. It was not until the energy crisis of 1973, and the boom in offshore energy exploitation of the North Sea, that pipeline contractors began to publicise their automatic welding developments. Apart from the shortage and uncertainty of supply of skilled 'stovepipe' welders using cellulosic electrodes, the objective of introducing automatic welding was to increase the daily laying rate by at least 30% and to increase the quality of joints (and thus reduce costly repairs) by exercising machine tool control over the welding parameters, and to impose realistic and repeatable welding procedures. With the increasing wall thicknesses and strength grade of sea-pipeline, coupled with the aggressive service environment of the North Sea, safety and the protection of the ecology became uppermost in people's minds. Against this background, a joint venture between the French oil company Companie Francaise des Petroles (CFP/Total) and the French pipeline group ETPM, commissioned the development of an automatic pipe welding machine for use on a laybarge. After the construction of a welding simulator and experimental testing (Saturne 01), upon which the feasibility of the MIG welding system could be established, a series of four industrial machines (Saturne 02) was commissioned in 1977. Before being used offshore these machines were installed in a purpose-built development shop, where the conditions on a barge could be simulated and practical conditions were commenced in order to qualify the procedure under API 1104 and Lloyds regulations and to train operators. In 1979 a new series of six machines (Saturne 03) was constructed to provide facilities for welding pipelines from 22 in to 42 in in diameter. The Company called Serimer (Societe d'Exploitation des Richesses de la Mer) was formed to commercialise and further develop the 'Saturne' machines - so called because of the four orbiting MIG-$CO_2$ heads which revolve around the pipe.

Fig. 1. 'Saturne' automatic welding machine.

## PIPELINE CONSTRUCTION FROM A LAYBARGE

The basic difference between automatic pipeline welding operations on a barge, offshore, is that the pipe is moving past the different welding stations, whereas on land the welding machines move down the line to complete the next joint. There are also other differences and space restrictions on a barge mean that it is impossible to have much distance between welding stations whereas on land space is no problem. The rate of laying a pipeline on land is determined by the unit length of the pipes whereas offshore the pipes are coated with concrete which makes handling more difficult and limits the unit length. On a large barge such as ETPM 1601 four stations are used for double jointing coated pipes 12 m long with elements of 24 m, which are then joined in the welding tunnel by the four 'Saturne' orbiting machines. (Fig. 2)

Double jointing operations are carried out by submerged arc welding of 2 x 12 m coated pipes, using four passes with the pipe being rotated on turning rolls. One pass is made from the outside onto the pipes aligned and held in a special clamp. The second pass is made on the second station from the outside, after removal of the clamp. The double jointed pipe is then transferred to the third station where an internal head makes the inside weld. Welding is completed with the fourth pass being made on the outside. X-ray examination is performed on the completed joint and, after examination and acceptance, the double jointed pipe is transferred to the feeding rack.

Fig. 2. One of the four 'Saturne' automatic welding stations on board the laybarge ETPM 1601.

The work on the pipeline itself takes place in the welding tunnel, which comprises four 'Saturne' 4-head welding stations, an inspection station and a repair station. The organisation of the line is quite conventional and starts with machining of the special bevels to provide the narrow gap edge preparation. (Fig. 3)

The pipes are placed in the 'line' and the central line-up clamp, umbilical control cables attached and the 'Saturne' machine lowered onto the pipe to make the root pass. After withdrawal of the internal clamp, visual inspection is carried out and any necessary repairs are made with manual metal arc electrodes. The second pass is made with this first welding station. The barge then advances to present the joint to the second welding station and the 'Saturne' machine which is already preprogrammed to make some of the filling passes as required. The barge then advances to present the joint to the third welding station and more filler passes are added. If required, preheating can then take place before the last capping pass is made at the final welding station at its preprogrammed welding conditions. (Table 1)

The barge advances to present the joint for X-ray examination, when the exposure, development, drying and interpretation of the film takes place within minutes. If the radiograph is accepted, the barge advances; if not, a repair is made manually and re-X-ray examination is performed.

To achieve success with so many operations underway at the same time requires thorough training and good work organisation. A remarkable degree of skill can be developed with a good team and a production rate of some 80 butt joints/day is commonly achieved.

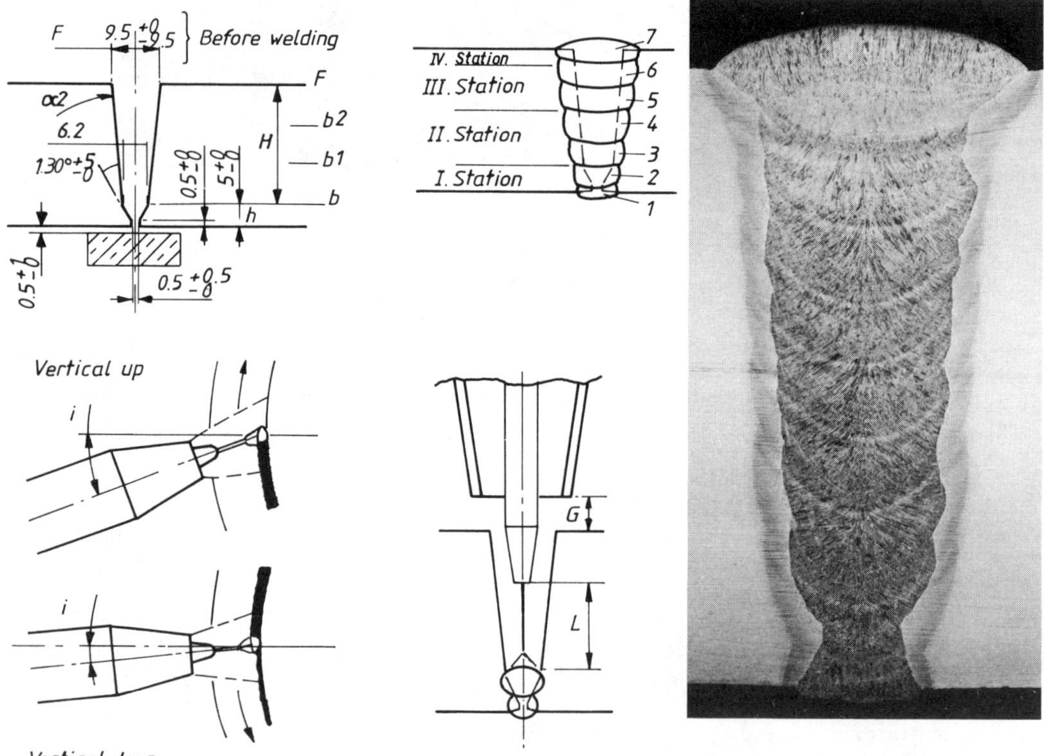


Fig. 3. Edge preparations for 'Saturne' automatic pipe welds on a laybarge.

TABLE 1 Typical preprogrammed welding conditions for 'Saturne' welding of pipeline (1mm diameter wire, 36 in, 22 mm wall, x65)

| Welding parameters | I | I | II | II | III | III | III | IV |
|---|---|---|---|---|---|---|---|---|
| Pass | 1U | 2D | 3D | 4D | 5D | 6D | 7D | 8D |
| Wire grade | SG3 | SG3 | SG3 | SG3 | SG3 | SG3 | SG3 | SG3 |
| Gas | $CO_2$/A | $CO_2$/A | $CO_2$/A | $CO_2$/A | $CO_2$/A | $CO_2$/A | $CO_2$/A | $CO_2$/A |
| Mixture | 50/50 | 50/50 | 50/50 | 50/50 | 50/50 | 50/50 | 50/50 | 50/50 |
| Stickout, mm | 13 | 13 | 13 | 13 | 13 | 17 | 13 | 10 |
| Oscillation, n/min | 0 | 160 | 130 | 13 | 130 | 130 | 140 | 180 |
| Width, mm | 0 | 2 | 25 | 3 | 4 | 5 | 6 | 7.5 |
| Weld speed, cm/min | 60 | 40 | 38 | 38 | 38 | 42 | 46 | 20 |
| Wire speed, n/min | 10 | 9 | 11 | 11 | 11 | 10.5 | 9 | 5.5 |
| Amps, A | 230 | 220 | 240 | 240 | 240 | 240 | 220 | 180 |
| Arc volt, V | 22.5 | 23.5 | 24.5 | 24.5 | 24.5 | 23.5 | 23.0 | 19.0 |
| Energy, J/cm | 5.2 | 7.5 | 9.2 | 9.2 | 9.2 | 7.6 | 6.9 | 9.7 |

## 'SATURNE' WELDING SYSTEM

The pipe dimensions of diameter and wall thickness represent the controlling factors in planning the production rate of a pipeline. With orbital MIG welding, the physical laws governing liquid steel under the forces of gravity strictly limit the size of the weld bead and the welding speed. Solidification processes and the thermal treatment of the pipes to be welded both play a part in the development of welding procedures used in an automatic welding system such as 'Saturne'. Geometrical considerations have resulted in a computer programme (described later) for predicting productivity and such approaches are based on the assumption that engineering tolerances in machining, welding and pipe manufacture are within tight control.

As mentioned previously, the 'Saturne' system was developed some years ago to achieve optimum productivity on a large laybarge handling a thick pipe with a wall size of >17 mm and a diameter from 22 in to 42 in, coated with concrete. The machine has already achieved success on two major offshore pipeline undertakings and is currently engaged in the North Sea. (Table 2)

TABLE 2 Pipeline operations undertaken by 'Saturne' welding system

| Year | Place | Diameter, in | Thickness, m | Steel | Number of joints |
|---|---|---|---|---|---|
| 1981 | Zakum | 24 in | 17 | x52 | 1,600 |
| | Persian Gulf | 24 in | 29 | x70 | 2,000 |
| 1982 | Woodside, Australia | 40 in | 24 | x65 | 5,600 |
| 1983 1984 | Statpipe, North Sea | 36 in | 22 | x65 | 14,000 |

The 'Saturne' system consists of a horseshoe frame which carries the supports (two rear and one front), clamping and earthing shoes, port and starboard welding carriages, tracking arms for upper and lower torches. The welding machine is clamped onto the pipe and the frame positions itself accurately over the joint by means of tracking arms which carry wheels located in the joint edge preparation. The horseshoe frame lowers onto the pipe and is closed when location in the joint is secured. The four torches are initiated from the start command, with each carriage starting after the other with a short delay. Welding with the four torches proceeds simultaneously, under the control of two operators.

To achieve satisfactory joints in a pipe in all positions the torches must describe a rotation around the pipe very accurately, and this is achieved simply with the tracking wheels which give a torch precision of $\pm 0.5$ mm. Each arm is equipped with an oscillator giving sinusoidal movement, the speed being nil when the edges of the joint are reached, thus giving good side wall penetration. A constant wire feed is most important to provide good weld quality and with the 'Saturne' system 200 W electrically controlled motors are used to select and maintain an accurate wire feed speed.

The execution of the first pass vertical-up represents an innovation for automatic orbital welding. The weld is made vertical-up with a torch angle of $15\text{-}20^{o}$ from the normal to the pipe, with no oscillation. This technique of vertical-up welding for the first pass eliminates the problem of lack of fusion in the root and also compensates for a pipe mismatch of up to 2 mm and this corresponds to the normal performance of a manual welder using a cellulosic electrode.

## WELD QUALITY

Although two operators are capable of handling the 'Saturne' system, it is common with larger pipe sizes for a third man to be positioned beneath the lateral welders. A fourth operator completes the barge team so that the various welding positions can be occupied successively and provision made for rest periods. The role of the operator-welders consists of setting-up and checking the torches before each weld using the two manual controls on the head, pressing the buttons to start the welding station and to observe the arc and weld pool. If it becomes necessary, the operator can intervene manually to centralise the oscillation or modify the arc length, but the overall adjustment of the equipment is fixed to close limits and cannot be changed. Slight adjustments to the torch position are needed to compensate for the out-of-roundness and mismatch of pipes. Although the 'Saturne' system is a fully automatic machine, the present team of trained and qualified welders substitutes for the lack of positive 'feedback' from the arc and weld pool, which would make 'Saturne' a true robot. Various codes exist for the acceptance of quality of land pipelines, such as the American AP 1104 and the UK BS 4515, and deal with typical workmanship defects such as porosity, inclusion and lack of fusion. The Norwegian Det Norske Veritas code was developed in 1981 for pipelines being laid in the North Sea and represents very severe requirements. These codes are being met by the 'Saturne' system. The requirements involve the measurement of the mechanical properties of joints including tensile and toughness properties, particularly in the root and below the surface. The development of the 'Saturne' system for offshore pipe welding of pipe in the North Sea has been achieved as a result of long periods of evaluation of welding procedures, and a special workshop, which simulates conditions on a barge, has been used for this purpose. This workshop is equipped with the necessary metallurgical and engineering facilities, and is ideally placed to provide the client with the results of months of tests to develop approved and reproducible procedures for a large offshore contract. For each contract in pipe laying, the procedures are developed in conjunction with a well established system of training the teams of operator-welders and electro-mechanical technicians who will work on the barge.

Parallel teams are trained to carry out the necessary quality control (quality engineers) to ensure that the welds are made according to the established procedure and to monitor the consumables and ancillary operations to welding.

Because of the automatic nature of 'Saturne', the maintenance of quality on the barge is to a high standard. There is, however, a limited occurrence of the standard MIG-$CO_2$ defects such as lack of fusion (50% of all defects) porosity and inclusions (30% of all defects) microfissures (10% of defects) and other defects (irregular shape, etc.). These defects are normally detected by the radiographic and ultrasonic inspection and repaired on the barge, and a repair rate of only 3% has been achieved. The acceptance of defects is sometimes based on the consideration of fracture mechanics principles and benefits are often obtained for client and contractor from the realistic acceptance of flaws based on the CTOD design curve.

## PRODUCTIVITY WITH THE 'SATURNE' SYSTEM

Two aspects of productivity on a large barge with an automatic welding system should be considered. There is the theoretical maximum productivity and the actual average productivity which can be achieved under practical conditions. The theoretical maximum productivity depends principally on the total weight which must be deposited to fill a joint, the number of passes which are necessary, the number of welding stations and the number of torches which can work simultaneously on one station. When these factors are considered from geometrical/welding technology points of view, the theoretical minimum time can be established. The actual average productivity can be established from operating experience on the barge with the 'Saturne' system, which is part of a chain of operations executed simultaneously and successively on the same joint. An interruption affecting one operation can affect all the others. The best average performance can be computed from a statistical consideraion of interactions between the different processes and the frequency of incidents which affect the various operations. Serimer has established a computer programme simulating the presumed operating conditions and the conditions affecting the welding of pipes with the 'Saturne' system, for different pipe diameters and wall thicknesses. (Fig. 4) The programmes have been prepared to provide a method of estimating commercial possibilities with specific contracts. Certain conclusions are already clear and it can be proved, for example, that for the barge ETPM 1601 the 'Saturne' system is optimum with four welding stations with pipes 24 m in length. The opportunity now exists to simulate productivity times with various models, including the prefabrication of pipes on land, number of welding stations, NDT stations and other welding processes for the future, e.g. flash welding, selectron beam or laser welding.

*Theoretical productivity of Saturne system*

*Influence of diameter and thickness*

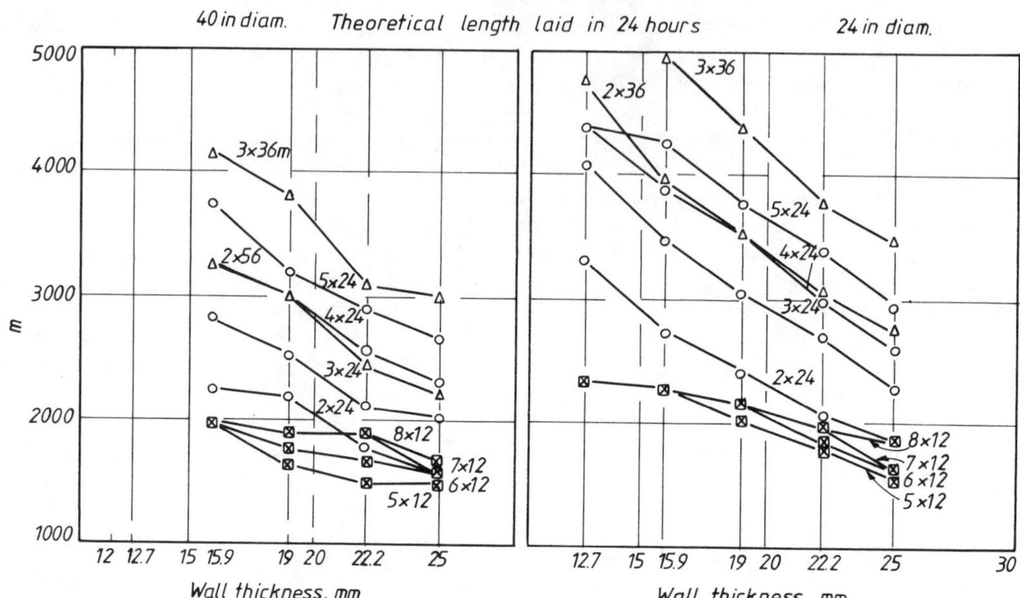


Fig. 4. Theoretical productivity curves for 'Saturne' system.

## CONCLUSIONS

The technical press describes many researches, principally in the USA and Japan, concerning intelligent welding robots which can position the torches and alter the welding parameters according to the conditions existing 'in-flight' in the weld pool. Further developments are expected to improve the present version of 'Saturne'.

Other welding processes such as electron beam and flash welding are in development and the impact of these developments will affect the design of barges for pipe laying.

Serimer is pursuing a programme of development to improve all the ancillary equipment such as clamps, alignment equipment, thermal treatment equipment and radiography. As far as the 'Saturne' system is concerned many details in improved design will be included in future machines without altering the fundamental principles.

Serimer is also pursuing a programme of studying the fundamentals of welding and particularly the effects of heat transfer in order to respond to the new steels which are being developed by the steel industry.

## ACKNOWLEDGEMENTS

The authors wish to thank their colleagues in ETPM/Serimer Dasa for the provision of information, particularly Mr. E.A. Minkiewicz, Chief Engineer and inventor of the 'Saturne' system.

# QUALITY OF WELDS IN LARGE DIAMETER PIPE JOINED BY FLASH BUTT WELDING

S.I. Kuchuk-Yatsenko*, B.I. Kazymov*, V.G. Krivenko*, D.L. Turner**

*E.O. Paton Welding Institute*
*Kiev, U.S.S.R.*
***McDermott Incorporated*
*New Orleans, Louisiana*

ABSTRACT

This paper shows that substantial deviations from previously proven welding conditions can cause discontinuities to be formed in flash butt welded joints, and that nonmetallic inclusions or nonmetallic films of various thickness are frequently found at these fusion line discontinuities. The various defects which can be encountered in flash butt welding are characterized and their causes are discussed in detail.

KEYWORDS

Flash butt welding, nonmetallic inclusions, weld discontinuities, gray spots, nick-break tests, weld testing, oxide films.

## INTRODUCTION

More than 35000 km of various size pipelines are in service in the Soviet Union; the largest of these, permitted by flash butt welding developments at E.O. Paton Welding Institute, is 1.42m (56 inches) in diameter.

Since 1977 the E.O. Paton Welding Institute and McDermott Incorporated have performed a number of investigations to develop welding technology and equipment for more efficient joining of offshore oil and gas pipelines.

Comprehensive tests of flash butt welded joints have proven their high and consistent quality which is provided by certain methods of welding process control. Many years of service experience of pipelines, flash butt welded in various climatic zones of the Soviet Union, testify the same; there have been no cases of weld joint failure. This is explained by the fact that weld defects appear only as a result of considerable deviations of main welding parameters from the preset values, these deviations being caused by a failure of separate units of the welding machine or control equipment. Such cases are immediately registered by a strip-chart recorder at the operators control. This initiates a signal for interruption of welding so that the equipment malfunction can be remedied. For these reasons the probability of defective joints in the welded pipeline string is negligible, and the application of subsequent non-destructive testing reduces this to an even lower level. Confidence in weld joint integrity is even further increased when the fracture mechanics approach is applied to assessment of very small weld discontinuities which are permitted to remain.

This paper deals with the classification of defects occurring in pipe welds from the standpoint of their physical nature, interrelation with deviation of main parameters of the welding process, and the affect on weld joint quality, and ability to be revealed by nondestructive testing (NDT).

## EVALUATION OF WELD QUALITY

As mentioned in a previous paper (Welding Journal, April 1982), radiography has been found to be very effective for disclosure of fusion line discontinuities such as voids and nonmetallic inclusions. Figure 1 shows such inclusions in an unacceptable flash butt weld. In every case it has been possible to confirm by subsequent nick-break testing that all significant discontinuities of this type are revealed by radiography. Conventional ultrasonic testing, however, is not as reliable.

Two weld discontinuities not usually detected by NDT are lack of fusion and gray spots. Lack of fusion can be prevented by weld parameter control along with strip-chart monitoring, and gray spot fracture is precluded by post weld heat treatment which toughens the weld fusion line by a factor of 10.

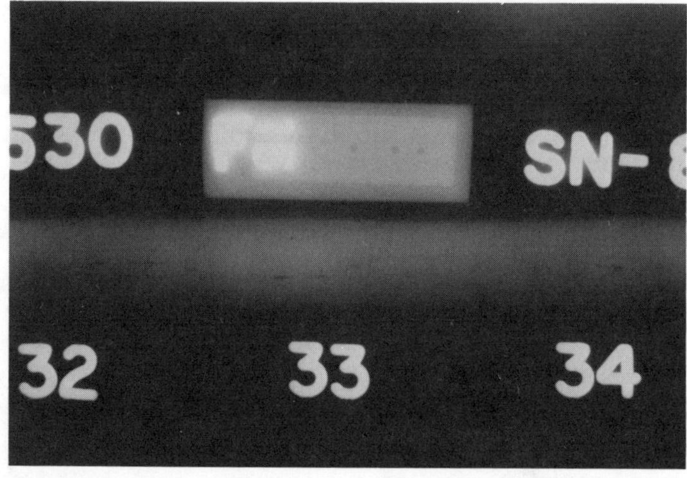


Fig. 1 Nonmetallic (slag) inclusions revealed by radiography (unacceptable)

Transverse tensile testing with or without weld reinforcement removal is not effective in examining weld discontinuities because such specimens almost always break in the pipe material. Side bend specimens reveal more, but only show discontinuities at or near one transverse plane.

Far more effective than tensile or side bend tests for weld quality evaluation are multiple nick-break specimens, because they reveal almost any fusion line discontinuity when tested in the as-welded condition. Less satisfactory is testing in the post weld heat treated condition because fracture then rarely occurs 100% on the fusion line.

To further facilitate weld quality evaluation, a nick-break specimen wider than specified by API Standard 1104 has recently been utilized with good success. Figure 2 shows a drawing of this specimen.

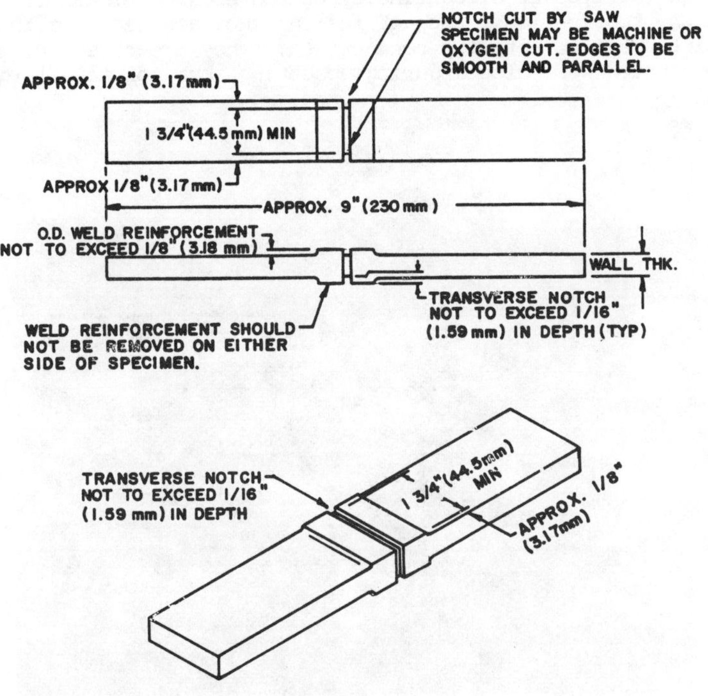


Fig. 2 Two inch type nick-break specimen

## WELD DISCONTINUITIES AND THEIR CAUSES

In June, 1983, a series of 20 flash butt welds was made in API 5LX Grade 60 pipe in order to produce clear-cut examples of flash butt weld discontinuities, and to further explore welding parameter tolerances. The first 10 welds were made with various flashing times; the second 10 welds were made with various upset strokes. All other welding parameters were held practically constant.

Each joint was radiographically inspected after welding, and 2 inch nick-breaks were taken from zones with radiographic discontinuities, if there were any. From 8 to 12 nick-breaks were taken from each weld and tested without post weld heat treatment to facilitate fusion line fracture. For each nick-break specimen there was excellent correlation with radiographic inspection. Following is a discussion of the various discontinuities found and their causes.

### Lack of Fusion

Figure 3 shows one type of discontinuity encountered on nick-break specimens during the June test series. Lack of fusions such as these are those areas of the weld joint where no metallic bond exists. They are caused by arc craters on the pipe ends which were closed during upsetting, but not metallurgically bonded.

Fig. 3 Lack of fusion discontinuities resulting from insufficient upset stroke.

Depending on their location on the fusion line, lack of fusion discontinuities can have a dark or light appearance. If they are open to the O.D. or I.D., surface of the pipe immediately after welding, they become oxidized, and therefore dark. If they are not connected with the pipe surface during weld cool-down their surface is most often light.

An exaggerated form of lack of fusion is the void. These defects are incompletely closed arc craters, and only occur with grossly insufficient upset stroke. A good example of such discontinuities is shown in Fig. 4 which was made with only 3.2 mm of stroke.

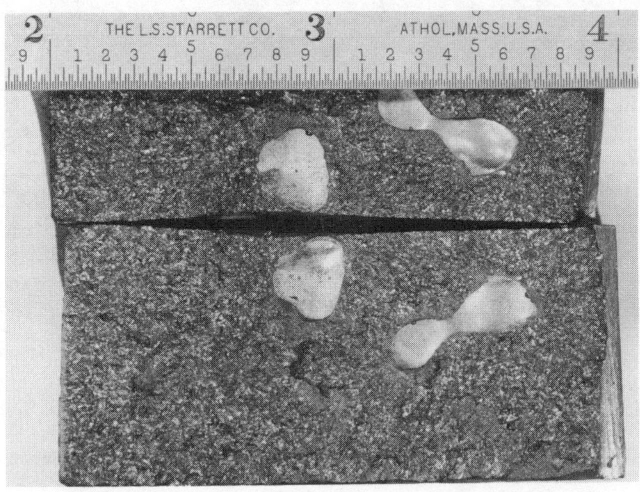


Fig. 4 Voids resulting from grossly insufficient upset stroke.

## Pipe Segregations

Another type of discontinuity which appears on the fracture face of nick-break specimens is shown by Fig. 5. The segregations on the right hand side of the photograph were caused by heavy banding or even laminations within the pipe itself. It is possible for these to be flattened on the fusion line during upsetting and to subsequently appear on radiographs. Fortunately the metallurgial cleanliness and thru-thickness properties of line pipe have been improved in recent years, and this type discontinuity is infrequent and usually less than 6.35 mm (1/4 inch) in length.

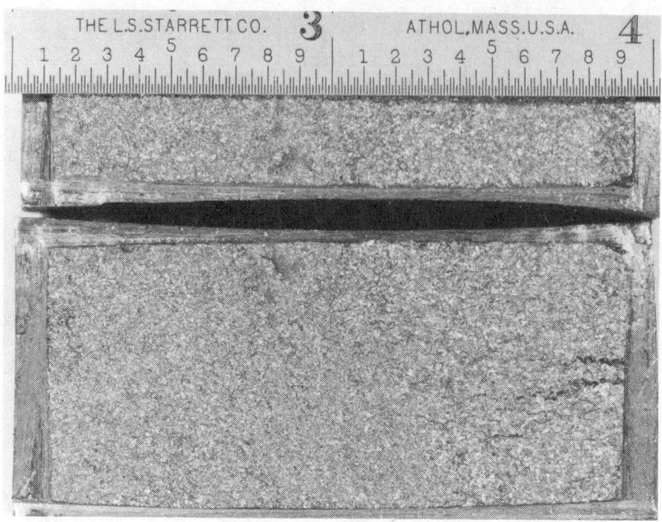


Fig. 5 Pipe segregations revealed by nick-break test specimen

Nonmetallic Inclusions

The most common discontinuity encountered in flash butt welds is the nonmetallic inclusion. These result from momentary interruption of the flashing process prior to upsetting, or from heavy pipe segregations. Figure 6 shows such a discontinuity. Nonmetallic inclusions are always black and friable, and can be physically lifted away from the fracture face. Microchemical analysis using energy dispersive spectra techniques (EDS) has revealed that the black material is composed of oxides of manganese, silicon, and iron. These inclusions are caused by welding slag trapped in arc craters on the fusion line at the end of the welding cycle and then flattened during the upset stroke. Such discontinuities are consistently and effectively revealed by radiography. The proposed new section of API Standard 1104 permits a maximum length of 3.17 mm (1/8 inch) and a maximum cumulative length of 12.7 mm (1/2 inch) in any 304.8 mm (12 inches) of weld for these inclusions.

Fig. 6 Nonmetallic inclusion revealed by nick-break specimen

## Gray Spots

Nonmetallic inclusions in the form of thick oxide films mentioned above usually appear during momentary interruption of the flashing process before upsetting. Sometimes very thin films occur and then are broken up by the upsetting operation. These become finely dispersed on the fusion line (the white etching zone shown by Fig. 7). When these microscopic discontinuities occur, the metallic bond is only partially ensured between the pipes. Areas on nick-break specimens which contain these microscopic inclusions are called gray spots because of their characteristic appearance which is shown by Fig. 8.

Scanning electron microscope (SEM) images of gray spots confirm their non-homogeneous nature. Figure 9 reveals a ductile mode of fracture and microscopic inclusions existing in the fracture dimples. At identical SEM magnifications the portion of the as-welded nick-break specimen without gray spots displays a completely different mode of fracture (cleavage). The microscopic inclusions found in a gray spot are not usually larger than 30 microns in size, and are generally rich in manganese and silicon.

It should be clearly understood that gray spots do not cause fusion line failure after post weld heat treatment. This is because the fusion line becomes more homogeneous, as well as considerably tougher during the normalizing treatment.

Fig. 7 Deeply etched cross-section of typical flash butt weld in API 5LX, Grade 60 pipe (as welded)

Fig. 8 Gray spot revealed by nick-break specimen

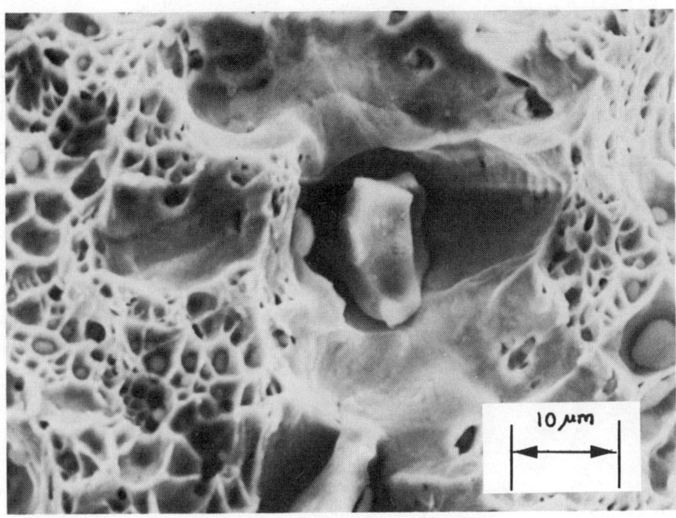


Fig. 9 Scanning Electron microscope image of nonmetallic inclusions on a gray spot

TABLE I
UPSET STROKE VS. WELD QUALITY

| WELD NO. | UPSET STROKE DEVIATION (%) | MECHANICAL TESTS | RADIOGRAPHIC DISCONTINUITIES TOTAL LENGTH (MM) | RADIOGRAPHIC DISCONTINUITIES MAX.INDIV.LENGTH (MM) |
|---|---|---|---|---|
| 532 | +41 | F | 1.52 | 1.57 |
| 529 | +26 | OK | 8.63 | 2.36 |
| 526 | +13 | OK | 7.11 | 2.36 |
| 515 | +02 | OK | 3.93 | 2.36 |
| 534 | -10 | OK | 7.11 | 3.17 |
| 535 | -15 | OK | 5.58 | 2.36 |
| 536 | -24 | F | 7.87 | 3.17 |
| 537 | -36 | F | 58.67 | 8.71 |
| 539 | -52 | F | 81.78 | 7.92 |
| 541 | -64 | F | 399.28 | 13.48 |

## CONTROL OF WELDING DISCONTINUITIES

Table I shows the correlation between radiographic and mechanical tests obtained when data from the June weld test series were analyzed. All nick-break specimens were acceptable as long as radiographic discontinuities were small, or not too numerous. Table I also shows the affect of various upset strokes on weld quality. For this specific pipe material and set of welding parameters, deviations as great as +26% and -15% resulted in fully acceptable welds.

Using these data it is possible to construct bar charts like Fig. 10 which graphically display individual weld parameter tolerances when all others are held near optimum. Such data can then be used as a guide to establish realistic tolerances for welding procedure control. Due to the statistical natúre of weld parameter development, however, a large number of welds must be made to establish a welding procedure which is different from those previously proven.

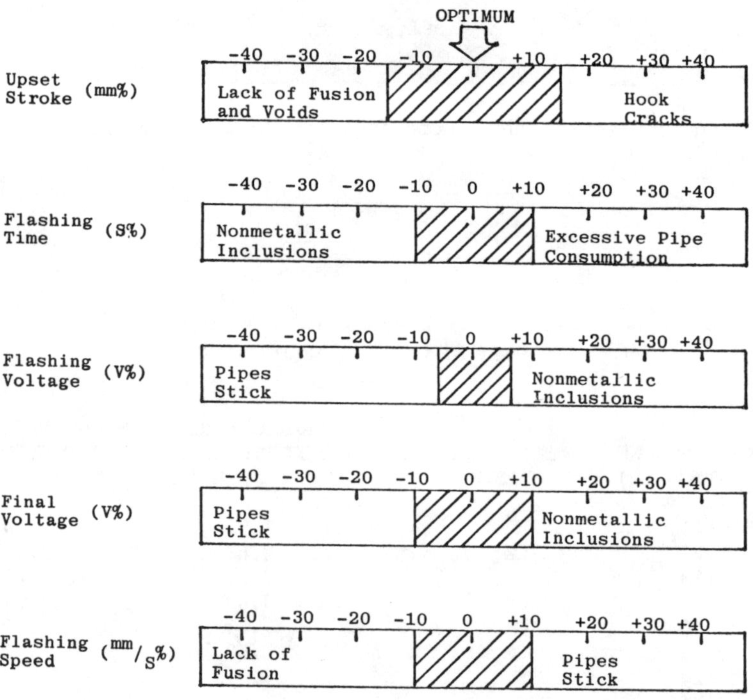


Fig. 10 Typical tolerance for flash butt welding parameters

# RESOURCES-SAVING TECHNOLOGY OF POSITION AUTOMATIC FLUX-CORED WIRE WELDING OF GAS PIPELINES WITH A FORCED WELD FORMATION

I.K. Pokhodnya, V.Ya. Dubovetsky, V.N. Shlepakov, L.N. Orlov, A.N. Kutovoy

*E.O. Paton Welding Institute of the Ukrainian Academy of Sciences*

A.G. Mazel, M.Z. Sheinkin

*Ministry of construction of enterprises of oil gas industries*

ABSTRACT

The method of self-shielding flux-cored wire welding with a forced weld formation has been described. The technology of the position automatic flux-cored wire welding of 1220-1420 mm dia. pipe circumferential joints with the root weld, made with coated electrodes, is given. The assembly-welding installation "Styk", designed for the realization of this method of welding, is also described.

KEYWORDS

Flux-cored wire, position circumferential welding, forced weld formation, welding technology, assembly-welding installation "Styk".

At present position welding of main pipelines is performed principally by manual arc process. The shortcomings of this technological process are low efficiency, long training of highly qualified welders, need in special electrodes, cost and labour consumption of which are high.

A volume of welding operations in pipeline construction constantly rises, this leading to a deficit of highly skilled welders for overhead welding, whose number at present inhibits the accomplishment of working programs.

An elimination of the above drawbacks and an essential increase of efficiency at the expence of forced weld formation were attained in welding with self-shielding flux-cored wire (1), the technological process of this welding method and welding equipment having been developed at the E.O.Paton Institute in collaboration with VNIIST and Kiev branch of SKB "Gasstroymachina".

The scheme of this process of pipeline welding with the forced weld formation is given in work (2) and determined by a presence of melting space formed by pipe edges, preceding layer surface and the forming device.

Flux-cored wire is fed into the melting space, and an electric arc is burning between the ends of the wire and the molten pool. Due to the arc heat and that of the welding pool the pipe edges are fused. The slag formed in the wire melting prevents the crystallized metal of the welding pool from sticking to the forming device.

As the weld is crystallized the forming device together with the welding machine are displaced upwards along the perimeter of the butt.

Welding of right and left semiperimeters of butts of large diameter pipes is performed simultaneously by two heads.

At the begining of welding by the first head a metallic insert usually of electrode wire (Fig.1a) is employed as a bottom of the melting space.

The tack weld can also serve as the bottom of the melting space. Welding by the second head (Fig.1b) is started from the weld previously made by the first head and carefully protected by a polishing device.

A lock weld in the top part of the pipe is made by transferring the process from the forced mode into the semiforced one by means of the angular corrector. When finishing the welding process by the first head the weld thickness is gradually reduced at the expence of increasing the welding speed. Welding by the second head is finished by overlapping the preceding weld.

In the process of position welding of butts of pipes a position of welding pool changes in space. Though welding of vertical spot of the butt can be performed rather easily, the weld formation in overhead and flat positions is difficult due to flowing out of the pool metal on the forming devices.

The pool is prevented from being flowed out using gas dynamic arc pressure which is determined by the current intensity, the wire diameter and by the angle of its inclination.

Besides, when welding in flat and overhead position, a non-uniform penetration of the pipe edges as to depth takes place. In the overhead welding those edges are melted more intensively which are adjacent to the inner surface of the pipes, while in flat position the edges adjacent to the outside one sustain more intensive melting. To eliminate this non-uniformity the electrode wire is displaced during welding as to the melting space height from the outside pipe surface at the beginning of welding to its inside one in the end of the process.

At present the organizations of Minneftegasstroy, especially Glavyuzhtruboprovodstroy, apply the technology of automatic welding with flux-cored wire using manual welding up of weld root on 1220 and 1420 mm dia. pipes with standard edge preparation. In this case, as in one-sided submerged arc welding using pipe welding machines of ПАУ type, welding of the main and the most labour-cunsuming part of weld section, is automated. Simultaneously, the studies on improving the means of weld formation from the side of inner pipe surface and of the technology of automatic welding the root weld are carried out.

The automatic welding is performed at direct current of reverse polarity using 2.3 mm dia. flux-cored wire with 40-45 mm stick-out under the conditions shown in Table 1.

Depending on the pipe wall thickness, welding is accomplished in several passes (Table 2). When the width of the groove is more than 14 mm, transverse oscillations are imparted to the electrode wire with 4-5 mm less swing than the groove width and at frequency of 30 - 120 Hz depending on the welding speed. The thickness of each layer is controlled by the height of the forming shoe tooth ente-

TABLE 1 Conditions for automatic position welding of butts with flux-cored wire in forced weld formation on manual welding up.

| Name of pass | Arc voltage, V | Welding current, A | Average welding speed, m/h |
|---|---|---|---|
| The first filling one | 25-28 | 260-320 | 12-14 |
| The second filling one | 25-28 | 260-400 | 10-20 |
| The facing one | 26-32 | 260-450 | 10-20 |

TABLE 2 Amount of layers in welding with flux-cored wire in forced weld formation on manual welding up.

| Wall thickness | 10-14 | 15-22 | 22-25 |
|---|---|---|---|
| Amount of layers | 1 | 2 | 3 |

ring the groove (Fig.2a). The reinforcement shape is determined by the groove sizes on the shoe for welding of facing layer (Fig.2b). The shoe groove for welding of the facing layer should be 10-15 mm wider than the groove between edges.

Each welding head is completed with the set of the shoes required.

The shoe design ensures their replacement in 15-20 s. The shoes are connected with the welding head by console-lever system providing the shoe being pressed to the pipe in the process of welding at 40-45 kg. force.

Besides the displacement around the pipe of the forming shoe cooled with water or antifreeze, the welding head fulfils all other functions usual for arc welding, i.e., provides feeding of welding wire, current supply, adjustment of stick-out value, correction of electrode position. Unlike other welding methods, the wire is fed into the melting space tangentially to the pipe surface and, if necessary, at an angle. The same features distinguish the system controlling the welding head. The necessity of having the arc position adjusted as to the height to the melting space is met by a possibility of stopping the welding head without interrupting the wire feed and arc burning and also without increasing the speed of movement. In the first case the arc and the welding pool are lifted to the front end of the shoe, in the second - lowered into the melting space. The devices controlling these functions are mounted on an operator's panel handle.

For welding of position butts of 1220 and 1420 mm dia. pipelines the set of equipment "Styk" was developed by experimental design technical bureau of the E.O.Paton Institute and the Kiev branch of construction design bureau "Gasstroymachina".

Each welding machine of "Styk" set is completed by two welding heads which are mounted on the opening rail track of hingegrab type. The system of rail track mounting on the pipe and of its locking is hydraulically controlled.

The rail track is suspended from a boom of self-propelling welding unit mounted on the base of caterpillar tractor TT-4. In the body of the power set a power unit is mounted, consisting of the 100 kW source of three-phase alternating current, three welding rectifiers, cooling station of the forming shoes, a control cabinet for the welding heads.

For auxiliary operations a non-self-propelling movable shop was developed, in the closed body of which the auxiliary equipment is arranged for welding preparation, such as: oven for calcinating of electrode wire and a device for winding it into cassettes, drilling facility, fitter's bench. For displacing along the track the shop is secured on a rigid rod to one of the power units. "Styk" sets are produced at Kakhovsky works of electrowelding equipment in the modifications (Table 3) which mainly differ in a number of welding units.

The modification is chosen depending on diameter and pipe wall thickness, organization of the works, location relief and other conditions.

All existing forms of work organization valid at present are based upon the technology usually in service for assembly of pipe butts and welding root weld and hot pass with organic electrodes. The amount of butts assigned by the program and prepared for a subsequent welding of butts is determined by a number of welders in the main team.

In its turn, this determines a number of welders who weld filling and facing layers synchronously with the main team. The replacement of welders making filling and facing layers is done from the estimation that one power unit of "Styk" installation, serviced by two operators, substitutes 4-5 welders. Such substitution provides synchronious operation with the main team.

TABLE 3 Complete set of "Styk" installation

| Version of performance | Diameters of pipes welded in mm | Quantity of units in the set, pcs | | | |
|---|---|---|---|---|---|
| | | inner aligner | power unit | welding head | movable shop |
| Styk | 1420 | 1 | 1 | 2 | 1 |
| | 1220 | 1 | 1 | 2 | 1 |
| | 1420 | 1 | 2 | 4 | 1 |
| | 1220 | 1 | 2 | 4 | 1 |
| | 1420 | 1 | 3 | 6 | 1 |

TABLE 4 Technical-economical indices for replacing the manual position welding of butts by an automatic welding with flux-cored wire using installations "Styk".

| | 120 mm dia. pipes 4 welders in the main team | | 7 welders in the main team | | 1420 mm dia. pipes 4 welders in the main team | | 8 welders in the main team | |
|---|---|---|---|---|---|---|---|---|
| | Manual welding | Instal. "Styk" | Manual welding | Instal. "Styk" | Manual welding | Instal. "Styk" | Manual. w. | Instal. "Styk" |
| Number of workers in the team | 22 | 19 | 34 | 24 | 29 | 21 | 44 | 29 |
| Number of welders | 10 | 4 | 17 | 7 | 14 | 4 | 24 | 8 |
| Number of operators | - | 4 | - | 4 | - | 4 | - | 6 |
| Number of mechanisms, pcs. | 9 | 8 | 12 | 9 | 11 | 8 | 15 | 9 |
| Efficiency, welds/shift. | 30 | 30 | 40 | 40 | 22 | 22 | 32 | 32 |
| Output per 1 welder/operator, welds/shift | 3.0 | 3.8 | 2.4 | 3.6 | 1.6 | 2.8 | 1.4 | 2.3 |
| Output per 1 worker, km/year | 8.0 | 9.2 | 6.9 | 9.3 | 4.4 | 6.1 | 4.25 | 6.45 |
| Labout consumption for welding of one butt, man/days | 0.73 | 0.63 | 0.85 | 0.63 | 1.32 | 0.96 | 1.37 | 0.91 |

It is seen from the Table that for 1220 mm dia. pipes the replacement of 10 welders by the installation "Styk" with four operators ensures welding of40welds per shift,the number of the workers in the team and of mechanisms having been reduced for about 25 %.

For 1420 mm dia. pipes the substitution of 16 welders for installation "Styk" with 6 operators provides the reduction of the number of workers in the team and of mechanisms for about 35%. Technical and economical indices of automatic welding of 1220-1420 mm dia. pipes with flux-cored wire (Table 4) testify 15-50% increase of the output per member of the team and 15-35% reduction of labour consumptions per weld as compared to the manual arc welding.

As compared to welding in shielding gases and resistance welding, which provide high rates of pipelines construction, the welding with flux-cored wire possesses a high versatility. If is easily usable for any kind of work and enables to weld turning angles, crane junction welds, short pipelines for earth structures and also pipelines in marshaland regions.

## REFERENCES

Pokhodnya I.K., Suptel A.M., Shlepakov V.N. Welding with flux-cored wire, Kiev, Naukova Dumka, 1972.

Paton B.E., Pokhodnya I.K., Dobovetzky V.Ya. at al. Automatic position welding of large dia. pipe butts with self-shielding wire. "Stroitelstvo truboprovodov", 2, 1981, pp.22-24.

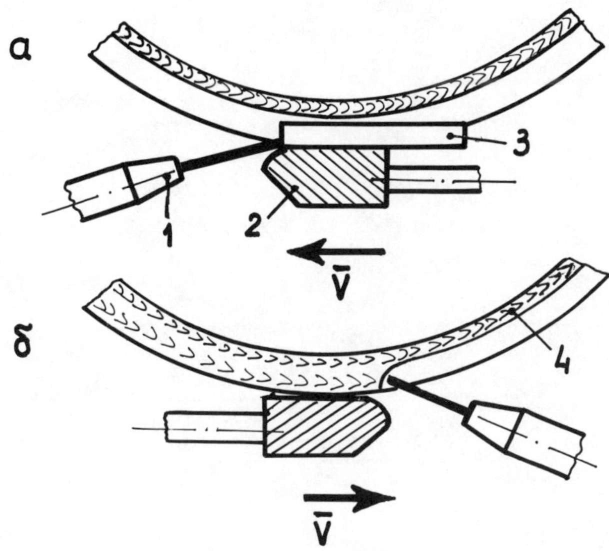


Fig.1. Scheme of beginning the process of welding of a - right and semiperimeter; b - left and semiperimeter.
1-welding head; 2-water-cooled shoe; 3-metallic insert;
4-zone of beginning the welding by the 2nd automatic machine.

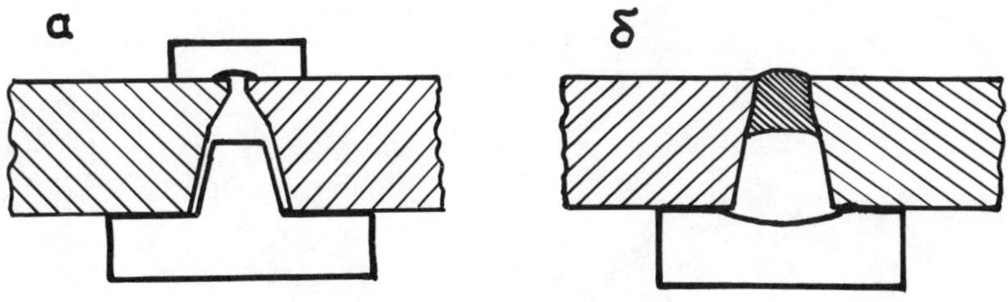


Fig.2. Shape of forming shoe. a-for welding the filling layer; b-for welding the facing layer.

# AUTOMATED WELDING DEVELOPMENT FOR ELECTRICAL POWER PLANT MAINTENANCE AND CONSTRUCTION

M.J. Tinkler

*Ontario Hydro Research Division*
*800 Kipling Avenue, Toronto, Ontario*

ABSTRACT

Welding situations are often encountered in nuclear power plant construction, and both nuclear and conventional plant maintenance and repair, which require that welding be done in areas where access is extremely poor and/or where a high radiation field is likely. Mechanized welding equipment is currently in use for some of those welding applications, particularly for circumferential welds around tubing and piping, but the orbital weld heads generally used require an operator to be present to visually monitor and fine-tune the operation in order to cope with unpredictable (and generally unavoidable) weld path deviation related to fit up and alignment tolerances.

An extensive research program is underway towards the development of a computer-controlled remote automated welding capability. The principal requirement is the incorporation of closed-loop feedback controls which enable the welding system to define weld path trajectory and change welding process parameters when necessary to account for any unexpected situation, without the presence of a human operator in the immediate vicinity of the welding location. Work is being directed towards two primary applications:

- orbital tube and pipe welding, construction or repair;

- repair of defects in heavy wall vessels.

In addition, the final closure welding of containers for spent fuel disposal is considered an important subset in orbital welding development.

In the first phase of the program, a computer-controlled opto-electric seam tracking device and a weld path record/playback capability have been developed for incorporation into an existing mechanized cold-wire gas-tungsten-arc orbital welding head. To achieve the required accuracy and reliability, this has necessitated considerable mechanical and electronic hardware and computer software development. With the incorporation of a microprocessor-based transistorized power supply, and experimental work to generate quantitative data concerning welding parameter/weld bead shape relationships, an automated closed-loop feedback system is envisaged where optimum welding parameters can be used for every point along the defined weld path.

The requirements for an automated system for heavy wall vessel repair are somewhat more complex because of the various activities needed to complete a remote repair (i.e. inspection and flaw characterization, flaw removal and cavity preparation, cavity fill by welding, and final inspection), which must be integrated into a single computer-controlled package. Substantial progress has been made in each aspect of automated repair welding but considerable work remains to develop a portable, field-compatible system.

From a metallurgical viewpoint, the temper-bead welding technique is the method of choice for heavy section repair welding, since it eliminates the need for post-weld heat treatment. Only manual, shielded-metal-arc temper-bead welding procedures are currently available, however, which cannot be considered for automation. Therefore, an additional major requirement is the development of effective gas-metal-arc (GMA) temper-bead welding procedures with the added complication that welding must be possible in all positions (i.e. horizontal, overhead). Current research suggests mixtures, can satisfy these requirements.

# THE POTENTIAL USE OF NON-ARC WELDING PROCESSES IN ENERGY RELATED FABRICATIONS

J.D. Russell

*The Welding Institute, Research Laboratory, Abington Hall, Abington, Cambridge CB1 6AL, UK*

## ABSTRACT

The technical features of EB, Laser and Friction Welding processes are reviewed in relation to their suitability for use in energy industry related fabrications. In addition special equipment developments are referred to for friction welding in order to adapt the process for pipe welding.

## KEYWORDS

Electron beam; Laser; Friction Welding; Steels; Joint Properties; Process development.

## INTRODUCTION

The welded products associated with energy related industries cover a vast range in terms of overall size, section thickness and type of material. It is therefore impossible to survey the whole field in a single contribution whilst covering the relevant aspects in sufficient detail to be meaningful.

It is therefore proposed to confine this review to recent developments in deep penetration processes (EB and laser) and friction welding (since flash welding is covered elsewhere in this conference).

The features which these processes have in common and which provide many advantages as well as some disadvantages are:-

1. Fast joining rates since a single pass operation is involved, resulting in short welding times.
2. Autogeneous processes i.e. very few consumables involved.
3. Highly mechanised with little reliance on operator skill.

The deep penetration and forge welding processes differ in a fundamental and important way which significantly affects their mode of exploitation. This arises from the different welding mechanisms involved. In the case of deep welding, the process is only limited by the depth of weld which can be achieved with the power and power density available i.e. once a given depth has been achieved then the weld can be of any length, limited only by practical restrictions on component handling.

However, with forge welding all the available power is dissipated in a single short operation, therefore the processes are area limited. In practice this means that a given size of machine can join a maximum area of joint and this can be of either a solid or tubular section. The sections also need to be fairly regular for even heating.

Thus there are basic differences between the non arc high joining rate processes which determine the application areas to which they are best suited.

In the following review the various process and materials features of these processes will be discussed from the points of view of their application to steel components of a tubular nature related to the energy industries.

## DEEP PENETRATION PROCESSES

### Basic Process Features

Both laser and electron beams, when focused, have sufficient power intensity to achieve deep welding by a vaporisation process which is now well documented and understood. However, there are detail differences between the various types of beam process which can have a significant effect on technical performance.

In simple terms the energy beam must have sufficient intensity to produce localised metal vaporisation in a controlled way. The vaporised material induces a 'keyhole' in the intended joint area by a 'rocket' mechanism and the depth of the keyhole, and hence depth of weld penetration, is controlled by the beam power and speed of translation relative to the workpiece. The heat transfer and metal plasma formation mechanism inside the keyhole can have a profound effect on welding performance.

The most common process, the VEB, is capable of producing the deepest weld penetration, Fig. 1. Here it can be seen that single-pass welds can be produced to depths in excess of 150mm, and the same equipment can be used for welding over a wide range from a few millimetres up to this level.

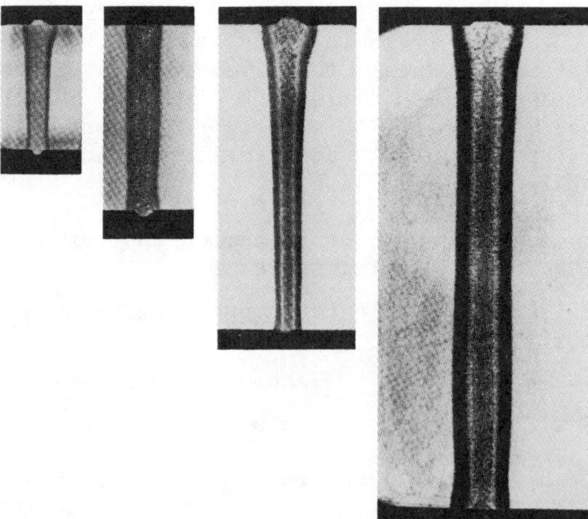

Fig. 1 Typical EB welds in 25, 50, 100 and 150mm thick C-Mn steels.

For comparable beam powers the NVEB process can achieve much less penetration (typically one-third). This stems from the substantial power loss in the formation of plasma and beam scattering at atmospheric pressure, even in a helium atmosphere. The degraded beam is then less able to produce so deep a weld because of the increased weld width. The advantage of welding in a nonvacuum environment, however often outweighs the disadvantage of reduced penetration per kilowatt.

The laser beam also operates in an open air environment and is capable of producing welds very similar to the VEB process, although with a much lower ultimate penetration capability.

## ELECTRON BEAM WELDING

The VEB process is far more widely used industrially compared to the NVEB process and has received much more development effort. Its wide penetration range capability has lead to developments being carried out for applications as diverse as pipeline butt welds,(as has been successfully demonstrated by the Total Company,) to nuclear vessel fabrication.

Work at The Welding Institute has been mostly concerned with the thicker section applications covering steels in the range 50-150mm.

Early work examined some of the technical questions which needed to be answered before EB welding could be considered for heavy steel fabrication. These questions were concerned with:-

(a) The influence of steel composition on weld quality and properties, particularly toughness.

(b) The effect of process variables on weld quality and properties.

(c) The ability of the process to be adapted to cope with heavy engineering situations involving poor joint fit up.

Since weld metal composition is dependent on base metal composition it was important to know how the normal commercial steel grade compositions would react. Two main weld metal defects were found to be possible with normal commercial steel grades. These are solidification cracking and porosity, and typical examples are shown in Fig. 2. Elimination of these defects was found to be influenced by both process and steel composition aspects. Both defects were minimised by low travel speeds, which needed to be typically 100-150mm/min for 50-150mm thick steels. Porosity occurrence was found to depend principally on oxygen content. The tolerable oxygen levels varied with steel thickness. In general very thick steels required an oxygen level of <50 ppm, typical of vacuum processed steels, whereas for thinner steels conventional Al killed steels were acceptable.

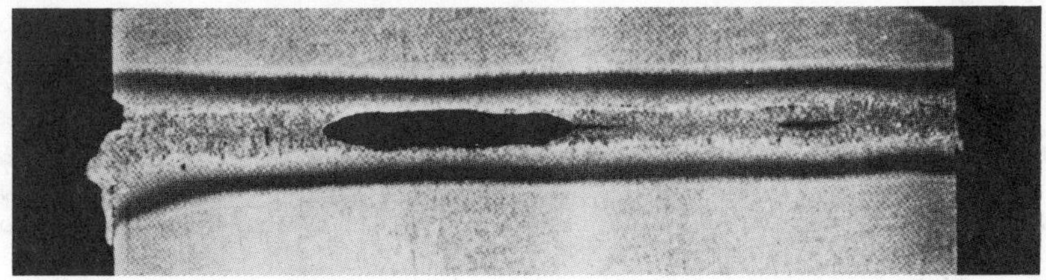

Fig. 2 EB weld in 100mm thick alloy steel showing porosity and solidification cracking

Solidification cracking was found to be influenced by the usual elements associated with weld metal cracking in arc welding. However, a statistical experiment examining the effects of C, S, P and Mn showed that small differences in individual element effects were apparent and a cracking severity index formula was devised as follows:-

$$\text{Cracking index} = 52.6C + 1972S + P(4268C-285) - Mn(1135S-9) - 21.2$$

This formula was used to characterise a number of different steel compositions and related to actual cracking observed during welding tests. Figure 3 illustrates one method of using the formula. This shows that both weld width and travel speed influence cracking for a given composition. Thus for crack sensitive compositions it is desirable to use low travel speed with a narrow weld.

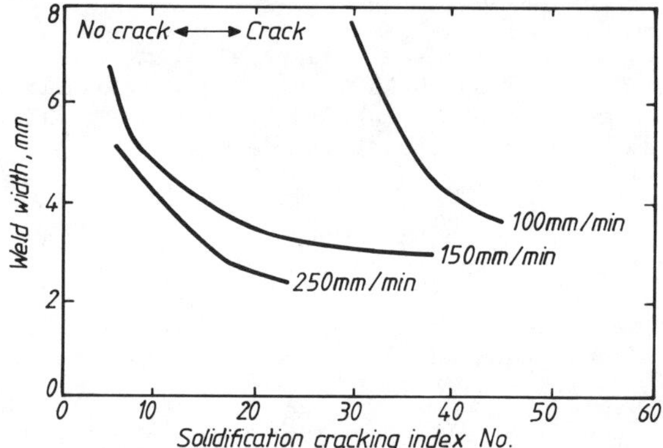


Fig. 3 Use of index to compare weld width and speed on incidence of cracking

For many practical situations the need to use a wide fusion zone, say 5mm could be in conflict with this and therefore if bad joint fit up was likely, a good quality steel would be preferred.

From a weld metal toughness point of view again interactions were found between welding procedure, and steel composition. The situation is very complex and is not yet fully understood. In general low C steels are very desirable together with low levels of residual elements. These are important from two points of view due to their influence on transformation and also solidification structures.

For high carbon grades e.g. >0.1%C a post weld tempering treatment is usually necessary in order to achieve good toughness. Other composition effects can be observed due to their influence on transformation structure e.g. alloying elements which increase hardenability can be beneficial. Figure 4 illustrates three 150mm thick steels giving vastly different Charpy V properties in the post weld stress relieved state.

For a steel typical of offshore platform construction such as BS 4360 50D the effects of welding speed and weld width have been compared for a steel of 0.18C, 1.5Mn and 75mm thick. The results are shown in Fig. 5, and confirm the desirability of a post weld heat treatment with minor benefits from using a narrow weld.

In general it can be said that sufficient work has now been done on factors affecting the acceptability of autogeneous EB welds in energy transporation and generation applications to justify the major step of development and installation of purpose built plant for specific applications. Although futher detailed investigations are necessary for some materials and applications.

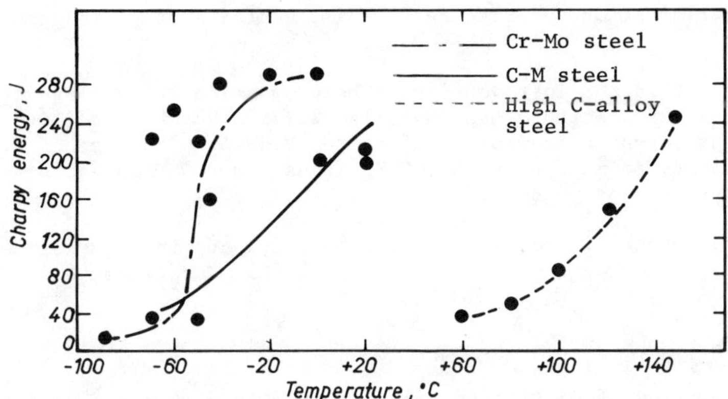


Fig. 4 Comparison of toughness (Charpy V) of three weld metals in 150mm thick steel after stress relief PWHT. - CMn steel; high C steel; Cr-Mo steel.

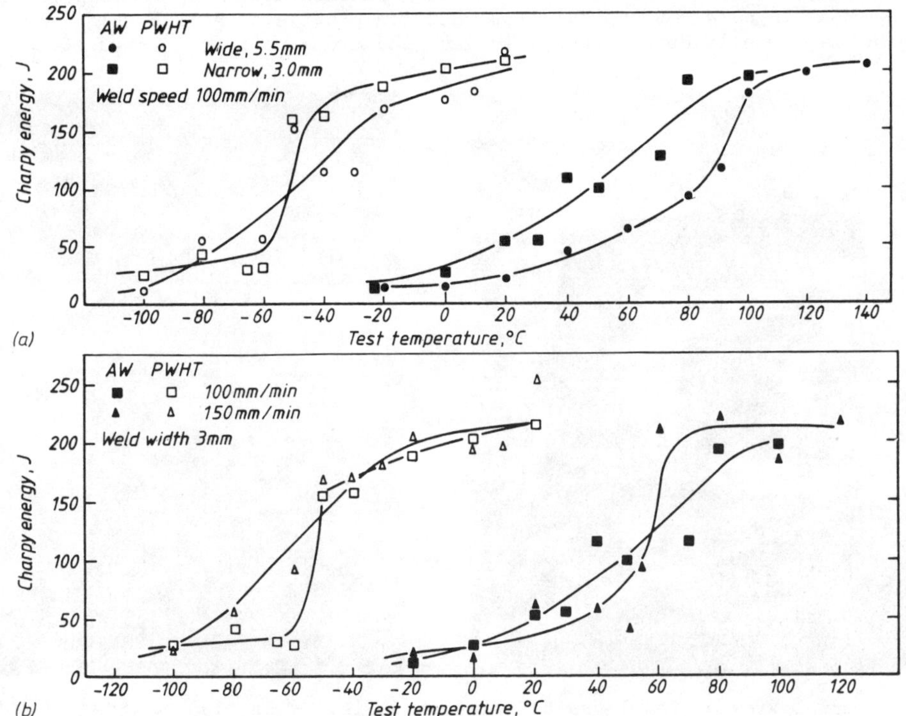


Fig. 5 Effect of process parameters on toughness of C-Mn steel weld metal (a) weld width, (b) travel speed.

## *Applications*

Many potential applications exist in the energy related fabrication field which could benefit both technically and economically from the use of EB welding. Specially designed equipments already exist to minimise the vacuum disadvantage that the process suffers from. Examples of these can be found in the girth welding of line pipe and in local and mobile vacuum systems for shop welding of main seams and circumferential seams in pressure vessels, containment plate, and platform tubulars.

In order to assist in the introduction of heavy section EB welding into appropriate industries the Institute has installed a large chamber facility for development work and prototype component manufacture, Fig. 6. Its size of 7.5 x 3.5 x 3.5M and a capacity to manipulate 10,000Kg loads under CNC operation should enable realistic demonstrations to be carried out.

The economic advantages increase with section thickness increase since apart from machining costs the welding cost is the same for 25mm and 150mm thick plates.

## NON VACUUM EB WELDING

The nonvacuum version of EB welding has until now been confined to the high speed welding of relatively thin materials (<6mm). However, work at Westinghouse Research Laboratory (USA) and, more recently at The Welding Institute, has shown that this process has the potential to achieve penetration of at least 50mm using beam powers up to 60kW. A typical cross-section profile of a NVEB weld in 30mm C-Mn steel is shown in Fig. 7 this weld was produced at a travel speed of 500mm/min.

Fig. 6 General view of the large EB welding equipment at The Welding Institute

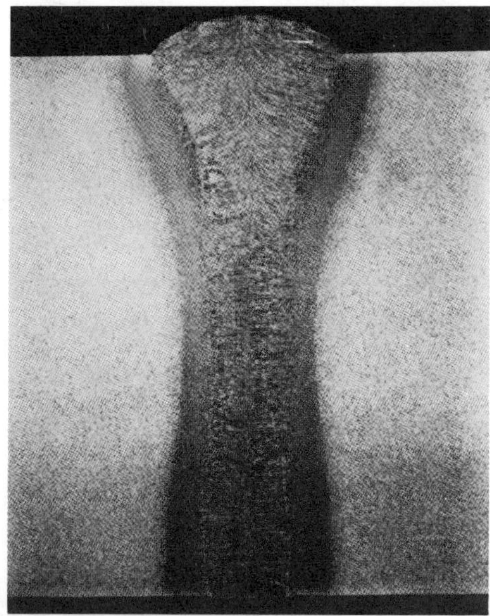

Fig. 7 Section of a NVEB weld in 30mm thick C-Mn steel, beam power 35kW, 500mm/min travel speed

Recent work has also shown that the gun-nozzle-work standoff distance can be increased to 25mm if necessary. However, for optimum performance in terms of minimum weld width and maximum penetration, this distance should preferably be 10-12mm.

The extension of NVEB welding into thick section materials opens up new areas of exploitation, particularly in pipe-welding. The longitudinal welding of pipe sections in the mill is an obvious area for investigation and with a mobile orbiting gun system, circumferential butt welds are a possibility.

Elimination of the vacuum enclosure, however, still leaves some problems: X-rays are produced and local protection against this hazard is required. Another factor to be taken into consideration is the need for a backing material support, when welding in the flat position. However, the larger width of the NVEB weld leads to less stringent demands on joint fit up.

### LASER WELDING

The development of the CW $CO_2$ laser into a high power system with capabilities up to 25mm penetration in steels opens up many new areas for laser exploitation.

There is an obvious difference in ultimate penetration performance per kilowatt between EB and laser beams. This arises as a result of the different beam/material interaction mechanisms associated with the electron and photon beams. In the keyhole process, metal vapour is formed which becomes ionised by interaction with the beam. In EB welding the ionised metal vapour/shielding gas is almost transparent to the high voltage electron beam. Conversely, with the laser the ionised material is almost opaque to the photon beam (at 10.6μm wavelength) and substantial energy loss occurs by intensification of the plasma. This applies to plasma within the keyhole and more visibly to plasma about it. The plasma problem has two main effects:

1. It absorbs energy which would otherwise be used to melt metal.

2. It transfers part of its energy to the workpiece giving rise to an increased fusion zone width, particularly at the top. The plasma/beam problem increases with decreasing travel speed and becomes serious at speeds less than about 700mm/min for 15kW lasers and about 300mm/min for 5kW lasers. Thus, low travel speeds are not practical, which is the main reason why electron and laser beam have different ultimate penetration performances.

From a materials point of view laser welds are very similar to EB welds although porosity problems tend to be more comparable with arc welds. Cracking, however, is similar to EB welding and unfortunately the laser suffers from the disadvantages that it cannot readily be used at low travel speeds due to the plasma problems, which are sometimes necessary to avoid solidification cracking in commercial steel grades. Therefore successful autogeneous laser welding is more dependent on good steel quality than EB welding. In order to minimise this problem, recent work has been concerned with the use of filler additions by wire feed in order to control weld metal chemistry as well as give more tolerance for bad joint fit up. These developments have led to the serious consideration of a purpose built laser system for line pipe girth welding. Figure 8 shows a typical laser weld section in 15mm steel.

An attractive feature of the laser process, in addition to its open air aspect, is its ability to be manipulated by mirrors. This enables a stationary power source to have its power distributed to a welding location which could be tens of metres away and then for the beam to be focused and manipulated along the joint by further mirror systems.

From a toughness point of view laser welds can be expected to be similar to EB welds for the same steel i.e. low C grades are preferable.

### *APPLICATIONS*

The main application in the present context for which laser welding is being developed are longitudinal and girth welding of line pipe. For these applications it is particularly important to use low C steel compositions within the various pipe grades in order to avoid the need for post weld heat treatment. The open air nature of the process gives it a great advantage over EB welding. However, the problems of manipulating beams over very long distances by long optical benches should not be minimised in view of the high degree of beam/joint alignment requirement by the deep welding processes.

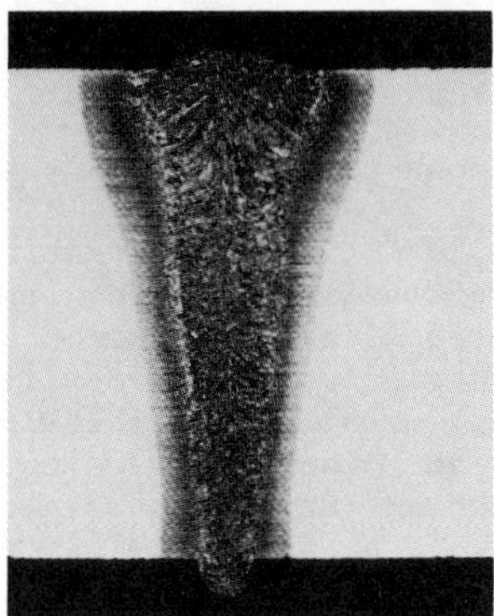

Fig. 8 Section of laser weld in 12.5mm C-Mn steel using 6kW laser power, travel speed 360mm/min.

## FRICTION WELDING

Friction welding has been used for 25 years for a variety of joining applications involving both similar and dissimilar combinations. Its most frequent use is in the mass production industries for repetitive joining of simple shapes. In these industries the process has acquired a reputation for consistently producing good quality welds once the correct welding procedure had been set up.

More recently friction welding in both continuous drive and inertia forms has been used for attachment of end fittings to drill pipe. In all current friction welding applications service performance requirements in terms of weld quality and properties are maintained by process control and monitoring systems which measure selected process parameters during welding. This is an important aspect for the future of friction welding in a quality concious industry since the solid phase weld cannot be assessed by conventional NDT methods. Thus quality assurance must be via process control and material selection.

In order to assess the influence of process variables on joint quality and properties a programme has been carried out at The Welding Institute on C-Mn steels in both solid bar and tubular form.

If was found that once conditions for the production of sound, defect free welds were established that variations in parameters such as rotation speed, force and burn off had little effect on weld metal toughness.

As with EB and laser autogeneous welds, the influence of steel quality was found to be important, but from a different point of view. With friction welds, steels with high inclusion contents tend to have poor impact properties. This is due to reorientation of the inclusions at the joint interface during the friction and forge stages. Therefore, low S, low $O_2$ steels are preferred. Even with good quality steels it is sometimes necessary to carry out a post weld temper and occasionally a full normalise and temper treatment in order to achieve good weld toughness.

Figure 9 compares the toughness of friction welds in 25mm dia. C-Mn steel bar (0.21C, 1.5Mn) in different states of heat treatment. Whereas Fig. 10 compares this steel with two better quality steels in the as welded state.

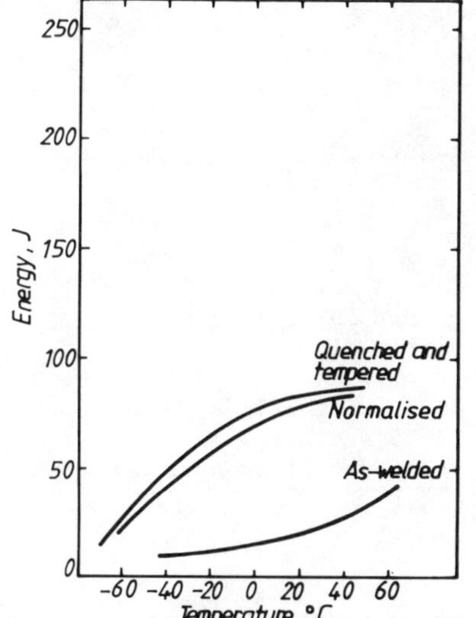


Fig. 9 The effect of postweld heat treatment on the impact properties of continuous drive friction welds in C-Mn steel (A).

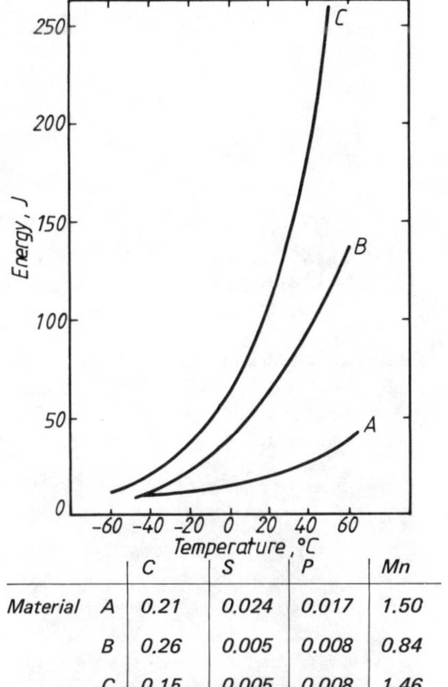


| | | C | S | P | Mn |
|---|---|---|---|---|---|
| Material | A | 0.21 | 0.024 | 0.017 | 1.50 |
| | B | 0.26 | 0.005 | 0.008 | 0.84 |
| | C | 0.15 | 0.005 | 0.008 | 1.46 |

Fig. 10 Impact properties of as-welded continuous drive friction welds for 3 different steel qualities.

For line pipe steels for which friction welding is currently being developed the necessary steel quality can be achieved to give adequate as welded toughness.

## *PROCESS DEVELOPMENT*

The necessity to rotate one side of a friction welded joint has been a serious inhibition to the application of the process to the joining of long lengths of pipe found in oil and gas transporation lines. Although the necessary heavy engineering developments are not impossible, a more elegant solution was developed at The Welding Institute a few years ago which is now known as Radial Friction Welding. In addition to avoiding the need to rotate large members, the internal bore flash is also avoided. The technique is illustrated schematically in Fig. 11. The pipe ends are machined to provide a root face of between 0 and 1mm and bevelled to produce a 100$^o$ included angle when they are butted together. A solid ring of suitably comparable material, but with a smaller bevel of $\sim$90$^o$, is located between the pipe ends. The ring geometry assists metal flow from the base of the weld preparation and also provides less cross-sectional area during contact, thus reducing the initial power demand. The pipes are firmly clamped to resist axial and rotational movement, and a mandrel is inserted in the bore to support the pipe walls and prevent metal penetration. Frictional heat generation and accompanying metal displacement occur when the ring is rotated and subjected to uniform radial compression. To complete the weld sequence after adequate heating, the ring is arrested, and the radial compressive load is either maintained or increased to consolidate the joint.

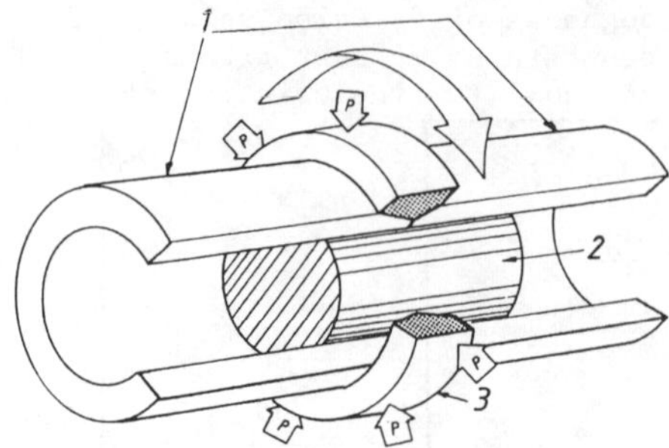


Fig. 11 Schematic arrangement for radial friction welding. 1 - stationary clamped pipes; 2 - expanding plug; 3 - consumable ring rotated and compressed radially.

A prototype machine was designed for joining 50mm OD, 6mm wall pipe and is shown in Fig. 12 and was capable of producing joints in 12-15 sec. of weld time. A section of a completed weld is shown in Fig. 13 illustrating the clean bore and complete bond formation at the two interfaces.

Fig. 12 Top view of prototype radial friction welding machine for 50mm diameter pipe.

Fig. 13 Half section of a radial friction weld showing clean bore.

The radial process has since been extended to 100mm OD pipe and current work with a licencees is extending this further to 300mm OD. The main initial applications for which the equipment is intended is for flow lines. For this requirement steel chemistry can be controlled and good joint properties can be obtained.

## CONCLUDING REMARKS

The single pass and forge welding processes such as friction welding have undergone significant developments in recent years with the aim of improving their acceptability in heavy engineering, and energy generation and transportation systems. The better understanding of materials and process factors which affect joint quality and properties should enable potential users to proceed to the next stage of large scale production plant installation in order to take advantage of the substantial technical and economic advantages which these processes offer in comparison with multi pass arc welding methods.

# MINIMUM HOLDING TIME FOR NORMALIZING OF ELECTROSLAG WELDMENTS

Sathish Rao Bala
*Research and Technology Centre, AMCA International Ltd.*
*P.O. Box 13160, Kanata, Ontario, K2K 1X4, Canada*

## ABSTRACT

To accomplish the normalizing of an approximately 75mm wide band of electroslag weld, the conventional practice is to normalize the whole can or the vessel in a furnace. It is a common practice in the industry to specify holding times up to one hour per 25mm of thickness for normalizing. However, because of higher fuel cost and distortion there is an incentive for fabricators to keep the holding time to a minimum or resort to local normalizing. If local normalizing is adopted, the heat loss can be substantial due to insulation limitations and hence there is also an incentive to keep the holding time for local normalizing to a minimum.

In view of the above, an investigation leading to the minimum holding time required for normalizing which would give adequate toughness of the electroslag weldments was carried out. For this, electroslag butt welds in A516 Gr 70 steels supplied by three sources were normalized. Each weldment was normalized at 900°C for 5, 15, 45 and 180 minutes and stress-relieved. The impact and tensile properties determined for the normalized and the stress-relieved weldments showed that they meet the code requirements and that there is no real advantage in furnace normalizing the electroslag welds for more than 5 minutes. The holding time as low as 5 minutes would, thus, save on fuel, reduce distortion and make local normalizing a viable process.

KEYWORDS

Normalizing, electroslag weld, heat-affected zone, impact energy, yield strength and ultimate tensile strength.

## INTRODUCTION

Two of the limitations of the electroslag welding process have been the occasional presence of grain boundary fissures, and the rather limited weld metal and heat-affected zone (HAZ) toughness due to the coarse grain size. It is now believed, as a result of the recent investigation (Lowe, 1981; Thibau, 1983) on Hydrogen in Electroslag Welds, that the former problem occurs only under extreme conditions of moisture or restraint, and is therefore, controllable. To overcome the latter problem, in critical applications such as pressure vessels for the nuclear industry, the ASME Code requires that the electroslag weldments be given a "grain refining treatment" (ASME Boiler and Pressure Vessel Code Sect. III, 1980). Metallurgically this requires that electroslag weldments be normalized. To accomplish the normalizing of an approximately 75mm wide band of electroslag weld, the conventional practice is to normalize the whole can or vessel in a furnace by attaching a few thermocouples onto the surface of the vessel.

It is well recognized that at temperatures greater than the upper critical, the time required for the phase transformation to austenite is quite small. Calculations based on heat transfer (Kreith, 1973) show that if a 100mm thick plate is heated to 600°C and then the plate surface is instantaneously brought to a normalizing temperature of 900°C, the centre of the plate reaches 900°C in about 10 minutes. In spite of this, it is common practice in the industry to specify holding times up to an hour per 25mm of thickness. In practice, the vessels are gradually heated to the normalizing temperature so that the temperature gradients that exist between the surface and the centre are minimal.

If one 3.65m long can, 3.65m in diameter and 100mm in wall thickness, has one longitudinal electroslag weld seam, then the weld that needs to be normalized weighs ~ 1/145 of the total weight of the can. At the present day fuel prices, the energy expense for normalizing these cans is so large that cost benefits of the electroslag process are decreased. Also, as the ASME Code stipulates that the difference between the maximum and minimum inside diameters at any cross-section must not exceed 1% of the nominal inside diameter, long holding times are detrimental. The magnitude of the distortion caused by self weight alone can be quite large due to the long holding time at high temperature, making it almost impossible to maintain such a tolerance. The distortion due to self weight, which is a high temperature creep phenomena, could be reduced by shorter holding times.

To experience the cost benefits of the electroslag process, fabricators have considered the local normalizing of electroslag welds. It is reported that local normalizing has been successfully performed in Russia (Novikov, 1958; Kroshkin, 1973), Japan (Suzuki, 1967) and European countries (Shakleton, 1975; Eichhorn, 1970; Wachtmeister, 1961) even though there is no knowledge of its application in North America. If local normalizing is adopted, the heat loss can be substantial due to insulation limitations and hence again there is an incentive to keep the holding time for normalizing to a minimum.

In view of the above, an investigation to determine the minimum holding time required for normalizing which would give adequate toughness of the electroslag weldments was carried out.

## EXPERIMENTAL PROCEDURE

### Materials

ASTM A516 Gr 70 steel plates used in this study were supplied by three mills. The chemical compositions and the thicknesses of these steel plates are given in Table 1 on the following page.

To perform the electroslag butt welds in A516 Gr 70 steels supplied by Mills A and B, Linde 40 wire (typical composition in wt %: C 0.15, Mn 2, Si 0.03, S 0.024, P 0.017, Mo 0.53), 3.2mm in diameter, and dry Arcos BV flux were used as the consumables. For the steel supplied by Mill C, a combination of Armco W-19 wire (typical composition: 3.5% Ni), 3.2mm in diameter, and dry Arcos Bv flux were used to obtain the butt welds. The Arcos BV flux was dried at 200°C for at least 24 hours and only removed from the oven just before commencing the welding.

The slag cap height and the electrode stickout were maintained constant at ~30 and ~50mm respectively. The actual welding set-up is shown in Fig. 1.

Fig. 1: Conventional electroslag welding set-up.

Furnace Normalizing of the Electroslag Welds.

A 255mm wide weldment region containing the weld metal and the HAZ was flame-cut symmetrically on either side of the weld parallel to the long axis of the weld, and was cut into 300mm long pieces. Two thermocouples were attached to each piece: one at the surface and the other at the geometric centre.

To simulate the local normalizing heat treatment, the weldment piece was brought up to a uniform temperature of 540 ± 25°C (preheated), in a box furnace kept at that temperature, and then transfered into another box furnace, kept at 900 ± 25°C. When the temperature at the centre of the weldment piece reached 900°C, in 45 to 70 minutes, it was normalized at that temperature for one of the desired holding times of 5, 15, 45, and 180 minutes and air cooled. In practice, 180 minutes are used as the holding time for furnace normalizing of electroslag welds in 100mm thick plates.

Stress-Relieving of the Normalized Weldment Pieces.

The normalized pieces were stress-relieved at 620 ± 25°C for 7.5 hours. As per the ASME Code (Boiler and Pressure Vessel Code Sect. VIII, 1980), the time of 7.5 hours included at least 3 stress-relief heat treatments for the steel thickness used in this study.

Mechanical Tests

Each stress-relieved weldment piece was saw cut to obtain 16 standard Charpy specimens in the weld metal from each of T/4 and (T/2 - 8mm) locations as shown in Fig. 2. Also, 16 HAZ Charpy specimens were obtained from (T/2 - 8mm) locations as shown in the same Figure. The rolling direction was perpendicular to the length of the Charpy specimens. The tensile specimens, 4.5mm in diameter and 25mm in gauge length were obtained from the T/2 location of the weld metal and the HAZ with their axes parallel to the rolling direction of the plate. These were tested at room temperature (21°C) in a Baldwin tensile testing machine with an initial strain rate of 0.01/min.

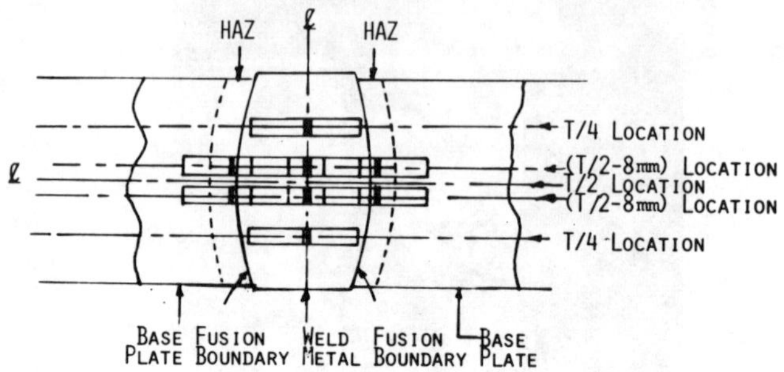


Fig. 2: Locations of Charpy specimens

For Charpy impact tests, V-notches were machined in the specimens on the side parallel to the thickness direction of the blocks perpendicular to the rolling direction. For HAZ specimens, notches were put such that the root of the notch coincided with the fusion boundary or 1mm away from the fusion boundary into the HAZ. Impact tests were carried out at test temperatures, in the range of -60°C to room temperature.

## Metallography

Thin slices were removed from the T/2 locations of the weldments. These slices were cut to obtain surfaces parallel to the thickness side of the base plate at locations such as the centre of the weld, the HAZ approximately 1mm away from the fusion boundary and the base material. These surfaces were polished to 1μm diamond and etched with 2% nital. Microstructures were obtained using an optical microscope.

# RESULTS

## Mechanical Properties

The mechanical properties such as yield strength (YS) and ultimate tensile strength (UTS) of the furnace normalized and stress-relieved weldments are given in Table 3.

Plots of CVN values vs. test temperatures were made for the weldments normalized for different holding times followed by stress-relieving. Smooth transition curves were drawn through the lower CVN energy points of the scatter, and are given in Fig. 3. It should be noted that when plotting the transition curves for the HAZ, the fractured specimens were examined after etching, to see whether the fracture surface was in the HAZ or the weld metal. If the fracture surface was in the weld metal, the test was discarded.

TABLE 1: Chemical Compositions of A516 Gr. 70 Steel Base Plates and the Respective Weld Metals

| Elements | MILL A wt.% | | MILL B wt.% | | MILL C wt.% | |
|---|---|---|---|---|---|---|
| | Base Plate | Weld Metal | Base Plate | Weld Metal | Base Plate | Weld Metal |
| C | 0.22 | 0.20 | 0.21 | 0.18 | 0.22 | 0.15 |
| Mn | 0.99 | 1.57 | 1.11 | 1.54 | 1.13 | 0.99 |
| Si | 0.17 | 0.10 | 0.25 | 0.06 | 0.24 | 0.14 |
| P | 0.01 | 0.01 | 0.01 | 0.01 | 0.01 | 0.01 |
| S | 0.022 | 0.018 | 0.008 | 0.008 | 0.014 | 0.008 |
| Cr | 0.07 | 0.08 | 0.01 | 0.03 | 0.22 | 0.14 |
| Ni | 0.02 | 0.07 | 0.26 | 0.20 | 0.28 | 2.17 |
| Mo | 0.01 | 0.23 | 0.01 | 0.24 | 0.08 | 0.06 |
| Cu | 0.014 | 0.04 | 0.25 | 0.22 | 0.21 | 0.24 |
| Nb | <0.003 | <0.003 | 0.03 | 0.01 | 0.003 | <0.003 |
| Al | 0.048 | 0.005 | 0.01 | 0.005 | 0.023 | 0.005 |
| N | 0.007 | 0.006 | 0.007 | 0.004 | 0.011 | 0.010 |
| O | 0.0084 | 0.0078 | 0.0021 | 0.0058 | 0.0023 | 0.0056 |

Conventional Electroslag Butt Welding of the Base Plates

Two welds were performed for each of the steels. The length and width of the plates were not less than 920 and 620mm, respectively. A constant gap width of 30mm was obtained by welding a sump at the bottom and two strongbacks on one side of the plates. The stationary water-cooled copper (cooling) shoe was located on the side of the weld gap where the strongbacks were welded. The butt welding was performed, with rolling direction of the base plate parallel to the length of the weld (ASME Boiler and Pressure Vessel Code Sect. III, 1980), using the conventional electroslag welding technique.

To start the weld, steel wool was placed in contact with the electrode wire and the sump. About 100g of the dry flux was poured into the weld gap and an arc was struck to melt the flux. Welding was performed automatically to obtain a weld of a minimum of 915mm long by using the welding parameters given in Table 2.

TABLE 2: Welding Parameters Used for the Conventional Electroslag Welding

| Base Plate Source | Mill A | Mill B | Mill C |
|---|---|---|---|
| Thickness of plate, mm | 100 | 70 | 76 |
| Electrode polarity, DC | +ve | +ve | +ve |
| Current, A | 800 | 525 | 525 |
| Voltage, V | 33 | 34 | 34 |
| Oscillation, mm/s | 4 | 4 | 4 |
| Dwell, s | 3-4 | 3 | 3 |
| Stroke, mm | 70 | 20 | 25 |
| Number of electrodes | 2 | 1 | 1 |
| Distance between electrodes, mm | 35 | - | - |
| Weld travel speed, mm/s | 0.14 | 0.14 | 0.13 |

TABLE 3: YS and UTS at T/4 Location of the Base Plate and T/2 Location of the Normalized and Stress-Relieved Weldments

| Base Plate | | | Holding Time For Normalizing | Weld Metal | | HAZ | |
|---|---|---|---|---|---|---|---|
| Source | YS MPa | UTS MPa | s | YS MPa | UTS MPa | YS MPa | UTS MPa |
| Mill A | 258 | 442 | 300 | 326 | 478 | 274 | 482 |
| | | | 900 | 312 | 465 | 268 | 493 |
| | | | 2700 | 309 | 471 | 269 | 475 |
| | | | 10800 | 290 | 450 | 277 | 514 |
| Mill B | 319 | 474 | 300 | 391 | 529 | 332 | 502 |
| | | | 900 | 381 | 521 | 320 | 487 |
| | | | 2700 | 364 | 496 | 318 | 469 |
| | | | 10800 | 344 | 480 | 294 | 456 |
| Mill C | 347 | 516 | 300 | 396 | 523 | 359 | 539 |
| | | | 900 | 396 | 524 | 343 | 531 |
| | | | 2700 | 398 | 526 | 340 | 598 |
| | | | 10800 | 373 | 528 | 337 | 515 |

### Microstructure

Several optical micrographs were compared to observe the microstructural changes occurring during the heat treatments. Some of the typical micrographs, at a magnification of 200 times, are shown in Fig. 4 for the weld metal and the HAZ in both the as-welded and the normalized plus stress-relieved conditions.

## OBSERVATIONS AND DISCUSSIONS

### YS and UTS

The YS and UTS values of the weld metal and the HAZ in the normalized and stress-relieved weldments decrease with the holding time, in most cases. However, the weld metal in the steel from Mill C shows virtually no variation, and the HAZ in the steel from Mill A shows no systematic effect of holding time on YS and UTS (see Table 3).

The weldments homogenized for 5 minutes show optimum YS values in the weld metal and the HAZ after stress-relieving. The YS and UTS values of the weld metal and the HAZ, normalized for 5 minutes and stress-relieved, are comparable to or better than those of the respective base plates which are stress-relieved (Bala, 1983). Refer to Table 3.

### Charpy Impact Energy Values

The impact energy transition curves plotted in Fig. 3 show that the energy values of the weld metal are not necessarily inferior at (T/2 - 8mm) location when compared with those at T/4 location for any length of "holding time". Generally, a holding time of 180 minutes for normalization yields better CVN values than other holding times for all the three weld metals. The lowest CVN values are observed for those weld metals normalized for 45 minutes. This is because during normalizing two simultaneous processes are taking place in the weld metal viz. homogenization of the cast structure and the austenite grain growth. Homegenization usually improves the impact energy, whereas austenite grain growth deteriorates it. The HAZ would undergo only the austenite grain growth during normalizing because of the initial homogeneity of the base plate. Thus, in the HAZ the impact property deteriorates as the holding time for normalizing increases from 5 to 180 minutes.

For the weld metal in steel from Mill C the CVN values are higher than the HAZ at all holding times. However, the welds from Mills A and B when normalized for 5 minutes and stress-relieved show better or comparable CVN values to those of the HAZ which was normalized for 180 minutes followed by stress-relieving. The CVN values of both the weld metal and the HAZ in steels from Mills A and B normalized for 5 minutes and stress-relieved are comparable to the CVN values of the respective stress-relieved, as-received or normalized base plates shown in Fig. 5 (Bala, 1983). However, the HAZ in steel from Mill C requires 45 minutes of normalizing followed by stress-relieving to observe the same effect.

If code requirements are considered, ASME Section III Sub-section NB (1980) requires a CVN value of 68J (50 ft. lbs) and a lateral expansion of 0.88mm (35 mils) at ~ 25°C (RT), assuming no break at -7°C for the drop weight tests generally observed, for the electroslag weld metal and the HAZ in a pressure vessel fabricated from A516 Gr. 70 steel. Similarly, CSA W59 Code (1977) requires an average of 20J (15 ft. lbs.) or a minimum of 14J (10 ft. lbs) CVN value at -18°C (0°F) in the weld metal and the HAZ of the same steel after normalizing and stress-relieving. Considering the code requirements, the marginal CVN values are observed only in those weld metals in steels from Mills A and B normalized for 45 minutes and stress-relieved. The rest of the welds and HAZ show better CVN values than those required by the codes.

Comparing the weld metal values to those of respective HAZ and base materials, it can be inferred that there is no real advantage in normalizing the weldments for more than 5 minutes. Thus, this study shows that the holding time for furnace normalizing of electroslag weldments can be considerably reduced from that of the conventional practice in order to save fuel and reduce distortion in the cans or the vessels. Also, shorter holding times make local-normalizing a viable and an economical process. Holding times as low as 1-2 minutes for local normalizing have been reported in the literature to give adequate toughness in the weld metal and the HAZ for some electroslag welds (Kroshkin, 1973; Suzuki, 1967; Shakleton, 1975; Eichhorn, 1970).

## Chemical Composition

Comparing the weld metal composition with the base material, Table 1 shows that the S contents in the weld metal as well as in the base material are lower in the case of steels from Mills B and C than that from Mill A. This partly explains why the CVN values are superior in the weldments of steels from Mills B and C. Also, weld metals in these steels from Mills B and C contain higher amounts of Nb and Ni, which have further improved the weld metal impact properties as seen in Fig. 3(a) and 3(b).

## Microstructures

The weldment microstructure clearly shows bainitic structure in the weld metal as well as the HAZ, in the as-welded condition (refer to Fig. 4). The bainitic structure disappears during normalizing and gives rise to two phase structures i.e. pearlite or degenerate structure. Even though no attempt has been made to determine the grain size in these weldments, it is expected that γ-grains grow at normalizing temperatures and is a time dependent process.

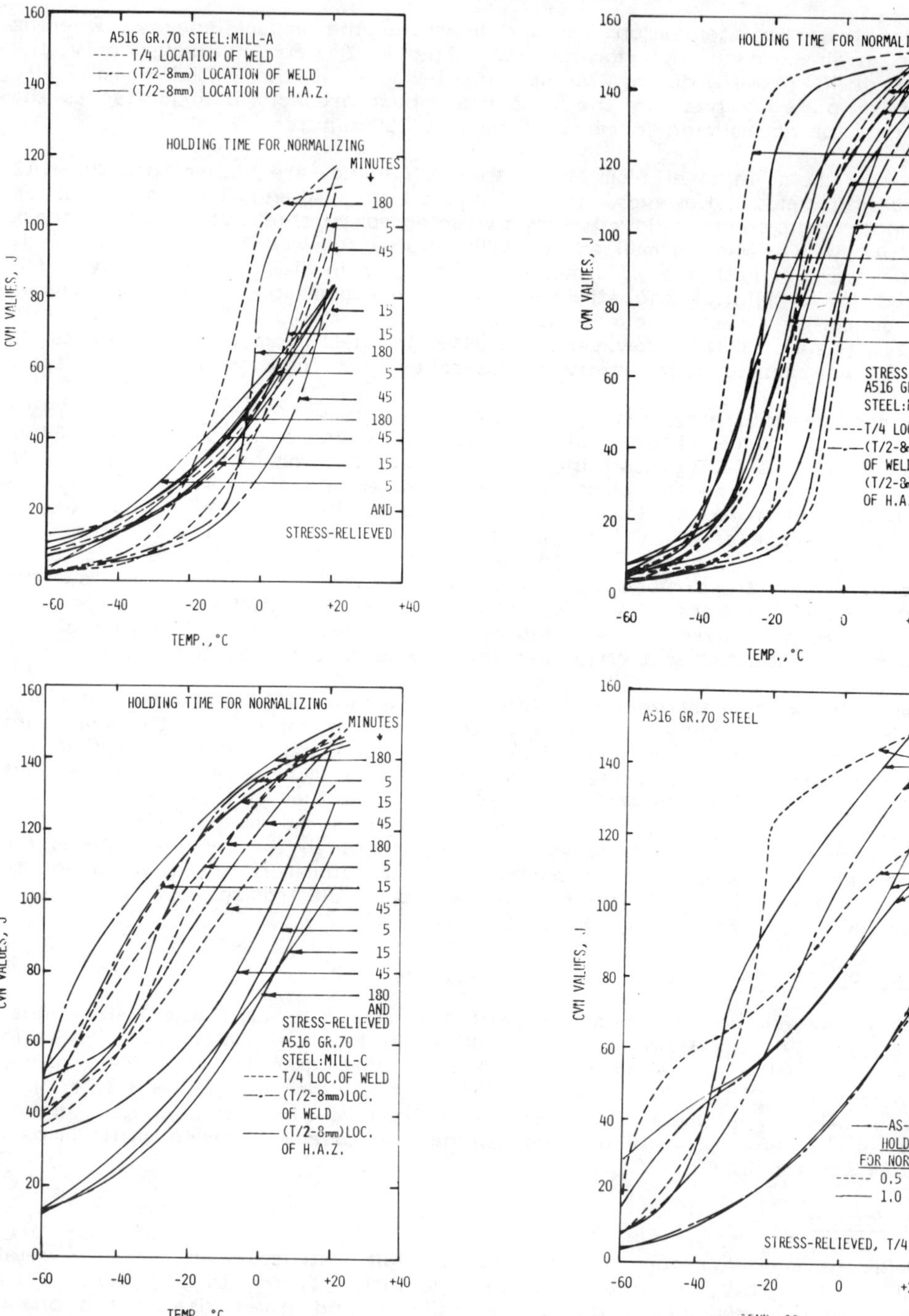


Fig. 3. The impact energy transition curves for the normalized plus stress-relieved weldments of A516 Gr. 70 steel from Mill (a)A, (b)B and (c)C.

Fig. 5. Impact energy transition curves of A516 Gr. 70 steel base plates from three different Mills.

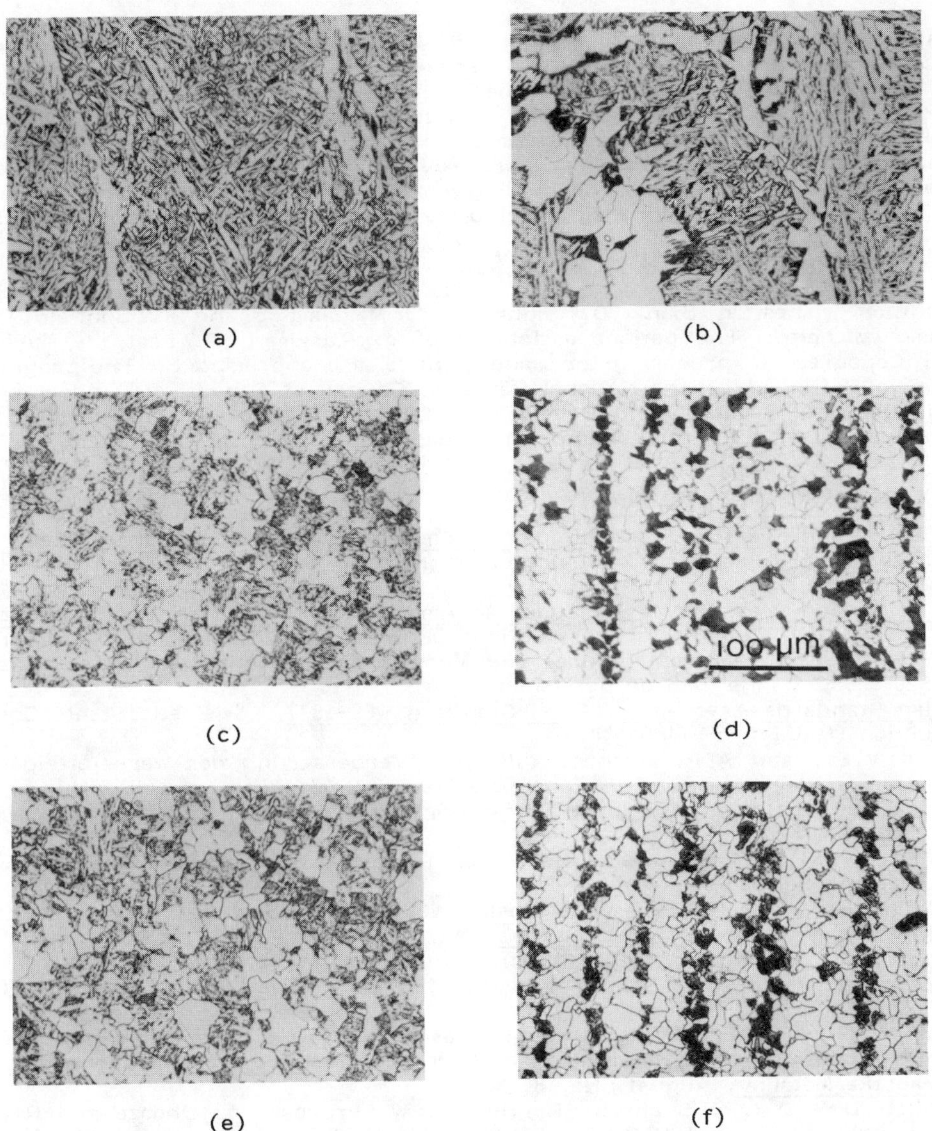


Fig. 4: The optical micrographs of the electroslag weld metal (a,c and e) and the HAZ (b,d and f) in A516 Gr. 70 steel from Mill B: (a,b) as-welded, (c,d) normalized for 5 minutes, and stress-relieved and (e,f) normalized for 180 minutes and stress-relieved. Mag. X200.

## CONCLUSIONS

Comparing the weld metal impact values and the tensile properties, which satisfied code requirements, to those of respective HAZ and the base material it could be inferred that there would be no real advantage in normalizing the weldments for more than 5 minutes. Thus, the 'holding time' for furnace normalizing of electroslag weldments could be considerably reduced from that of the conventional practice. This would save fuel as well as reduce distortion in the cans or the vessels. The holding time as low as 5 minutes for normalizing showed that local-normalizing would be a viable and economical process.

## ACKNOWLEDGEMENT

The author wishes to thank D. Skinner, Z. Maksimovic and S. Bonfield for experimental help. The patient assistance of J. Tackman, P. Last, J. Taylor and J. Capogreco in preparing the manuscript is also appreciated. The program was partially funded by the National Research Council of Canada and undertaken in the Corporate Research and Development Centre of AMCA International Ltd., who gave permission for publication of this paper.

## REFERENCES

ASME Boiler and Pressure Vessel Code, an American National Standard, Section III, Div. 1, "Rules for Construction of Nuclear Power Plant Components", 1980 Edition.

ASME Boiler and Pressure Vessel Code, an American National Standard, Section VIII, Div. 1, "Rules for Construction of Pressure Vessels", 1980 Edition.

Bala S.R. (1983). "Microstructures and Mechanical Properties of Steels Soaked at 620-950°C Range", to be published.

Canadian Standards Association (CSA) Standard W59-1977, "Welded Steel Construction (Metal Arc Welding)".

Eichhorn V.F., and A.R. Shaheeb (1970). "Verbesserung der Werkstoffeigenschaften in der WarmeeinfluB-Zone von Electro-Schlacke - SchweiBverbindungen durch ortliches kontinuierliches Normalisieren" VDI-Z 112, Nov., Nr. 21, 1461-1467.

Kreith F., (1973). "Principles of Heat Transfer", 3rd Edition, (Chemical Engineering Series).

Kroshkin V.A., et al, (1973). "Electroslag Welding with Simultaneous Indirection Normalizing", Avt. Svarka, No. 10, 48-51.

Lowe G., S.R. Bala and L. Malik (1981). "Hydrogen in Consumable Guide Electroslag Welds: Its Source and Signficance", The Welding Journal, 60, 258s-268s.

Novikov V.N., I.E. Tutov and A.I. Kondrashev (1958). "Localized (Induction) Heat Treating of Electroslag Welds", Translated from Metallovedenie i Obrabotka Metallov, August, No. 8, 38-43.

Shackleton D.N., A.R. Shaheeb, P. Hirsch, W. Provost, A. Dhooze and R.E. Dolby (1975). "Research Developments in Electroslag and Electrogas Welding", IIW Doc. XII-J-43-75.

Suzuki H., S. Fujimosi and Y. Nishio (1967). "Continuous Normalizing Electroslag Welding Process (First Report) IIW Doc. XII-I-3-67.

Thibau, R., and S.R. Bala (1983). "Influence of Electroslag Weld-Metal Composition on Hydrogen Cracking", The Welding Journal, 62, 97s-104s.

Wachtmeister I., (1961). "The Quality of Electroslag Welds in As Welded and Heat Treated Conditions", Undocumented, 79-96.

# INTEGRATED ROBOTIC WELDING THROUGH REAL TIME IR VISION

G. Bégin*, J.-P. Boillot** and D. Villemure**

*Industrial Material Research Institute, 75 De Mortagne Blvd., Boucherville (Quebec) J4B 6Y4
**Welding Institute of Canada, Quebec Centre, 75 De Mortagne Blvd., Boucherville (Quebec) J4B 6Y4

## ABSTRACT

Adaptive functions, whereby small changes in seam location, seam width and heat dissipation regimes are sensed and taken into account through real time regulation of torch coordinates and welding parameters, are required in robotic welding complex parts, especially when the latters are made of thin materials. Artificial vision and artificial intelligence thus become essential parts of any integrated welding system. It will be shown in the following paper why and when adaptive functions are required, and how they can be achieved through simple feature extraction of the thermal field surrounding the weld area, as perceived by an infrared detection system. A laboratory prototype able to achieve seam tracking and pool geometry control will then be described and its performances discussed.

## KEYWORDS

Real time infrared robotic arc welding control; intelligent welding robot, real time vision.

## INTRODUCTION

Figure 1 illustrates schematically what an integrated welding fabrication system should be. Inspired from N/C machining technology, it should incorporate off-line programmation from engineering drawings and a weld data base, in order to improve the robot duty cycle, and to optimize the sequence of events in the programme from off-line simulation of the time required to weld a particular piece. It should also be able to recognize and to cope with the inevitable variations associated with all the tolerances incorporated into a manufacturing scheme.

The latter requirement is adaptivity which essentially is the ability of the system to modify in real time and for each production unit, the rough programmation issued either from a «learn mode» on the first unit of a series, or from a more appropriate CAD file coupled to a welding data base.

In order to assume adaptivity, the following informations are required on a real time basis.

a - Joint location and gap size;
b - Pool geometry and location.

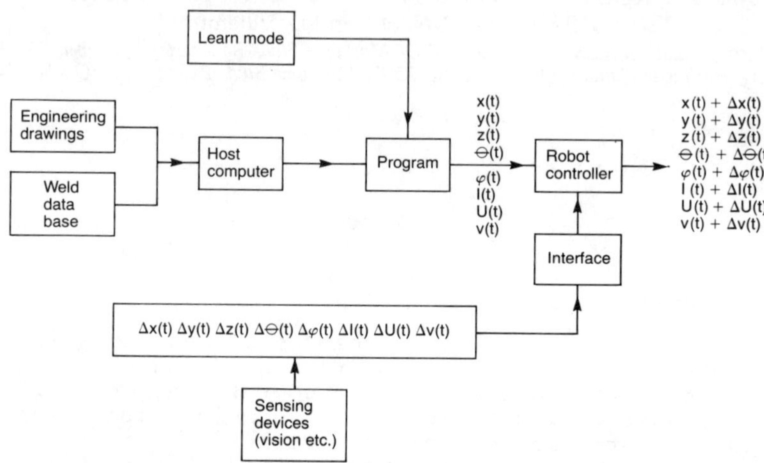


Fig. 1 Block diagram showing an ideal integrated automatic welding system

The first set of information leads to proper adjustments of spatial torch coordinates and mass balance adjustments, whereas the second one leads to heat balance adjustments. Table 1 resumes the adaptive functions just mentioned and displays the nature of the information necessary to perform those functions. Table 1 also shows what particular parameters an adaptive system should be able to change continuously in order to perform each of the adaptive functions.

TABLE 1

| ADAPTIVE FUNCTIONS | REQUIRED INFORMATION | CORRECTIVE ACTION TAKEN ON |
|---|---|---|
| Relative position of the torch and work piece | Joint location | Torch coordonates |
| Mass balance | Gap size | Welding speed/or wire feed rate |
| Heat balance | Pool size,dimension and location | Linear energy through welding current or welding speed |

## WHY ARE ADAPTIVE FUNCTIONS REQUIRED

### Seam Tracking

Seam tracking in term of corrective actions that changes the torch coordinates from those of the programmed path, is required whenever the cumulative imprecisions exceed the programmed path reproducibility, to the extent that welding defects occur. This occurence is of course a function of any particular assembly and is best illustrated by an example. Figure 2 shows to what extent misalignment between the electrode tip and the axis of the joint can be tolerated. For that particular case, when 5 mm thick plates, butt welded with a gap of 2 mm, using GMAW process in the short-circuited mode at 120 amperes and with a wire diameter of 1,2 mm, it is seen that, when the mismatch is or exceeds 1 mm, a defect occurs on the back side. This means that if all the cumulative sources of mismatch could lead to a total misalignment in excess of 1 mm, seam tracking is necessary.

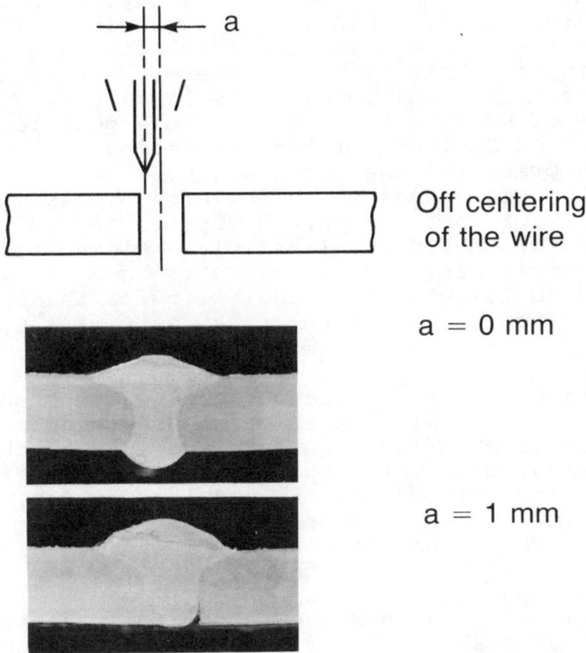


Fig. 2 Effect of wire position v/s gap axis (for GMAW)

In order to enhance the significance of that tolerance requirement in term of cutting, shaping, preassembly and positioning precision, let us look at the sources of imprecision of the system.

TABLE 2

| Source of imprecision | ± 2σ value or 95% confidence level |
|---|---|
| Programmed path imprecision | ± 0,2 mm |
| Programmed path reproducibility | ± 0,2 mm |
| Positionner reproducibility | ± 0,1 mm |
| Position of parts on positionner within the jig & fixture | ± 0,2 mm |

Table 2 adds up a potential misalignment of ±0,7 mm, whereas we have seen that for the particular case under study, misalignment up to ±1 mm could be tolerated. This leaves a margin of ±0,3 mm for the cumulative effect of parts dimensional tolerances associated with cutting and shaping. This is quite severe, particularly for limited production where mechanical shearing is largely used. For complex assemblies made of many parts, and where the position of a given joint is influenced by those of the preceeding ones, the severity on parts dimension precision increases further, and it becomes obvious that seam tracking is necessary unless high precision machining is implemented. If for instance part dimension reproducibility of ±0,5 mm (±2σ or 95% confidence level) is achieved through shearing sheet metal, then, in the most unfavorable case, as in an assembly of n parts in series, the reproducibility of the last one within a frame of reference locked on the first part is ± 0,5 n mm.

Table 3 shows, in term of confidence levels, the chances of making sound welds without a real time seam tracking module as n increases. The ±2σ confidence level to make the axis of the seam to coïncide with the programmed path incorporates the misalignments due to the programmed path reproducibility and imprecision, the positioner location error and the part location error in the fixtures on the positionner. Therefore, the ±2σ for the first joint is ±1,2 mm, ±1,7 mm for the second, ±2,2 mm for the third, etc.

Under these premisses, Table 3 shows that the probabilities of making soundwelds without real time seam tracking, as defined by the misalignment distribution contained within the ±1mm limits of weld defects occurence, is 89% in the most unfavorable case for an assembly containing one weld, 76% for an assembly containing two and 64% for one containing three, etc. It is thus seen that problems arise rapidly as assemblies become more complex.

TABLE 3

CONFIDENCE LEVEL FOR
MAKING SOUND WELDS WITHOUT SEAM TRACKING

| n | $\pm\sigma$ | $\int_{-1\ mm}^{+1\ mm} p(x)dx$ |
|---|---|---|
| 1 | ± 0,60 | 89% |
| 2 | ± 0,85 | 76% |
| 3 | ± 1,10 | 64% |
| 4 | ± 1,35 | 54% |
| 5 | ± 1,60 | 46% |

Figure 3 shows an example where the improper location of electrode tip lead to poor quality welds in robotic welding.

Another advantage resulting from the use of seam tracking modules, even with a high probability of sound welds consists in the optimization of weld metal quantities in fillet joints.

Fig. 3 Poor quality weld, due to improper torch location, in robotic welding

AWS code for prequalified joints specifies a leg length. In the case of 6,4 mm thick plates, the leg length should be 3 mm. Figure 4 shows the weld profile of such a fillet weld as the electrode tip is shifted from the bisecting position.

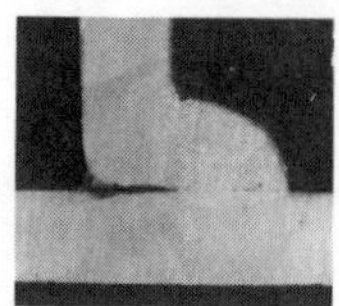
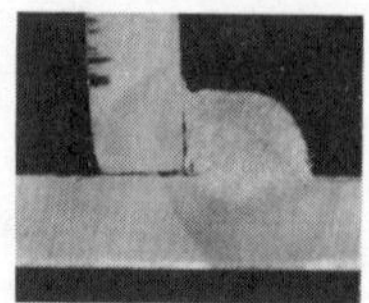
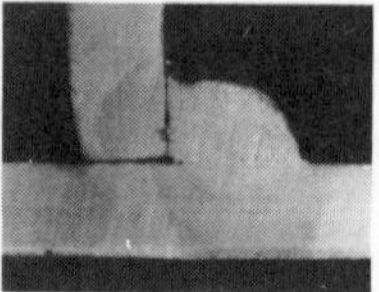

Fig. 4 Evolution of fillet weld profiles when shifting of the electrode tip from the bisecting position

## Heat and Mass Balance Control

Heat balance. In all welding operations, there are changes in the heat dissipation regime and a true steady state condition really never occurs. It is therefore necessary, in order to maintain a constant pool size, to adjust continuously the parameters controlling the heat input. In many cases, this is done by programming the linear energy either through welding current or welding travelling speed variations. This is illustrated in Fig. 5 where the linear energy is reduced by increasing the travelling speed at constant mean current in order to compensate for the continuous increase of the temperature of the work piece. In cases of complex parts where heat sinks or sources may occur along a weld line, adjustments of welding parameters are required to maintain a constant pool size and a good quality weld.

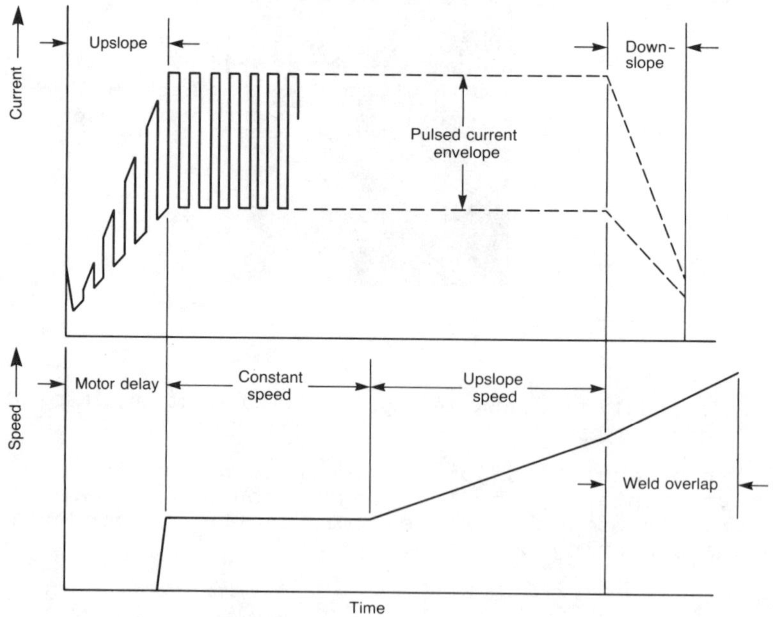


Fig. 5 Pipe welding program to take into account heat balance

Figure 6 illustrates an example where welding parameters and torch coordinates must be changed to comply with the constant pool size and appropriate pool location criteria.

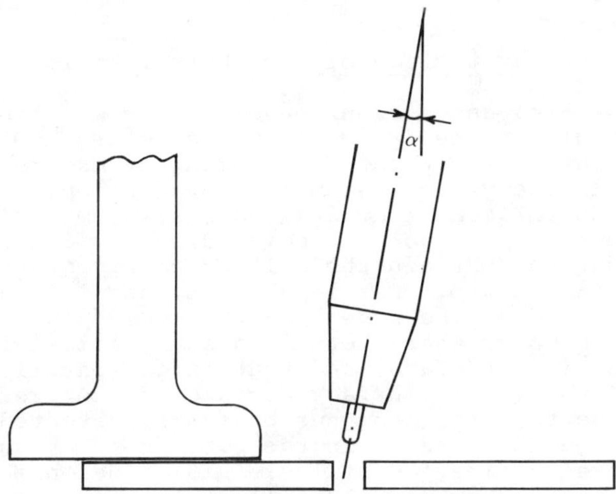


Fig. 6 Example where torch coordinates must be changed to comply with heat balance

Mass balance. Mass balance adjustments are required whenever gap variationsoccur. Especially in parts made of thin materials and characterized by small gap sizes, large relative variations of gap inevitably occur. They are caused by the misalignments associated with cutting, shaping and preassembly imprecisions. A design gap of 2 mm may vary from 1 to 3 mm if the parts tolerances are ±0,5 mm. That implies ±50% variations in weld metal requirements. That may lead to severe weld defects. For the 1 mm gap, with a weld metal deposition rate based on a 2 mm gap, we will have a lack of penetration. For the 3 mm gap, we will have a burn through or a collapse. In either cases, improper weld geometry occurs. It is thus seen that real time sensing of gap size with the appropriate corrective action in term of mass balance is necessary to achieve optimal quality welds in this material.

## HOW COULD THE ADAPTIVE FUNCTIONS BE IMPLEMENTED

Pool size and location, as well as gap size and gap axis position are the kind of information that can be obtained from real time vision. Real time vision is thus a prerequisite for adaptive welding. However, incorporating a real time vision system into a welding process is not an easy task. In order to do so, one is confronted with arc interference, fumes, and rapid processing time problems. In some instances, it is possible to overcome the problems. In seam tracking, feature extraction from a projected laser beam optically shaped into a thin plane has been used with some success. For pool size control however, direct vision of the pool in the visible range implies artifacts such as arc extinctions and/or complex image treatments involving many grey levels. In most cases, integrated systems for seam tracking and for pool geometry control are facing relatively high costs.

## Significant Feature Extraction of the Thermal Field

Because the pool area and its surroundings are emitting infrared radiations, vision within the IR spectrum can be used to perform the various adaptive functions, without the need for external illumination sources (laser in the video systems). In addition, the relevant geometrical features for seam tracking and pool geometry location and control are inherent parts of the emitted energy field. There is in fact a sharp contrast between the emitted energy at the edge of the plates and within the gap. For lap joints, significant differences in the power emitted by the upper and the lower plates can also be found. It also happens that there is a sharp emissivity discontinuity at the liquid solid interface, and that large spatial emissivity variations occurs at the liquid surface because of the presence of oxyde particles. These features can thus be easily detected on-line by properly sampling the IR emission in the pool area. Figure 7 shows schematically where significant profiles of IR emission should be taken, whereas Fig. 8 and 9 show actual on-line measurements of such profile, revealing joint location, gap size, pool location and pool width. Simple feature extraction in selected profiles within the IR emission field in the pool area can thus provide the necessary information to achieve adaptive welding. Figure 10 illustrates schematically how a simple 2-channel module can «see» the required features. This scheme has become the basis for the integrated system that we have developped and that will be described below.

## DESCRIPTION OF A LABORATORY PROTOTYPE

A laboratory prototype for seam tracking butt welds based on the previously mentioned principles was designed and tested in a joint IMRI-WIC venture.

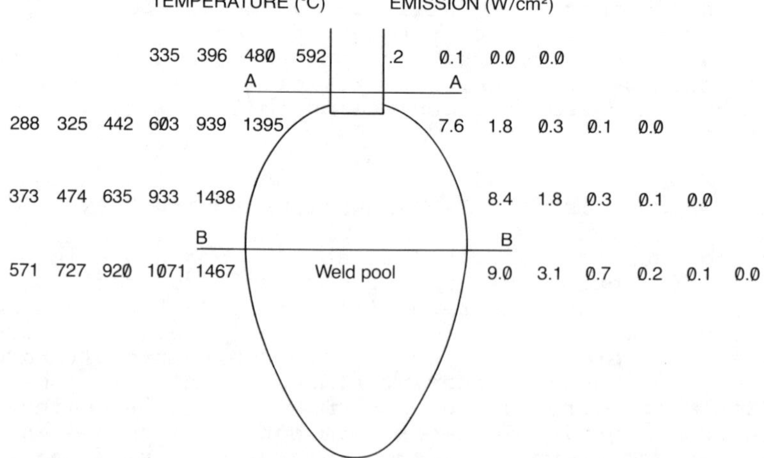


Fig. 7 Computer plot of the surface temperature and of the power emitted per square centimeter in the vicinity of the weld pool

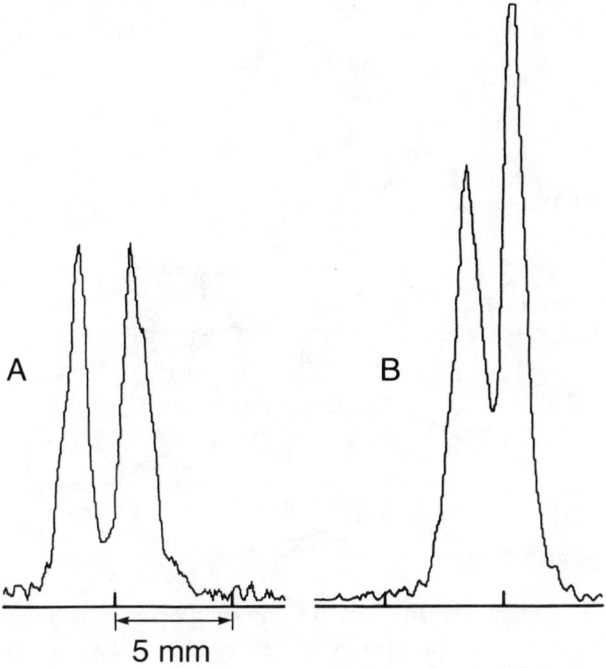


Fig. 8 Records of the emission profile during scanning ahead of the pool

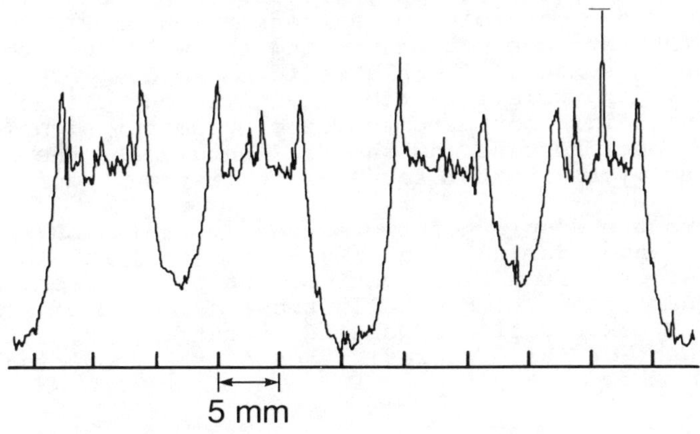


Fig. 9 Records of the emission profile during scanning across the pool

The process involved in this instance was GMAW operating in the short-circuited mode and the weld was a single pass full penetration one made on 4 mm thick steel plates.

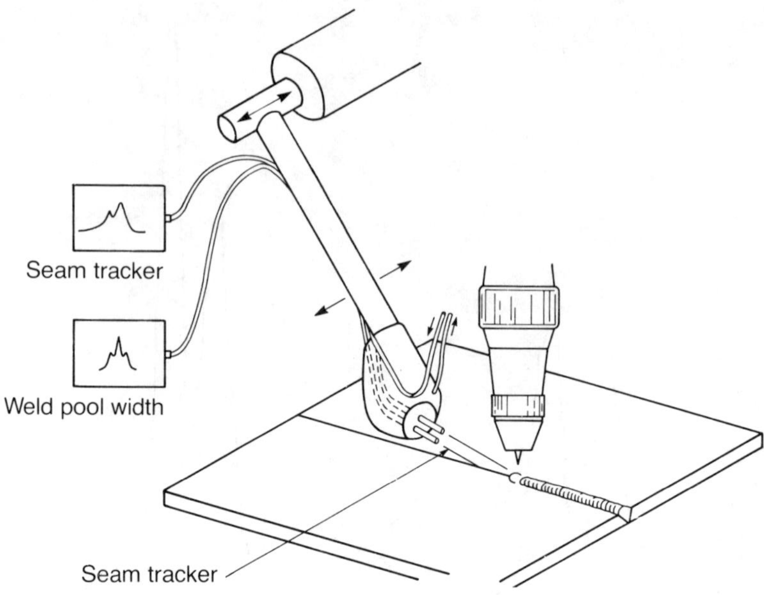


Fig. 10 Experimental set-up for preliminary measurements

The experimental set up is shown on Fig. 11. A single optical fiber focussed at the plate level was mechanically oscillated at 4 Hz in front of the pool on line aa. The signal, guided through an optical fiber was fed into a pyrometer located remotely from the welding area. The analog signal of the pyrometer was transformed into a digital one through an A/D converter, properly processed in a Z-80 microprocessor whose output fed the controler of a translation unit oriented perpendicularly to the carriage motion driving the welding torch and the attached scanning sensor. The plates to be welded were laid on·the translation unit and were intentionnaly given curvilinear shapes. For analysis and visualisation purposes, the pyrometer signal was also fed into a Nicolet oscilloscope and the data handled by the Z-80 fed and stored in a IBM microcomputer for on-line display and further analysis.

Figure 12 shows the display of two successive emission profiles detected by the sensor during a welding operation. As expected, absolute values vary largerly from time to time, but the edge of the preparation are always clearly seen. Figure 13 shows a record of all on-line corrective actions taken during a welding operation. Finally, Fig. 14 shows photographs of the preparation and of the resulting weld.

Another advantage of the IR vision scheme just described is that it also provides the necessary information for heat input control through pool proximity sensing. The vision scheme used to locate the gap can in fact sense pool proximity and can thus be used to control the linear energy input by reacting on welding speed for example. In the event of variations in the heat dissipation regimes, the weld pool surface changes and the relative location of the scanned line with respect to pool edge also varies. The sign and the rate of change of the average temperatures characterizing the profile is thus used to regulate the the welding speed. The next figure (Fig. 15) compares how the control loop reacted to insure constant pool size despite drastic changes in the heat dissipation regimes caused by the holes. The same figure shows that when the control system was turned off, large increases in pool size were experienced where the heat flow was restricted. This shows again the great versatility of the IR vision approach to perform adaptive welding functions.

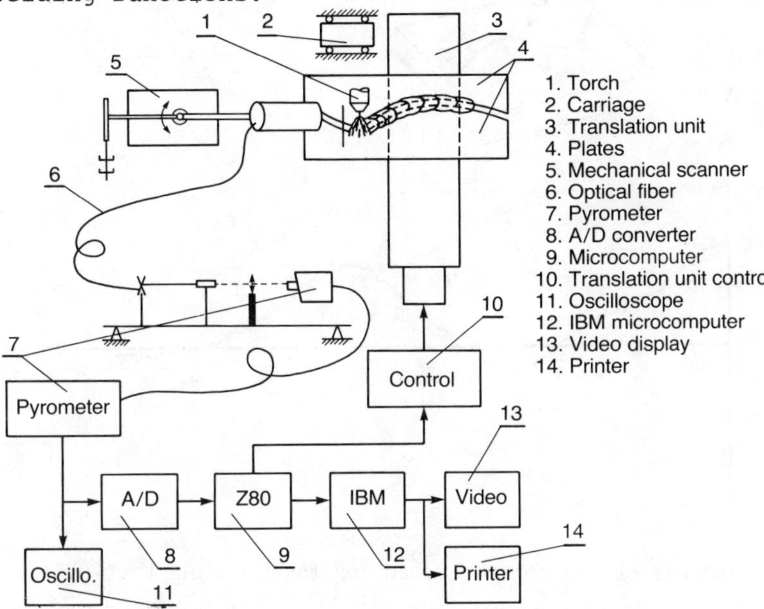


Fig. 11 Block diagram of an experimental seam tracker

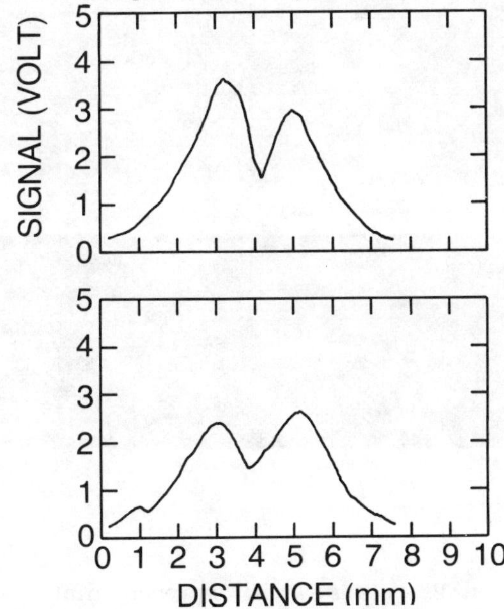


Fig. 12 Successive emission profiles in front of the pool taken during a welding operation

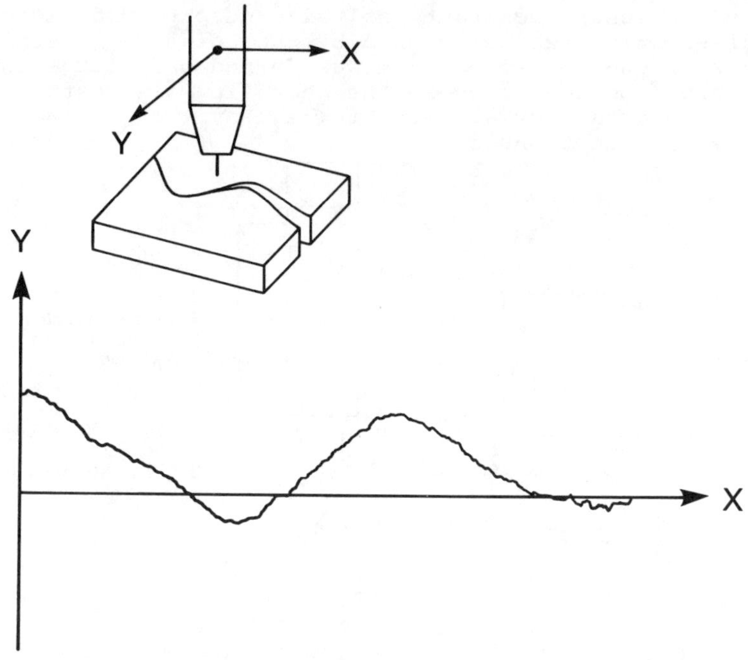


Fig. 13 Records of the corrective action taken during the welding operation

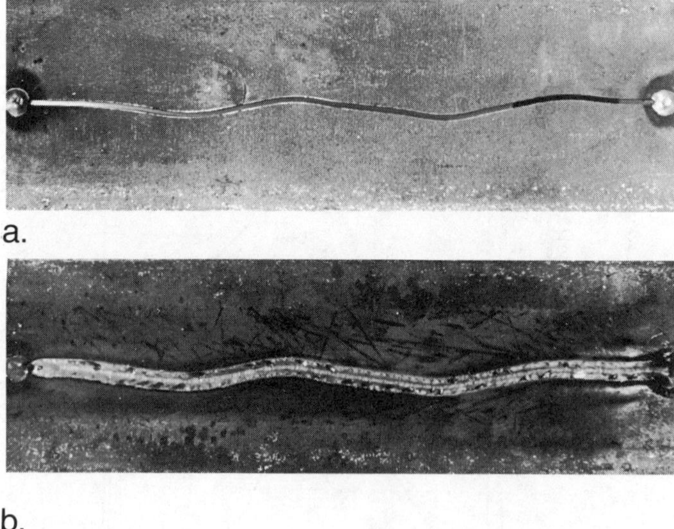


Fig. 14 a. Plate simulating improper joint preparation
b. Same plate after adaptive GMAW welding

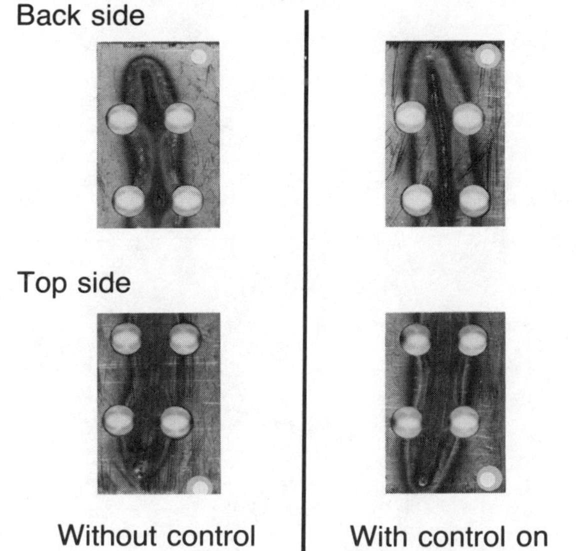


Fig. 15 Effect of infrared control on heat balance

## CONCLUSION

Unless costly equipments are implemented to reduce the cumulative imprecisions associated with parts cutting, shaping, jigging and positioning, robotic welding of parts made of thin material will not be possible without on-line modification of the programme to comply with misalignments between torch trajectory and joint position, gap variations and changes in heat dissipation regimes. Real time vision is thus a prerequisite for expanding robotic welding to complex parts involving large relative misfitt. The various actual video approaches used to incorporate vision to welding systems imply external illumination, arc extinctions and/or complex image treatment involving high cost and limiting welding speed for on-line corrective action.

The fact that the required geometrical features such as gap size and location, and the liquid-solid interface are contained in the emitted power coming from the weld area within the emission bandpass of simple IR detection systems opens the way to simple feature extraction at high speed and relatively low cost.

An integrated system based on gap and pool recognition through emission profiles at selected places in the pool area has achieved the welding adaptive functions required for on-line modification of welding robot programmes. This system works in real time, is based on a relatively simple technology, does not require external illumination, and can cope with a variety of situations to make robotic welding profitable to fabrication shops dealing with highly deversified products in limited quantities.

# ON LINE WELD PENETRATION MEASUREMENT USING AN INFRARED SENSOR °

J.L. Fihey, P. Cielo*, G. Begin*
*Production et utilisation de l'énergie*
*IREQ, Institut de Recherche d'Hydro-Québec*
*Varennes, Québec J0L 2P0*

## ABSTRACT

A new Approach to weld penetration monotoring in GTAW or PAW is proposed. Analytical as well as numerical calculations show that a strong correlation exists between the weld penetration and the solidification time of the weld pool. The physical origin of such a correlation is that the dissipation time of the heat of fusion is larger for a deeper molten pool. The experimental verification of the basic principle is carried out by means of a rugged and inexpensive fiber-optic sensor which probes the temperature distribution of the weld pool. A discussion is presented of the relevant parameters such as the spatial and temporal variation of the thermal gradients around the weld pool as well as the infrared emissivity of the surface.

## KEYWORDS

Welding, Penetration, Infrared sensing, Solidification

° Supported in part by the National Research Council of Canada through the Welding Institute of Canada.

* National Research Council of Canada - IMRI Montreal H4C 2K3

## INTRODUCTION

Automatic welding machines and more recently welding robots replace the human welders in many applications. One drawback of those machines is that they do not have the ability to react to unexpected events, e.g. variations in joint fit-up or weld pool wander because they are not equipped to sense those events. In order to improve the quality of the welded products we must develop welding machines with on line sensing and control capability. The most relevant parameters are the weld pool geometry and position and its thermal history (gradients, rate of cooling). The depth of penetration of the weld is one of the most difficult quantity to monitor. Systems involving measurements of radiations on the back face of the weld have already been implemented and work well for the detection of complete penetration when the back face is accessible (Boughton and co-workers, 1979; Rider, 1975). However, the back face is often not accessible. Moreover, back face sensors perform poorly in the detection of partial penetration. In this work we propose a new approach based on the measurement with an infrared sensor of the weld pool solidification time during arc interruption. An analytical solidification model is developped and yield a simple relationship between solidification time and depth of penetration. The model is successfully tested versus a more accurate numerical model and experimental results.

## ANALYTICAL SOLIDIFICATION MODEL

Consider the problem of weld pool solidification after arc extinction (Fig. 1). Heat is accumulated in the pool as latent heat of fusion and superheat. This heat will be dissipated at a rate which is a function of the thermal gradient in the solid metal surrounding the pool. This is a transient tridimensional heat flow problem which cannot be solved analytically unless major approximations are made. The problem is greatly simplified with the four following assumptions:

- No superheat in the liquid pool
- The thermal gradient in the solid at the liquid-solid interface is independent of position and time during solidification.
- The weld pool has the shape of a portion of sphere.
- The divergence of the heat flow as well as the finite thickness of the plate can be neglected.

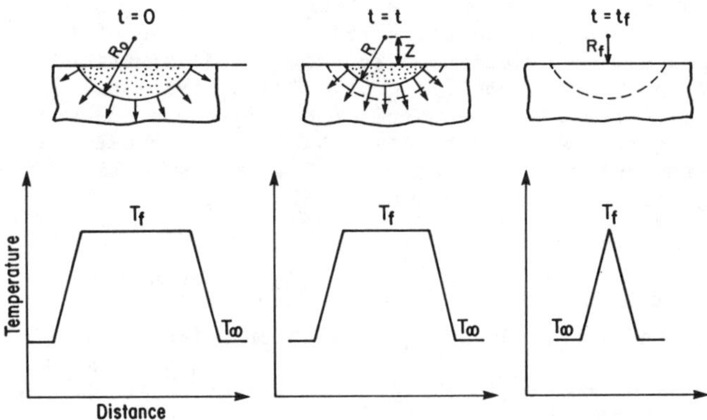


Fig. 1: Weld pool solidification model

The heat flow leaving the interface in the solid state equals the heat flow produced at the interface. The applicable differential equation is then:

$$A k_s \left(\frac{dT}{dR}\right)_s = \rho_s H_f \frac{dV}{dt}$$

Where:

A is the interface area
V is the weld pool volume
$k_s$ the thermal conductivity of the solid metal
$\left(\frac{dT}{dR}\right)_s$ the thermal gradient in the solid phase at the interface
$\rho_s$ the density of the solid
$H_f$ the heat of fusion of the solidifying metal

Since dV = AdR, the equation may be simplified:

$$\frac{dR}{dt} = \frac{k_s}{\rho_s H_f} \left(\frac{dT}{dR}\right)_s \qquad (2)$$

During solidification the interface remains perpendicular to the heat flow direction, thus the solidification process may be simulated by decreasing the sphere radius down to $R_f$ as shown schematically in Figure 1. Integration between $t = o$ and $t = t_f$ yields:

$$t_f = \frac{D \rho_s H_f}{k_s \left(\frac{dT}{dR}\right)_s} \qquad (3)$$

The solidification time, $t_f$, is directly proportionnal to the penetration depth and inversely proportionnal to the thermal gradient in the solid at the interface. According to this equation the depth of penetration can be deduced from measurements of both the solidification time and the thermal gradient in the solid. However because a number of approximations have been made we must test the validity of this equation by comparison with a numerical calculation as well as with the experimental results.

## NUMERICAL MODEL

Because of its power and versatility, the numerical approach is today the primary method of solution of multidimensional heat-transfer problems (Thomas, 1980). Welding applications of such numerical models have already been reported in the literature (Paley and Hibbert, 1975; Westby, 1968). We thus used a numerical model to test our analytical expression for different weld pool sizes and shapes, variable heat input and plate materials of different thermal parameters.

The three-dimensional finite-difference model used in our computations uses the explicit iterative approach to solve the thermal diffusion equation:

$$\frac{\partial T}{\partial t} = \alpha \left(\frac{\partial^2 T}{\partial x^2} + \frac{\partial^2 T}{\partial y^2} + \frac{\partial^2 T}{\partial z^2}\right) \qquad (4)$$

by subsequently applying the Fourier equation:

$$\underline{Q} = k_s \text{ grad } T \qquad (5)$$

and the energy conservation equation:

$$\rho_s C_p \frac{\partial T}{\partial t} = - \text{div } \underline{Q} \qquad (6)$$

where $\alpha \equiv k_s/\rho_s C_p$, $C_p$ is the specific heat at constant pressure, and $\underline{Q}$ is the thermal flux. This two-step approach simplifies the boundary conditions, while being more exact in this case where the material properties are temperature-dependent and thus spatially non-uniform (Cielo, 1973).

The simple case of an autogenous weld on a thick plate was assumed. The model covers a 20mm x 20mm x 9mm volume with an elementary cell $(dx)^3 = 1$ mm$^3$. The temporal iteration is pursued for a period of typically 10 seconds with time increments dt satisfying the stability condition:

$$\frac{\alpha \, dt}{(dx)^2} \leqslant \frac{1}{6} \qquad (7)$$

The 9mm-thick plate is assumed of infinite extent. Continuity conditions of the form:

$$\frac{\partial T}{T\partial x} = \text{constant} \tag{8}$$

were imposed at the boundaries of the 20mm x 20mm modelled area. Heat losses by radiation, air convection and partial vaporization were assumed at the upper surface of the plate. The expressions for such losses, as well as the values of the thermal parameters for the two cases of carbon steel and stainless steel are listed in Table 1. The expression for the vaporization losses is obtained from the published data for iron (Quigley and co-workers, 1973), which are the only data available. Condensation was simulated by re-injecting 75% of the vaporization loss on a 1 $cm^2$ area of the surface around the weld pool.

The thermal power of the arc was injected over a 20$mm^2$ surface on the center of the weld pool. The injected heat distribution had a nearly gaussian shape. The travel speed of the arc was in all cases 1mm/s, and the frame of reference was fixed with respect to the arc. A radial convective flow was assumed at the surface of the liquid pool, both during welding and after the arc extinction until solidification. the hydrodynamic forces producing such currents are the arc pressure and the resultant of the surface tension components, which depend on the temperature gradient, surface topography and chemical composition of the surface layer. As such parameters are hardly predictable, we used a semi-empirical approach by using a simplified version of the Navier-Stokes equation at the weld pool surface:

$$\frac{d\sigma}{dT}\frac{dT}{dx} = \mu\,\frac{v}{l} \tag{9}$$

where $\sigma$ is the surface tension, $\mu$ is the temperature-dependent viscosity of the liquid metal, v is the flow velocity, and l is the approximated thickness of the flowing layer. For most metals, $\frac{d\sigma}{dT}$ is temperature-independent, while l is nearly constant over the weld pool surface. The following expression was thus used in our model for the velocity of the surface flow:

$$v = \frac{K}{\mu}\frac{dT}{dx} \tag{10}$$

where K is a constant whose value was determined empirically by comparing the sectional shape of the modelled pool with the real shape obtained by after-welding metallographic inspection.

The calculated weld pool boundaries after arc extinction are plotted in Figure 2 a, b and c for a stainless steel plate. In this case a W/D ratio of 2.8 was aimed. The complete surface temperature profile along the X X' axis is plotted in Figure 3. The temperature under the arc is very high (≃ 2500 °C); this value is however a function of the heat losses by vaporization which in turn depend on the degree of contamination of the surface. The temperature drops very rapidly after arc extinction, then remains approximately constant during the solidification period and drops again after complete solidification. The termal gradient at the liquid-solid interface decreases as the solidification progress. Furthermore the gradient is a function of the position along the weld pool boundary as shown in Table 2. Nevertheless the thermal gradient on both sides of the weld pool at the electrode level (lateral thermal gradients) remains a good approximation of the mean thermal gradient. The relationship between the depth of penetration and the solidification time has been computed both with the numerical and analytical model for three types of weld pools: two stainless steel weld pools with a W/D ratio of respectively 2.8 and 6.0 and one carbon steel weld pool with a W/D ratio of 3.2. the parameters used for the analytical model are drawn from Table 1. The thermal conductivity at the fusion temperature was used. The analytical equations are then:

Stainless Steel: $$D = \frac{t_f}{7411} \left(\frac{dT}{dR}\right)_s \qquad (11)$$

Carbon Steel: $$D = \frac{t_f}{5180} \left(\frac{dT}{dR}\right)_s \qquad (12)$$

Where D is expressed in cm, $\left(\frac{dT}{dR}\right)_s$ in °C/cm and $t_f$ in seconds.

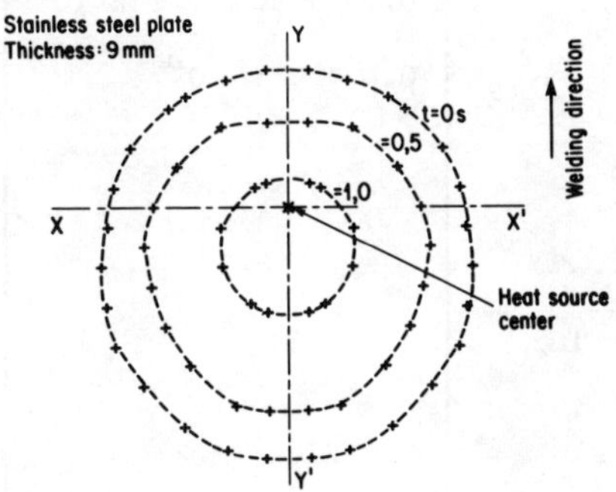


Fig. 2-a: Computed weld pool boundaries after arc extinction - Surface.

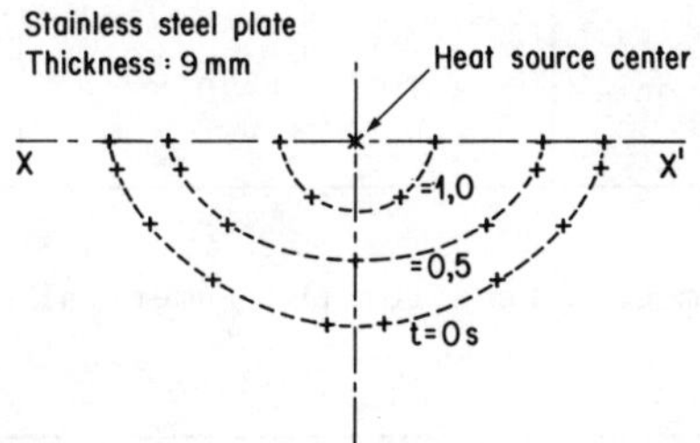


Fig. 2-b: Longitudinal cross-section along plane XX'

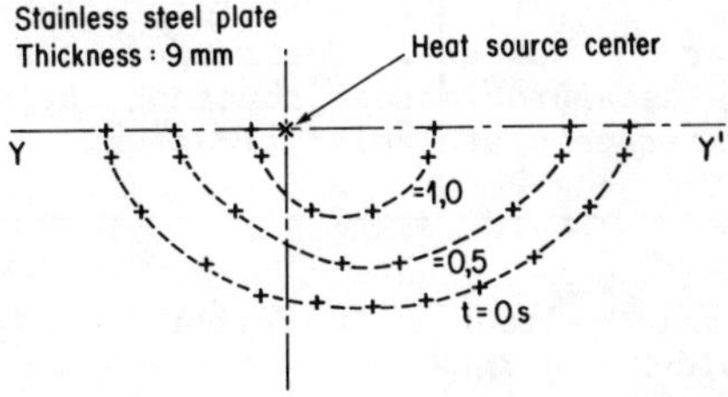


Fig. 2-c: Longitudinal cross-section along plane YY'

| Parameter | Carbon Steel | Stainless Steel |
|---|---|---|
| Temperature of fusion (°C) | 1480°C | 1450°C |
| Thermal conductivity (watt/cm °C) | From 0 to 900°C: $0.6 - T \times 2.8 \times 10^{-4}$ above 900°C: $0.24 + T \times 1.2 \times 10^{-4}$ | $0.16 + T \times 10^{-4}$ |
| Specific heat (joules/g °C) | $0.4 + T \times 10^{-4}$ | $0.4 + 9.2 \times T \times 10^{-5}$ |
| Specific Mass ($g/cm^3$) | 7.8 | 8.0 |
| Heat of allotropic transformation ($\alpha - \gamma$) (j/g) | 23 | - |
| Heat of fusion (j/g) | 270 | 270 |
| Heat loss by vaporization ($w/cm^2$) | $10^{-4}$ exp $(6.7\ 10^{-3}\ T)$ | $10^{-4}$ exp $(6.7\ 10^{-3}\ T)$ |
| Heat loss by radiation ($w/cm^2$) | $5.67\ 10^{-3} \times \varepsilon (T^4 - 293^4)$ | $5.67 \times 10^{-3} \times \varepsilon (T^4 - 293^4)$ |
| Heat loss by convection ($w/cm^2$) | $5 \times 10^{-4}$ (T-20) | $5 \times 10^{-4}$ (T-20) |

TABLE 1: Parameters Used for the Numerical Calculation.

| Position | | Thermal gradient (°C/cm) |
|---|---|---|
| Surface | Front | 4 000 |
| | Rear | 1 700 |
| | Side | 2 700 |
| underbead | | 2 500 |

TABLE 2: Thermal Gradient at Different Locations along the Liquid-Solid Interface.

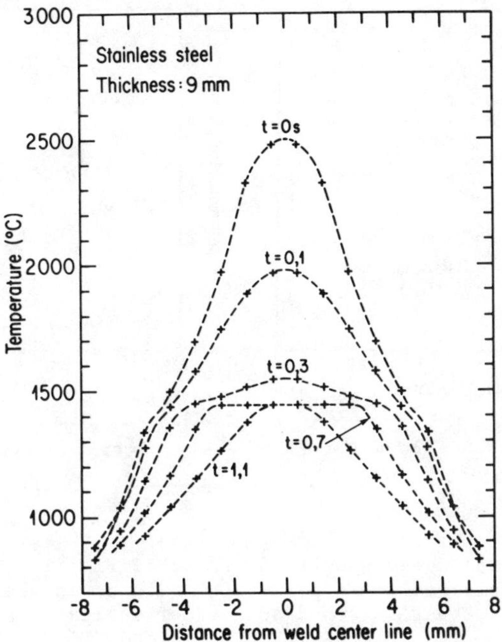


Fig. 3: Computed surface temperature profile along the XX' axis after arc extinction.

The computed and calculated values of D are plotted versus solidification time in Figures 4a, b and c. There is a good agreement between the two curves up to a solidification time of 0.6 to 0.8 s. For longer solidification time the computed curve lies above the analytical one. This discrepancy can be related to the error introduced in the analytical model by the assumption of a constant thermal gradient at the liquid solid interface during all the solidification period. A more complex quadratic expression with coefficients derived empirically from the numerical values could be used. However, we feel that the simple analytical expression given above is a satisfactory approximation for most practical situations.

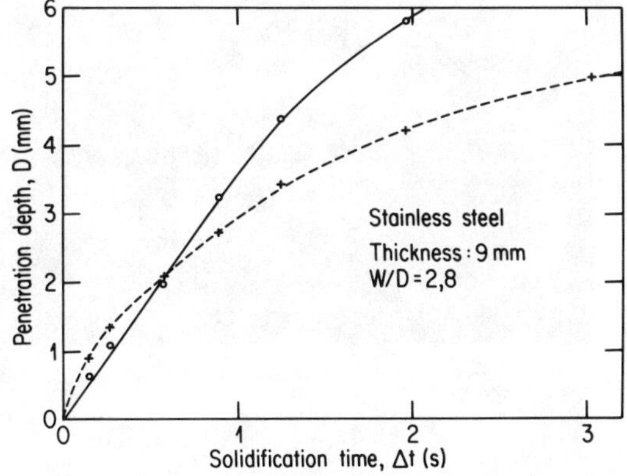


Fig. 4-a: Weld penetration depth versus solidification time. —— : analytical model; ---- : computed. Stainless Steel, W/D ≈ 2.8

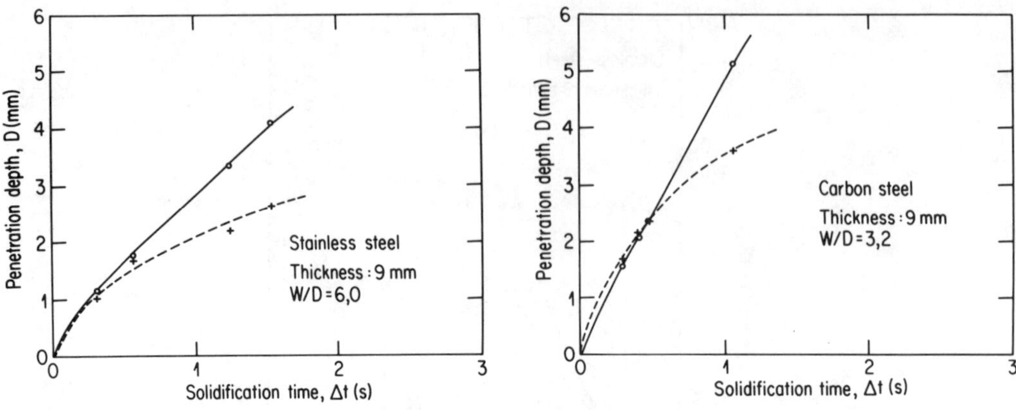


Fig. 4-b: Stainless Steel, W/D ≈ 6.

Fig. 4-c: Carbon Steel, W/D ≈ 3,2.

## EXPERIMENTAL TECHNIQUE

A series of experimental measurements of the solidification time of carbon steel and stainless steel weld pools was performed. The experimental setup is shown in Figure 5, the workpieces were clamped on a water cooled copper plate. A fusion

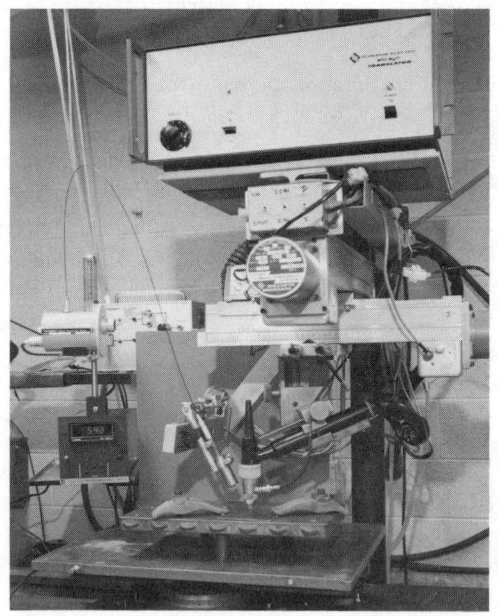

Fig. 5: Experimental set-up.

path was made with the GTAW process. The transistorized source was controlled by a microprocessor. The torch could be driven in the X and Y direction by two stepping motors. The heart of the set up is the infrared sensor. The radiation emitted by the pool is focused on the tip of an optical fiber and transmitted through a chopper blade and a 2.2 μm slot filter to a PbS detector. The signal generated is amplified and linearized such as the output of the pyrometer is proportional to the temperature. The following relationship holds in the temperature range close to the fusion temperature

$$V \propto \sqrt[4]{\sigma \varepsilon A F} \quad x\ T \tag{13}$$

where $\sigma = 5.67\ 10^{-8}\ W \cdot m^{-2}\ K^{-4}$

$\varepsilon$ is the emissivity of the workpiece

A is the spot size

The response time of the pyrometer is 0.1s for 90% of the signal. The sensor was pointed at a distance of 50 mm from the workpiece. The angle between the sensor axis and the surface was 60°.

## EXPERIMENTAL RESULTS

With the infrared sensor two types of measurement have been performed. First the transverse temperature profile was measured with the sensor scanning the pool, then the solidification curve of the weld pool after arc extinction was measured with the sensor aiming toward the weld pool center. When the arc is on the infrared emission of the arc is scattered by the surface and eventually adds up to the thermal emission of the weld pool. The temperature of the arc (≃ 10 000 K) is much higher than the temperature of the weld pool (≃ 1800 K) but the emissivity of the arc is very low. It has been calculated (Rider, 1979) that for wavelengths greater than 2 μm the weld pool radiations dominate. Thus eventhough the arc interfers with the weld pool temperature profile this interference is restricted to the arc root or the anode spot and should not impede temperature gradient measurements at the weld pool edge.

A serie of measurements were performed on carbon steel (P3) and stainless steel weld pools (P4, P5). All the plates were 9mm thick, two different heats of stainless steel 304L were used, P4 had a better penetration profile (W/D ≈ 2.8) than P5 (W/D ≈ 5). This behavior is related to a variation in sulfur content between the two heats and has been reported elsewhere (Fihey and Simoneau, 1982). For all the experiments pure argon shielding gas was used, the welding speed was 1mm/s and the arc length 3 mm. The current was varied between 75 and 180 A. Two identical fusion pass were made for each settings. The first pass for thermal gradient measurement and the second pass for solidification time measurement after arc extinction and simultaneous torch travel interruption. The solidification curve was recorded on an X-Y recorder. Figure 6 compares an experimental curve and a computed one for a weld pool on plate P4, the current was 150 A. The two peaks in the solidification plateau are due to emmissivity variations. There is a good agreement between the two curves.
The lateral thermal gradients before arc extinction may be measured experimentally, however the values obtained are not very accurate because they depend on the exact emissivity of the surface. Furthermore, the relatively low response time of the pyrometer sets a limit to the higher values obtainable for a given sweeping rate. All the experimental values are reported in Table 3 for a serie of 10 experiments. The bead width and depth were measured on a cross-section of each weld, the time for complete solidification was calculated from the solidification curves and the mean value of the lateral gradients deduced from the recorded scans are also reported. The computed values of the depth of penetration for a given

solidification time and W/D ratio are included in the table as well as the computed lateral gradients. Finally the depth of penetration was calculated using the analytical model and the computed thermal gradients.

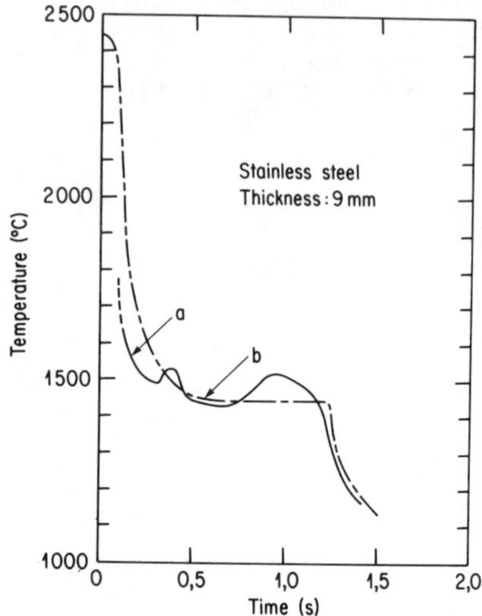


Fig. 6: Solidification curve, a) experimental; b) computed.

## DISCUSSION

The experimental results reported in Table 3 cover a narrow range of penetration depth (0.17 cm - 0.35 cm) nevertheless this range is typical of autogenous welding with the GTAW process. The computed depth of penetration correlates well with the experimental results in this range, the error being always less than 10 %, there is some discrepancy however between the computed and experimental thermal gradient. This is not surprising owing to the difficulties encountered in our measurements of this parameter: low response time of the pyrometer, uncertainty about the emissivity. The more accurate computed thermal gradients were used for the calculation of D with the analytical relationship. The agreement remains good as far as the solidification time does not exceed about 0.8 s.

We should point out that the analytical model does not include explicitely the W/D ratio in the calculation of D. The variation of W/D however is embodied in the thermal gradient. The stainless steel plate P5 behaves abnormally with respect to the heat input, the depth of penetration does not increase with heat input. Nevertheless the analytical model is able to discriminate between the increase in solidification time due to an increase in W rather than D. It is thus essential to measure more precisely the thermal gradient during welding. To do so we will have to add an emmissivity correction to the pyrometer reading and to increase the response time both of the pyrometer itself and of the scanning method. In order to increase the scanning speed and avoid an impractically fast mechanical scan, we are now developing a sensor comprising an array of optical fibers coupled with electronically scanned photocells.

Penetration measurement with an infrared sensor may be used for applications where it is possible to momentarily extinguish or transfer the arc. One application could be for pulsed GTA welding where complete solidification between pulses is seeked (Lucas and Males, 1979). In this case the depth of penetration would be measured after each pulse. The welding speed would also have to be pulsed. The low intensity background arc could be momentarily transfered to an auxiliary electrode during the measurement. Another approach would be to rely on the bead width for continuous adaptive control and check only occasionnally the penetration at the beginning of the weld and after each unexpected event.

| Identification | Experimental parameters | | | | | | Computed | | analytical model |
|---|---|---|---|---|---|---|---|---|---|
| | I (A) | $t_f$(s) | D(cm) | W(cm) | W/D | gradient (°C/cm) | D(cm) | gradient (°C/cm) | D(cm) |
| P3 | 125 | 0.34 | 0.21 | 0.63 | 3.0 | 2270 | 0.20 | 2650 | 0.17 |
| P3 | 150 | 0.38 | 0.21 | 0.67 | 3.2 | 2125 | 0.21 | 2650 | 0.19 |
| P3 | 175 | 0.50 | 0.24 | 0.76 | 3.2 | 1910 | 0.25 | 2600 | 0.25 |
| P4 | 103 | 0.55 | 0.21 | 0.49 | 2.3 | 2060 | 0.20 | 2800 | 0.21 |
| P4 | 125 | 0.98 | 0.27 | 0.76 | 2.8 | 2020 | 0.29 | 2550 | 0.34 |
| P4 | 150 | 1.18 | 0.35 | 0.93 | 2.7 | 2000 | 0.32 | 2500 | 0.40 |
| P5 | 95 | 0.45 | 0.17 | 0.59 | 3.5 | 2100 | 0.17 | 2800 | 0.17 |
| P5 | 110 | 0.45 | 0.17 | 0.75 | 4.4 | 2100 | 0.16 | 2600 | 0.16 |
| P5 | 135 | 0.56 | 0.17 | 0.88 | 5.2 | 2050 | 0.16 | 2450 | 0.19 |
| P5 | 180 | 0.80 | 0.17 | 1.13 | 6.6 | 2000 | 0.18 | 2100 | 0.23 |

TABLE 3: Experimental parameters and comparison between the experimental computed and calculated values of the depth of penetration. For all the experiments: welding speed 1mm/s, arc length 3mm and pure argon shielding gas.

## CONCLUSION

Real-time thermographic inspection during the welding process is a very powerful technique and one of the most promising approaches for automatized process monitoring. In this paper we propose a novel approach to depth penetration monitoring by a time-resolved thermographic inspection method. An analytical as well as numerical model of heat propagation during pool solidification is presented. A strong correlation is obtained between the solidification time and the depth of penetration of the weld pool.

An experimental verification of this method is carried out, and the agreement with the theoretical model is quite satisfactory. A fiber-optic sensor has been developed for surface temperature monitoring without relying on a delicate and relatively expensive infrared camera. Such a new approach for weld penetration monitoring is particularly suitable for on-line control of pulsed GTA welding.

## ACKNOWLEDGEMENT

The authors are grateful to Raynald Simoneau for valuable discussions and suggestions. They also wish to thank A. Di Vincenzo for assistance in the experimental work and J. larouche for the metallographic work.

## REFERENCES

Boughton, P. and G. Rider (1979). Advances in Welding Processes, Welding Institute, London, 203.

Cielo, P. (1983). Analysis of pulsed thermal inspection. 14th Symposium on N.D.E., San Antonio.

Fihey, J.L. and R. Simoneau (1982). Weld penetration variation in GTA welding of some 304L stainless steels. International Conference on Welding Technology for Energy Applications, 139.

Lucas, W. and B.O. Males (1979). Pulsed TIG welding in the fabrication of nuclear components and structures. Welding and Fabrication in the Nuclear Industry, BNES, London, 343.

Paley, Z. and P.D. Hibbert (1975). Computation of temperature in actual weld designs. Welding Journal, 54, 11, 3855.

Quigley, M.B.C. and co-workers (1973). Heat flow to the workpiece from a TIG welding arc. J. Phys. D., 6, 2250.

Rider, G. (1975). Low Light and Thermal Imaging Techniques, IEE Conference.

Rider, G. (1979). On line Measurement of weld pool surface size. Welding and Fabrication in the Nuclear Industry, BNES, London, 351.

Thomas, L.C. (1980). Fundamentals of Heat Transfer, Prentice Hall, 158.

Westby, O. (1968). Temperature distribution in the workpiece by welding. Technical University of Norway.

# MICROPROCESSOR TECHNOLOGY IN AUTOMATIC INTEGRATED WELDING SYSTEMS

C. Wilson
*Irco Industries Limited,*
*Brantford, Ontario, Canada*

ABSTRACT

A machine is described which illustrates the successful use of microprocessor-based technology in welding. Designed specifically for weld build-up of locomotive main bearing supports, the machine achieved a productivity gain of 500% over the previously used manual process. A Teach/Weld system was developed which allowed direct-entry of weld control data, thereby eliminating the need for operator programming and increasing the flexibility of the control system. The paper also examines the underlying principles involved and goes on to suggest further applications of the system. Maintenance operations requiring weld build-up and hard surfacing processes are thought to be suitable areas for this technology, where its special advantages would have the greatest impact on productivity.

KEYWORDS

Microprocessor-based technology; teach/weld system; teaching head; direct entry of weld control data; digital image; programmable controller; hard facing; locomotive engine; welding productivity.

INTRODUCTION

<u>Opening Remarks.</u> "Necessity is the Mother of Invention." This sixteenth-century proverb most accurately characterizes the circumstances under which the control system described here was developed. It did not come about as the end-product of a planning process, nor was it in any sense "discovered" during the course of research. Instead, it emerged as the only way to solve a particular weld control problem, and was used accordingly. At the time, its development was entirely goal-oriented, and only after the technique had been successfully demonstrated did its full potential become apparent. This paper therefore seeks not only to give an account of the system as developed, but also to examine its potential for further application in connection with welding.

Background. The useful lifespan of a diesel locomotive is in the order of 25 years, and apart from the downtime required for maintenance, much of this represents continuous duty. Major overhauls are made every 5 or 6 years. One of the operations concerned is the checking for, and possible rectification of engine block sag, which develops under the sustained conditions of heat, stress and vibration. Traditionally, the rectification involved a manual weld build-up process of the main bearing support surfaces, followed by a machining operation which re-established the true centre-line. Using simple weld positioning equipment, the manual welding process alone required about 10 hrs.

As part of a modernization program, a major U.S. locomotive repair shop gave Irco an order to design and build an automatic machine specifically for this welding operation. The proposal called for a head and tailstock arrangement to support the engine block, and a sidebeam and carriage to move the welding torch over the entire length of the engine. Stopping and starting of the welding arc, it was assumed, could be controlled through the use of cams and limit switches located on the sidebeam. However, with mechanical construction virtually complete, a problem was found with the engine main bearing dimensions: The spacing varied, not only between engines of different sizes, but also between engines of the same size. In other words, the weld control data was unique to each engine block. Consequently, the proposed method of weld control could not have worked. Exactly why this discovery was made at such a late stage need not be discussed here, but it is worthwhile to note that this problem became the starting point for the development of the Teach/Weld technology, which is now the main subject of this paper.

## MECHANICAL CONFIGURATION

Machine Components. Major mechanical items were all based on standard weld positioning equipment. The engine block was supported by the head and tailstock in such a way that it could be rotated about its crankshaft axis. The tailstock was truck-mounted to accommodate different sizes of engine. (See Fig. 1)

The sidebeam and carriage assembly supported the welding torch and wire feed unit, allowing the torch to travel the length of the engine, and on into the "parking zone." A pneumatically operated vertical slide provided torch travel of about 30 inches, which was necessary to clear the tailstock. The carriage was kept in the parking zone during loading and unloading of engine blocks.

Teaching Head and Welding Torch. Both had a common mounting bracket so that the teaching head sensed the same point that the torch would be welding if it were in position. The torch was offset at about $15^{o}$ to the vertical to facilitate access to the complete bearing surface. It could be pivoted about the vertical axis to give two positions, corresponding to the two directions of rotation of the engine block. (See Fig. 2 )

The mounting bracket was itself attached to a mechanical oscillator which superimposed a lateral oscillation on the longitudinal motion of the carriage. The amplitude was adjustable, and the net result was a fine "zig-zag" weld deposition path. During a teaching run, the oscillator was inoperative, being stopped in the centre position.

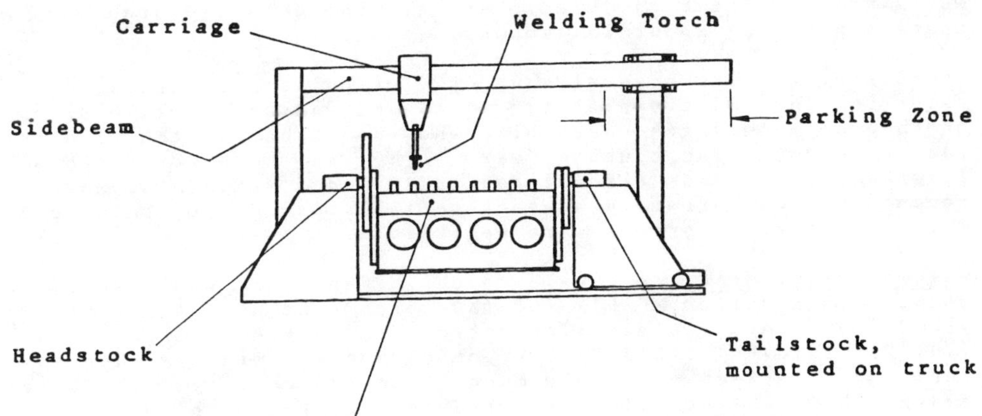


Fig. 1. Automatic Engine Welding Machine showing mechanical details. Locomotive engine block is inverted for welding operation.

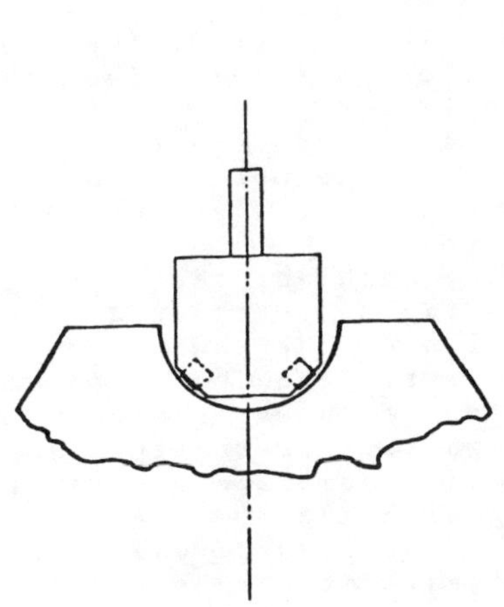

Fig. 2a. Teaching Head. Engine in 0° position.

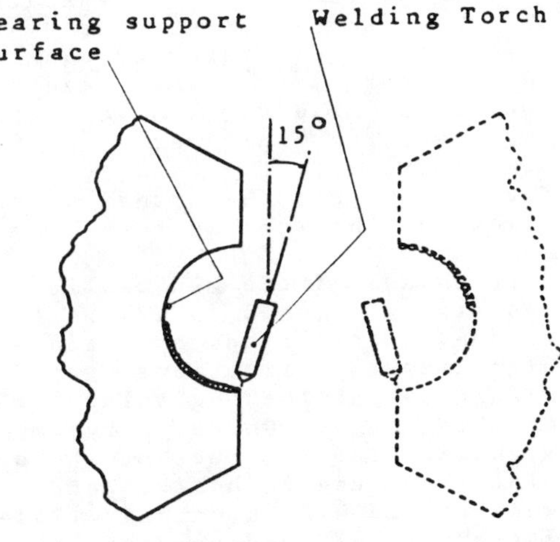


Fig. 2b Torch Mounting. Engine in 90° position.

Illustration shows first quadrant in fully welded condition.

Changing the teaching head over to the welding torch was a manual operation taking about 20 seconds.

Encoders. Two incremental encoders, or pulse generators, were used to determine the positions of the torch and engine block. One was mounted on the carriage drive assembly and the other on the main spindle of the headstock. Resolution was 1/16" in both directions, (i.e. longitudinally and circumferentially) and since the motion of the torch was always from headstock to tailstock during both teaching and welding, backlash errors were negligible.

Welding Equipment. A single-torch Mig process was used. The power source was a Miller MP 65E and the torch a Bernard #1332 Model C. The wire feed unit was a Miller-matic, and National Standard NS101 wire (.035") was used with a 95/5 argon/oxygen gas mixture. The torch, wire reel and wire feed unit were mounted on the carriage, and power source and gas bottles were floor-mounted next to the machine.

## CONTROL SYSTEM

General Principles. Since all welding processes involve relative motion as well as arc control, it follows that the data required to control the process must have a spatial (or positional) component as well as the welding arc parameters. By keeping track of the accumulated pulse-count on both axes, the position of the torch relative to the engine block can be determined. It amounts to a coordinate system defining the two-dimensional space containing the bearing support surfaces, which in this case would have been in the shape of a trough. The principle behind the Teach/Weld system may be better understood if one vizualizes the coordinate system as a "map" of the welding area, onto which is written the welding parameter information. The teaching run is where the information is picked up by the system and transferred to the map. Then, during the welding operation, the torch moves sytematically over the weld area, reading the welding parameter information off the map as it goes. Significantly, the weld control data is acquired directly during a "dry run" of the actual welding cycle, and is more of a "digital image" of the welding task than a programmed set of instructions.

In the case of the Automatic Engine Welding Machine, the welding task was simply to weld on all parts of the bearing support surface and not to weld in the spaces. Since the location of the surfaces varied significantly from one engine to the next, each one had to be "learned" at the beginning of the cycle. This was achieved by the teaching head which, by means of magnetic proximity switches, signalled the presence or absence of a welding surface as it passed along the engine block. Each time the signal changed, the pulse-count was stored in a memory table. Two tables were used: one for "arc start" values and one for "arc stop" values. At the beginning of a welding run the first "arc start" value was retrieved from memory and compared with the running total of the pulse-count. When equality occurred, the arc was automatically switched on and the first "arc stop" value retrieved from memory. Again, this was compared with the running total of the pulse-count, and when equality occurred, the arc was switched off, and the second "arc start" value retrieved. By selecting successive alternate start and stop values, and comparing them with the running total of the pulse-count, arc control was achieved for the entire length of the block.

The flow chart in Fig. 3 shows the logical basis of the control program to do the teaching run. "N" is the pulse-count, and the flow chart loop would be executed each time a pulse was received from the encoder. The "status bit" corresponds to the teaching head signal in a teaching run, and would control the arc-start relay when welding.

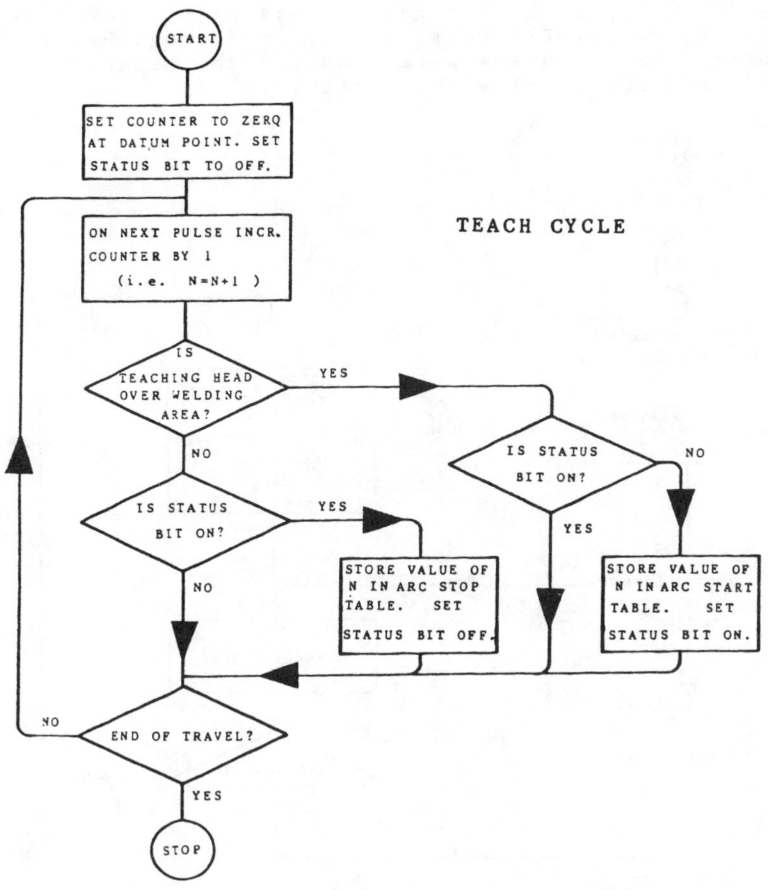


Fig. 3. Teach cycle flow chart.

Both the teaching and welding runs should be regarded as sub-routines in the overall machine control program. The flow chart in Fig. 4 illustrates this and shows the other major parts of the complete program. For example, the "Initial Positioning" sub-routine is concerned with getting the torch (or teaching head) to the beginning of its run. Note that the program would execute the loop containing the welding sub-routine several times, corresponding to the number of passes required to weld up one quadrant.

Hardware. An Allen-Bradley PLC/2 programmable controller was used at the heart of the system. Utilizing the Z80-A microprocessor (running at 4.0 MHz) it had 1K of user memory. Eight operator-initiated inputs, 21 machine-initiated inputs and 9 outputs connected it with the real world. The two encoder signals required fast-response DC input modules and the two teaching head signals required regular DC input modules. All other inputs and outputs were 110 V AC. Extensive use of the Allen-Bradley I/O terminal was made together with the tape storage facility. BEI encoders were used and the proximity switches in the Teaching Head were supplied by Allen-Bradley.

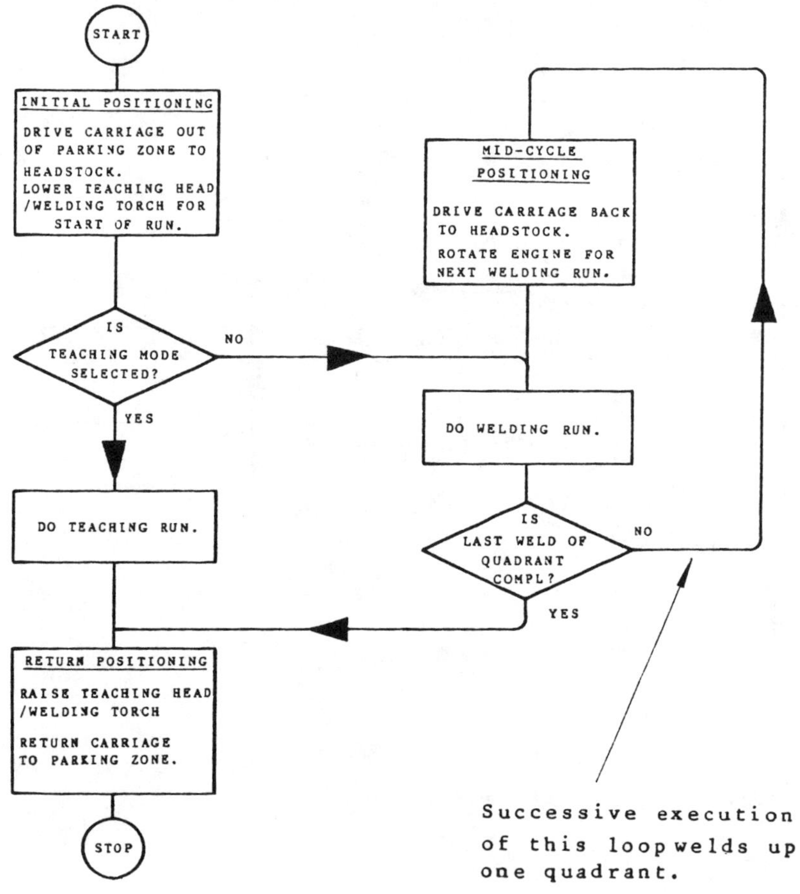


Fig. 4. Overall machine cycle flow chart.

<u>Control Program Details.</u> The entire program had to be written in terms of relay-logic ladder diagrams, which is the "language" of many programmable controllers. For the Teach and Weld sub-routines, a "look-up table" structure had to be devized, and writing and reading the table values was achieved through use of the "PUT" and "GET" statements. In its final form, about two-thirds of the memory was taken up with data handling and the two memory tables, with the balance being used for general machine control. The system could accommodate up to 10 bearing surfaces, corresponding to a 16 cylinder engine.

Some other features were included, one of them being an automatic oil-hole skip routine, which interrupted the weld when an oil hole was encountered. Another one was the automatic reversion to the manual mode whenever the emergency stop button was pushed. This was for safety reasons and the operator would have to drive the carriage back to the parking zone before the automatic mode could be re-activated. Also, welding could be interrupted and then resumed without the need for re-teaching.

A cardboard model, capable of all the movements of the real machine was made for testing and debugging the program. This proved invaluable in reducing the amount of real-time testing. Thorough documentation was prepared for the customer, which together with in-depth training sessions at the time of acceptance trials, resulted in a smooth commissioning period at the customer's plant.

## OPERATION AND PERFORMANCE

<u>Operating Cycle.</u> With bearing caps removed, the engine block was inverted and loaded into the machine. The operator would select the "teach" mode, install the teaching head and initiate the cycle. Automatically, the carriage would drive out of the parking zone, along to the headstock end, lower the teaching head and start the teaching run. Within two minutes, the carriage would be back in the parking zone, having completed the teaching run. Next, the operator would remove the teaching head and insert the torch in either the "A" or "B" position. Restarting the cycle would cause the carriage to drive to the headstock end, lower the torch and start the first welding run. On completion, the carriage would return to the headstock end, the engine block would rotate a pre-programmed amount and welding would resume. After the first quadrant was welded, the carriage would return to the parking zone where the operator would swing the torch round to the other position. Restarting the cycle would initiate welding for the other quadrant. The entire cycle was automatic except for the installation and setting of the teaching head and welding torch.

Through use of the Automatic Engine Welding Machine, the cycle time was cut to less than two hours. Weld quality, in the form of consistency, was improved, and the more even distribution of heat (resulting from welding all the bearing surfaces a bit at a time) helped minimize distortion. A productivity gain of at least 500% was achieved.

The customer subsequently developed a special jig to hold the bearing caps and these were welded up in exactly the same manner. Figures 5 & 6 show a comparison between the manually welded surface and one welded on the Automatic Engine Welding Machine.

Fig. 5. Manually welded bearing surface.

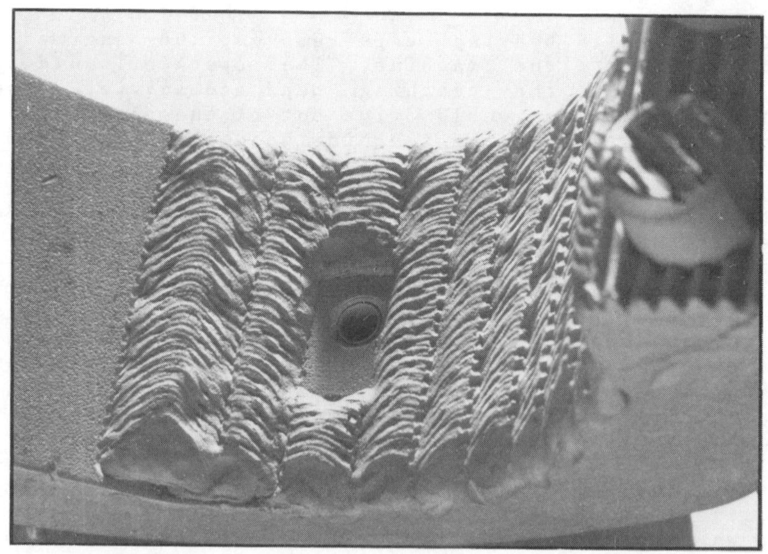

Fig. 6 Bearing surface welded on Automatic Engine Welding Machine. The oil-hole "skip" was exaggerated on this test piece to reveal the oil-hole clearly.

## DISCUSSION

Method of Data Storage. It will be recalled that the term "digital image" was used to describe the form and content of the data required to control the welding process. What distinguishes it, in the author's opinion, from a conventional program is that its "sense" does not depend on the order in which it is written or read. It is a self-sufficient body of information, so to speak. This has two practical advantages: Firstly, it allows considerable flexibility for changing the sequence of a welding task without having to change the weld control data itself. Secondly, the weld control data is stored in a form much more akin to the way a person tends to visualize it, and this would facilitate the operator's understanding of how the system works.

On the other hand, such an approach is inherently memory-intensive, and there is the risk of using valuable memory capacity to store redundant information. Therefore, the most promising applications are likely to be those where the weld control data varies from job to job, or at least from one small batch to another. The increased cost of a large memory capacity would be more than offset by the benefits of flexibility and adaptability.

Method of Data Input. In the teaching run, the machine goes through the motions of the welding cycle while the weld control data is fed into memory. For the machine described above, the data was acquired directly by means of a sensor. This is the simplest method, but it relies on there being some easily detected physical feature which corresponds with the area to be welded. In many applications, this is unlikely to be the case, and operator-initiated input is the practical alternative. One possible configuration would involve a simple pointer in place of the teaching head. The operator would have a push-button control with which he would signal when the pointer was over an area to be welded. The teaching run would still be similar to that with the teaching head, except the operator would initiate the signal instead of the teaching head. Also, data input could be extended to include welding parameters such as welding speed, wire feed speed, current setting etc. The operator would simply select pre-programmed sets of parameters from a memory table and assign them to the appropriate areas as the pointer passed over them. Thus a fairly complex digital image could be built up with this direct-entry method without requiring any operator programming skills.

Further Aplications. An example of where this Teach/Weld technology might well be used is the maintenance of crusher hammers. Restoration to the original profile through weld build-up is followed by hard surfacing. Since crusher hammers are of similar shape but may come in many different sizes, the weld control data is likely to be different for each batch.

Hammers would be placed in a simple jig capable of rotating them about their mounting-hole centres. (Figure 7 illustrates this.) The welding torch would be carried by a sidebeam and carriage arrangement similar to one described above. It would require the additional feature of controlled vertical motion of the welding torch. Data acquired in the teaching run would establish the location of each hammer and also contain a measure of wear for each one, assigned by the operator. Perhaps a wear classification system would be used, with the most worn hammers being class 5, for example.

The welding run would begin with building up only those hammers of a class 5 wear value, until they were equivelent to class 4. Then, all the class 4's would be built up to class 3's, and so on until all hammers were restored to true profile. The final welding runs would be with the hard facing electrode.

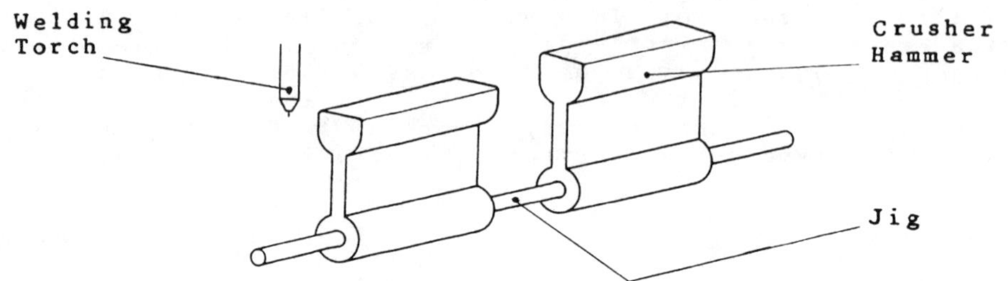


Fig. 7. Rebuilding and hard surfacing of crusher hammers.

Advantages of such a machine would be the uniformity of welds, minimization of heat build-up, low unit labour content (most of the cycle would be entirely automatic), and ease of adaption to different sizes of hammer. Also, no special programming would be required by the operators.

## CONCLUSIONS

The Automatic Engine Welding Machine described in this paper has demonstrated the successful use of a microprocessor-based Teach/Weld system. Utilizing direct-entry techniques for weld control data and combined with a "digital image" approach to data storage, the system offers the advantage of increased flexibility while eliminating the need for programming by the operator. It is suggested that a more sophisticated system based on the same principles could find application in manufacturing and maintenance operations where weld build-up and hard facing processes are employed. Although inherently memory-intensive, the system's special advantages would nevertheless increase productivity where job to job dimensional variations would normally demand reprogramming on conventional systems.

## REFERENCES

Am. Soc. for Metals, (1979). Metals Handbook 8/e, Vol. 6, 152-166.

## ACKNOWLEDGEMENT

The author wishes to express his gratitude to Radio Shack Division of Tandy Electronics Limited for asssistance in the final preparation of this paper.

# PROTECTION SYSTEM AGAINST ELECTRICAL SHOCK HAZARD FOR SMAW WELDERS

J.M. Pelletier* and R. Simoneau*

*Institut de Recherche d'Hydro-Québec
Varennes, Québec, Canada

ABSTRACT

Following extensive work done at IREQ, an automatic electronic system has been designed to protect welders working in damp environments; the system controls the open circuit voltage at the electrode and insures a safe condition in all welding conditions. With the implementation of a failsafe scheme, it has proven to be reliable and no field malfunction have been noticed at this time.

KEYWORDS

SMAW Welding; shock protection system; automatic protection system; damp environments; failsafe system.

## INTRODUCTION

Due to the nature of the welding process involved in SMAW welding as well as the unavailability of proper protection devices, it is one of the few conditions where safety standards permits a high voltage source ($\sim$80V) with high power capacity (100 to 300A) to be left unprotected and at the welder's reach (the electrode tip); this high voltage is necessary to insure reliable arc starting and reduced voltage sources exhibit poor performance. Although electrical shocks from welding machines do not oftenly cause serious injuries, they are common to welders, unpleasant and can cause secondary injuries in some cases.

The open circuit voltage delivered by the welding supply is usually in the range from 70V to 100V; as the welding is initiated, it drops to 15-30V. To alleviate the hazard associated with high open circuit voltage, various types of contactors, manual and semi-automatic, have been investigated; the purpose of this contactor is to disconnect the welding electrode from the power source, reclosing the circuit when manually triggered by the welder. Usually mechanical devices suffer from poor reliability with slow response time and it explains the limited use of these devices.

With the advent of power transistors and SCR'S, all solid state switches, very reliable and fast, are now available, which do offer a better solution; their price is accessible and, properly rated, can have a lifetime much greater than a contactor.

## OBJECTIVES

Devices commercially available suffered from major drawbacks, namely

- poor reliability of the power contactor
- manual or semi-automatic operation which involved welder's assistance

The objectives of the project were to design a system overcoming those problems; being automatic, welders could not bypass it and it would act as if "nothing" had been added between the power source (welding machine) and the electrode. The overall cost should be acceptable and the system should be failsafe, if possible.

## STUDY OF DYNAMIC SKIN IMPEDANCE

The overall concept of the system is to add a "normally open switch" between the power source and the electrode and to design a circuit that automatically closes the circuit. As the welder touches the plate with the electrode, the circuit senses this condition and closes the switch; as he resumes his welding, the switch reopens the circuit. Should the welder touch his body, the circuit senses this condition and the switch remains open.

Dynamic electrode impedance was found to be the most usefull variable to establish a significant difference between body impedance and the plate impedance; other sensing variables did not yield good results. As a result, a preliminary literature survey was done.

Current flow is the factor that causes injury; that is, the severity of electrical shock is determined by the amount of current flowing in the body (Ref. 1, 2). The fatal current level is around 500 mA, the lost of control level around 76 mA and the perception threshold around 5 mA in DC: we should therefore limit the current to a level below the perception threshold.

Knowledge of body impedance is essential to establish triggering levels; from reference 1 and from our tests, the following results were compiled:

| | |
|---|---|
| -- dry skin | 100,000 to 600,000 ohms |
| -- wet skin | 1,000 ohms |
| -- internal body (hand to foot) | 400 to 600 ohms |
| -- ear to ear | 100 ohms |

The minimum resistance of human body with wet skin is 2,000 ohms (2 skins); our triggering point should be set accordingly.

## DESIGN AND OPERATION OF THE CONTROL CIRCUIT

From the measurements and literature survey made on skin impedance, we know that it never falls below 2,000 ohms; we also know that current flowing in the body should be limited to less than 5 mA.

The block diagram of the complete system is as follows (for DC applications):

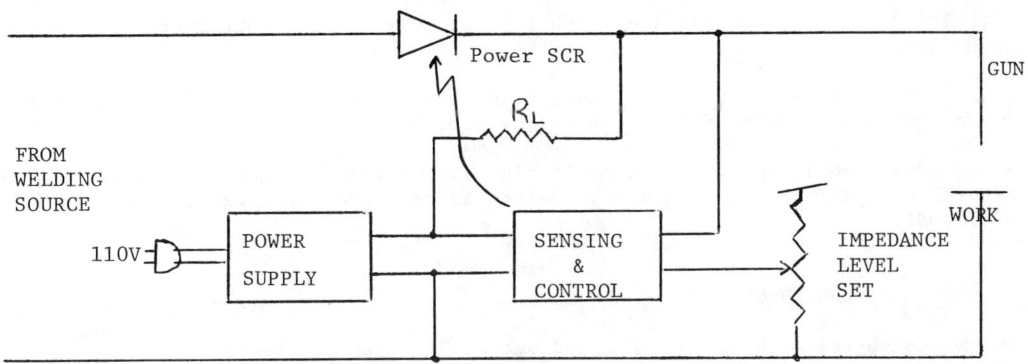


Fig. 1: BLOCK DIAGRAM OF THE SYSTEM

A power SCR (normally open switch) is placed in series between the welding supply and the electrode; it is normally open and this isolates the electrode (GUN) from the source. A small DC voltage derived from the power supply is applied to the electrode and the current level limited by the value of resistance $R_L$: this voltage is used to measure the dynamic impedance between the electrode and the work. The sensing and control circuit measures this impedance and compares it to a pre-adjusted value: if it is lower, the power SCR is triggered and full power applied to the gun; as the welder resumes its operation, this power SCR re-establishes its non-conducting state. If the welder touches any part of his body, the power SCR remains off, thus preventing the welder from a possible shock. Adding another power SCR parallel to the first SCR (but in opposite direction) extends the range of usefullness to AC welding sources.

## FIELD TESTS

Three prototypes were mounted and experienced field testing for 3-4 months period; they performed satisfactorily in situations where welders were subjected to damp environments, namely in shipyards and construction sites.

From these tests, some additional capabilities were added (such as "pulse stretchers", etc...) to improve welding startup with difficult electrodes (7018).

One of the limitations of the present system found from experiments is its inability to protect welders if high frequency is added to the output of the welding source (one option provided to welders to initiate an arc if the working plate is heavily rusted); no action can be taken to overcome this problem since high frequency "pass" through the main switch (power SCR) and reaches the electrode. Should manufacturers of power SCR provide us with SCR having a better behaviour in presence of high voltage high frequency spikes, this problem will be overcome.

## FAILSAFE SCHEME

One of the major requirements of the system is to be failsafe, i.e. if some components goes bad, the system must be designed to be safe to the welder.

One of the limited ways to achieve this goal is to build a system where each and every of the "initial" components is duplicated, both components being requested to agree. Therefore, assuming an active circuit performs a function, another circuit is connected in parallel with the main circuit and both must agree on the same function.

This feature has been fully implemented in the latest versions of the system and has enhanced the security level of the system.

As an additionnal precaution, a monitoring circuit, independant of the sensing and control circuit, triggers an audible alarm if the output voltage reaches a dangerous level.

## CONCLUSION

An extensive research project carried out at IREQ has led to the development of an automatic electronic system for the protection of welders working in damp environments; field tests of the system confirmed the positive results expected.

Although this system has limitations when high frequency is added to the welding source, its range of usefullness is acceptable and safety regulations should emphasize the use of this system.

## REFERENCES

1. Accident Prevention Manual for Industrial Operation, National Safety Council, 7th edition, Chicago 1974, Chapter 41, Electrical Hazards, P. 1255.

2. Dalziel, C. F. (1941). Effect of Electric Current on Man, Electrical Engineering. (1947). Scientific Facts concerning Electrical Hazards, National Safety News. (1961). Electricity-Good and Faithfull Servant, National Safety News.

# FOUR ENERGY-RELATED CASE STORIES OF USING STAINLESS WELDING CONSUMABLES

Martin Crowther
*Sandvik AB, Sandviken, Sweden*

ABSTRACT

Four energy-related application areas have been selected where stainless welding consumables play an important role.

Armoured flexible conveyors (AFC's) built up of line pan units are used in modern underground longwall mines. The line pan is constructed from carbon steel plates and Hadfield steel castings. This dissimilar metal joint has been traditionally welded using AWS 316 covered electrodes. Major manufacturers have now switched to MIG welding using an 18Cr-8Ni-7Mn consumable. Weld mechanical properties have improved, hot cracking of the weld deposits has been eliminated and the overall welding economy has been vastly improved.

Carbon steel flowlines in corrosive oil and gas wells are successively being replaced by stainles steel systems. Conventional austenitic stainless steels cannot be used due to their susceptibility to chloride induced stress corrosion cracking. A special duplex stainless steel, SAF 2205 (22Cr-6Ni-3Mo), is used with success. The parent metal properties are suited to this application and the weldability is as good as for austenitic stainless steels. High strength and good corrosion properties are attainable in the welds using matching welding consumables.

The British Advanced Gas-cooled Reactors (AGR) have a concrete pressure vessel containing both the reactor core and the four main steam boilers. The steam penetrations through the vessel wall are of modified ASTM 316 austenitic stainless steel. Fabrication is carried out by use of narrow-gap TIG welding and orbital TIG welding. The fully automatic narrow-gap welding process described here has to date given defect-free welds at 100% X-ray inspection on 144 units.

Hydroelectric power is usually harnessed by the use of Kaplan or Francis turbines. Kaplan turbine blades and wicket gates are surfaced with 17% Cr steel strip electrodes and a special flux to give austenitic-martensitic-ferritic deposit with high resistance to cavitation and erosion corrosion. The wicket gate support rings for Francis turbines are also surfaced with 17% Cr steel strip electrodes, in order to give a surface that will not "stick" to the wicket gates, which in this case were surfaced with austenitic stainless steels.

Spent fuel from nuclear reactors is transported both on site and from the nuclear facility to treatment plants in special fuel transport flasks. These thick-walled vessels weighting up to 76 tons are weld overlayed internally and externally with austenitic stainless steel deposits of the 308L or 347 type. Submerged-arc strip surfacing is the preferred method of overlaying the vessels due to the high productivity and high-quality deposits that can be obtained.

KEYWORDS

Armoured flexible conveyors, flowlines, AGR penetrations, hydroturbines, nuclear waste flasks, MIG-welding, TIG-welding, narrow-gap TIG-welding, submerged-arc strip surfacing, stainless steels, weldability, weld metal properties.

## INTRODUCTION

Five application areas have been selected where stainless welding consumables play an important role in the construction and function of equipment. Two examples refer to power generation and the other three are components for material transportation. They all represent areas throughout the whole power generation cycle, from the extraction of fuels to the disposal of fuel wastes. Welding procedures and properties of deposits are presented.

## ARMOURED FLEXIBLE CONVEYORS (AFC) FOR COAL MINING

The main competitor to oil, both as a fuel and as a raw material for the production of a wide range of chemicals, is, of course, coal. In order to make coal extraction more cost effective, new methods are being continually developed. Coal extraction is divided into two fields - opencast mining, and underground mining.

In the latter case an extraction method gaining rapid acceptance worldwide is long-wall mining. This method utilizes power loaders and armoured flexible conveyors (AFC) which involves a high degree of mechanization. An AFC is as long as the coal-face on which it is installed, usually between 160 and 200 metres.

The AFC consists of a series of heavy-duty steel trays, called line pans (Figure 2), which are flexibly connected together. These line pans are connected at one end to a drive unit and at the other end to a return unit. Coal is transported to the drive end by a series of scraper bars which are connected together by chains. The chains are normally attached to the ends of the scraper bars and lie in channels at the sides of the line pans.

The chains pass along the length of the conveyor, then return along the underside thus forming two continous loops. The chains are driven by powered sprockets mounted in the drive unit pulling the chains, scraper bars and their load towards the delivery end. The sprockets are powered by one or more electric motors of up to 250 HP.

On the face side of the conveyor there are ramp plates to scrape the floor clean, and on the goaf side a spill plate is attached to prevent coal from falling off the conveyor. Hydraulic rams which are used to push the conveyor forward are connected between the roof support and the ramming bars on the spill plates. The power loader, which extracts the coal from the face, rides on top of the AFC. It is propelled either by a rack-and-pinion system or by chains. Typical output from a 200 metre face is 1500-2000 tonnes per shift.

Fig. 1 Assembled AFC with power loader and hydraulic roof supports

Fig. 2 Line pan and ancillary attachments. A = deck plate, B = side section, C = connection piece

The AFC is a strong, heavy-duty construction. It must carry not only the static weight of the power loader and coal, but it is also subjected to great stresses both as the loader travels along the AFC and when the AFC is pushed into the coal face.

An AFC system generally has a life span of about one year, i.e. the time it takes to work out a coal face. At the end of this period, the drive and return units are dismantled and the roof supports removed. The line pans are left in the disused face. Consequently, about 200 line pans/per coal face/per year are used. In the UK, for example, this means that about 80 tonnes of welding consumables per year are used to weld the line pan components together. Although the service life is so short, it is imperative that no failures occur in any of the components as it is a difficult and costly job to replace line pans. The properties thus required are typical of a "fit for purpose" situation.

The line pans are usually constructed from two or three materials. In the majority of cases the deck plate (A) must have a good wear resistance and is of a relatively high carbon steel or a low-alloyed steel of about 20-25 mm thickness. The sides are hot-rolled sigma-shaped sections (B) generally of a similar material.

The location and connection pieces (C) are subjected to heavy impact when the AFC is being repositioned on the coal face and when the power loader traverses the AFC. The end connector has a complicated shape which is produced by casting. The material most extensively used is Hadfield steel (1.2% C, 13% Mn). This is a fully austenitic steel which work hardens rapidly by impact.

For many years an AWS ER316 (18Cr/12Ni/2.5Mo) welding consumable has been used to join the Hadfield steel end castings to the low-alloy and carbon steel sigma profiles and plates.

In tests run by a leading manufacturer of mining equipment it was found that by the use of Sandvik 18.8.CMn (18Cr/8Ni/7Mn) as filler metal the connector strength was increased from 40 to 100 tonnes without change of section size. The reason is the higher tensile strength and superior hot cracking resistance of 18.8.CMn weld deposits compared with those obtained with the previously employed 316 type of filler metals. An added bonus is that 18.8.CMn has a lower price than AWS ER316 filler metal. Depending on the size of the line pan between 1 and 2 kg of welding consumables are used per pan.

In the interest of production efficiency a single J-preparation is used. This is the simplest form of preparation, as the Hadfield steel is cast, and no machining of the groove faces is necessary.

Naturally, the welding process has to give a high deposition rate, ease of operation and sound weld deposits. Most fabricators have switched from MMA welding to MIG welding using either a 1.2 or 1.6 mm wire electrode with a shielding gas of Ar + 5% $CO_2$. This shielding gas gives an optimum balance between weldability, bead profile, weld properties and cost.

Typical mechanical properties for weld metal in the as-welded condition are

| | |
|---|---|
| Yield strength, $R_{p0.2}$, N/mm$^2$ | 380 |
| Tensile strength, $R_m$, N/mm$^2$ | 670 |
| Elongation, A5, % | 48 |
| Impact Strength, Charpy V, R.T., J | 140 |
| Hardness, HB | 220 |
| Hardness, HRC | 15 |

The weld deposit rapidly work hardens upon impact to about 40 HRC. The ductility of the weld deposit in conjunction with freedom from defects is also of prime importance. Samples of production welds between the Hadfield steel end castings and the carbon steel sigma profile have been taken and have passed face, root and side bend tests (Figure 3).

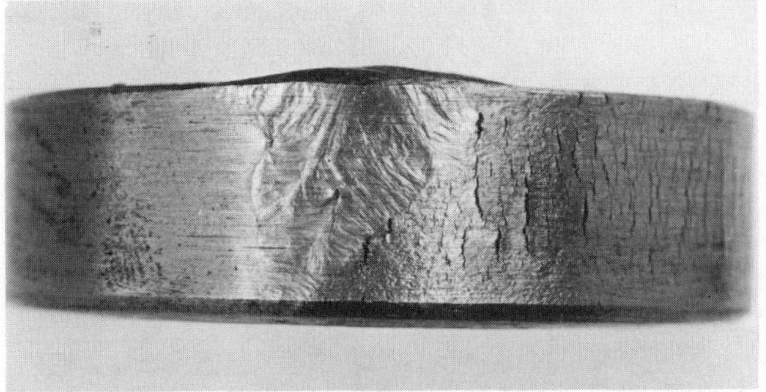

Fig. 3 Side bend specimen

The chemistry and microstructure of the weld deposit are important in determining its properties. The weld deposit obtained by using an 18Cr/8Ni/7Mn welding consumable between the two line pan components is fully austenitic with a nominal composition of 0.2 C, 0.4 Si, 6 Mn, 15 Cr, 7 Ni. Austenitic weld deposits are often considered as being extremely sensitive to hot cracking, and this was a great problem with the welds produced with ER 316 fillers. 18Cr/8Ni/7Mn filler metals, due to their high manganese content, have been found to have a superior hot cracking resistance.

It can be concluded that the 18.8 CMn filler has successfully replaced AWS ER316 in Europe as it gives better weld metal properties at lower cost.

## FLOWLINES IN OIL AND GAS PRODUCTION

Many of the oil and gas wells being operated today have a product which is highly corrosive due to the presence of water, hydrogen sulphide and/or carbon dioxide often together with chlorides.

Carbon dioxide dissolved in water produces an acid solution which results in a high corrosion rate for carbon steels, especially at elevated temperature. Hydrogen sulphide dissolved in water also produces an acid solution, which is liable to produce sulphide stress cracking. This form of cracking is regarded as being a type of hydrogen embrittlement and is most prevalent in high-strength steels containing martensite, but it also occurs in ferritic steels. Sulphide stress cracking is most pronounced at room temperature.

Chlorides which also frequently occur in oil and gas wells, usually in the form of sodium chloride, accelerate corrosion in carbon steels. In conventional stainless steels chlorides are responsible for localized attack in the form of pitting, crevice corrosion and stress corrosion cracking.

If enhanced recovery methods, such as hot-water and steam injection, are used to stimulate oil production, the corrosive conditions are further aggravated by both the additives and the high temperatures involved.

Corrosive oil and gas resevoirs are often located at great depth and under high pressure, thus requiring high-strength piping in the wells.

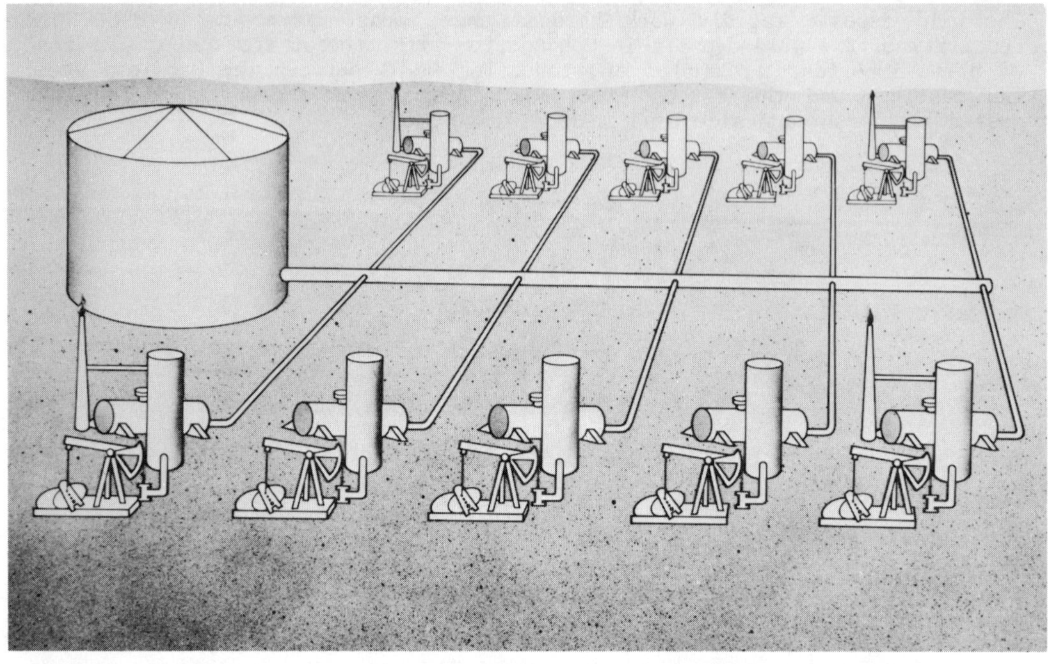

Fig. 4 Oil flowlines of carbon steel

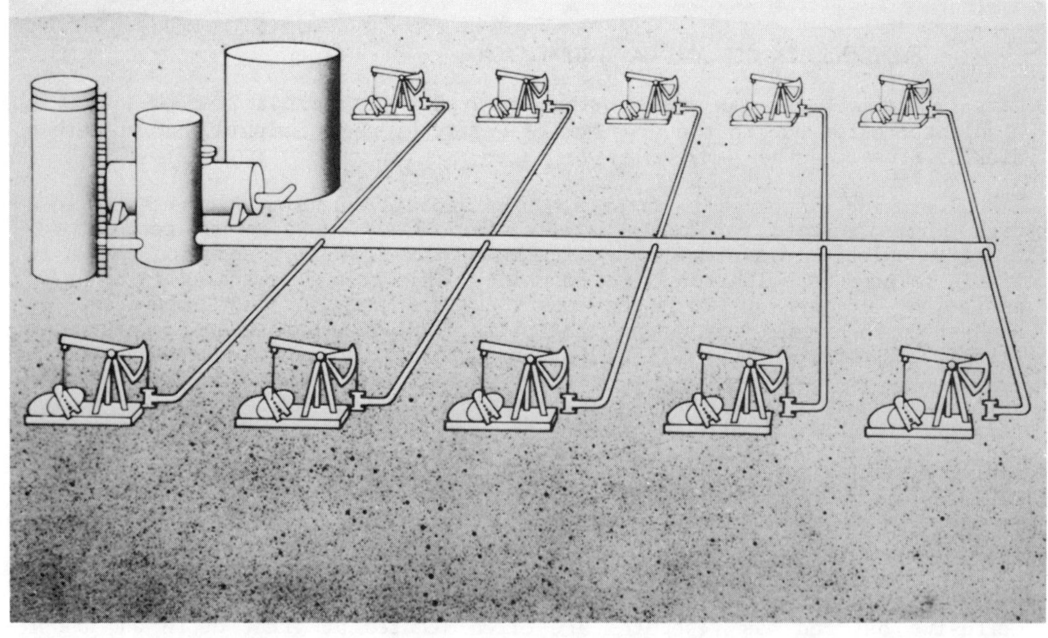

Fig. 5 Oil flowlines of stainless steel SAF 2205. Centralized purification.

One interesting area where special stainless steels have been utilized and where great savings in plant investment have been made is at the wellhead of production wells. Flowlines connecting the wellhead to the pipelines or storage tanks are made of SAF 2205 which is a ferritic-austenitic stainless steel containing 22 Cr, 5.5 Ni and 3% Mo.

This steel has been found to solve corrosion problems where previously the use of inhibitors or organic coatings failed. Standard stainless steel grades, such as AISI 316/316L, cannot be used. Although high in chromium and molybdenum and therefore rather resistant to pitting and crevice corrosion, they are very prone to chloride induced stress corrosion cracking.

Consequently, the purification plants normally found at each well can be replaced by one central unit.

SAF 2205 is a low-carbon, nitrogen stabilized ferritic-austenitic stainless steel. It belongs to the second generation of ferritic-austenitic steels, developed from the early AISI 329 type of steel. It possesses not only very good corrosion resistance and mechanical properties but also very good weldability and good properties after welding. The composition of SAF 2205 is given below together with AISI 316L and AISI 329.

Table 1 Chemical composition, (percentage by weight)

| Grade | C max | Si max | Mn max | P max | S max | Cr | Ni | Mo | N |
|---|---|---|---|---|---|---|---|---|---|
| SAF 2205 | 0.030 | 0.8 | 2.0 | 0.030 | 0.020 | 22 | 5.5 | 3.0 | 0.14 |
| AISI 316L | 0.03 | 1.00 | 2.00 | 0.045 | 0.030 | 17 | 12 | 2.5 | - |
| AISI 329 | 0.20 | 0.75 | 1.00 | 0.040 | 0.030 | 25.5 | 3.8 | 1.5 | - |

The mechanical properties of SAF 2205 are compared in Table 2 with those of 316L. As can be seen, the yield strength of SAF 2205 is more than double that of AISI 316L at a relatively high ductility.

Table 2 Mechanical properties

| Grade | 0.2% yield strength $N/mm^2$ min | 0.2% yield strength ksi min | Elongation A5 % min | Tensile strength $N/mm^2$ min | Tensile strength ksi min | Vickers hardness 30 kg approx. |
|---|---|---|---|---|---|---|
| Quench-annealed | | | | | | |
| 316L | 195 | 28 | 45 | 500 | 73 | 150 |
| SAF 2205 | 450 | 65 | 25 | 680 | 99 | 260 |
| Cold worked | | | | | | |
| SAF 2205 | 900 | 130 | 10 | 965 | 140 | 330 |

The corrosion properties of SAF 2205 can be summarized in the following diagrams.

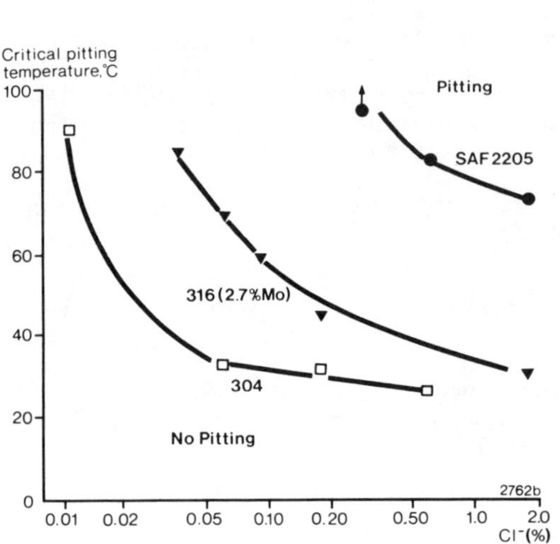


Fig. 6 Critical pitting temperatures for SAF 2205, AISI 304 and 316 at different sodium chloride concentrations. Potential vs SCE = 300 mV; pH = about 6.0. Owing to the low oxygen content in gas and oil wells the potential there is much lower and allows much higher critical pitting temperatures.

Fig. 7 Results of SCC tests in 40% $CaCl_2$ at $100^oC$; pH = 6.5. Time to failure as a function of applied stress, divided by the tensile strength. Quench-annealed specimens.

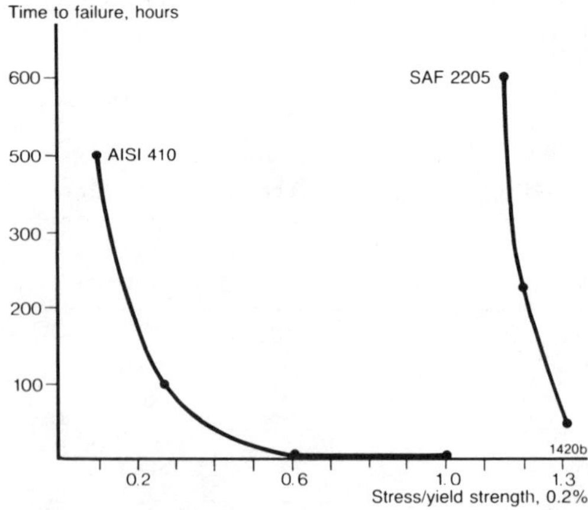


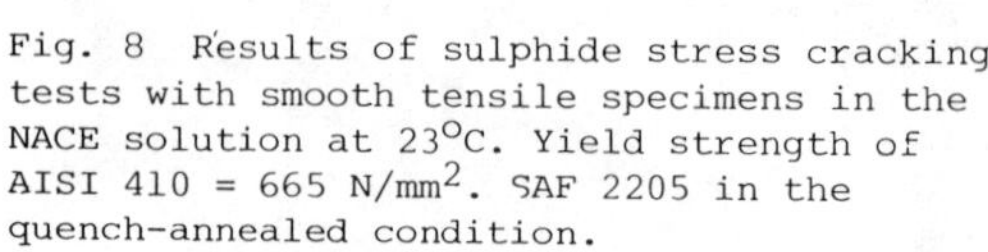
Fig. 8 Results of sulphide stress cracking tests with smooth tensile specimens in the NACE solution at $23^oC$. Yield strength of AISI 410 = 665 $N/mm^2$. SAF 2205 in the quench-annealed condition.

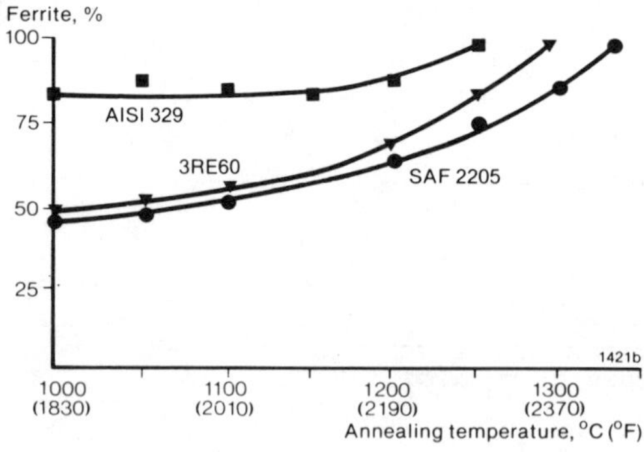


Fig. 9 Effect of temperature on austenite transformation in AISI 329 and SAF 2205. Annealing time 1 h.

As can be seen, SAF 2205 offers many advantages over the other materials commonly used in wellhead applications.

A chain is as strong as its weakest link, and in conventional ferritic-austenitic materials this is often the weld. In many cases a dissimilar filler metal is used and, consequently, some properties of the weld joint can be inferior to those of the parent metal.

When welding AISI 329, the high-temperature heat affected zone (HT-HAZ) becomes almost completely ferritic and coarse-grained. This causes low toughness and sensitivity to intergranular corrosion. Due to its higher nickel equivalent/chromium equivalent ratio, and the content of nitrogen, reformation of austenite occurs readily in the HT-HAZ of SAF 2205, implying good mechanical properties and corrosion resistance.

During the cooling portion of the weld thermal cycle the austenite will recover. The rate of recovery is dependent on the choice of austenitizing elements. Although a slow cooling rate would aid the recovery of austenite it would also, due to the high Cr content, promote sigma-phase. The austenitizing effect in SAF 2205 is achieved by using an element which diffuses rapidly, namely nitrogen. Nitrogen accounts for nearly 40% of the austenitizing effect.

Table 3 Effect of alloying elements on austenite stability.

| Element | Content % | Ni-equivalent[1)] | Austenitizing effect % |
|---|---|---|---|
| Ni | 5.5 | 5.5x1 = 5.5 | 49 |
| N | 0.14 | 0.14x30 = 4.2 | 38 |
| C | 0.02 | 0.02x30 = 0.6 | 13 (C + Mn) |
| Mn | 1.8 | 1.8x0.5 = 0.9 | |
| | | 11.2 | 100 |

[1)] According to De Long

Nitrogen has a diffusion coefficient 2500 times greater than nickel at 1400°C and 600 times greater at 1000°C.

Based on a cooling rate of about 120°C/sec in the temperature range 1400°C-900°C, nitrogen will diffuse about 250 µm whereas nickel will only diffuse about 1.7 µm.

As a consequence of the weld thermal cycle, the grain size of the HAZ of ferritic-austenitic stainless steels is about 100-150 µm, i.e. the maximum diffusion distance to reach a grain boundary is 50-75 µm. Obviously the recovery of austenite is attributed mainly to nitrogen diffusion.

The difference in HAZ structure of AISI 329 and SAF 2205 can be clearly seen in the microphotos shown in Figures 10 and 11.

A dissimilar filler metal is used to identify the fusion line.

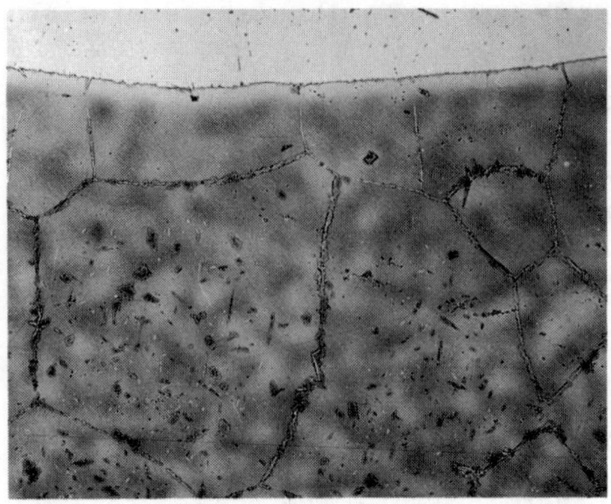

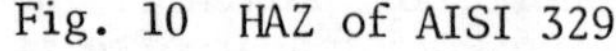

Fig. 10 HAZ of AISI 329 x250

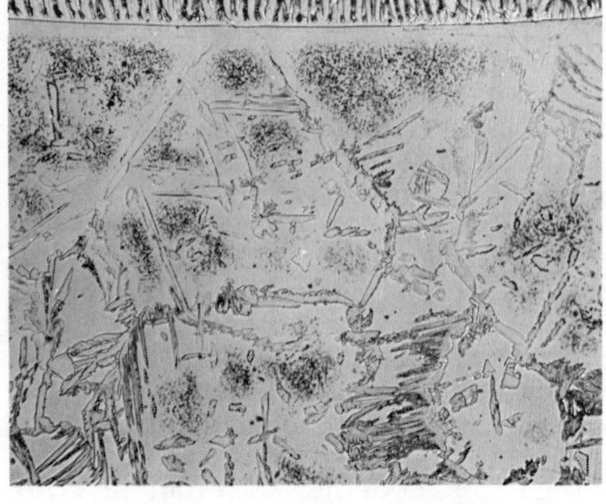

Fig. 11 HAZ of SAF 2205. Austenite is clearly visible in the grain boundaries. x250

Due to the microstructural stability possessed by SAF 2205 it can be welded with matching consumables. Welding tests have been carried out with different thicknesses of parent metals and with autogeneous TIG, TIG with filler and manual metal arc welding. Details of the welding procedures used and the results obtained are indicated in Tables 5 and 6. The chemical compositions of base and filler metals are given in Table 4.

Table 4

Chemical compositions of base metals and fillers used in the welding tests.

| | C | Si | Mn | P | S | Cr | Ni | Mo | N |
|---|---|---|---|---|---|---|---|---|---|
| SAF 2205[1)] melt No 762065 | 0.018 | 0.38 | 1.62 | 0.011 | 0.006 | 22.04 | 5.57 | 3.01 | 0.13 |
| SAF 2205[1)] melt No 745929 | 0.013 | 0.40 | 1.60 | 0.013 | 0.015 | 22.01 | 5.48 | 2.96 | 0.13 |
| 22.6.3.L[1)] melt No 762915 | 0.016 | 0.35 | 1.67 | 0.011 | 0.009 | 22.12 | 5.61 | 3.01 | 0.14 |
| AISI 1518[2)] | 0.18 | 0.3 | 1.4 | - | - | - | - | - | - |
| 22.9.3.LR[2)] | 0.025 | 1.0 | 0.8 | - | - | 22.0 | 9.0 | 3.0 | 0.10 |

1) melt analysis

2) nominal analysis

Table 5 Test welding procedures used for SAF 2205

| Item | | |
|---|---|---|
| I | Type of joint: | Closed square butt joint |
| | Base metal: | SAF 2205 tube O.D. 114.3 x 3 mm from melt No. 762065 |
| | Welding method: | Automatic TIG-welding without filler metal |
| | Welding parameters: | 120-130A/10V/125 mm/min |
| II | Type of joint: | Butt joint: single V (60°) |
| | Base metal: | SAF 2205 tube O.D. 168.3 x 12.7 mm from melt No. 762065 |
| | Filler metal: | Bare wire 22.6.3.L from melt No. 762915<br>Covered electrode 22.9.3.LR, diam. 3.25 mm |
| | a Welding method: | Manual TIG-welding with filler metal |
| | Welding parameters: | Root run: 190A/1.3kJ/mm<br>1st filling run: 150A/1.7kJ/mm<br>2nd " " : 175A/2.2kJ/mm<br>3rd " " : 175A/2.4kJ/mm<br>4th " " : 175A/2.5kJ/mm |
| | b Welding method: | Manual TIG-welding with filler metal for the root run. The remainder of the joint was manual metal arc welded. |
| | Welding parameters: | Root run: 190A/1.3kJ/mm<br>1st filling run: 105A/0.8kJ/mm<br>2nd " " : 105A/1.2kJ/mm<br>3rd " " : 105A/1.2kJ/mm<br>4th " " : 105A/1.4kJ/mm |
| III | Type of joint: | Dissimilar butt joint, single V (60°) |
| | Base metal: | SAF 2205, plate 3 mm thick from melt No. 762065 and AISI 1518 (CMn-steel), 3 mm thick |
| | Filler metal: | Bare wire 22.6.3.L from melt No. 762915 |
| | Welding method: | Manual TIG welding with filler metal |
| | Welding parameters: | Root run: 35A/0.2kJ/mm<br>Filling run: 35A/0.3kJ/mm |
| IV | Type of joint: | Dissimilar butt joint, single V (60°) |
| | Base metal: | SAF 2205, plate 7 mm thick from melt No. 745929 and AISI 1518 (CMn-steel), 7 mm thick |
| | Filler metal: | Covered electrode 22.9.3.LR, diam. 2.5 mm |
| | Welding method: | Manual metal arc welding |
| | Welding parameters: | Root run: 90A/1.1kJ/mm<br>1st filling run: 90A/1.0kJ/mm<br>2nd " " : 90A/1.0kJ/mm |

Table 6 Results of welding trials with SAF 2205.

| Item | Transverse weld tensile strength N/mm$^2$ | Bend test[2] transverse weld | Impact (CVN), weld metal Nm | Hardness (HV3kp) weld metal top/root | Intergranular corrosion test ASTM A262/E | Pitting corrosion test in 2% $CuCl_2$ at 20°C, (1+3+3) x 24 h |
|---|---|---|---|---|---|---|
| I | 716/weld metal[1]<br>761/weld metal | Satisfactory<br>Satisfactory | Not tested | 247-261 | Satisfactory | No pitting |
| IIa | 769/weld metal[1]<br>764/weld metal | Satisfactory | 135,150,168 (0°C)<br>54,64,65 (-20°C) | 254-280/272-284 | Satisfactory | No pitting |
| IIb | 804/weld metal[1]<br>805/weld metal | Satisfactory | 53,57,63 (0°C)<br>48,51,55 (-20°C) | 220-276/272/284 | Satisfactory | No pitting |
| III | 610/CMn-steel[1]<br>600/CMn-steel | Satisfactory | - | 241-272/195-254 | - | - |
| IV | 560/CMn-steel[1]<br>570/CMn-steel | Satisfactory | - | 229-257/- | - | - |
| Base metal SAF 2205 t = 3 mm melt 762065 | 840<br>850 | - | - | 257-265 | - | - |
| Base metal SAF 2205 t = 7 mm melt 745929 | 760<br>770 | - | - | 238-257 | - | - |
| Base metal AISI 1518 t = 3 mm | 570<br>570 | - | - | - | - | - |
| Base metal AISI 1518 t = 7 mm | 570 | - | - | - | - | - |

1) Place of fracture
2) Top and root side of the welds were tested. The radius of the former was 2 x t and the specimens were bent 180°

Pure argon was used as shielding gas for TIG welding. All welding was carried out in the flat position. From the welded joints specimens for mechanical and corrosion testing were taken out. The results of the tests are summarized in table 6. They were all satisfactory and agreed with the requirements on the parent metal.

The greater ductility of an austenitic weld implies that it can tolerate slag inclusions better than a ferritic-austenitic weld. Since TIG welding gives welds virtually free from slag inclusions, a matching filler like Sandvik 22.6.3.L can be used.

With covered electrodes, on the other hand, one has to allow for a certain amount of slag inclusions and, therefore, in order to improve ductility by more austenite in the weld, an electrode with 9% nickel is used. Typical weld deposit structures are shown in Figures 12 and 13.

SAF 2205 shall be welded without preheating and normally PWHT is not necessary. The thermal expansion is lower and the thermal conductivity is higher than for austenitic stainless steels and, consequently, distortion and residual stresses due to welding are lower.

Fig. 12 Weld metal from covered electrodes 22.9.3.LR. Etched in Murakamis etchant. x200

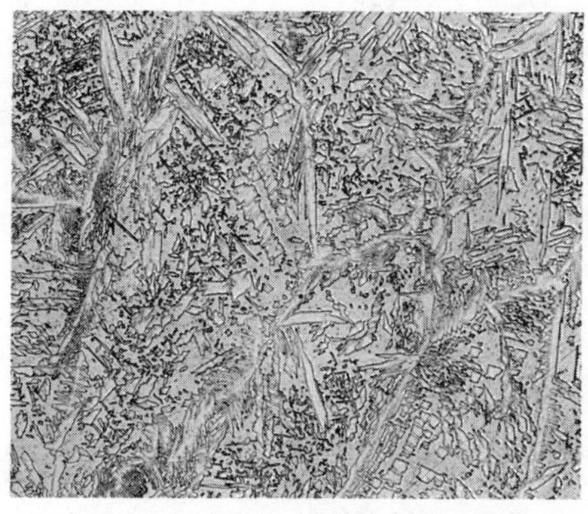

Fig. 13 Weld metal from TIG filler 22.6.3.L. Etched in Murakamis etchant. x200

## AGR PENETRATIONS

Nuclear power in Britain is produced mainly by gas-cooled reactors of either the Magnox or AGR types. The latest twin 660MW AGR's are being constructed at Heysham and Torness, and it is the welding of certain components in these reactors that will be presented here.

Briefly the advanced gas-cooled reactors use slightly enriched U235 in stainless steel canning tubes located in a graphite moderator. The cooling medium $CO_2$ is pumped through the core and conducts away heat at about 600°C to the four boilers located in the boiler annulus between the gas baffle and the pressure vessel in prestressed concrete.

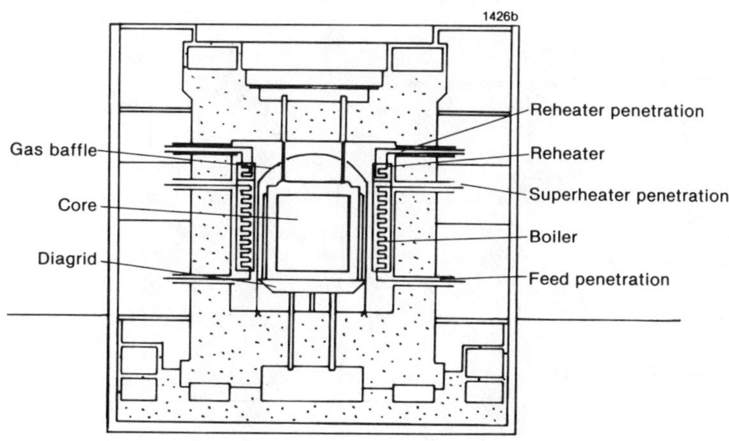


Fig. 14 Cross section of AGR.

The four boilers are separated into quadrants by division plates, and each boiler comprises three separate factory assembled units of rectangular cross section. Each unit is supported from below on beams attached between the pressure vessel and the gas baffle cyclinder. The reheater units are supported from the pressure vessel roof with sliding joints between the main boiler and reheater casing. The boilers are of the once-through type to minimize the number of pressure vessel penetrations.

The design and materials used for the steam and water penetrations differ considerably. The feed water penetrations, of which there are two per unit, consist of carbon steel liners with 1Cr/1.5Mo tubes. Of the superheater and reheater outlet penetrations there are three and two per unit respectively. They are constructed from 316 type austenitic stainless steel (both the liners and tubes). Narrow gap TIG welding is employed to fabricate the liners and orbital TIG welding is employed for butt welding of tubes to themselves and to the tube sheets. Some aspects of the automatic narrow-gap TIG-welding used for the steam outlet penetrations will be presented.

The penetration consists of a combination of pipes and forgings as shown in Figure 15. The locations of the narrow gap TIG welds are also indicated.

Fig. 15 Sketch of steam outlet penetration.

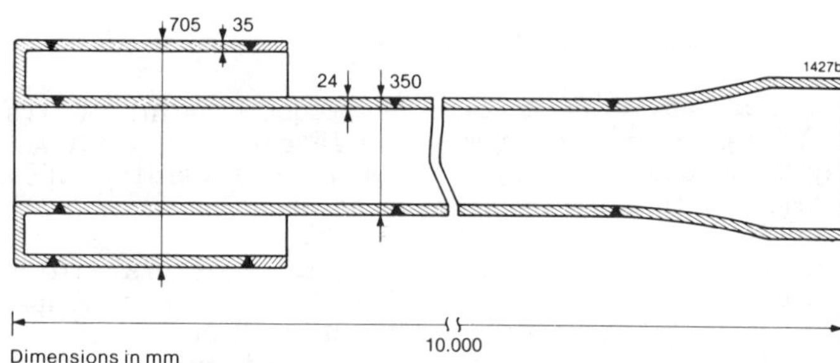


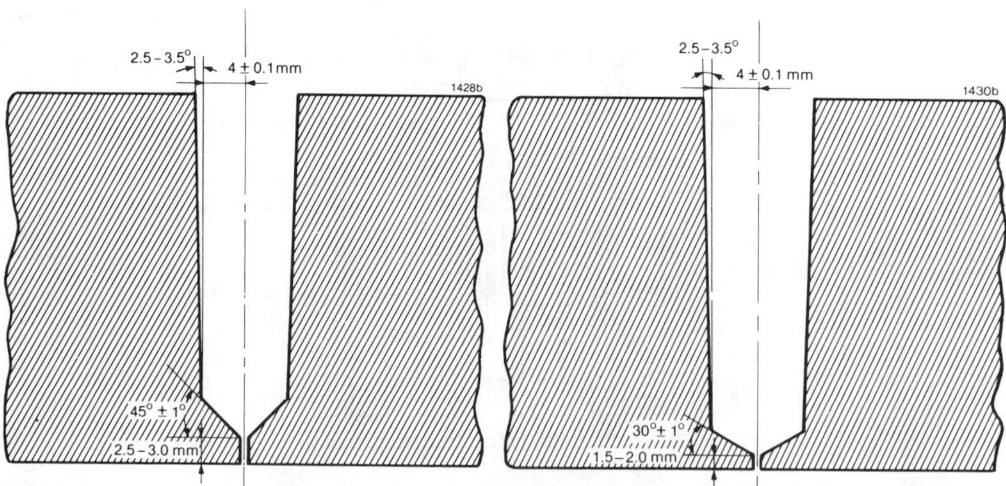

Fig. 16 316-316 joint preparation.

Fig. 17 316-carbon steel preparation.

The material thicknesses used are 24 and 35 mm. The same joint preparation is used for both thicknesses when both parent metals are 316. The transition joint between 316 and carbon steel has a slightly different preparation. Both are shown in Figures 16 and 17. The widest part of the joint is about 10 mm before welding but towards the end of the welding cycle this has shrunk to about 8 mm and the sides of the preparation have become nearly parallel.

Once the welding has been initiated it progresses continuously until the weld is completed. The welding parameters are preprogrammed on a Babcock MkVII Digital Weld Programmer. Constant arc voltage control is employed and there is an additional facility where the operator can move the electrode laterally and follow the position of the arc on a samll screen. The picture is obtained via a fibre-optical link.

Fig. 18 General view of welding area.

Fig. 19 Close-up view of welding station.

The welding cycle is divided into three parts - root run, first filler pass and remaining passes.

The typical parameters are:

| Pass | Current A | Voltage V | Travel speed mm/min | RPM |
|---|---|---|---|---|
| Root | 190 | 11 | 96 | 0.29 |
| 1st filler | 170 | 12 | 105 | 0.3 |
| Rest | 230 | 12 | 105-132 | 0.3 |

The torch shielding gas used is 95% Ar+5% $H_2$ and the inside of the pipe in the vicinity of the weld root is purged with 100% Ar. For the 316-carbon steel transition joints 100% AR is used as torch shielding gas.

A special 316L filler wire (1.6 mm diam.) is used for the 316-316 joints. It was produced to suit the special requirements of this customer and has the following composition.

| C | Si | Mn | Cr | Ni | Mo | FN |
|---|---|---|---|---|---|---|
| $\leqslant$0.020 | 0.4 | 1.8 | 19.4 | 13.2 | 2.2 | 7 |

Its designation is: Sandvik 19.12.3.L.

The welding consumable used for the 316-carbon steel transition joints is ER NiCr3 (1.6 mm diam.).

The weld deposit has an austenite matrix with a samll amount of ferrite. All production welds are, of course 100% X-rayed, and all of the 144 penetrations welded to date have been defect free.

Fig. 20 Microstructure of completed weld, 316-316, 35 mm thick.

The productivity of the process is very good and the time taken to weld the 350x24 mm pipes is about one hour. The alternative procedure of one TIG root run, two MMA passes and the remainder being welded by SAW, would take about three hours.

The total time taken to complete all the narrow-gap TIG welds is about 12 hours per penetration. The next stages of fabrication are to weld the tube bundles to the tube sheet and then the tube sheet onto the forged cone at the end of the penetration liner.

The above is a short account of one example where narrow-gap TIG wlding has shown great saving in both welding time and consumables. Simultaneously it increased the quality level of the final product.

## WICKET GATES IN HYDROTURBINES

One of the renewable sources of electric power is water power. About 350 GW of hydroelectric power is generated today. This is about 15% of the world potential for hydroelectric power generation.

Electricity is generated by hydroturbines which can be of various types. However, the predominating types for large-capacity turbines are the Kaplan turbine for low to medium heads (20-50 m) and the Francis turbine for medium to high heads (50-700 m). See Figures 21 and 22.

The turbine blades and wicket gates are often fabricated by forging the major components and welding them together. The welding consumables used in the fabrication are partly for joint welding of constructional and ferritic-martensitic-austenitic stainless steels, partly for surfacing of carbon or low-alloy steels with stainless steel deposits of the martensitic-ferritic type.

Two specific applications for submerged arc strip surfacing of turbine components will be described here.

But for the fact that a strip electrode is used, the principle of submerged arc strip surfacing is the same as for submerged arc wire welding.

The strip surfacing process is characterized by:

- high weld metal quality
- high deposition rate and area coverage capacity
- high degree of reproducibility and reliability
- low dilution with parent metal
- no fumes or arc light

### Kaplan type turbine

For a Kaplan type turbine in Brazil a ferritic-martensitic-austenitic weld metal was specified for surfacing of the wicket gates in order to obtain adequate resistance to cavitation corrosion and erosion corrosion.

In the first layer a low-carbon 17% Cr steel strip electrode was used together with a Ni and Mo alloying flux to obtain the classic 13Cr/4Ni/Mo ferritic-martensitic structure. With the same strip electrode and flux greater corrosion resistance was obtained in the second layer of the weld deposit where a 15Cr/5Ni/Mo deposit was obtained.

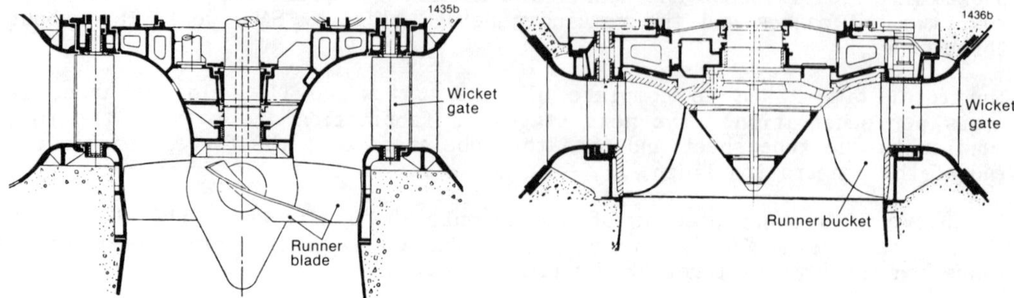


Fig. 21 Cross-section of Kaplan turbine.

Fig. 22 Cross-section of Francis turbine.

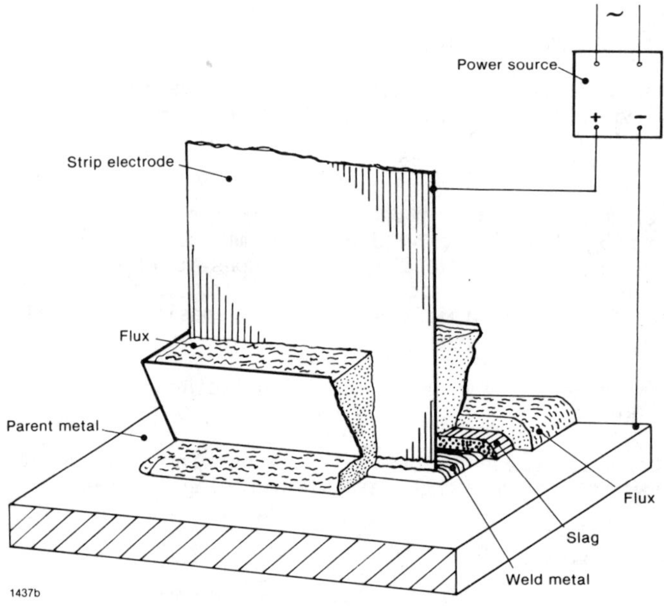


Fig. 23 Process principle for submerged-arc strip surfacing.

The welding procedure was

| | |
|---|---|
| Parent metal: | AISI 1518 |
| Strip electrode: | Sandvik 17.L 60x0.5 mm |
| Flux: | Soudometal Record 4B 410 NiTl |
| Welding current: | 750 A D.C.R.P. |
| Voltage: | 26 V |
| Travel speed: | 100 mm/min |
| Stickout: | 30 mm |
| Interpass temp.: | 150°C |

This procedure resulted in the following weld deposit compositions:

| Element | 17.L strip electrode | Weld deposit 1st layer | 2nd layer |
|---|---|---|---|
| C | 0.013 | 0.045 | 0.027 |
| Si | 0.33 | 0.63 | 0.64 |
| Mn | 0.64 | 0.68 | 0.62 |
| Cr | 17.2 | 13.7 | 15.5 |
| Ni | - | 4.0 | 5.0 |
| Mo | - | 0.36 | 0.44 |

The agglomerated alloyed flux gives a first-layer deposit with a pronounced nickel content and a noticeable pick-up of molybdenum. A fairly low silicon pick-up is noted.

The wear resistance depends on chemistry and structure of the deposit. It is difficult to assess, other than empirically, but guidance is given by hardness and toughness of the deposit:

| Weld deposit condition | Property | 1st layer | 2nd layer |
|---|---|---|---|
| As welded | Hardness, HB | 387 | 342 |
| After 8 hours at 600°C (cooling in the furnace) | Hardness, HB | 204 | 264 |
| | Tensile strength, N/mm$^2$ | 864 | 865 |
| | Elongation, A5, % | 15 | 14 |
| | Impact strength, J (Charpy V 0°C) | 34.4 | 24.8 |
| After 3+3 hours at 520°C | Hardness, HB | 342 | 319 |
| | Tensile strength, N/mm$^2$ | 1008 | 980 |
| | Elongation A5, % | 13 | 13 |
| | Impact strength, J (Charpy V 0°C) | 28.8 | 17.6 |

## Francis type turbine

The other surfacing application for turbines concerns the wicket gate support rings of Francis turbines built in Sweden.

Three vertical Francis turbines with a diameter of approximately 4 m were constructed recently in Sweden. In order to prevent the wicket gates from getting stuck because of fretting corrosion, it was decided that the support rings should be surfaced with stainless steel. However, since the wicket gate blades were made of austenitic stainless steel it was decided that the support rings could not be surfaced with an austenitic stainless steel owing to the risk of seizing between two austenitic stainless steels.

For this reason a ferritic stainless steel weld deposit was chosen.

Three welding procedures were investigated before the production welding procedure was adopted. A low-carbon 17% Cr strip electrode, as in the previous example, was used together with a readily available chromium compensating flux. The trial welding procedures were as follows:

| | |
|---|---|
| Parent metal: | AISI 1518 (90 mm thick) |
| Strip electrode: | Sandvik 17.L, 60x0.5 mm |
| Flux: | ESAB OK 10.92 |
| Preheating: | None |

| Procedure | A | B | C |
|---|---|---|---|
| Welding current, A | 675 | 750 | 825 |
| Strip feed, mm/min | 420 | 510 | 540 |
| Voltage, V | 27-28 | 27-28 | 27-28 |
| Speed, mm/min | 90 | 100 | 110 |
| Overlap, mm | 12 | 10 | 12 |

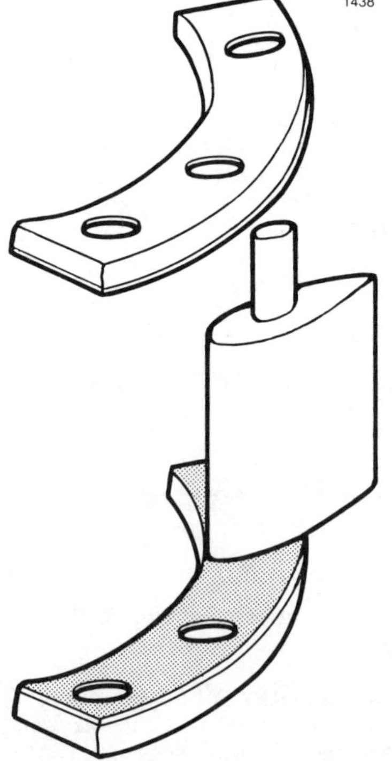


Fig. 24 Sketch of wicket gate and ring assembly.

The main requirement on a ferritic weld deposit is that it is free from cracks and sufficiently ductile to withstand service conditions. Both ductility and the corrosion resistance of welded ferritic chromium steel can be restored by a short heat treatment at about 750°C. Such a heat treatment is not suitable for mild steels and, consequently, it cannot be used for a surfaced component. However, many surfaced components are stress relieved but usually at temperatures below 600°C. A comprise was made and a temperature of 620°C was chosen for different holding times to simulate the heat treatment for different parent metal thicknesses.

The weld metal properties before and after heat treatment were as follows:

| Condition | Property | A | B | C |
|---|---|---|---|---|
| | Microstructure | Ferrite + martensite | Ferrite + martensite | Ferrite + martensite |
| As-welded | Hardness, HV<br>Bend test | 251<br>- | 253<br>Failed | 242<br>Failed |
| Heat treatment 620°C 20 min | Hardness, HV<br>Bend test | -<br>- | 236<br>Passed | -<br>- |
| Heat treatment 620°C 1 h | Hardness, HV<br>Bend test | -<br>- | -<br>- | 214<br>Passed |
| Heat treatment 620°C 5 h | Hardness, HV<br>Bend test | 225<br>Passed | -<br>- | -<br>- |
| Heat treatment 620°C 8 h | Hardness, HV<br>Bend test | 210<br>Passed | - | - |

Based on these trial procedures, the alternative B was chosen for giving the best results in terms of slag removal, tie-ins and bead edge appearance.

After the rings had been surfaced and heat treated at 620°C for 3 hours, holes were drilled for the wicket gate shafts and the surface was machined to the correct finish.

# NEW WELDING FILLER METALS FOR THE WELDING OF GIRTH WELDS ON PIPELINES OF CORROSION-RESISTANT CrNiMoN-DUPLEX STEELS

E. Perteneder, J. Tösch, G. Rabensteiner

*Vereinigte Edelstahlwerke Aktiengesellschaft (VEW), formerly Bohler Comp. 8605 Kapfenberg, Austria*

## ABSTRACT

The ferritic-austenitic duplex steel X 2 CrNiMoN 22 5, material number 1.4462, because of its good resistance to stress corrosion cracking, is used today for pipelines for the transport of chloride-containing crude oil or natural gas (acidic gas). For this reason, welding filler metals which satisfy the requirements set in regard to strength, toughness, and corrosion resistance, were developed especially for the welding of pipe girth welds. These are the coated rod electrode BOHLER FOX CN 22/9 N, and the rod electrode CN 22/9 N-IG for TIG welding. A special characteristic of the coated rod electrode are the welding properties adapted to the difficult conditions of the field laying of pipes made of steel 1.4462. The electrode FOX CN 22/9 N was used in the Netherlands with good result in the practical laying of pipes made of this steel, with the dimensions Ø 101 mm and 256 mm, and wall thicknesses of 3,25 mm to 7 mm.

## KEYWORDS

Line pipe welding; duplex stainless CrNiMoN-steel pipes; corrosion resistance; weld metal structur; mechanical properties.

## INTRODUCTION

The ferritic-austenitic duplex steels unite the good resistance against chloride-induced stress corrosion cracking which rust-resistant ferritic steels exhibit, with the good resistance to hydrogen-induced stress corrosion cracking (HSCC) which the austenitic steels possess. As a consequence of the properties obtained through the $\delta/\gamma$ composition structure, their economical range of application above all lies where high mechanical stress exists in addition to chemical stress (Herbsleb, 1980; Koren, 1982; Oppenheim, 1982; Oredsson, 1983).

Novel transport methods in crude oil and crude gas extraction led to the use of the steel X 2 CrNiMoN 22 5 (material number 1.4462; VEW designation; A 903) for transport pipelines, tanks, shut-off valves, as well as chemical manufacture of apparatus for the construction of clarifying plants, gas filters, and heat exchangers. Decisive for its employment are the cases where natural gas, in connection with chlorides, puts strongly corrosive stress on the material, and where other construction materials are eliminated on technical or cost grounds.

In the present report, a coated rod electrode and a TIG welding wire which have both been recently developed for the steel 1.4462 are presented (Tösch, 1982; Perteneder, 1983).

## TEST RESULTS

### Influence of the alloy structure and welding technique on the mechanical-technological properties of all weld metal

For the development of a suitable welding filler material, the optimization of the mechanical properties were prominent at first. The basis for this was the requirement that the electrodes, along with good welding behavior, should correspond in mechanical-technological and corrosion respects to the base metal. In order to determine the influence of the structural condition and of the ferrite content in this respect, three promising weld metals were so adapted as alloys that a distinct ferrite gradation occured. Fig. 1 contains the chemical composition and the ferrite content of the all weld metal of the test alloys.

| Designation of specimen | Chemical composition [wt%] | | | | | | | | | Ferrite content | |
|---|---|---|---|---|---|---|---|---|---|---|---|
| | C | Si | Mn | P | S | Cr | Mo | Ni | $N_2$ | Förster [Scale interv.] | Metallogr. [%] |
| German-Standard Nr.:1.4462 | ≤0,03 | ≤1,0 | ≤2,0 | 0,03 | 0,02 | 21,0 to 23,0 | 2,5 to 3,5 | 4,5 to 6,5 | 0,08 to 0,20 | — | — |
| 1 | 0,034 | 0,63 | 0,58 | 0,019 | 0,009 | 20,83 | 2,71 | 9,15 | 0,15 | 17 - 23 | 17±3 |
| 2 | 0,036 | 0,83 | 0,74 | 0,019 | 0,010 | 22,21 | 2,82 | 8,51 | 0,13 | 29 - 36 | 31±5 |
| 3 | 0,034 | 0,70 | 0,61 | 0,018 | 0,009 | 22,79 | 2,84 | 8,21 | 0,13 | 43 = 55 | 46±6 |

Fig. 1. Chemical composition of the all weld metal test specimen

The variable ferrite contents were, on the one hand, obtained by grading the Cr-content under the use of the specification limits of material 1.4462; and, on the other hand, obtained by adjustment of the Ni-content. Along with the C-content, with C=max.o,o4%, the Ni- and N-contents were so adjusted that, in light of the high cooling speed of the weld metal, a $\delta/\gamma$ relationship at room temperature exists that permits the attaining of adequate toughness values. As the following results show, this is only to be obtained in a weld metal with Ni-contents distinctly higher in comparison to the base material. On the basis of the imbalance condition of the weld metal, there occured in this the undesirably high ferrite content of substantially over 5o Vol.-%, with Ni-contents of 4,5-6,5%, as present in the base material. For the weld metal, a Ni-content of 8-9% can, however, be regarded as a favorable range. The ferrite contents were measured magnetically (by Förster instrument) and metallographically.

Fig. 2 shows the mechanical values and the notched bar impact work in the RT of the three test alloys as a function of the ferrite content of the all weld metal. For this, DIN 1913 tests were employed. One can see from the diagram that the notched bar impact work distinctly declines with an increasing ferrite portion in the structure. If one starts from a requirement for the all weld metal of ≥ 47 J at 2o°C (≥ 4o joule at -1o°C) with the ISO-V test, then the ferrite content is to be reduced unconditionally as stated above. At the lowest, the boundary is to be considered as at least 25 Vol.-%, since otherwise no adequate stress corrosion cracking resistance is guaranteed. The restrictions stated require a careful and exact use of analysis. Accordingly, alloy 2 (rod electrode BOHLER FOX CN 22/9 N) is considered to be the most favorable and is drawn upon as the base alloy for all further investigations.

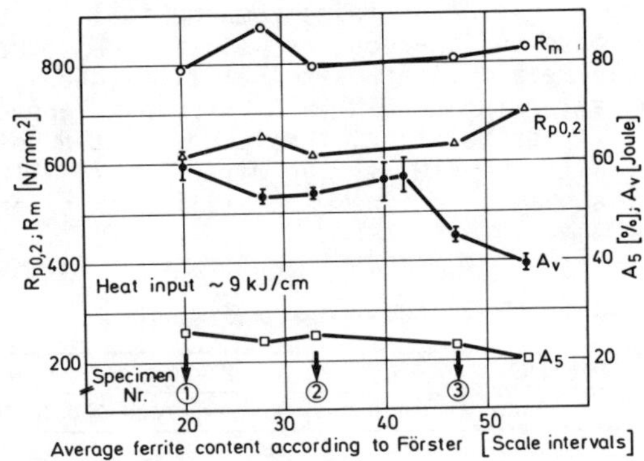


Fig. 2. Mechanical properties of the all weld metal of type X 3 CrNiMoN 22 8 3

In ferritic-austenitic weld metals, the welding technique and composition structure, in addition to the ferrite content, also exercise a distinct influence on notched bar impact work. Fig. 3 shows this dependence by means of DIN 1913 tests, which were also welded with FOX CN 22/9 N. As a result, the weave bead technique works positively in comparison with the string bead technique. The difference, measured with ISO-V probes, amounts to about 1o joules, and is thus quite distinct. As Fig. 3 further shows, the welding technique employed has practically no influence on the weld metal analysis. The analysis samples were also extracted from the specific compound.

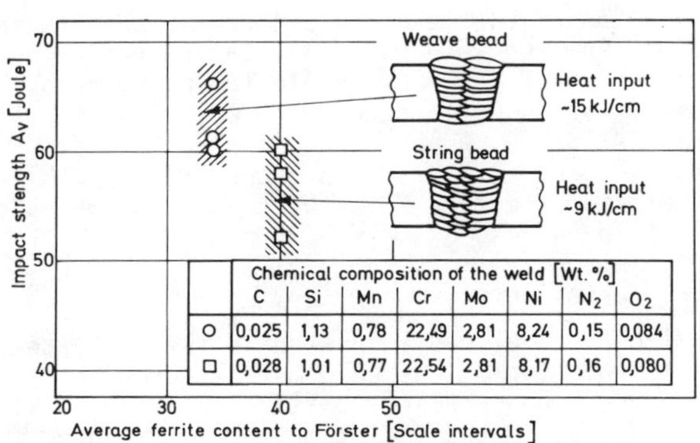


Fig. 3. Influence of welding technology on the impact strength of the all weld metal X 3 CrNiMoN 22 8 3.

The reasons for the higher toughness of weave beads in relation to string beads are certainly to be sought in the different arrangement of structure and in the somewhat slighter ferrite content of the oscillated weld metal.

Presupposing the chemical composition of FOX CN 22/9 N, further experiments with a same-type TIG welding wire (CN 22/9 N-IG), ∅ 2,o mm, were carried out according to the TIG process under Ar. Fig. 4 contains the chemical composition of the wire; the welding parameter observed thereby; as well as the mechanical-technological values for the all weld metal. It appears that the mechanical-technological values of the all weld metal according to DIN 1913 correspond approximately to that of the coated rod electrode. As against that, the values of the notched bar impact work are distinctly higher with the TIG welding wire CN 22/9 N-IG than with FOX CN 22/9 N. These results fall easily into line with similar experiences with austenitic or soft martensite CrNi(Mo) weld metal.

| Chemical composition [wt.%] | C | Si | Mn | P | S | Cr | Mo | Ni | $N_2$ |
|---|---|---|---|---|---|---|---|---|---|
| CN 22/9N-IG | 0,037 | 0,49 | 1,67 | 0,012 | 0,014 | 22,52 | 3,10 | 8,31 | 0,14 |

Welding process : TIG
Wire diameter : Φ 2,0 [mm]
Current : 210 [A]
Voltage : 15 [V]
Welding speed : 130 [mm/min]
Heat input : ~15 [kJ/cm]
Shielding gas : Ar
Specimen welded according to DIN 1913
Ferrit content according to Förster : 35 [Scale intervals]

| Tensile test [N/mm²]; [%] | | | | | Impact test ISO-V [Joule] | |
|---|---|---|---|---|---|---|
| $R_{p0,2}$ | $R_{p1,0}$ | $R_m$ | $A_5$ | Z | +20 [°C] | −10 [°C] |
| 637 | 691 | 802 | 28 | 64 | 137 | 128 |

Fig. 4. Mechanical properties of CN 22/9 N-IG

## Corrosion Resistance

The range of application of the material 1.4462 already mentioned demands special corrosion properties from the weld material. The corrosion tests were undertaken with all weld metal of the FOX CN 22/9 N electrode. Four-layered surfacings as well as welded joints were used in accordance with DIN 1913. In the Huey test, in accordance with ASTM A 262-79, Practice C, an abrasion rate of o,29 g/m$^2$.h was detected. By comparison, abrasion rates of about 8,o g/m$^2$.h were found in the Huey test with a weld metal of the type X 3 CrNiMo 19 12 3, which corresponds to approximately a 25-fold values.

## Pitting Corrosion

The pitting corrosion behavior was inspected in the FeCl3-test and in the NaCl-test. As is evident from Fig. 5, the weld metal in the FeCl3 test (1o%, acqueous FeCl3 solution), posesses an excellent pitting corrosion resistance. This result is the more noteworthy, since the test temperature amounted to +3o$^o$C at a test length of 24 hours. In comparison, the weld metal of the type X 3 CrNiMo 19 12 3, which according to this test should be used, is strongly susceptible to pitting corrosion.

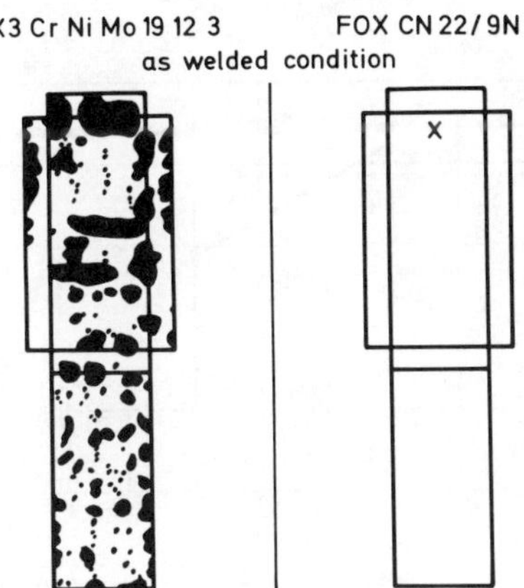


Fig. 5. Pitting test in aquous 1o% $FeCl_3$ solution
Test temp.: 3o$^{o}$C
Duration: 24h

The examination in the NaCl test was carried out at +6o$^{o}$C and +9o$^{o}$C in an $H_2S$-saturated, acqueous solution with 9o g of NaCl. In this test, no pitting corrosion could be detected, but only a general abrading corrosion. The abrasion rates detected lay at o,o1 and o,o2 g/m$^2$.h respectively.

## Stress corrosion cracking resistance

The resistance against chloride- and hydrogen-induced stress corrosion cracking are especially important requirements for the filler metal for the steel 1.4462. The following tests were accordingly carried out, and the following results obtained:

Examination according to NACE TM-o1-77. Before the examination, the samples were electrolytically polished, so that a layer of 1oo μm was abrades. Fig. 6 gives the result. Within the limits of the investigation of the weld metal, the length of the examination was increased above the required 72o hours (1 month) to the max. 362o hours (5 months). This was done in order to obtain as accurate a determination of the actual edge stress as possible. As is evident from the course of the curve, the axial tensile stress of 7oo N/mm$^2$ at the beginning was gradually lowered, and finally an edge stress of 3oo N/mm$^2$ was obtained. During the experiments, a steady fumigation with $H_2S$ took place.

Examination according to DIN 5o9o8. This test was carried out by means of a loop test piece. It appears that the required service life of at least 5oo hours to the tearing of the loop could be observed. For comparison, the weld material AWS E 316 L-16 was examined; and this yielded a significantly greater susceptibility to cracking as well as more rapid progress of the crack.

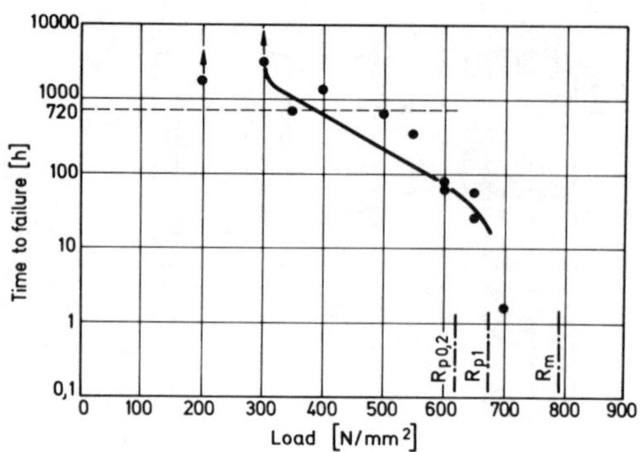


Fig. 6. FOX CN 22/9 N
SCC-test according to NACE TM-o1-77

<u>Examination in the crust test.</u> In this test, the weld metal was examined in the form of girth tensile tests in 3% NaCl solution under crust formation. The upper shaft of the tensile probe was electrically heated, so that for this purpose a crust of NaCl formed at the water line. This test represents an especially acute examination method for chloride-induced stress corrosion cracking, and makes possible a very good differentiation of the material behavior. The results are diagrammatically represented in Fig. 7. For purposes of comparison, the austenitic steels X 5 CrNi 18 9 (VEW A 5oo) and X 2 CrNiMo 18 1o (VEW A 2oo) are recorded, in addition to the duplex steels X 2 CrNiMoSi 19 5 (VEW A 9o1) and X 2 CrNiMoN 22 5 (VEW A 9o3). The service life of the weld metal of FOX CN 22/9 N lies in the dispersion range of the analogous base material (Kohl 1981).

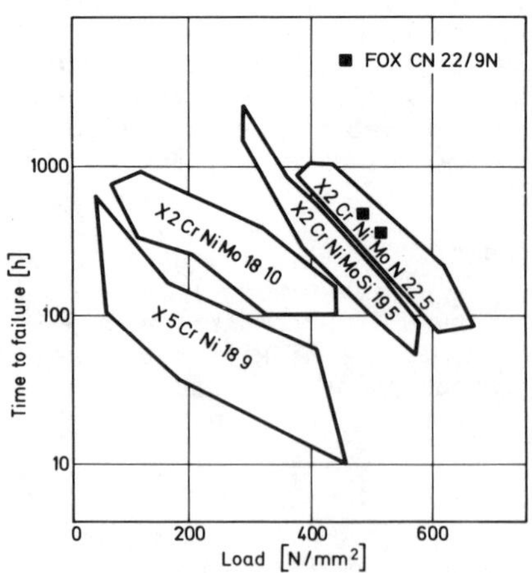


Fig. 7. SCC in 3% NaCl solution with crust formation

Welding of pipes made of material number 1.4462

In the context of the practical testing of the coated rod electrode, girth welds were welded to pipeline conduits made of steel 1.4462. This demanding special task of the economical welding of pipe girth welds has become necessary today through the building of pipelines made of this steel.

The weldings were first carried out in the lab under simulated field conditions. The pipes had a diameter of 15o mm and a wall thickness of 3.5 mm. The demands on the weld properties of such coated rod electrodes are manifold, and the attaining of the favorable compromise represents a difficult optimization problem. The rod electrodes, particularly for a successful working under field conditions in the pipes, must satisfy the following conditions:

- stable, non-deflectable arc;
- rapid draining of the slag out of the root opening;
- good weldability in the overhead position;
- good climbing ability and gap bridgeability;
- high pores security;
- constant good weldability in a wide range of current.

Samples were removed from a pipe connection welded with such kinds of rod electrodes Ø 2,5 mm. The reason for the selection of the electrode diameter of 2,5 mm was that this represents, for the wall thickness of 3,5 mm, an upper limit, and, in relation to mixing and structure modulation, represents intensified conditions. Therefore, the relatively slight wall thickness of 3,5 mm in connection with the relatively great rod electrode diameter of 2,5 mm, leads to a large thermal modulation of the base material in the HAZ.

For the elucidation of the base material modulation, Fig. 8 shows the macro-section of a pipe weld joint with an 11-fold magnification. Cross sections are depicted in the six-o'clock and the twelve-o'clock position. The macro-surveys let variably wide zones of coarsened ferrite be detected. The HAZ in the root area of the twelve-o'clock position is the most pronounced. The metallographic ferrite evaluation according to the point-count method yielded in this zone of coarsened ferrite a portion of about 7o Vol.-%, as opposed to the about 6o Vol.-% in the unmodulated base material.

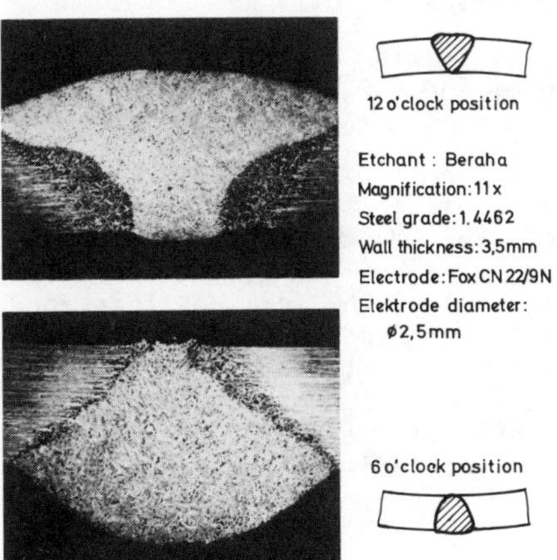


Fig. 8. Macrographs of specimens taken from pipe welds

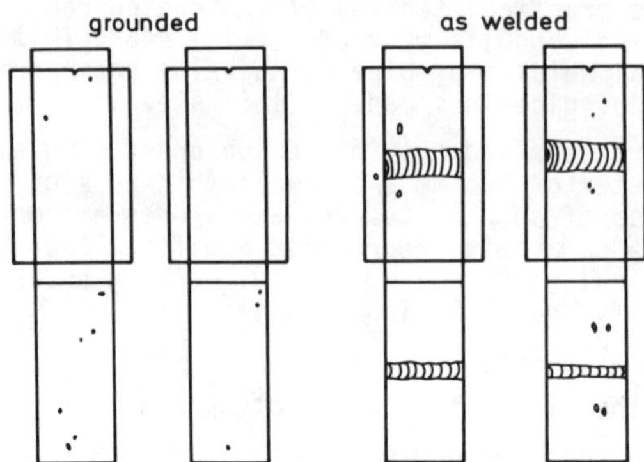


Fig. 9. Pitting test in aquous lo% $FeCl_3$ solution
Test temp.: 3o$^{o}$C
Duration: 24h

The results of the traction and bending experiments corresponded to the desired specifications; that is, the increase of the amount of ferrites as well as the grain coarsening of this phase in the HAZ led to no distinguishable impairment of the mechanical-technological properties of the base material.

Of interest, too, was the examination of the pitting corrosion resistance of samples from the pipe weld joint. Fig. 9 shows the total surface of the samples extracted after an examination period of 24 hours in acqueous lo% $FeCl_3$ at 3o$^{o}$C. Only the base material partially showed a slight attack. The weld metal exhibited manifestations of corrosion. These positive results show that it is today possible to produce weldable CrNiMoN-duplex steels, which, even with the relatively larger thermal modulation of the HAZ, display no structural alterations impairing the mechanical-technological and chemical properties. The statements met are also valid for greater wall thickness. It thus appears that, as a result of the more rapid heat dissipation in thicker sheets, the recognizable structural impairment is naturally slighter in welding in the HAZ.

Furthermore, the large technical testing of the FOX CN 22/9 N rod electrodes took place in the dimensions Ø 2,o mm and Ø 2,5 mm, with the field laying of pipes made of the steel 1.4462, with the dimensions of:

Ø 4" (1o2 mm) x s = 3,25 mm
up to
Ø 1o" (254 mm) x s = 5,o mm

The laying of pipes was carried out by the firm HAK/Isselmonde in the Netherlands. In particular, the welding of the electrodes to the direct current-negative pole met the test. In comparison to the direct current-positive pole, a significantly altered burn-out behavior results, which expresses itself in a doubling of the number of short circuits per unit of time. An especially steady root formation, as well as an improved top position weldability in an overhead position, were attained in that way.

## CONCLUSION

For the welding of the duplex steel, material number 1.4462 (X 2 CrNiMoN 22 5; VEW designation A 9o3), welding filler metals in the form of coated rod electrodes and TIG rods were developed. The most essential difference with the base material is the higher Ni content of the all weld metal necessary for reasons of toughness.

The ferrite content of the weld metal lies at 35 Vol.-%. The all weld metal of the welding filler metals described, FOX CN 22/9 N and CN 22/9 N-IG, fulfill the mechanical-technological toughness and corrosion resistance requirements for the base material 1.4462.

The suitability of the rod electrode for the vertical up-welding of pipes for the use in pipeline construction, is to be mentioned as a special feature. The new electrodes FOX CN 22/9 N and CN 22/9 N-IG are already in use, both for laying pipelines and for the chemical manufacture of apparatus in general.

## ACKNOWLEDGEMENT

The authors would like to thank the research promotion fund of the Austrian Federal Chamber of Trade and Industry for the support they have received in their R&D work.

## REFERENCES

Herbsleb, G., and R.K. Poepperling (198o). Corrosion Properties of Austenitic-Ferritic Duplex Steel AF 22 in Chloride and Sulfide Containing Environments. Corrosion-Nace, Vol. 36, 611-617.

Kohl, H., and W. Wedl (1981). Stress Corrosion Cracking Test with Crust Formation. Proc . 8th ICMC, Vol. I, 524-529.

Koren, M., and G. Hochörtler (1982). Eigenschaften des ferritisch-austenitischen Stahles X 3 CrMnNiMoN 25 6 4. Stahl und Eisen 1o2, 5o9-513.

Oppenheim, R., and G. Chlibec (1982). Eigenschaften des nichtrostenden Stahles X 2 CrNiMoN 22 5 (Remanit 4462) im Vergleich mit anderen ferritisch-austenitischen Stählen. Thyssen Edelst. Techn. Ber. 8, 187-193.

Oredsson, J., and S. Bernhardsson (1983). Performance of High Alloy Austenitic and Duplex Stainless Steels in Sour Gas and Oil Environments. Materials Performance, Vol. 22, 35-41.

Perteneder, E., J. Tösch, H. Schabereiter, and G. Rabensteiner (1983). Neuentwickelte Schweißzusatzwerkstoffe zum Schweißen korrosionsbeständiger CrNiMoN-legierter Duplexstähle. Österr. Schweißtechnik 37, 83-86, 1o2-1o4.

Tösch, J., H. Schabereiter, E. Perteneder and G. Rabensteiner (1982). Nieuwontwikkelde lastoevoegmaterialen voor het lassen van voestvaste CrNiMo-gelegeerde Duplex-staalkwaliteiten. Lastechniek 48, 1o6-116.

# WELDABILITY OF DUPLEX STRUCTURE 12Cr-(Mo,W) STEELS

E.J. Vineberg*, T. Wada*, T.B. Cox*, and C.C. Clark**

*Climax Molybdenum Company of Michigan
Ann Arbor, Michigan
**AMAX Tungsten
Greenwich, Connecticut

ABSTRACT

Low carbon and 12Cr-(Mo,W) steels have been developed and their weldability evaluated with y-groove cold cracking tests. Continuous cooling transformation diagrams have been determined for these steels to characterize the transformations occurring in the heat affected zone (HAZ) of weldments. The steels, which have ferrite-martensite duplex microstructures, are low carbon alternatives to the martensitic 12Cr-1Mo steels such as DIN X20 CrMoV 12 1, commonly used for superheater tubing. The elevated temperature properties of these new duplex structure steels are equivalent to or higher than those of the 12Cr-1Mo steels. The most promising steel has a nominal composition of 12Cr-1.5Mo-1W and exhibits creep resistance significantly higher than that of DIN X20 CrMoV 12 1. As the HAZ of these duplex structure steels does not completely transform to austenite during welding, its hardness remains lower than 400 HV1, while fully martensitic 12Cr-1Mo exhibits a single-pass HAZ hardness higher than 500 HV1. The lower HAZ hardness eliminates the need for pre-heating.

KEY WORDS

Chromium-molybdenum steels, chromium-molybdenum-tungsten steels, elevated temperature steels, duplex structure steels, weldability, mechanical properties, creep resistance, continuous cooling transformation diagrams, steam tubing, power generation.

## INTRODUCTION

Martensitic 12Cr-1Mo-V steels such as DIN X20 CrMoV 12 1 are used extensively in Europe in steam tubing applications at electrical power generating stations. These steels require typically pre-heat and post-weld stress relief to avoid HAZ cracking. North American boiler manufacturers have historically preferred to use austenitic stainless steels, such as Type 304, for superheater and reheater tubing thereby avoiding the higher fabrication costs associated with preand post-weld heat treatments. However, the austenitic steels are less efficient for heat transfer than ferritic steels. Also, operating experience indicates that the welds joining dissimilar metal austenitic and ferritic components are susceptible to premature failure during service (Dooley and co-workers, 1982). These operating problems could be avoided if all ferritic construction were used; thus, there is a need for a ferritic steel with sufficient strength that is at the same time readily weldable.

A research program was undertaken at the Climax Molybdenum Company of Michigan Research Laboratory to develop a 12% chromium steel with the strength of the martensitic 12Cr-1Mo-V steel, which typically contains 0.2% carbon, and the weldability of Type 304 stainless steel. To achieve the desired weldability, the carbon level was decreased from 0.2% to a range of 0.07 to 0.11%. The accompanying decrease in strength was found (T. Wada and Co-Workers, 1983) to be offset by increasing the molybdenum content from 1% to 2%.

The present paper describes additional progress in the 12Cr-2Mo steel development effort, particularly in the area of weldability. It was reasoned that replacing a part of the molybdenum with an equivalent amount of tungsten on an atomic basis would increase the elevated temperature strength. The steel in this class with the most attractive properties nominally contains 12Cr-1.5Mo-1W, and much of the data in this paper are on this composition. The steel exhibits a duplex microstructure of ferrite plus martensite, and exhibits the excellent weldability previously demonstrated for the duplex structure steels.

## EXPERIMENTAL PROCEDURES

### Materials

The experimental steels were prepared from pure metals and ferroalloys either as 57 kg (125 lb) heats induction melted in air under an argon cap or as 25 kg (55 lb) heats induction melted under vacuum. An argon plus nitrogen atmosphere was introduced to the latter after meltdown of the charge materials and prior to the addition of the alloying elements. The compositions of the steels are listed in Table 1. The steels were forged and rolled to 16 mm (5/8 in.) thick plate and

TABLE 1 Compositions of the Test Steels

| Heat Number | Element, % | | | | | | | | | | | (ppm) |
|---|---|---|---|---|---|---|---|---|---|---|---|---|
| | C | Mn | Si | Cr | Mo | W | Ni | V | Nb | P | S | Al | N |
| 6274[a] | 0.079 | 0.68 | 0.25 | 12.02 | 2.01 | - | 1.06 | 0.25 | - | 0.016 | 0.017 | 0.040 | 350 |
| 6455[a] | 0.070 | 0.56 | 0.22 | 11.87 | 1.50 | 1.04 | 1.46 | 0.18 | 0.045 | 0.019 | 0.013 | 0.012 | 315 |
| P2322[b] | 0.10 | 0.72 | 0.28 | 11.97 | 2.02 | - | 0.60 | 0.22 | 0.028 | 0.012 | 0.007 | 0.004 | 263 |

a) Air melt
b) Vacuum melt

then heat treated. The air melted heats were held at 1050 C (1920 F) for 0.5 hour and air cooled, followed by a one hour temper at 705 C (1300 F) and air cooling. The vacuum melted heat was held 1 hour at 1050 C (1920 F) and air cooled, then tempered 2 hours at 730 C (1345 F) and air cooled.

### Metallography

Specimens for metallographic examination were mounted, mechanically polished, and etched using either Vilella's martensitic etchant (1% picric acid and 5% hydrochloric acid in methanol) or an electrolytic process using a 50% nitric acid solution with the specimen as the cathode. The microstructures were examined by both optical and scanning electron (SEM) microscopy. Ferrite fractions were determined by standard point counting techniques.

### Weldability Tests

The cold cracking susceptibility of the test steels was determined using 16 mm (5/8 in.) thick y-groove plate assemblies. (T. Wada and Co-Workers, 1983) The two plates were anchor welded at both ends of the groove to provide restraint during the test welding. The anchor welds were made using a low-hydrogen 9Cr-1Mo electrode material. The test welds were made without pre-heating of the plates using either the 9Cr-1Mo electrode or an experimental 12Cr-1.6Mo-2Ni electrode material and the following conditions: direct current, reverse polarity, 150 to 170 A, 18 to 23 V, 14 cm/min. (5.5 in./min.) travel speed, and 14 kJ/cm (37 kJ/in.) heat input. The welded plate assemblies were allowed to cool to room temperature and, after 48 hours, were sectioned for hardness measurements and inspected for cracks.

Welded joints for mechanical testing were prepared from 16 mm (5/8 in.) thick plates measuring 75 by 350 mm (3 by 14 in.). The weld joint was made along the 350 mm (14 in.) edges using a double v-groove geometry with an included angle of 70 degrees and a 3 mm (0.125 in.) root face. The plates were tack welded with a spacing of about 3.5 mm (0.138 in.) at the root. The welding passes were made without preheating using conditions similar to those used in welding the y-groove plate assemblies: 160 A, 23 V, a 13 cm/min. (5 in./min.) travel speed and a 17 kJ/cm (43 kJ/in.) heat input. The second pass was placed on the side opposite the first pass, after grinding the root portion of the first pass. The third and fourth passes were made over the first pass, and the fifth and sixth were made over the second pass. One portion of the weldment was stress relieved at 700 C (1290 F) for 1 hour and the balance evaluated in the as-welded condition.

### Determination of Continuous Cooling Transformation Diagrams

Transformation studies were performed using a quenching dilatometer. The instrument employs induction heating of a 5 mm (0.197 in.) diameter by 10 mm (0.394 in.) long specimen and either gas quenching or programmable radiative cooling. The lower critical temperature ($Ac_1$) and temperatures for carbide transformation ($T_C$) and completion of carbide dissolution were determined from the heating curves. Using average cooling rates between 800 and 500 C (1470 and 930 F) ranging from 0.02 to 314 C/sec (0.036 to 565 F/sec), a complete continuous cooling transformation (CCT) diagram for material initially austenitized at 1050 C (1920 F) was constructed. A CCT diagram for a simulated welding condition was constructed from the results of four specimens heated separately to 1300 C (2370 F) for 1 minute and cooled at rates ranging from 1 to 40 C/sec (1.8 to 72 F/sec).

### Tensile and Impact Tests

Room temperature tensile tests were performed using duplicate specimens 6.35 mm (0.250 in.) in diameter with a 25 mm (1 in.) gauge length. The specimens were cut from the normalized and tempered plates parallel to the rolling direction. The strain rate was 0.3%/min. in the elastic range and 5%/min. in the plastic range. Impact toughness was determined using standard Charpy V-notch specimens. The specimens were cut from the normalized and tempered plates parallel to the rolling direction.

### Phase Stability

To accelerate the changes in microstructure which occur during service exposure and to determine the maximum temperature for creep rupture testing, a bar of the

12Cr-1.5Mo-1W steel was cold rolled 50% and aged for 1,000 hours in a gradient furnace at temperatures ranging from 540 to 830 C (1000 to 1525 F). After aging, the bar was sectioned for metallographic examination at specific locations corresponding to aging temperatures ranging from 600 to 800 C (1110 to 1470 F) in 25 C (45F) increments.

Creep Rupture Tests

Creep rupture tests were performed in air at temperatures ranging from 565 to 677 C (1050 to 1250 F) on specimens with gauge diameters and lengths of 8.25 and 33 mm (0.325 and 1.3 in.), respectively. Tests were performed using specimens machined from the normalized and tempered plates and also from stress relieved welded sections in the case of one steel (Heat P2322).

Oxidation Tests

The 12Cr-1.5Mo-1W steel was tested for resistance to cyclic and continuous oxidation by suspending specimens over small crucibles (to catch any scale that spalled) and heating in air at 650 C (1200 F). In the cyclic tests, the specimens were held at temperature for 22 hours, removed from the furnace and allowed to air cool to room temperature, after which the cycle was repeated. Weight change measurements were performed after each 5 to 10 cycles for a total of 67 cycles, or a total of 1,500 hours at temperature. In the continuous tests, the specimens were weighed at the end of each 1,000 hours of exposure and were run for four such intervals, for a total exposure time of 4,000 hours.

## RESULTS AND DISCUSSION

Microstructure

The microstructure of the 12Cr-1.5Mo-1W duplex structure steel in the normalized and tempered condition, shown in Fig 1, consists of ferrite and tempered martensite. To understand how this microstructure develops, it is useful to consider a schematic equilibrium phase diagram for the alloy system, as shown in Fig. 2. The nickel and low carbon contents of these 12Cr-(Mo,W) steels allow ferrite to be stable at all temperatures up to the liquidus. Thus, when the steels are heated during processing or in welding, they are never fully austenitic, but rather they

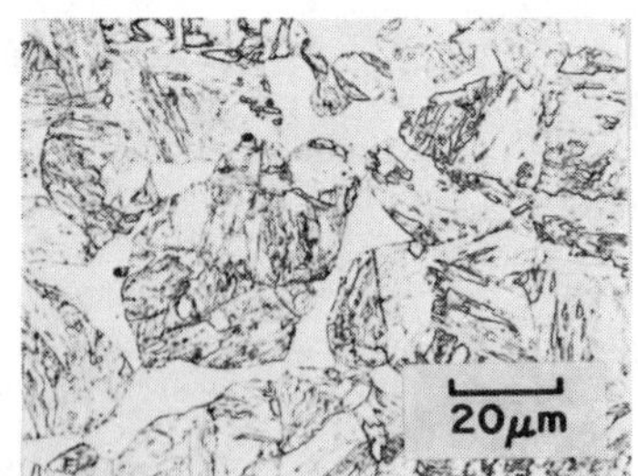


Fig. 1. Microstructure of the normalized and tempered 12Cr-1.5Mo-1W steel (Heat 6455), etched in 1% picric acid and 5% HCl in methanol.

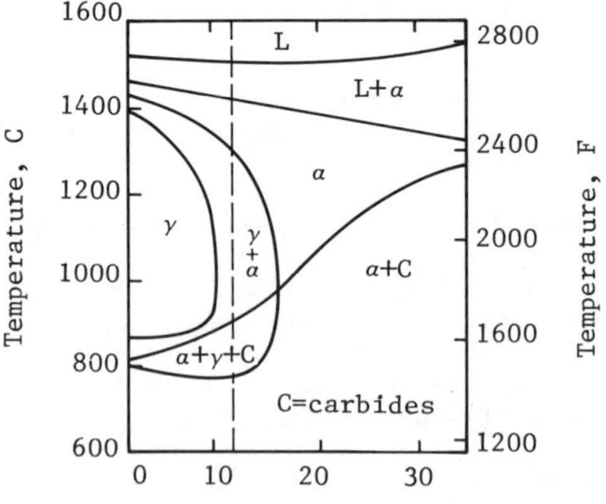


Fig. 2. Cr-Mo-W steel phase diagram.

contain varying amounts of austenite. As seen in the phase diagram, at the temperatures reached in a weld heat affected zone the microstructure is predominantly ferritic with a relatively small fraction of austenite, whereas at the normalizing temperature, 1050 C (1920 F), the ferrite fraction is 15-25%.

Weldability

There was a total absence of HAZ cracking in the y-groove tests of the duplex structure steels even when welded without pre-heating and post-weld stress relief. In contrast, the martensitic 12Cr-1Mo-V steel welded and examined under the same conditions exhibited 100% HAZ cracking. Pre-heating the martensitic steel at 150 C (300 F) reduced the level of HAZ cracking, but 20% cracking was still observed. (T. Wada and Co-Workers, 1983) The martensitic 12Cr-1Mo-V steels are normally (Sandvik Ht9-Tech Info; N.G. Persson, 1980) welded with pre-heating at about 430 C (805 F).

Hardness traverses of representative sections of duplex structure 12Cr-2Mo and 12Cr-1.5Mo-1W steels are shown in Fig. 3. Both steels exhibit peak HAZ hardness

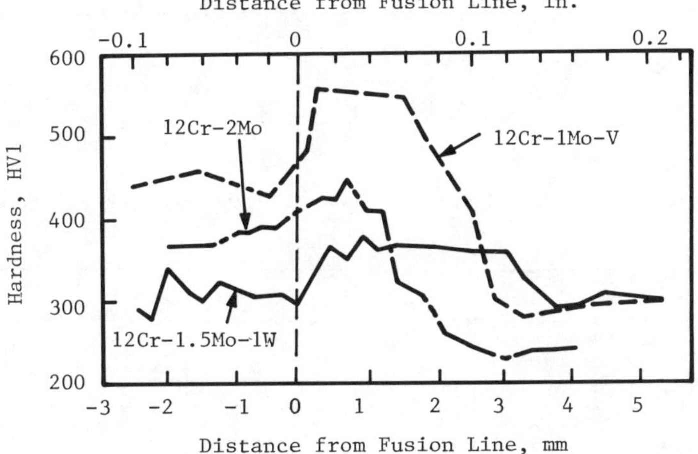


Fig. 3. Hardness profiles of y-groove weldments.

levels of 350 to 375 HV1, compared to weld metal and base plate hardness values in the ranges of 300 to 350 HV1 and 250 to 300 HV1, respectively. The HAZ hardness of the duplex structure steels differ dramatically from the 550 HV1 HAZ hardness of the fully martensitic steel reported earlier. (T. Wada and Co-Workers, 1983) Pre-heating has little affect in reducing the HAZ hardness level of the martensitic steel; only with post-weld tempering does the HAZ hardness of the martensitic steel approach the base plate hardness level.

CCT Diagram and Microstructures of HAZ

The CCT Diagram for a simulated welding condition in which the CCT specimens were heated to 1300 C (2370 F) for 1 minute and rapidly cooled is shown in Fig. 4. This diagram may be compared with the CCT diagram for the steel heat treated at 1050 C (1920 F), which is shown in Fig. 5. The $M_s$ and $M_f$ temperatures for cooling from 1300 C (2370 F) are typically 260 and 80 C (500 and 170 F), respectively, each 20 to 40 C (36 to 72 F) lower than the corresponding values for the

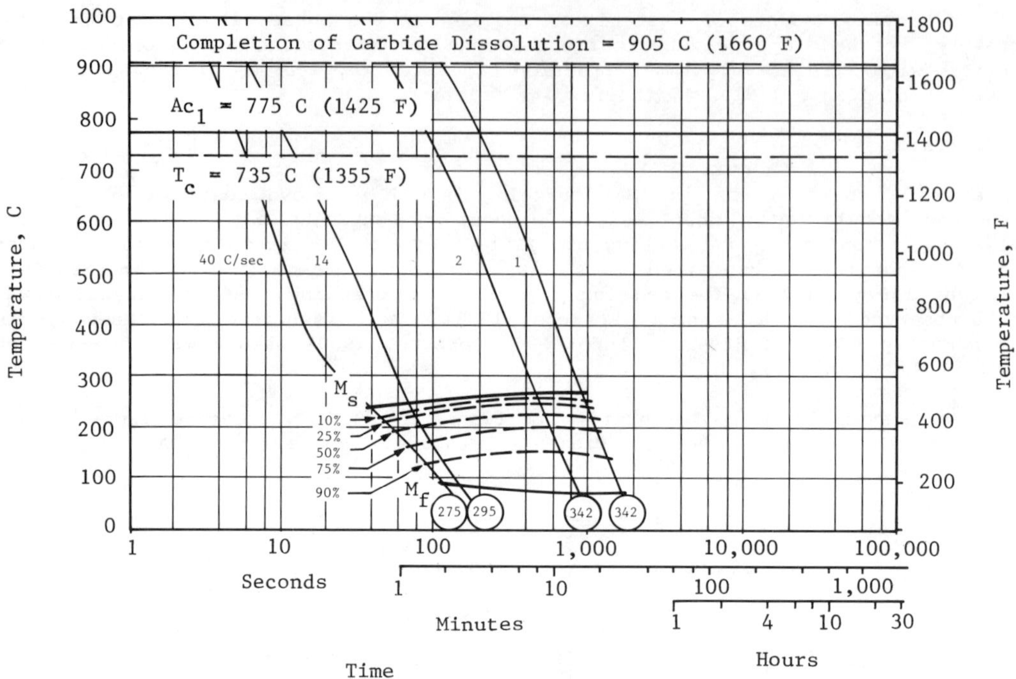


Fig. 4. CCT diagram of the 0.08C-12Cr-2Mo-1Ni-0.3V steel cooled from 1300 C (2370 F)

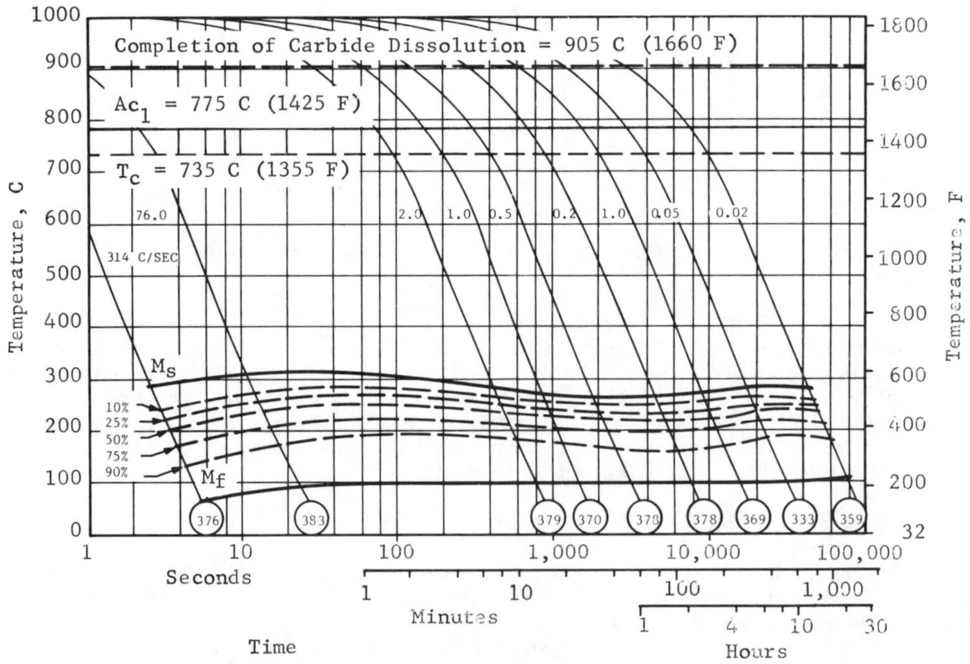


Fig. 5. CCT diagram of 0.08C-12Cr-2Mo-1Ni-0.3V steel austenitized at 1050 C (1920 F)

steel austenitized at 1050 C. The hardness levels of 275 to 340 HV1 in the simulated welding condition are also significantly lower than those in the heat treated condition as seen from the hardness values appended to the cooling curves. These hardness values in the simulated welding condition indicate that the peak hardness in the HAZ of the duplex structure 12Cr-2Mo and 12Cr-(Mo,W) steel welds is inherently low in the region near the fusion line where the peak temperature is relatively high.

Microstructures of CCT specimens cooled at various rates from 1300 C (2270 F) are presented in Fig. 6. These microstructures are a mixture of ferrite and marten-

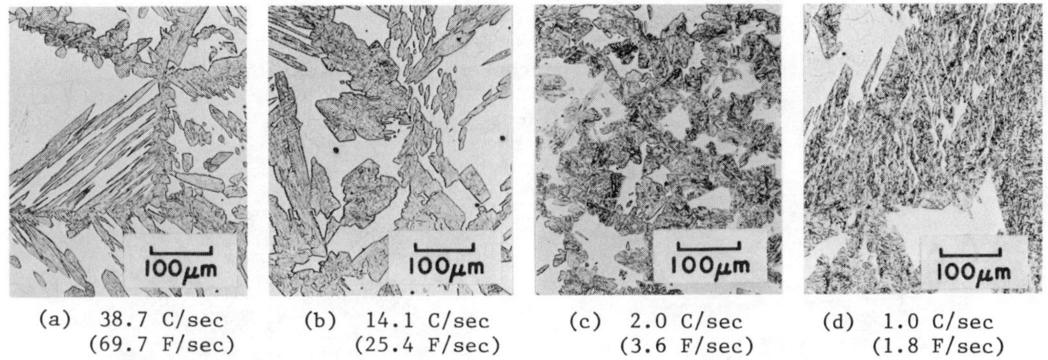


(a) 38.7 C/sec (69.7 F/sec)  (b) 14.1 C/sec (25.4 F/sec)  (c) 2.0 C/sec (3.6 F/sec)  (d) 1.0 C/sec (1.8 F/sec)

Fig. 6. Simulated HAZ microstructures for the 12Cr-2Mo-V steel (Heat 6274) continuously cooled at various rates from 1300 C (2370 F), etched in 1% picric acid and 5% HCl in methanol.

site, but are quite different from the normalized and tempered structure shown in Fig. 1. The differences are a consequence of both the phase balance at 1300 C (2370 F) compared to that at 1050 C (1920 F) and the varying cooling rates.

The microstructure consists primarily of ferrite with approximately 10% austenite. At lower temperatures, the ferrite fraction decreases (austenite fraction increases) such that between 1000 and 1200 C (1830 and 2190 F), the microstructure consists of nominally 20% ferrite and 80% austenite. The amount of austenite that forms during cooling from 1300 C (2270 F) depends on the cooling rate. With decreasing cooling rate, an increasing fraction of ferrite transforms to austenite as shown in Fig. 6. Also, as the cooling rate decreases, the nature of the austenite changes from an acicular morphology shown in Fig. 6a to the more equiaxed morphology shown in Figs. 6c and d. The austenite transforms to martensite below the $M_s$ temperature.

The microstructure of the HAZ in a 12Cr-2Mo y-groove test specimen is shown in Fig. 7. A high volume fraction of ferrite is observed in the HAZ near the fusion line, decreasing with distance into the base plate. Acicular martensitic islands are also observed at the fusion line as shown in Fig. 8a, taken from a multi-pass weld. Fig. 8b shows the microstructure in the HAZ somewhat distant from the fusion line where the peak temperature was appreciably lower than that at the fusion line. The original martensitic region in the base metal transformed to austenite without significantly changing the morphology. After cooling to room temperature, ferrite is observed in the prior austenitic region, but no carbide precipitation is observed.

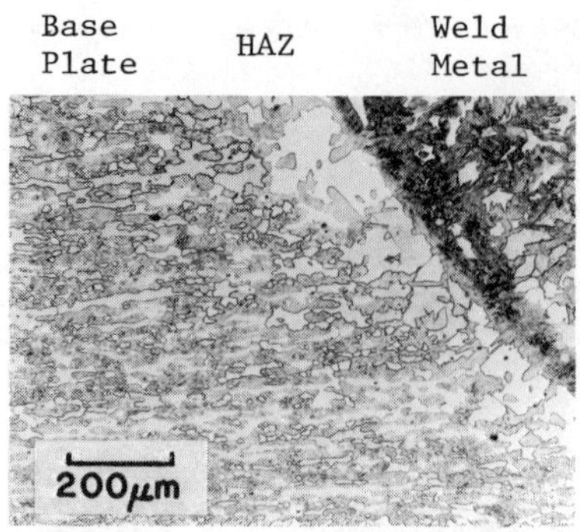


Fig. 7. HAZ microstructure of a duplex structure 12Cr-2Mo-V steel, etched in 1% picric acid and 5% HCl in methanol.

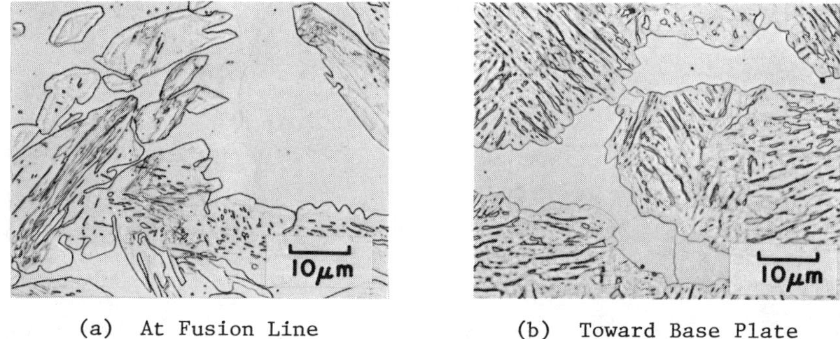


(a) At Fusion Line (b) Toward Base Plate

Fig. 8. Microstructures of the HAZ of a multi-pass weld of a 12Cr-2Mo-V steel, etched electrolytically in 50% $HNO_3$ with reverse polarity.

Tensile and Impact Properties

The room temperature tensile properties and the impact toughness of the experimental duplex structure steels are presented in Tables 2 and 3, respectively. Compared with the martensitic 12Cr-1Mo-V steels in the same tempered condition, the

TABLE 2 Tensile Properties at Room Temperature

| Heat Number | Steel | 0.2% Offset Yield Strength, MPa | (ksi) | Tensile Strength, MPa | (ksi) | Elongation % | Reduction of Area, % |
|---|---|---|---|---|---|---|---|
| 6274 | 12Cr-2Mo-V | 761 | (110) | 894 | (130) | 22 | 62 |
| 6455 | 12Cr-1.5Mo-1W-V-Nb | 799 | (116) | 927 | (135) | 21 | 64 |
| P2322 | 12Cr-2Mo-V-Nb | 671 | (97) | 833 | (121) | 19 | 51 |

TABLE 3 Impact Toughness Properties

| Heat Number | Steel | Room Temperature Energy, J (ft-lb) | Room Temperature Shear, % | Upper Shelf Energy, J (ft-lb) | FATT C (F) |
|---|---|---|---|---|---|
| 6274 | 12Cr-2Mo-V | 85 (63) | 100 | 130 (96) | 0 (30) |
| 6455 | 12Cr-1.5Mo-1W-V-Nb | 70 (52) | 62 | 115 (85) | 20 (70) |
| P2322 | 12Cr-2Mo-V-Nb | 44 (32) | 50 | n.d. | 20 (70) |

n.d. - not determined

duplex structure steels typically exhibit somewhat lower room temperature tensile strength and markedly higher room temperature and upper shelf impact energies. These effects are all attributed to the combination of the presence of ferrite in the microstructure and the lower carbon level of the tempered martensite of the duplex structure steels as compared to the fully martensitic steel. The room temperature mechanical properties of the 12Cr-1.5Mo-1W and 12Cr-2Mo steels are quite similar.

Phase Stability

Microstructures of the cold rolled 12Cr-1.5Mo-1W steel aged at various temperatures are shown in Figs. 9a-d. The microstructure is stable with increasing aging

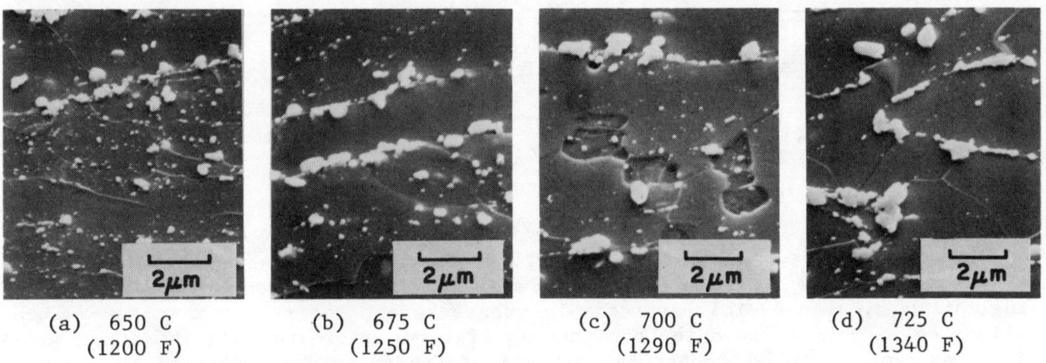


(a) 650 C (1200 F) (b) 675 C (1250 F) (c) 700 C (1290 F) (d) 725 C (1340 F)

Fig. 9. Microstructures of the 12Cr-1.5Mo-1W steel (Heat 6455) aged 1000 hours at various temperatures, etched electrolytically in 50% $HNO_3$ with reverse polarity.

temperature up to 700 C (1290 F). Some dissolution of the intragranular precipitates and coarsening of the grain boundary precipitates are observed at 700 C, but no gross microstructural changes occur up to this temperature. The precipitates were identified by x-ray diffraction as primarily $M_{23}C_6$. At 725 C (1340 F), both recrystallization and precipitate coarsening are observed, both effects accelerating with increasing temperature.

Creep Rupture Properties

The creep rupture strength and minimum creep rate of the duplex structure steels are presented in Figs. 10a and b, respectively, and are compared to the corresponding properties for the martensitic 12Cr-1Mo-V steel. The properties of the

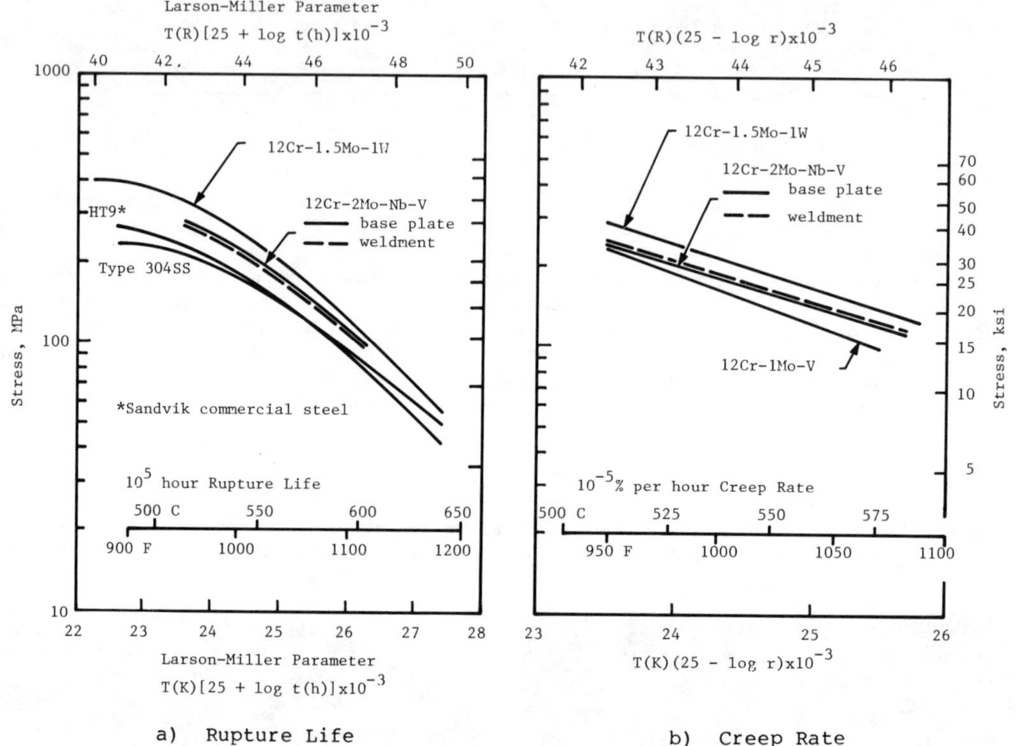


a) Rupture Life b) Creep Rate

Fig. 10. Rupture properties

12Cr-2Mo-V-Nb steel weldments, Heat P2322, creep tested in the as-welded condition are also indicated. The duplex structure steels in general and the 12Cr-1.5Mo-1W steel in particular exhibit appreciably greater creep resistance than the martensitic steel. This is seen in both the higher creep strengths and the lower minimum creep rates. The weldments exhibit 90% of the rupture strength of the corresponding base plate material and reduced minimum creep rates, indicating that welded sections of the duplex structure steels will not pose problems in terms of reduced creep resistance.

Oxidation Resistance

In neither the cyclic (1,500-hour) nor continuous (4,000-hour) oxidation tests at 650 C (1200 F) in air did the 12Cr-1.5Mo-1W steel exhibit significant weight change. It thus appears that the 12% chromium duplex structure steels offer excellent oxidation resistance.

## CONCLUSIONS

The improved weldability of duplex structure 12Cr-2Mo and 12Cr-(Mo,W) steels over that of the martensitic 12Cr-1Mo-V steels is attributed to the presence of large fractions of ferrite in the heat affected zone, and the consequent reduction in HAZ hardness. The ferrite in these steels results from the reduced carbon content and increased level of molybdenum or molybdenum plus tungsten. A steel of nominal 12C-1.5Mo-1W composition exhibits particularly attractive mechanical properties and creep behavior, with higher creep strength and lower minimum creep rates than either the martensitic 12Cr-1Mo-V steels or austenitic Type 304 stainless steel at temperatures as high as 650 C (1200 F). Weldments of the duplex structure steels exhibit creep resistance nearly equal to that of the base plate.

## REFERENCES

1. Sandvik HT9 - Technical Information on Seamless Tubing and Bar Materials.

2. N.G. Persson, "Characterization of Hardenable 12% Cr Steels for Energy Conversion Systems," published in "Alloys for the Eighties," pp. 143-150, Climax Molybdenum Company, 1980.

3. R.B. Dooley, G.G. Stephenson, M.J. Tinkler, M.D.C. Moles and H.J. Westwood, Welding Research Supplement, February 1982, pp. 45s-49s.

4. T. Wada, P.J. Grobner and E.J. Vineberg, "Duplex Structure 12Cr-Mo Steels with Improved Weldability," published in "Ferritic Steels for High-Temperature Applications," Ashok K. Khare (editor), ASM Conference Proceedings, 1983.

# WELDING 304L STAINLESS STEEL TUBING HAVING VARIABLE PENETRATION CHARACTERISTICS

I. Grant*, M.J. Tinkler**, G. Mizuno* and C. Gluck***

**Atomic Energy of Canada Ltd., Sheridan Park Research Community, Mississauga, Ontario*
***Ontario Hydro Research Division, 800 Kipling Avenue, Toronto, Ontario*
****Ontario Hydro Generation Projects Division, 700 University Avenue, Toronto, Ontario*

## ABSTRACT

Following difficulties with heat-to-heat variation in weld penetration enountered during GTA welding of stainless steel tubing, a program was undertaken to identify the cause of the problem and to develop remedial techniques. The variation in weld penetration was attributed primarily to differences in sulphur and, to a lesser extent, oxygen content, in broad agreement with a published model for such effects based on surface tension. The incorporation of copper alloy heat sinks into the standard welding head was successful in overcoming the problem. Both argon - 1% oxygen shielding gas and multipass welding procedures have been demonstrated to be promising alternative solutions, but these techniques have not been fully developed. To prevent the recurrence of the problem in future construction, a limit of 100-200 ppm sulphur has been included in the material specification for tubing purchases.

## KEYWORDS

Stainless steel; 304L; tubing; Gas Tungsten Arc; variable weld penetration; sulphur; oxygen; heat sinks; argon-oxygen mixtures; multipass.

## INTRODUCTION

Type 304L stainless steel tubing is used extensively in instrumentation systems for the CANDU (Canadian Deuterium Uranium) nuclear electricity generating stations. The systems in a typical reactor unit require some 50,000 m of tubing in sizes ranging between 6 mm OD x 1.2 mm wall thickness to 25 mm x 2.4 mm. Installation of this tubing involves about 15,000 butt welds. The welds are produced at site using a portable orbital Gas Tungsten Arc (GTA) welding machine which fuses the tube ends, without the addition of filler, inside a protective chamber flooded with argon. The welding current, rotation speed, and gas flows conform to a preset cycle, which is standardized for each tube size, under the control of the welding power supply (Delaney, 1978). Table 1 lists a typical standard procedure and also gives details of some of the experimental techniques described below. Many of the instrumentation systems contain reactor primary coolant, and lack of fusion defects in the welds are not acceptable.

This method was used to join instrument tubing during the past construction of the 4-unit Bruce 'A' and Pickering 'A' generating stations in Ontario. Experience at that time was very satisfactory, with low weld defect rates. Recently, however, during the construction of several new generating stations, difficulties have been experienced with variable and irregular weld penetration causing lack of fusion defects in the welds. These effects, which occurred although the standard welding procedures were being used, were observed only when joining certain batches of tubing. The increased weld reject rates which resulted, and the need for increased inspection of welds and sorting of tube material according to weldability, delayed and added substantially to the costs of tubing installation.

Therefore, a research and development program was conducted with the objective of identifying the nature of the material-related variation in tube weld quality, and developing remedial techniques for field application. The results of this program are outlined in the following sections.

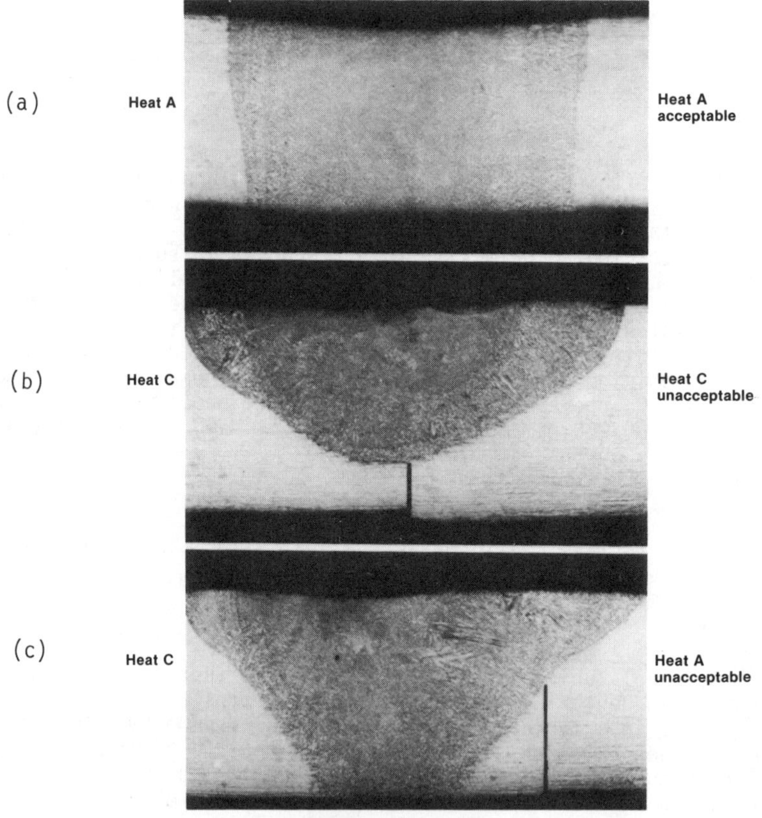


Fig. 1 Base material effects on GTA weld cross-section in 25 mm x 2.4 mm 304L tubing.

## THE NATURE OF MATERIAL-RELATED VARIATION IN TUBE WELD QUALITY

Field experience had been that the differences in welding behaviour occurred between heats of tubing, and this was confirmed by subsequent investigation. Some heats of tubing would give wider, shallower welds than normally expected using the standard welding procedures, as illustrated in Figs. 1(a) and (b). Different heat treatment lots, or even different tube sizes, within a heat exhibited similar behaviour. Heats of tubing were thus classified broadly as having a high ratio of weld penetration depth/width (D/W) or a low D/W. The D/W ratio has been found to be a useful indicator of the fusion characteristics of a material, for instance (Metcalfe & Quigley, 1977).

Of further, major concern was the fact that when high D/W tubes were welded to low D/W tubes, the weld paradoxically melted more of the low D/W tube. This was termed 'weld puddle shift' and is illustrated in Fig. 1(c). The skewed weld cross-section again often caused lack of joint fusion. When weld puddle shift occurred, the weld pool assumed an asymmetric, kidney-shaped appearance in contrast with the normal oval shape, as illustrated in Fig. 2. In this case, the deepest weld penetration occurred off-centre in the pool, towards the rear on the low D/W tube side.

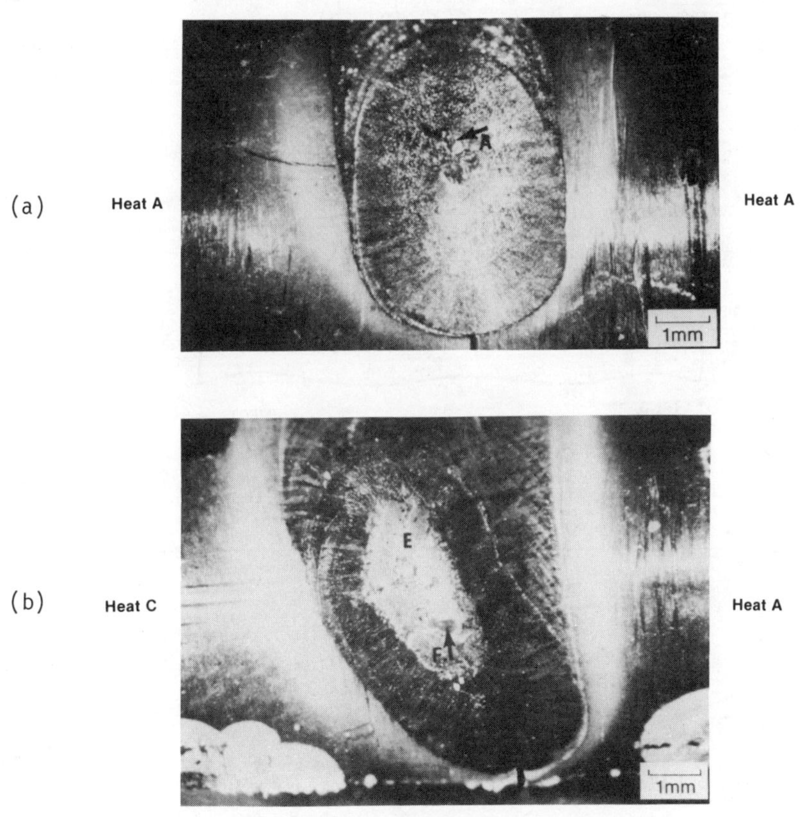


Fig. 2 Weld pool shape when joining a) similar materials, and b) dissimilar materials.

The tubing in the construction inventory was studied in order to identify factors which correlated with welding behaviour. A strong link between weld D/W or weld puddle shift and chemical composition, primarily the sulphur content, emerged from this study. Heats of tube with a low sulphur content, less than 40 parts-per-million (ppm), were those which gave reduced or irregular penetration when welded to themselves and which attracted the weld when joined to tubes having a high sulphur content, greater than 80 ppm. For instance, in Fig. 1, heat 'A' had a sulphur content of approximately 90 ppm and heat 'C' less than 30 ppm.

The link between the sulphur content and weld puddle shift was further explored as follows. Samples from 19 heats of 9.5 mm 304L tubing in the construction inventory, with sulphur contents in the range 15 to 140 ppm, were welded in turn to tubes at the high and low ends of this range. The welds were then sectioned axially and prepared metallographically to enable precise measurement of the weld cross-section with respect to specially-made witness marks on the tube outer surface. Fig. 3 provides a key to weld dimensions defined by this technique. The change in weld location, in terms of centre-line shift, with sulphur content is illustrated graphically for the two cases by Figs. 4(a) and (b).

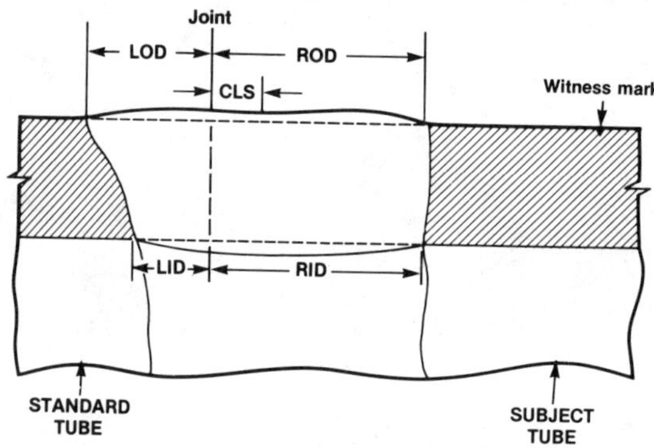


Fig. 3 Key to weld dimensions.

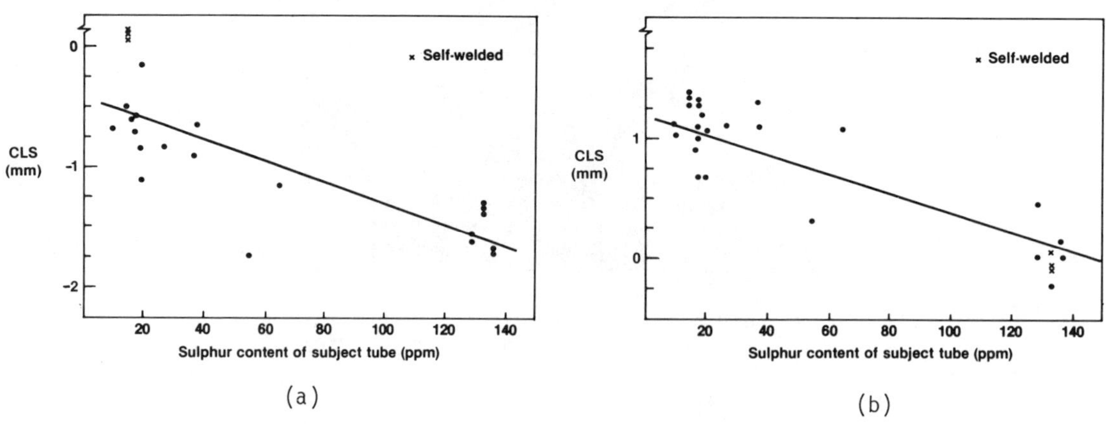


Fig. 4 Variation of CLS in 9.5 mm tubing when subject tubes are welded to: a) a 'low-sulpur' standard, S=15 ppm, and b) a 'high-sulphur' standard, S=133 ppm.

Other compositional factors could not be ruled out, however. Workers elsewhere have observed recently that variation in the oxygen content of austenitic stainless steels appears to cause changes in weld penetration (Fihey and Simoneau, 1982; Heiple and Roper, 1982a). A test for effects of oxygen was therefore carried out. Four heats were selected from the previous 19 which had sulphur contents of either 20 or 65 ppm and oxygen contents of approximately 40 or 80 ppm, forming a complete factorial experiment with sulphur and oxygen at two levels. These tubes were welded to a high-sulphur standard, then sectioned and measured as before. The results are illustrated in Fig. 5. Oxygen was observed to have effects similar to, but somewhat weaker than, sulphur.

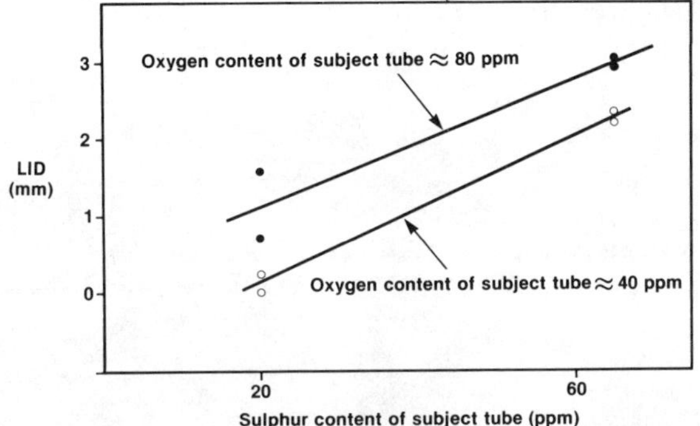


Fig. 5 Variation of LID with sulphur and oxygen content in 9.5 mm tubing.

It has been suggested also that the level of alkali or rare-earth elements in austenitic stainless steels can affect welding characteristics (Ludwig, 1968). According to the tube supplier, rare-earth treatments were not used at any time during steel refining and, consequently, the levels of these elements generally would be expected to be less than 10 ppm. There was no evidence of significant differences in the calcium content of heats having different penetration characteristics. However, there was some indication from the field trials that low sulphur heats having slightly increased levels of aluminum, titanium, or boron exhibited more extreme low D/W behaviour than other low sulphur heats. The results with low sulphur tubing generally were more variable and less well-explained on the basis of sulphur alone, as is seen clearly from the scatter in Figs. 4(a) and (b).

## DEVELOPMENT OF REMEDIAL TECHNIQUES FOR FIELD APPLICATIONS

A welding problem which can be attributed to variations in material composition may be dealt with at that level. The material specification for the purchase of stainless steel tubing for instrumentation systems has been revised to limit sulphur content to the range 100-200 ppm, and to prohibit any new alloy or trace element additions. This should prevent a recurrence of the problem in future.

It was, however, necessary to deal with the existing tubing inventory. Interim guidelines for field welding were defined, prohibiting the joining of high sulphur (>80 ppm) tubing to low sulphur (<40 ppm) tubing. This approach did allow instrument tube welding to progress, but was slow and costly. Thus the development of more suitable long-term solutions remained a high priority. Of some seventeen techniques evaluated, the three described below provided sufficient promise in terms of effectiveness and ease of implementation in the field to merit further investigation and development.

Copper Alloy Heat Sinks

One control concept was to limit the extent of tube melting on either side of the weld joint by positioning heat sinks around the tubes being welded. Accordingly, prototype heat sinks were manufactured from a copper alloy in the form of relatively thick cylindrical sleeves machined to fit around 9.5 mm diameter tubing. This technique was found to be extremely successful, reducing weld puddle shift to negligible values even for the most difficult to weld tube combinations. Therefore, a welding procedure was developed (see Table 1) and copper alloy heat sinks fitted into the weld heads for use in field construction, as illustrated in Fig. 6. These modified weld heads have now been used in production for some time with weld quality and reject rate as good as, or better than, the levels attained prior to the emergence of the weld puddle shift problem.

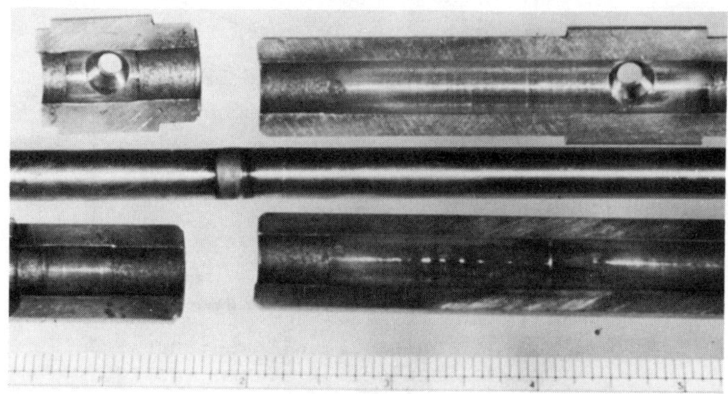

Fig. 6 Copper heat sinks for 9.5 mm tubing.

The heat sink design was complicated by several factors. The thick section heat sinks used in the preliminary trials could not be fitted into the low profile weld head for small diameter ($\leq$12.6 mm) tubing, which is required to deal with areas of restricted access. Clearance within this head limited heat sink thickness to 1.8 mm. With thinner section heat sinks, problems were quickly encountered when bending of the longer heat sink (see Fig. 6) led to poor surface contact and loss of effectiveness. Even the use of a high-strength heat-treated beryllium copper alloy did not alleviate this problem, so relatively minor modifications were finally made to the existing weld heads to allow the incorporation of heat sinks

TABLE 1 - WELDING CONDITIONS FOR 9.5 mm TUBING

| Technique | Start Current (A) | Finish Current (A) | Fixture Rotation Speed (RPM) | Arc Length (mm) | Shield & Backing Gas | Other |
|---|---|---|---|---|---|---|
| Std. Procedure | 35 | 26 | 4.6 | 1 | Argon | |
| Copper Heat Sinks | 54 | 54 | 1.7 | 1 | Argon | Heat sink gap = 5 mm |
| Argon - 1% Oxygen | 38 | 25 | 4.0 | 1 | Argon - 1% $O_2$ | |
| Multipass | Pulsed Current 54 peak 33 background | | 8.5 | 1 | Argon | 7 passes (continuous) |

4.6 mm thick. This increased cross-section provided the necessary heat absorption and mechanical stiffness characteristics.

An additional concern noted during the development program was the occurrence of occasional copper contamination of the weld surface. Although this was mostly innocuous mechanical transfer onto the solidified weld or tube surface, a few isolated instances of unacceptable fusion line defects were observed. A related issue was the fact that performance of the heat sinks in controlling weld puddle shift could be adversely affected by wear, which would reduce surface contact. Wear-resisting nickel plating and titanium carbide coatings were evaluated. The titanium carbide coating performed extremely well, and was consequently included in the heat sink design. Coated areas are characterized in Fig. 6 by their darker appearance.

### Shielding Gas Mixtures and Multipass Procedures

Two other techniques for controlling weld puddle shift were sufficiently promising to merit more detailed study. Previous workers have reported that small additions of oxygen or hydrogen to the argon shielding gas improves weld penetration in difficult to weld stainless steel tubing (Carrick & Paton, 1982; Fihey & Simoneau, 1982). Three shielding gas mixtures were tested: argon - 0.1% oxygen, argon - 1.0% oxygen and argon - 5% hydrogen. To find appropriate welding procedures for the three gas mixtures, systematic experiments were employed following a response surface methodology (Wang & Ramussen, 1972), with welding start current, end current, and welding fixture travel speed being the factors varied.

The test results indicated that the argon - 1.0% oxygen mixture significantly reduces weld puddle shift using the welding procedure given in Table 1. Surprisingly, the argon - 0.1% oxygen and argon - 5% hydrogen mixtures were found to provide no significant benefit. Due to the success of the copper heat sink solution, however, no further effort was made to develop the argon - 1% oxygen technique for field use.

The concept of using a multipass welding technique for instrument tube welding stemmed from the observation that for the standard welding procedure, the small overlapping weld bead formed during current slope-down was typically much better centred than the main weld bead itself. In the preliminary test program, a seven pass pulsed current welding procedure was identified which provided promising results. This procedure, given in Table 1, was then used to make about 50 identical welds with each of three commercially available tube welding systems. Although weld puddle shift was substantially reduced on average, the phenomenon could not be controlled consistently and some cases of lack-of-fusion were observed. It is possible that the welding equipment may have been at fault in being unable to provide satisfactory repeatability of programmed welding parameters under relatively severe working conditions. Since no attempt had been made to optimize the multipass welding procedure prior to consistency testing, however, it is more likely that the procedure used may have been unnecessarily sensitive to relatively minor variations in the set welding parameters. Again, due to the success of the heat sink technique, no attempt was made to further develop the multipass procedure for field application.

## DISCUSSION

### Causes of Material-Related Variation in Tube Weld Quality

Base material effects on GTA weld penetration have been reported in journals and conference proceedings for the past 25 years. The problem illustrated here displays many of the symptoms described in the literature, namely: weld D/W varies between heats of material and, when different heats are joined, the weld is attracted or skewed towards the low D/W heat, as shown previously by Moisio & Leinonen (1980). It is this variability that is of practical significance and which leads to unacceptable results in some cases when standard welding procedures are applied.

The variability of weld penetration is generally believed to be due to differences in the residual or minor alloying content of the material, since it occurs mainly between heats of material or between materials made by alternative refining routes, such as air melted or vacuum arc refined steels. Previous studies have not, however, produced widespread agreement on which specific elements are the cause. On aggregate, the results have simply emphasized the lack of understanding of the basic mechanisms (Glickstein & Yeniscavich, 1977).

Heiple and co-workers (1980, 1981, 1982a) recently conducted a well-controlled series of experiments which showed that sulphur, oxygen, selenium, tellurium, cerium, and aluminum additions to austenitic stainless steels affect fused zone shape. Heiple & Roper (1982b) proposed a model to explain these results based on the premise that fluid flow is generally the major factor influencing weld pool shape. In particular, the model states that surface tension gradients across the surface of the weld pool cause fluid flow known as 'Marangoni Convection', and that the surface tension gradients are sensitive to small concentrations of surface active elements in the weld pool, as illustrated schematically in Fig. 7.

Our investigation indicates that the weld D/W characteristics of heats of 304L tubing, irregular weld penetration, and weld puddle shift are related to the chemical composition of the tubing. Material with relatively low sulphur content exhibits low D/W ratios whereas material with higher sulphur content exhibits higher D/W ratios. When joining high to low sulphur material, the weld shifts towards the low sulphur material. Varying oxygen content has effects similar to, but somewhat weaker than, those of sulphur and the effects of the two elements appear to be additive over the range tested. Sulphur and oxygen are surface active in liquid iron, therefore our results may be interpreted directly in terms of the model above (Figs. 1 and 7). Using the data for surface tension given by Gupt and others (1976) it is possible to show approximately, following the simple analysis of Andersson (1974), that the surface tension forces are of a sufficient order of magnitude to cause convective motion. The poorer correlation between sulphur and weld puddle shift with lower sulphur tubing may be due to other elements assuming greater importance as the level of the principal surface active component, sulphur, is reduced. Variation in oxygen content may well account for much of the scatter in Fig. 4. The possible effect of aluminum, titanium, and boron at low sulphur levels may be explained on the basis that, although they are not surface active, they would combine with oxygen, effectively removing the remaining surface active component from the system.

Alternative explanations for minor element effects on GTA fused zone shape will be found in the literature. Many of these invoke interaction between the base material and the welding arc, for instance as proposed recently by Fihey and Simoneau (1982) and by others previously (Bennet & Mills, 1974; Ludwig, 1968; Metcalfe & Quigley, 1977; Savage & co-workers, 1977). The useful heat from the GTA arc is produced mainly in the current carrying region on the work-piece surface known as the 'anode spot' due to the work function of the surface and the anode voltage drop (Nestor, 1962; Quigley & co-workers, 1973; Schoek, 1963). According to explanations of this type, the physical nature of the arc is sensitive to small concentrations of certain elements in the workpiece, thereby affecting the heat generation in, or the location of, the anode spot. It is well known also that fluid motion in the weld pool may be caused by electro-magnetic induction, which can affect weld pool shape appreciably, depending on the electrode geometry and the current distribution on the surface (Andrews & Craine, 1978; Lawson & Kerr, 1976; Woods & Milner, 1971).

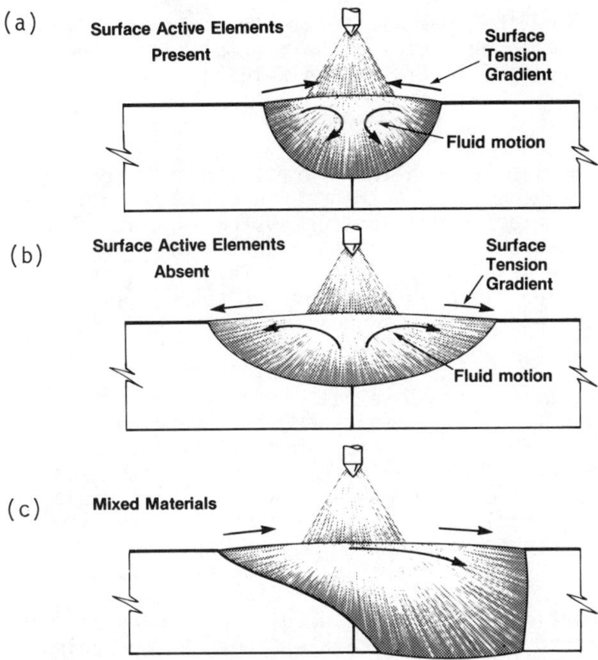


Fig. 7 Marangoni Convection model for minor element effect on GTA fused zone shape. (After Heiple and Roper 1982b).

However, Heiple and co-workers(1983) recently reported an experiment in which a defocussed laser was used to produce weld beads similar to GTA welds. Surface active elements were found to have effects on the laser weld D/W similar to those reported earlier for arc welds, in the absence of any possible arc/material interactions or electromagnetically-induced fluid flow. In view of this most recent work by Heiple and colleagues, we feel that our results with production heats of steel are best explained in terms of the surface tension model and, conversely, that our results lend support to this model as indicating the primary cause of base material induced variation in GTA fused zone shape.

Despite the broad agreement between our results and the model predictions, there remain several detail points which require elucidation, concerning especially the case of weld puddle shift when different heats are joined. From the shapes assumed by the weld pool, Fig. 2, the patterns of fluid flow must be more complex than that depicted in the simple model of Fig. 7(c). Further, it is unclear how compositional differences across the surface of the weld pool could persist strongly in view of the vigorous motion which they are supposed to cause, and therefore how weld puddle shift could be other than a transient phenomenon at the beginning of a weld bead. In connection with joint tracking sensors, Chin and colleagues (1983) showed using infra-red thermography that a slight offset of the arc heat source from the interface between two plates causes a marked perturbation in the classical symmetric pattern of isotherms around the moving weld pool. Preferential heat transfer to one side of the joint by convection in the weld pool may be viewed in effect as an offset of the heat source from the joint, and the thermal resistance of the interface would have similar effects on the temperature field around the weld pool. It may be that surface tension driven motion continues weakly throughout the weld due to the melting of fresh material at the front of the pool, serving to initiate weld puddle shift, which is then exacerbated by the effect of the interface on the thermal field. It would be interesting to verify this proposal experimentally by changing the thermal resistance of the interface, perhaps by diffusion bonding prior to fusion welding.

Practical Solutions

The heat sinks are effective because they permit a higher heat input to the weld pool, overwhelming imbalances in heat transfer due to convective motion. The sensitivity to surface contact conditions demonstrates that the mechanism is a thermal effect rather than a direct influence on the arc. The heat sinks involved only minor changes to the welding procedure and welding equipment, and have been successfully applied in the field to a variety of tube sizes. As an indirect benefit, the welding procedure now has greater tolerance to small changes in welding parameters.

The results with gas mixtures were somewhat contrary to those reported previously. Fihey and Simoneau found that 1% oxygen tended to reduce penetration whereas 0.1% oxygen improved it. This discrepancy is possibly due to differences in welding techniques, since we found the effectiveness of the 1% oxygen mixture to depend on the welding parameters. Argon - oxygen mixtures would be very easy to apply in practice, using premixed bottles. The results are also of academic interest because the increased penetration may be explicable in terms of the Heiple and Roper model. Clearly this is a candidate for further research.

The results with the continuous multipass welding procedure also indicated some, inconsistent, improvement. It is uncertain whether the mechanism in this case is homogenisation of the weld metal by successive passes or equilibration of the thermal field around the weld between subsequent passes. With modern equipment, multipass orbital welding is easily achieved, so that this technique also would be worth further investigation.

A longer term solution is to specify the content of surface active elements in order to ensure uniform weld penetration characteristics. We believe that this may be readily achieved by restricting the sulphur content to between 100 and 200 ppm. Tubing has been ordered to such a specification and the preliminary indications are of satisfactory weldability. The major problem with this route is controlling the unreported residual elements. The Heiple and Roper model predicts that if sulphur is 'high' then the effects of varying concentrations of oxygen or oxygen getters such as aluminum would be minimal. However, elements which combine with sulphur, such as cerium, might be expected to induce low D/W behaviour even if sulphur is high. All such elements and their effects have not been identified.

## CONCLUSIONS

Heat-to-heat variation in the welding characteristics of 304L tubing has been attributed to differences in the level of minor elements, primarily sulphur. Tubing with sulphur contents less than 40 ppm exhibits reduced or irregular penetration compared to tubing with higher sulphur contents and, when low sulphur tubing is joined to higher sulphur tubing, the weld shifts markedly towards the low sulphur side. The oxygen content appears to have additive effects similar to, but somewhat weaker than, those of sulphur, and there is some evidence that also the levels of aluminum, titanium, and boron affect fused zone shape when sulphur is low.

The results are in broad agreement with a published hypothesis for minor element effects on fused zone shape based on surface tension induced motion in the weld pool.

Copper heat sinks are a practical and effective means of controlling weld puddle shift and irregular penetration in tube welding. Argon-oxygen shielding gas mixtures and multipass welding may be potential alternative remedies, but require further work to demonstrate consistent improvements. Specifying a minimum limit as well as a maximum for sulphur content of 100 ppm and 200 ppm respectively will result in more uniform weldability of future material supplies.

## ACKNOWLEDGEMENTS

The authors would like to thank the managements of Atomic Energy of Canada Limited and Ontario Hydro for permission to publish this paper.

This paper previously appeared in the proceedings of the conference on 'The effects of residual, impurity, and micro-alloying elements on weldability and weld properties', published by The Welding Institute, Abington Hall, Abington, UK.

## REFERENCES

Andersson, D. (1974). Streaming due to a thermal surface tension gradient. Weld. Res. Abroad, 20, 55-64.

Andrews, J.R. and R. Craine (1978). Fluid flow in a hemisphere induced by a distributed flow of current. J. Fluid Mech., 84, 281-289.

Bennet, W.S. and G.S. Mills (1974). GTA weldability studies on high-manganese stainless steel. Weld. J., 53, 548s-553s.

Carrick, L. and A. Paton (1982). Materials and welding of small bore pipework for nuclear fuel reprocessing plant. Proc. Conf. Welding Technology for Energy Applications, American Society for Metals, Metals Park, Ohio. 281-289.

Chin, B.A., N.H. Madsen, and J.S. Goodling (1983). Infrared thermography for sensing the weld process. Oral presentation to the 64th American Welding Society Annual Convention, Philadelphia, Pennsylvania.

Delaney, M. (1978). Metal Constr., 10, 32-33.

Fihey, J.L. and R. Simoneau (1982). Weld Penetration Variation in GTA welding of some 304L stainless steels. Proc. Conf. Welding Technology for Energy Applications, American Society for Metals, Metals Park, Ohio.

Glickstein, S.S. and W. Yeniscavich (1977). A review of minor element effects on the welding arc and weld penetration. Bulletin No. 226, The Welding Research Council, New York.

Gupt, K.M., V.I. Yavoiski, A.F. Vishkarov, and S.A. Bliznukov (1976). Study of the effect of concentration and temperature on the surface tension of the Fe-O-S system. Trans. Indian Inst. of Metals, 29, 286-291.

Heiple, C.R., R.J. Cluley, and R.D. Dixon (1980). The effect of aluminum on GTA weld geometry in austenitic stainless steel. Proc. Conf. Physical Metallurgy of Metal Joining. The Metallurgical Society of the American Institute of Mining, Metallurgical, and Petroleum Engineers, Warrendale, Pennsylvania. 160-165.

Heiple, C.R., and J.R. Roper (1982a). Effects of minor elements on GTAW fusion zone shape. Proc. Conf. Trends in Welding Research in the United States, The American Society for Metals, Metals Park, Ohio. 489-520.

Heiple, C.R. and J.R. Roper (1982b). Mechanism for minor element effect on GTA fusion zone geometry. Weld. J., 61, 97s-102s.

Heiple, C.R., J.R. Roper, R.T. Stagner, and R.J. Aden (1983). Surface active element effects on the shape of GTA, laser, and electron beam welds. Weld. J., 62, 72s-77s.

Lawson, W.H.S., and H.W. Kerr (1976). Fluid motion in GTA weld pools: Part 1, Weld. Res. Intl. 6, No. 5, 63-77; Part 2, No. 6, 1-17.

Ludwig, H.C. (1968). Current density and anode spot size in Gas Tungsten Arc. Weld. J., 47, 234s-240s.

Metcalfe, J.C. and M.B.C. Quigley (1977). Arc and pool instability in GTA welding. Weld. J., 56, 133s-139s.

Moisio, T. and J. Leinonen (1980). The influence of minor elements on the welding arc and weld penetration in austenitic stainless steel. Proc. Conf. Arc Physics and Weld Pool Behaviour, The Welding Institute, Abington, Cambridge, 285-288.

Nestor, O.H. (1962). Heat intensity and current density distributions at the anode of high current inert gas arcs. J. Appl. Phys., 33, 1638-1648.

Quigley, M.B.C., P.H. Richards, D.T. Swift-Hook, and A.E.F. Gick (1973). Heat flow to the workpiece from a TIG welding arc. J. Phys. D.: J. Appl. Phys., 6, 2250-2258.

Savage, W.F., E.F. Nippes, and G! \ Goodwin (1977). Effect of minor elements on fused zone dimensions of Inconel 600. Weld. J., 56, 126s-132s.

Schoek, P.A. (1963). An investigation of the anode energy balance of high intensity arcs in argon. In W. Ibele (Ed.), Modern Developments in Heat Transfer, Academic Press, New York. 353-400.

Wang, K.K., and G. Ramussen (1972). Optimisation of inertia welding process by response surface methodology. J. Eng. for Industry, 94, 999-1006.

Woods, R.A. and D.R. Milner (1971). Motion in the weld pool in arc welding. Weld. J., 50, 162s-173s.

# WELD POOL GEOMETRY VARIATION IN GTA WELDING OF AUSTENITIC STAINLESS STEEL

J.-L. Fihey and R. Simoneau
*Production et Utilisation de l'Énergie*
*IREQ, Institut de Recherche d'Hydro-Québec*
*Varennes, Québec J0L 2P0*

ABSTRACT

A number of Weldability problems have been reported with automatic GTA welding of austenitic stainless steels. Variation in penetration from heat to heat and weld pool shift have raised concern in the welding community about the reliability of the GTAW process for these applications. The effect of slight variation in the composition of the base metal is now recognized. Our experience is that sulfur and oxygen content of the parent metal as well as oxygen content of the shielding gas may affect the penetration pattern. In this work the weldability characteristics of a low sulfur 317 stainless steel plate is investigated. A correlation is made between the measured thermal profile of the weld pool, the visual observation of the weld pool surface and the anode movement. A new experimental set-up was designed to monitor the anode position. The experiments show that any shift of the weld pool is accompanied by a smaller shift of the anode. It is suggested that the formation of a blue vapor on the weld pool surface plays a major role in the heat redistribution process.

KEYWORDS

Weldability, stainless steel, penetration, weld pool and anode shift, residual elements.

## INTRODUCTION

A number of weldability problems have been reported with automatic GTA welding of stainless steels. Variation in penetration from heat to heat and weld pool shift have raised concern in the welding community about the reliability of the GTAW process for these applications. According to the research work performed during the last years the effect of slight variation in the composition of the base metal is now recognized (Heiple and Roper, 1982 a; Fihey and Simoneau, 1982). It has been shown in particular that sulfur has a strong effect. In our laboratory we experimented with nine heats of 304 L stainless steel and one heat of 317 stainless steel. Five of them including the 317 had a sulfur content equal to or less than 0.003 wt %; the five remaining heats had a sulfur content equal to or greater than 0.010 wt %. The low sulfur heats exhibited constantly a poor penetration profile compared to the high sulfur heats. There is not yet a complete agreement as to the mechanism which controls this phenomena. Two main trends can be identified:

- Penetration controlled by convection in the liquid pool. the convection flow being set up either by electromagnetic forces or surface tension gradient (Heiple and Roper, 1982b).
- Penetration controlled by the anode phenomena (Ludwig, 1968, Fihey and Simoneau, 1982).

In this work a new test is developped in order to measure the anode movement during a GTA weld pool shift or instability. The thermal profile of the weld pool is also measured. All the experiments are performed on a 13 mm thick 317 stainless steel plate. The composition of the plate is given in Table 1.

## EXPERIMENTAL TECHNIQUE

### Anode Position Measurement Test

The test specimen, shown in Figure 1, consists of a 13 mm thick 317 plate with a saw cut in the middle leaving only a small bridge 2 mm thick and 0.5 mm wide between the two parts. Two copper leads brazed on both sides of the plate conduct the current to two 150A-50 mV shunts. The anode position on the test specimen is related to the current difference $\Delta I$, between both sides of the plate. In the vicinity of the bridge there is a linear dependency between $\Delta I$ and the anode position x. For a current of 145 A the experimental calibration with a moving arc yields the following relationship:

$$\frac{\Delta I}{\Delta x} = 2.76 \text{ A/mm}$$

| C | Si | Ni | Cr | Mn | P | S | N | O |
|---|---|---|---|---|---|---|---|---|
| 0.018 | 0.31 | 14.20 | 18.37 | 1.65 | 0.033 | 0.002 | 0.038 | 0.005 |

Table 1: Chemical Analysis of the 317 Stainless Steel Heat.

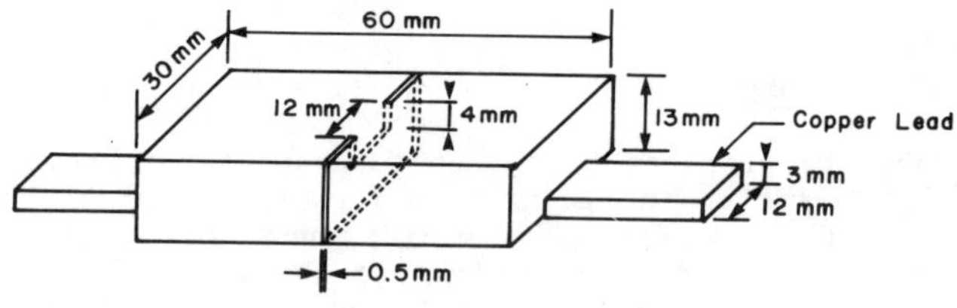


Fig. 1: Test specimen

When the arc is farther from the bridge, the temperature gradient between the two sides becomes too high and the resulting resistivity difference disturbs the measurement. All the experiments on the test specimen were conducted with a stationnary arc in the middle of the plate. The experimental set-up is schematically described in Figure 2. A GTAW torch is positionned above the specimen, the torch can be driven in the X and Y direction by two stepping motors. An alumina tube pointing towards one side of the weld pool could provide an auxiliary gas flow. The anode movement was monitored during an instability created either naturally or with the auxiliary gas.

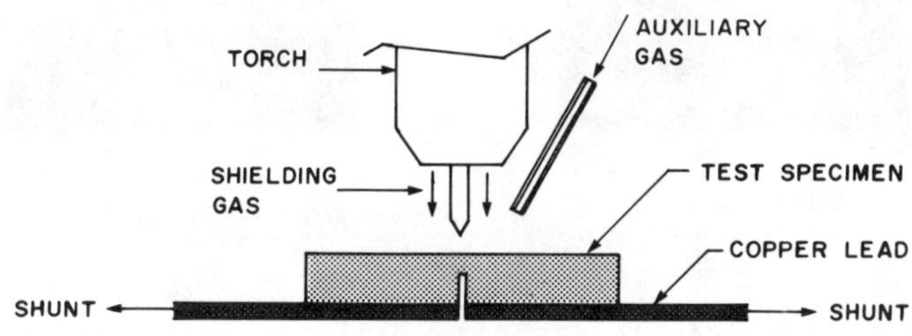


Fig. 2: Experimental set-up.

### Thermal Profile Measurement

The thermal profile was measured with an infrared optical fiber sensor which is an improved version of the pyrometer already described in another paper presented in this conference (Fihey and co-workers, 1983). The response time of this thermal system is 5 ms and the spot size 0.5 mm. The sensor was mounted on a driving mechanism which scanned the weld pool at a speed of 5 mm/s. The sensor position was simultaneously monitored.

## EXPERIMENTAL RESULTS

### Effect of Shielding Gas Composition

In order to test the weldability of the 317 plate a fusion path was made with the GTAW process. The current was 145 A, the thoriated tungsten electrode has a diameter of 2.4 mm with a 30° vertex angle, the arc length was 3 mm, the welding speed 1.5 mm/s. In a first experiment pure argon (99.999) shielding gas was used, the flowrate was 12 l/min. Under these conditions the weld pool is stable (very small oscillations) and a poor penetration profile is observed. For a bead width of 9.7 mm the depth of penetration is only 2.1 mm. In a second experiment with a mixture Ar - 0.1% $O_2$, the weldability changes dramatically. The weld pool is unstable (large oscillations) and the penetration increases (3.9 mm) whereas the weld pool width decreases (7.8 mm). The two different behaviour are well illustrated in Figure 3 which shows the weld bead surface for the two experiments. Note the weld pool oscillations in the second experiment.

The anode movement was then measured on a stationary weld pool on a test specimen. The results are reported in Figure 4. With the pure argon shielding gas the anode was stable and only a very small ripple is observed in the current, whereas with the Ar-0.1% $O_2$ mixture the weld pool instability is related to an instability of the anode. Any shift or excursion of the weld pool is accompanied by a shift in the same direction of the anode. The maximum current variation amounts to 1.1A corresponding to an anode displacement of 0.4 mm, whereas the maximum weld pool excursion is 2.6 mm (Fig. 5). After about 30 seconds the instability disappears almost completely.

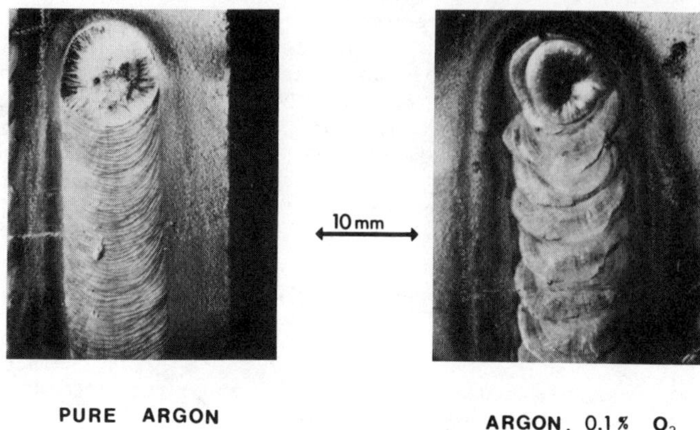


Fig. 3: Effect of shielding gas composition upon the weldability of a low sulfur 317 stainless steel plate. I = 145 A, welding speed 1.5 mm/s.

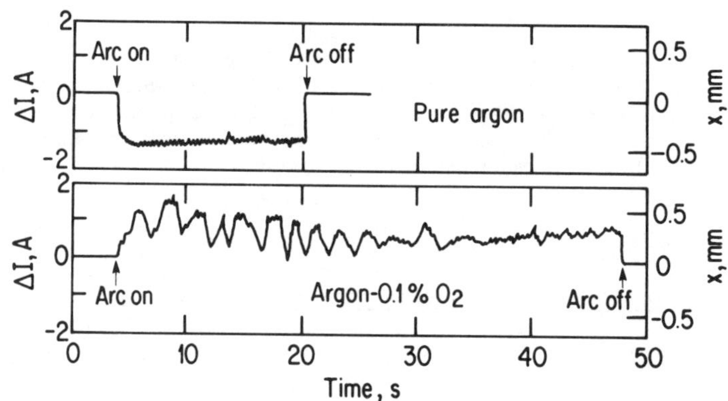


Fig. 4: $\Delta I$ as a function of time for the two types of shielding gases. Welding current 145 A.

The thermal profiles of the two types of weld pool have been recorded and are shown in Figures 6 a and b together with a schematic view of the weld pool surface. With pure argon shielding gas a blue ionized vapor covers most of the weld pool surface. The ripple in the thermal signal of the weld in the blue vapor zone (Fig. 6-a) as well as the sharp transition at the vapor edge suggests that this vapor rather than the liquid metal itself carries the energy outside the arc zone. With a mixture Ar −0.1% $O_2$ an unstable jet of blue vapor emerging from the arc zone extends over the liquid metal in the direction of the weld pool excursion. After about 30 seconds however the blue vapor disappears and the weld pool stabilizes. The thermal profile of a stabilized weld pool is shown in Figure 6-b. The thermal gradient in the liquid is now visible outside the arc zone. This gradient is quite flat compared to the gradient in the solid metal.

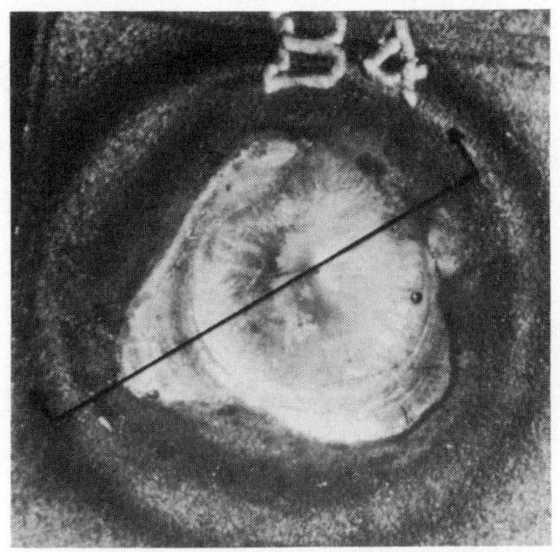

10 mm

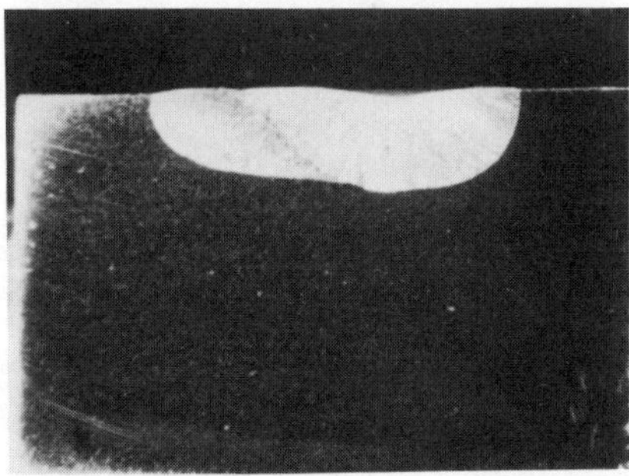

Fig. 5: Macrography of the surface and the cross-section of the weld fused zone showing the effect of the addition of 0.1% $O_2$ to the argon shielding gas on the weld pool stability.

## Weld Pool Shift Triggered by an Auxiliary Gas

In order to better understand the mechanism of the weld pool instability an experiment was prepared where a controlled weld pool shift could be triggered at will. This experiment involved the use of an auxiliary Ar-1% $O_2$ gas flow blown at a rate of 2 l/min towards the edge of a stable weld pool. A very large weld pool deviation could thus be created as shown in the macrographs of Figure 7. This phenomenon was again ephemeral with a lifetime of about 50 seconds. The weld pool center was shifted 3 mm to the left, away from the auxiliary gas. The shape of the weld pool was changed to an ellipsoidal configuration. The cooling action of the auxiliary gas as well as its mechanical blowing action were not involved since the phenomenon disappeared almost completely after a few tenths of seconds without interruption of the auxiliary gas flow.

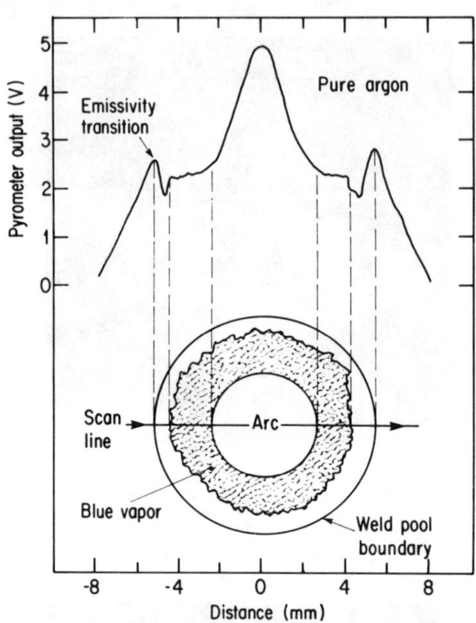


Fig. 6: Thermal profile and schematic view of the weld pool surface. a) pure argon shielding gas.

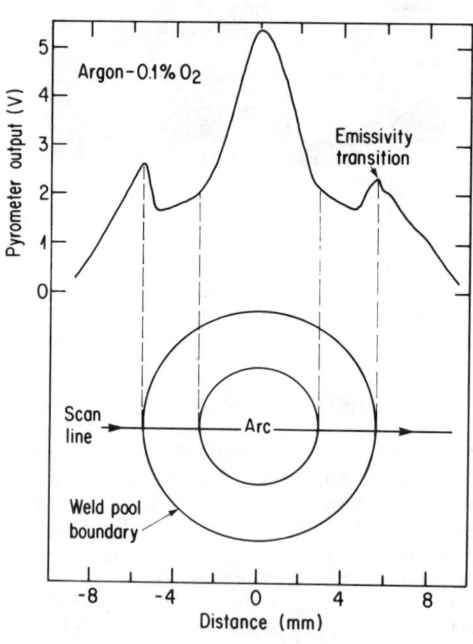


b) Ar-0.1% $O_2$.

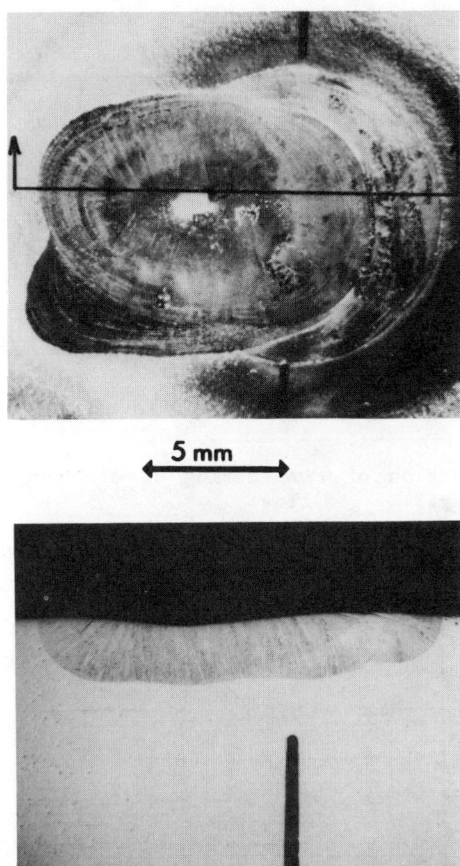


Fig. 7: Macrography of the surface and the cross section of the fused zone of a weld shifted to the left by an auxiliary argon -1% $O_2$ gas. Note the saw cut in the test specimen.

An anode shift of 1 mm in the same direction was simultaneously measured (Fig. 8). The blue vapor was again observed emerging from the arc root (Fig. 9-a and b). The arc position was recorded before and after the weld pool deviation. It was shifted about 0.9 mm in the direction of the weld pool deviation. The phenomenon occurring during this experiment is quite similar to the natural weld pool instability already observed and can be seen as a controlled amplification of this instability.

## DISCUSSION

In all the experiments described in this work any weld pool shift, either natural or triggered by an auxiliary gas, was accompanied by a shift in the anode position and an extension of the blue ionized vapor in the same direction. The anode shift

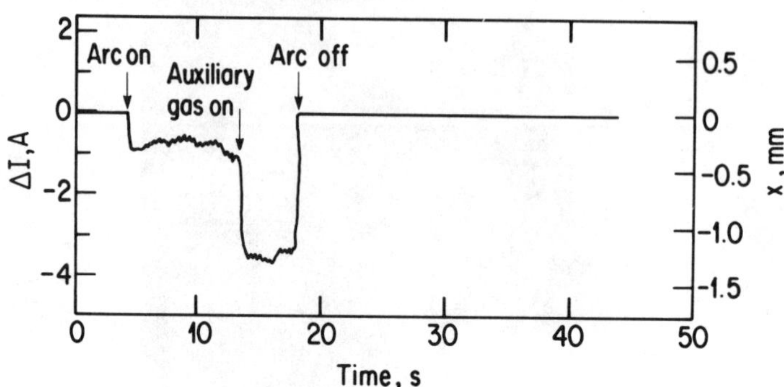


Fig. 8: ΔI as a function of time during a weld pool shift triggered by an auxiliary Ar-1% $O_2$ gas flow.

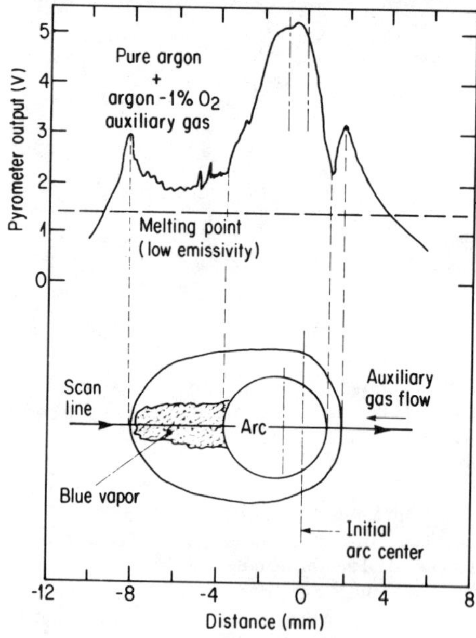


Fig. 9: Thermal profile and schematic view of the weld pool surface a) longitudinal scan

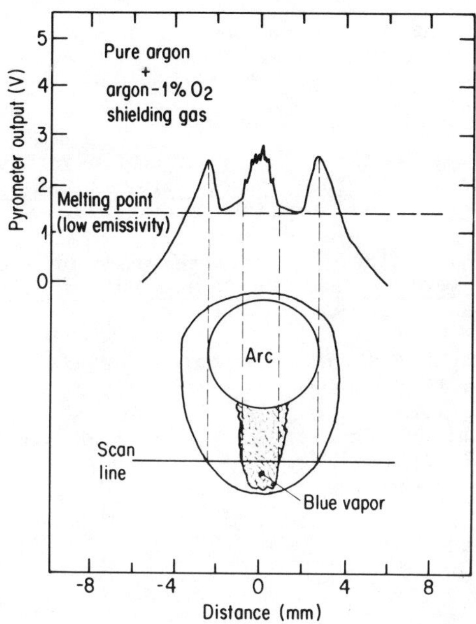


b) transverse scan.

is always smaller than the weld pool shift and may result only from a shift of the arc root as observed in Figure 9-a. The anode could thus be restricted to the arc root the heat being transported outside the arc zone through a secondary mechanism. It is not yet completely clear wether the energy is carried in the plasma or in the liquid pool, but the sharp transition in the thermal profile at the blue vapor edge suggests that this vapor carries the energy outside the arc zone. The blue vapor composition as well as its interaction with the weld pool must be thoroughly investigated for a complete understanding of the phenomenon.

## CONCLUSION

The GTA weldability of a low sulfur 317 stainless steel plate has been investigated. With pure argon shielding gas the plate exhibits a poor penetration profile. This penetration is increased with an Ar -0.1% $O_2$ mixture, however the weld pool may be unstable depending upon the welding parameters. The anode movement during an instability was monitored and the thermal profile was measured with an infrared sensor. The instability could be triggered at will using an auxiliary Ar -1% $O_2$ gas flow. Any weld pool shift is accompanied by a smaller shift of the anode in the same direction. A blue ionized vapor seems to carry some energy outside the arc zone thus playing a major role in the weld pool deviation and oscillation phenomena. More work has to be done in order to understand the interaction between this blue ionized vapor and the weld pool.

ACKNOWLEDGEMENT

The author wish to thank A. DiVincenzo and J. Larouche for assistance in the experimental work.

REFERENCES

Fihey, J.L. and R. Simoneau (1982). Weld penetration variation in GTA welding of some 304 L stainless steels. "Proceedings, International Conference on Welding Technology for Energy Applications" Gatlinburg, Tn, 139.

Heiple, C.R. and J.R. Roper (1982a). Effects of minor elements on GTAW fusion zone shape. "Trends in Welding Research in the United States", ed. S.A. David, Metals Park Ohio, 489.

Heiple, C.R. and J.R. Roper (1982b). Mechanism for minor element effect on GTA fusion zone geometry. Welding Journal, 61, (6), 97S.

Ludwig, H.C. (1968). Current density and anode spot size in the gas tungsten arc. Welding Journal, 47, 2345.

# NARROW GAP WELDS USING UNDER STRENGTH WELD MATERIAL

**Barry M. Patchett and Donald G. Bellow**
*University of Alberta, Edmonton, Alberta T6G 2G8*

## ABSTRACT

In the production of narrow gap welds cracking can often occur due to the low ductility of the high strength weld material as specified in the ASME Code. To overcome this problem a lower strength weld material with increased ductility is used in combination with various weld thickness to plate thickness ratios (aspect ratio). The purpose of this paper is to describe the experimental results which were obtained in evaluating the behaviour of under strength weld material in narrow gap welds in A516 Gr70 plate. Aspect ratios between 0.5 and 1.0 have been evaluated using 25 mm thick plate. The arc welding procedure used both an inert gas and mildly oxidizing gas. The strength of the weld was evaluated by tension tests performed on stress relieved specimens taken across the weld and along the direction of the weld. The results showed that the ultimate strength of the weld, which had an ultimate tensile strength 25% less than that of the base metal, actually increased by eight percent as the aspect ratio was reduced from 1.05 to 0.55.

## KEYWORDS

Narrow gap welds; understrength butt welds; aspect ratio; tensile strength; plastic flow stress.

## INTRODUCTION

Many pressure vessel fabrication codes, including the ASME Boiler and Pressure Vessel Code, require that the strength of the weld metal used in weld joints must exceed the minimum specified parent metal strength. Since the Code design criteria are based on ultimate tensile strength, an A516 Gr 70 joint must exceed 480 MPa (70 ksi) even after all heat treating cycles, including intermediate stress relief treatments and any extra heat treatments due to repairs. In very large pressure vessels, the number of intermediate heat treatments, including repairs, can lead to total heat treating times in excess of 10 hours. Many weld metals will experience a significant drop in yield and ultimate tensile strength during such heat treatments, often to the extent that the UTS drops below the minimum code criterion.[1] A higher strength electrode can be used to offset the loss of strength, but often at the expense of increased susceptibility to root pass cracking in restrained joints. It is therefore desirable to know if low strength or "undermatching" weld metal can be tolerated in fabrications, and is so, under what conditions of joint design and differential in both yield and ultimate tensile strength from base metal levels. Work by Satoh and Toyoda (1975, 1979) has shown that significant undermatching between weld metal and base metal strength can be tolerated in high-strength low alloy steels if narrow-gap joint configurations are used. This is due to the constraint on plastic deformation in the thin layer of weld metal by the stronger base metal which puts a triaxial stress on the lower strength metal. The early experiments by Satoh and Toyoda (1975) used a flash welding technique for most of their results, in order to obtain very low width to depth ratios in the weld metal, which they called the "relative thickness". They found that if the relative thickness is 0.25 or less, and the flat bar tensile test specimen width is five times the plate thickness, then the ultimate tensile strength of the complete weld joint will equal or exceed the specified base metal strength if the undermatching of the weld metal is 15% or less. To be specific, the low alloy base plate has an ultimate tensile strength of about 830 MPa - this strength was achieved in the joint using an E62016 (9016) electrode of about 660 MPa strength. Joint strength matching base metal strength can be assured if the strength disparity is 10% or less. Later work by Satoh and Toyoda (1979) showed that the lower strength weld metal not only minimized cracking in root passes but also allowed lower preheating temperatures by as much as 25°C. However, high-strength low alloy steels have very high yield strength, particularly as a percentage of ultimate tensile strength. The steel used by Satoh and Toyoda, for example, had a yield to ultimate ratio of 0.93, whereas typical C-Mn pressure vessel steels have ratios of about 0.55 to 0.70. C-Mn weld metals, on the other hand, have rather high yield to ultimate ratios which often exceed 0.75 (Wheatley and Baker, 1962). Therefore the behaviour of low-strength weld metals in low alloy steels is unlikely to be applicable to C-Mn steels and furthermore the high yield characteristics of C-Mn weld metals may improve performance in undermatching situations.

Narrow-gap welding procedures and processes have advanced rapidly in recent years. The most popular processes are the SAW and GMAW processes. The SAW process is most often used in an overlapping fillet weld (or dipass) procedure to fill a relatively narrow-gap of 20-25mm in width (Grist and Armstrong, 1980) but flux removal problems can occur in very narrow-gaps using a monopass (or one bead width) procedure. The GMAW process can be adopted to monopass procedures with gaps as narrow as 10mm in plates up to 300mm thick (Kimura et al 1979). The process has been used to fabricate a number of pressure vessels in C-Mn steels such as A533 GrB Cl 1 and A516 Gr70 and low alloy steels such as A387 Gr22 Cl 1 (Malin, 1983). However, these joints were all made with overmatching consumables, with no attempt to assess the effects of narrow-gap procedures on the tensile deformation characteristics of the weld joint.

Previous studies have therefore shown that narrow-gap welding processes are commerically viable, that C-Mn and low alloy steels can be fabricated into large pressure containing structures, and that undermatching weld metal can provide

---

[1] The loss in strength follows the Holloman-Jaffe parameter and can be predicted with reasonable accuracy if temperature and time cycles are known.

$$P = T(20 + \log t) \qquad T = {}^\circ K$$

$$t = \text{time in hours}$$

joints of tensile strength equal to the specified base metal strength in high-strength low alloy steels if a suitable weld aspect ratio (or weld gap to plate thickness ratio) is achieved. There is little knowledge available in the tensile deformation response of undermatching C-Mn weld metals in narrow-gap welds for thick plates of widely used pressure vessel steels such as A516 Gr70. The present work is a first step in assessing undermatching weld metals in C-Mn steels to see if their use is possible in large fabrications to minimize welding cracking difficulties while maintaining acceptable strength.

## EXPERIMENTAL PROCEDURES

Plates 25 and 50mm thick made of A516 Gr70 steel and approximately 0.5 x 0.7m were butt welded together by a gas metal arc using a mildly oxidizing gas. For the low strength welds an AWS EL12 electrode was used. One set of tests used a Kobe C-Mn twist wire which produced weld metal strengths exceeding the minimum strength specified for the base metal.

Before welding, the gap was set so that a range of aspect ratios could be evaluated. For an aspect ratio of 1.0 the welds were produced with three overlapping passes. For an aspect ratio of 0.75 two overlapping passes were used and for aspect ratios of 0.5 or less a single pass was used.

The tensile specimens were sawn out of the welded plate material and were stress relieved for one hour at 650°C. The specimens were milled to shape with the final surfaces being ground. Clip-on electrical resistance strain gauge transducers were attached to each tensile specimen. One transducer with a gauge length of 10mm was used to measure strains in the weld, one with a gauge length of 25mm was used to measure strains on the base metal and one with a gauge length of 50mm was used to measure strains across the weld which included a combination of weld metal and base metal strain behaviour. Specimens were cut in two widths, 38mm and 76mm, in order to determine the influence of specimen width on the test results.

The tensile specimens were loaded in a 400,000 lb. (1,780 kN) universal testing machine. The load was recorded simultaneously with the strain from each of the three transducers, digitized at 3.5 sec. intervals, and stored on a perforated tape. The tape was then analyzed by a computer which produced analog plots of stress vs log plastic strain and the log of the slope of the stress-strain curve vs plastic strain.

## RESULTS

In Figs. 1-3 the stress-strain behaviour of the weld metal is compared with that of the base material. Figure 1 shows that the base metal had a lower yield than that of the weld metal, which arises from the high yield-to-ultimate strength ratio in C-Mn weld metal, even though the ultimate strength of the weld metal was 25% below the base metal strength. On average the yield and ultimate stresses for the base metal and welds are given in the Table.

TABLE Base Metal and Weld Metal Yield and Ultimate Stresses

| Material | Yield (MPa) | | Ultimate (MPa) | |
|---|---|---|---|---|
| | longitudinal | transverse | longitudinal | transverse |
| A 516 Gr70 | 405 | 299 | 545 | 496 |
| weld metal | 363 | 321 | 410 | 420 |
| twist wire | - | 405 | - | 522 |

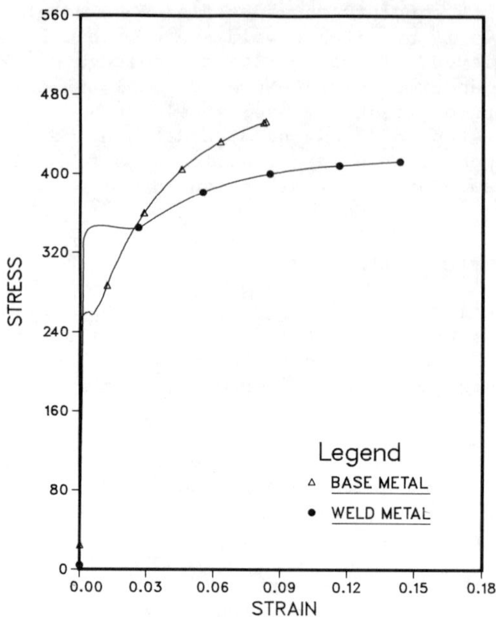


Fig. 1 Stress-strain curves of weld metal and base metal.

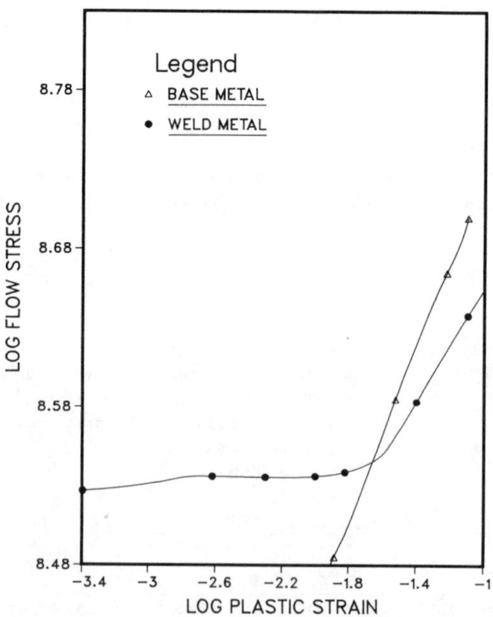


Fig. 2 Log flow stress vs log plastic strain for weld metal and base metal.

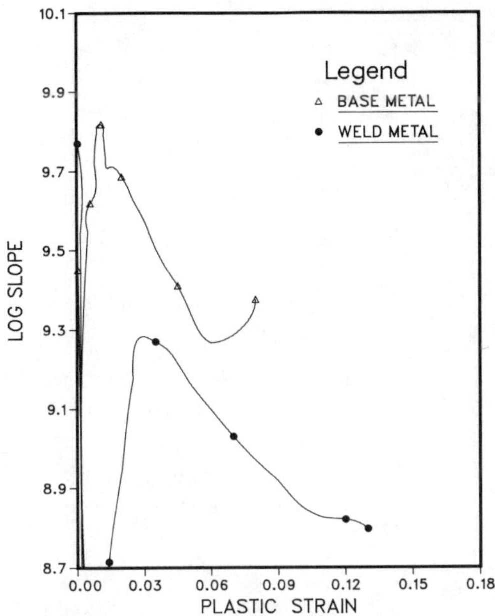


Fig. 3 Log slope of stress-strain curves vs plastic strain for weld metal and base metal.

Figures 2 and 3 illustrate the difference in the plastic strain behaviour between the base metal and the weld metals. It is seen that the base metal work hardens more than the weld metal as would be expected from the alloy content of the two metals. The work hardening rate, as shown in Fig. 3, indicates that the base metal had a peak log slope of 9.8 at a plastic strain of 1.0% whereas the weld metal had a peak log slope of 9.3 at a plastic strain of 3.0%.

A typical weld according to the ASME Boiler Code was obtained using a Kobe C-Mn twist wire. This weld was as strong as the base metal as required in the code. Figures 4 - 6 show the stress-strain behaviour of the "normal strength" weld for a gap of 13 mm. In Fig. 4 the strength of the base material (lower curve) was less than that of the weld material (upper curve) or across the weld as shown by the intermediate curve. These observations are amplified in Fig. 5 where the plastic flow stress was greater for the weld metal than that for the base metal. In Fig. 6 the degree of work hardening of the weld metal is seen to be 10.1 at a very low plastic strain which was greater than that for the base metal at approximately 9.8. It is also seen that as the plastic strain increased the slope for the weld metal fell more rapidly than it did for the base metal.

Similar graphs were plotted for the understrength narrow gap welds for which a typical set for the 13mm gap are shown in Figs. 7 - 9. Figure 7 shows that the base metal and weld metal yielded at 340 MPa and 330 MPa respectively. From Fig. 8 it is evident that the work hardening of the weld metal took place at a greater plastic strain than did the base metal but that log slope in Fig. 9 reached a peak of 9.55 indicating that the weld strength was approaching that of the base metal. For comparison, the peak log slope for the under strength weld metal for the 19mm gap weld tested was 9.4 and for the 25mm gap weld was 9.3, thus establishing the trend that as the weld gap decreased the degree of work hardening in the weld increased. Of course, part of this increase in weld strength as the gap narrowed

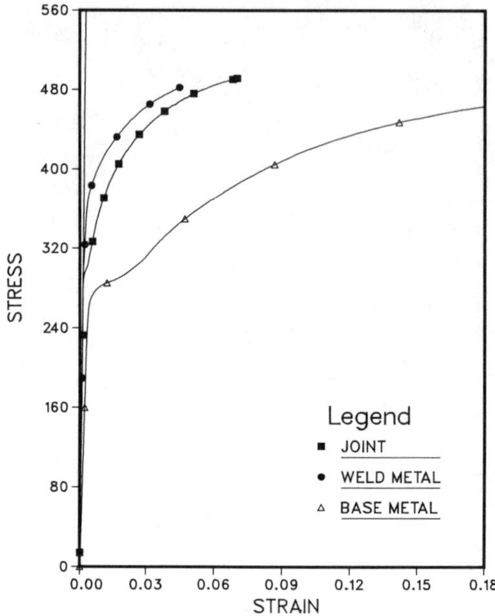


Fig. 4 Stress-strain curve for "normal strength" weld (twist-wire).

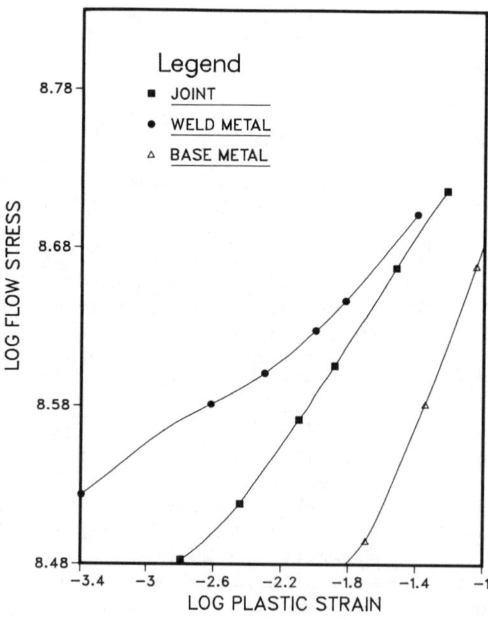


Fig 5. Log flow stress vs log plastic strain for "normal strength" weld (twist-wire)

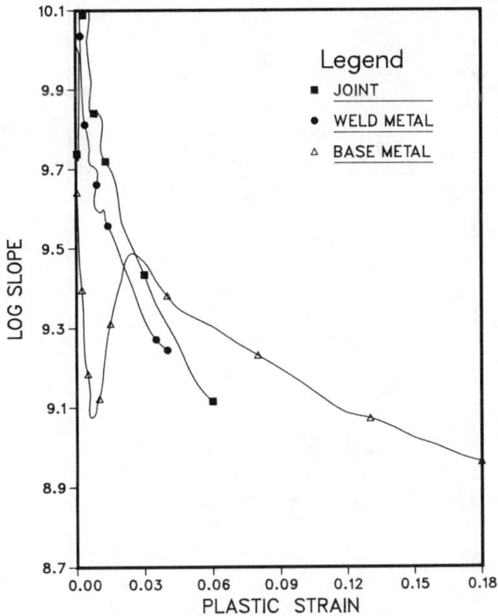


Fig. 6 Log slope of stress-strain curve vs plastic strain for "normal strength" weld (twist-wire).

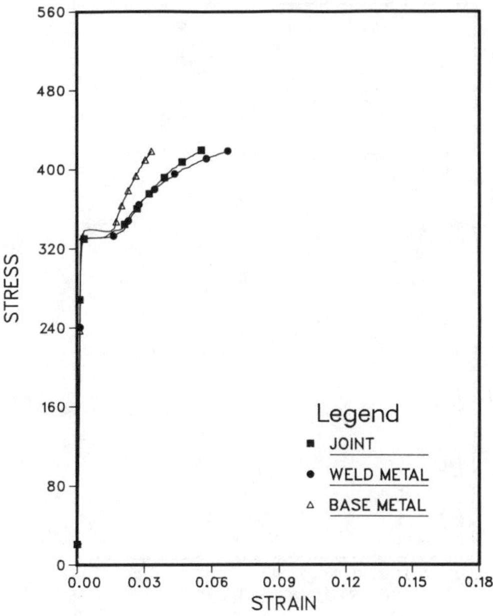


Fig. 7 Stress-strain curves for 13 mm narrow gap weld.

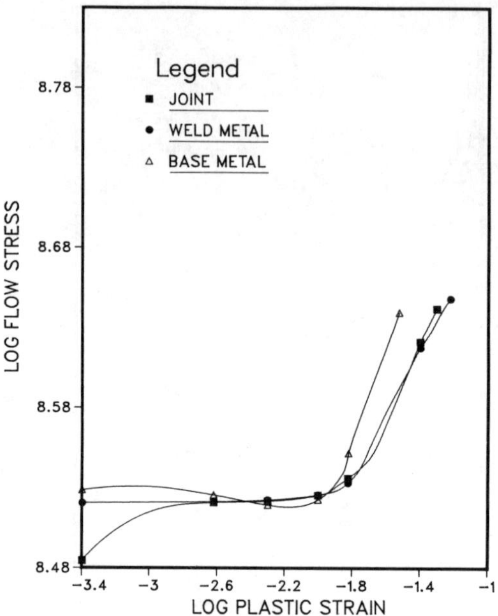


Fig 8. Log flow stress vs log plastic strain for 13 mm narrow gap weld.

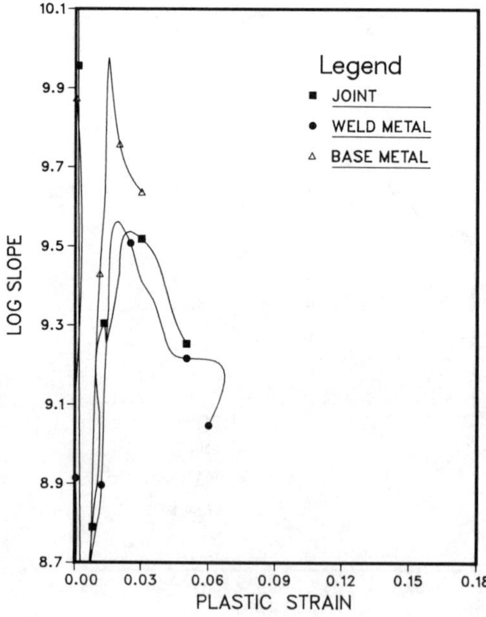


Fig. 9 Log slope of stress-strain curve vs plastic strain for 13 mm narrow gap weld.

can be attributed to dilution of the weld with the base metal, while the rest is due to the plastic constraint of the narrow gap weld metal imposed by the surrounding plate material.

The strength of the weld as a function of the aspect ratio has been plotted in Fig. 10. It is evident that as the gap of the weld decreased, for a given plate width, the ultimate strength of the weld increased and approached that of the base metal. The yield strength of the weld metal was relatively unaffected by changes in the aspect ratio.

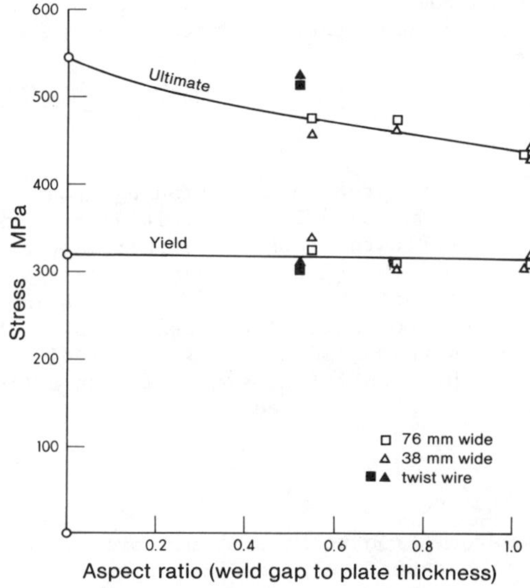


Fig. 10. Weld strength as a funtion of aspect ratio.

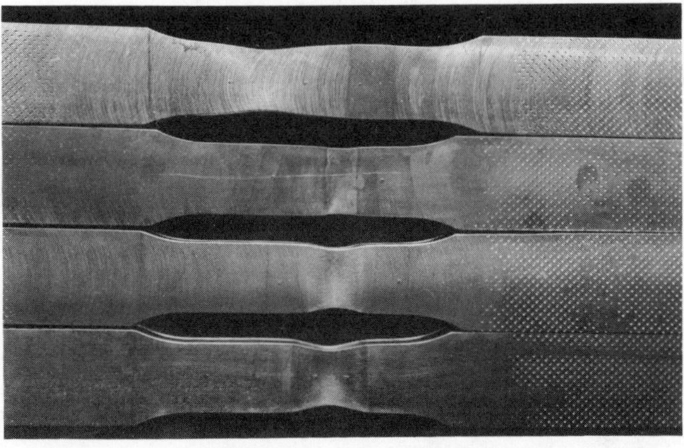

Fig. 11 Photograph showing pulled tensile specimens.

Although only two test specimen widths were evaluated, 38 and 76mm, in general there was an increase in strength of about three percent for the 76mm wide specimen compared with the 38mm wide specimen. This observation was consistent with that observed by Satoh and Toyada (1975) who reported that the strength of the weld increased up to a specimen width of five times the thickness and remained relatively constant for specimen widths greather than 5 x t.

A photograph of pulled welded tensile specimens is shown in Fig. 11. At the bottom of the photograph a weld gap of 25mm shows the yielding that took place in the weld whereas, as the two specimens above this show, as the weld gap narrowed the degree of yielding in the weld was decreased. For comparison, a "normal strength" weld is shown in the top picture where the base metal yielded whereas the weld retained its original shape.

## CONCLUSIONS

While this paper presents the first step in a continuing program to investigate the feasability of understrength narrow gap welds the initial results are encouraging. It was shown that the ultimate strength of the weld increased by eight percent as the aspect ratio was reduced from 1.05 to 0.55. Further improvements in weld strength are expected for aspect ratios less than 0.55 which implies that the benefits will be primarily associated with base metals of 50mm thickness or more. There is also scope for trials with a weld metal of intermediate strength between the present weld metal and the base material, e.g., with an ultimate strength drop of about 10 - 15% rather than the 25% achieved in this work.

## ACKNOWLEDGEMENTS

The authors express their appreciation to Messrs. G. Lorimer, C. Bicknell and T. Sharratt who assisted in the experimental work. Appreciation is also expressed for the financial assistance from NSERC through grant 2705-A and from D.S.S. (E.M.R) through contract 14 SU. 23440-0-9197.

## REFERENCES

Grist, F.J. and F.W. Armstrong (1980). Welding Journal, 59 (6), 30.
Kimura, S., I. Ichihara and Y. Nagi (1979). Welding Journal, 59 (7), 44.
Malin, V.Y. (1983). Welding Journal, 62 (4), 22 and 62 (6), 37.
Satoh, K. and Toyoda (1975). Welding Journal, 54 (9), 311.
Satoh, K. and Toyoda (1979). Welding Journal, 58 (2), 26.
Sawada, S., K. Hori, M. Kawahara, M. Takao and I. Asano (1979). Welding Journal 58 (9), 17.
Wheatley, R.M. and R.G. Baker (1962). British Welding Journal, 9 (6), 378.

# OFFSHORE STRUCTURES FOR THE NORTH SEA HAZ HARDNESS REQUIREMENTS AND PRACTICAL IMPLICATIONS

S. Tandberg
*A/S ESAB, Larvik, Norway*

## ABSTRACT

Experience from the fabrication of fixed structures for the North Sea are presented, as seen from the point of view of a welding consumables supplier who has been extensively involved as a consultant to the fabricating yards. Particular empasis is placed on the HAZ hardness problem, surveying specification requirements and how they have been fulfilled in practice. Based on a fitness for purpose/cost conciousness philosophy, indications are given as to what will be future developments.

## KEYWORDS

Offshore structures, HAZ hardness, offshore specifications, welding procedures, hardness control, hydrogen cracking, stress corrosion.

## INTRODUCTION

National standards, and codes by ASME, AWS, API etc. are not considered adequate for the welding fabrication of permanent North Sea structures. Separate fabrication specifications are written and issued for each large project, by the oil company and their engineering consultant. Welding, and weld quality requirements are important parts of these specifications, which as a rule are more stringent than the regulations laid down by classification societies and statutory authorities.

From a welding fabricator's point of view, specifications ought to be realistic, i.e. quality requirements in line with general knowledge on fitness for purpose, significance of defects etc. This creates respect for the rules, and efforts are made to produce overall high and even quality. Unrealistic requirements tend to favour concentration of efforts towards the checkpoints, while intermediate weld footage is taken less seriously.

Secondly, specifications ought to be uniform, particularly regarding quality criteria and acceptance limits. Today there are differences between projects, between countries, between oil companies. This tends to reduce the value of experience, new welding procedures are developed and qualified for each new project, cost becomes higher than necessary.

HAZ hardness is typical of a requirement that has been both unrealistic and variable. It has taken years to arrive at the relatively stable, but still not quite satisfactory situation of today.

## WHY HARDNESS CONTROL

HAZ hardness is being maximized for the purpose of avoiding various types of hydrogen induced cracking. Whether "ordinary" cold cracks or stress corrosion cracks depend upon circumstances and environment. One thing to remember is that cracking susceptibility is not proportional to hardness alone. Other parameters, like microstructure as such, stress intensity and in particular, hydrogen, are no less important. Nevertheless, hardness is the easy one to measure, and the one being checked during weld procedure testing and production.

## STRUCTURES IN AIR AND SEAWATER

"Ordinary" cold cracking during or shortly after welding is to be avoided. Known as underbead cracks, toe cracks, delayed cracks etc. these may occur in sensitive areas of the HAZ. To avoid them, specifications over the last five years have maximized hardness at 280, 300, 310, 325 and 350 HV, for modern structural steels with a minimum yield strength up to and including 355 N/mm$^2$.

Published work over the same period of time indicate that the actual crack sensitivity limit for these steels are probably 400 HV or higher provided hydrogen is controlled to below 10 ml per 100 g. (Fig. 1)
The immediate deduction is that 350 and may be 325 HV may be regarded as "reasonable" limits, while 280 HV is clearly "unreasonable". Typically, the 280 requirement in practice also led to expensive welding procedures, involving high preheat and rather artificial bead positioning, including a lot of grinding. This was the only way to get the procedures qualified and the production started, no one really knows how accurate these procedures were adhered to during the actual production.

A very similar structure, built later to a 325 HV max requirement has not shown any increase in hydrogen cracking frequency. Welding cost, however, was significantly reduced, also partly due to a new and lower carbon steel.

Interesting to note is that even people outside the fabricator's own ranks are influenced by the "unreasonable requirement" thinking. For instance test lab. technicians, who tend to be a little bit "kind" when results are close to the border line. A post-survey of hardness measurements made during fabrication of a large offshore structure illustrates this point. (Fig. 2) This frequency diagram is based on a total of 494 tests, partly weld procedures, partly production test coupons. The column covering the 270 - 280 HV range

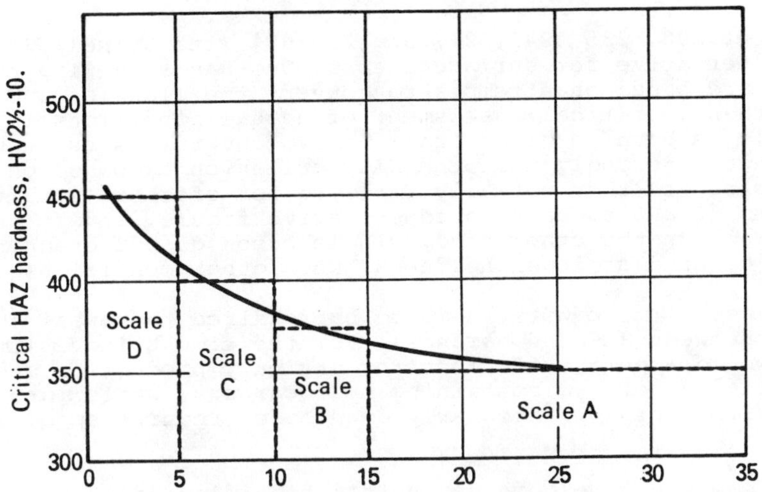


Fig. 1. A Welding Institute diagram, showing a relationship between hydrogen content and hardness to avoid cracking.

is remarkably higher than the 280 - 290 HV column, when compared to a normal distribution. The acceptance limit was 280 HV, one cannot help wondering whether some of the 281's and 282's may have been pushed across the border into safety, by a very slight misreading of the Vickers diagonals.

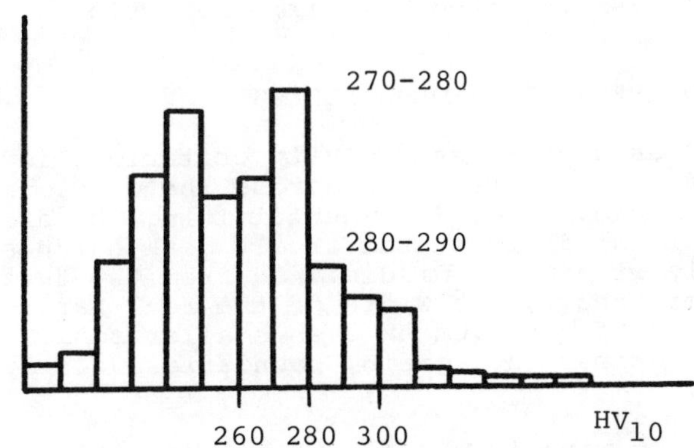


Fig. 2. Distribution of HAZ hardness test values, based on 494 measurements. Specified maximum, 280 $HV_{10}$.

## STRUCTURES IN SOUR SERVICE, HIGH $H_2S$ MEDIA

Here stress corrosion cracking may be a problem, and hardness is known to have a significant effect. Specifications have called for maximum at 200, 235, 243, 247 and 260 HV, with Brinell (200-235 HB) as an alternative for surface checks. One has a feeling that these figures are based on attempts to convert the classic 22 HRC by NACE, rather than a critical assessment of actual conditions in each case. It should be born in mind firstly that conversions as such are never recommended, secondly that the NACE criterion is based on steels of higher strength than normally employed for offshore structures. This points to 22 HRC as being a conservative figure, and it probably is, in general. On the other hand, single cases of SCC cracking is known to have occured at lower hardness, when other conditions are extreme.

In conclusion, more work seems to be required to reveal the full connection between SCC and hardness. For the time being a limit of 240 - 260 HV would appear sufficient for ordinary production, storage and transportation of $H_2S$ containing hydrocarbons, while more unpredictable environments, particularly down hole, require special investigation.

Fabricators would welcome any unification, including how and where hardness should be checked. And a more realistic approach to quenched and tempered steels, where HAZ hardness is sometimes maximized close to or even below that of the unwelded material.

## STRUCTURES IN LOW $H_2S$ MEDIA

Seawater with a small amount of $H_2S$ (rarely above 100 ppm) and high pH may create conditions favourable to stress corrosion. Published information and practical experience on subsea structures so far points towards the same limit as previously suggested. At below 350 HV hydrogen induced SCC is unlikely to occur in seawater, except maybe under high levels of cathodic (over) protection.

## HOW AND WHERE TO CONTROL HARDNESS

Hardness, when used as a crack sensitivity criterion, is normally controlled by a series of impressions across the weld, in a transverse cross sectional macro. (Fig. 3) When limiting the discussion to CMn and fine grain treated structural steel, peak hardness will occur in the HAZ, normally within 0,5 mm distance from the fusion line. This peak may be very sharp, the width of the real hard part of the HAZ only a few tenths of a mm. On an even smaller scale there may be hardness variations across each grain, boundaries usually being softer than the center.

With this knowledge in mind it should not be too difficult to specify a testing procedure fit for its purpose, described by:

> Impressions should be large enough to cancel out the variations across a grain, but small enough to pick up the peak of the gradient across the HAZ. One or more should be placed in the region where the peak has to be expected.

The obvious choice is the Vickers method, preferably with a 5 kg load. First HAZ impression to be placed as close as possible to the fusion line, next at a 0,5 mm distance. Rockwell and Brinell methods are not suitable, because of impression size.

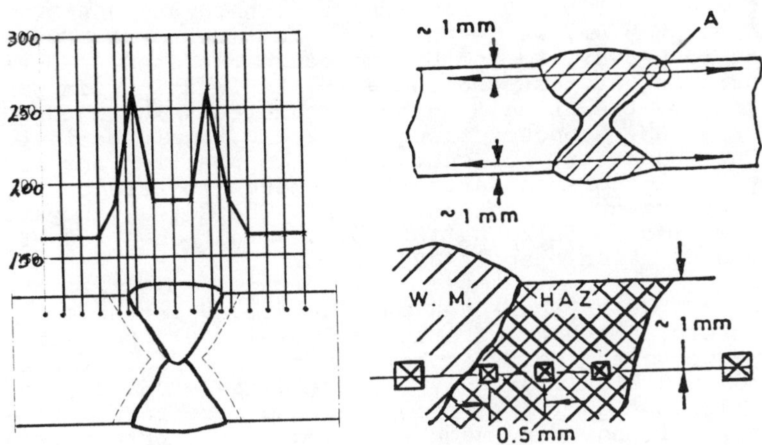


Fig. 3. Hardness profile across a welded joint, and detailed "where to measure" specification.

Today $HV_5$, located as above, is the preferred method, found in most specifications. But, over a period of say 5 years, a number of other variants have also been applied, creating some confusion among fabricators. A couple of examples will explain why:

Jackets, Decks and Modules have been controlled by $HV_5$ og $HV_{10}$, but impression positioning has varied. Some specifications call for a given number of impression to be placed "in the HAZ", no other details given. Others specify 1 mm distance between points, others again that the first point should be 1 mm outside the fusion line. In practice, neither of these will pick up the peak of the hardness curve. Consequently, welding procedures developed to meet these requirements are useless when the next contract requires hardness to be checked within the first half mm outside the fusion line.

In 1978, Metal Construction (UK) published a paper on "Economic welding procedures for demanding applications", based on the welding of BS 4360:50D (modified) steel commonly used for offshore structures. Carbon equivalent was 0,46 maximum. All common offshore requirements were shown to be fulfilled, including hardness which was measured at 235 $HV_5$ maximum.

Other fabricators, new to the offshore scene, who used these procedures as models for their own, were seriously disappointed. Although the steel now had a 0,43 $CE_{max}$, peak hardness values were frequently in the 350 - 400 HV range. Procedure tests were not approved, and had to be redone at double cost.

Reason, the new specifications wanted hardness checked close to the fusion line, while the British work had the first point quite a bit further out, losing the HAZ peak altogether.

Riser Pipe for $H_2S$-service was welded by a Norwegian yard, on licence from a US company who also supplied steel specification (HSLA) and welding procedures. D.n.V. required max. hardness 260 $HV_5$. The Americans said no problem, we do this all the time to 22 HRC, which is equivalent to 248 HV. So the US procedure was used, and values up to 300 HV found on the test coupon macro. Explanation, in USA hardness controll was by the Brinell method, measured values judged against a limit of 235 BHN, supposed to equal 22 HRC, again believed to equal 248 HV. The inspector, no doubt trying to position his indentor correctly, would cover an area large enough to effectively average out the hardness across the entire HAZ. The Norwegian yard had to develop new procedures to satisfy D.n.V., although 260 $HV_5$ is presumably an easier match than 22 HRC.

The offshore industry has been particularly vulnerable to incidents of this nature, a number of other examples could be referred to. This is partly due to the North Sea being a merging place for technology, habits and tradition from USA, UK, Scandinavia and continental Europe, partly to lack of qualified people during a certain build-up period. The situation is now very much improved, but fabricators still need to have a critical eye open when a new project specification turns up.

## PRACTICAL SOLUTIONS

Unrealistic or not, hardness requirements had to be met, to get procedures approved and production started. Experiments to find applicable welding procedures were based on the simple fact that hardness is determined by two main parameters.

- Steel analysis, with carbon content being the deciding factor, assisted by a number of other elements.

- Thermal history of the HAZ

Both can be reasonably well controlled under workshop conditons, while a third parameter, distribution of second-phase particles, is a research laboratory tool.

The real serious hardness-problem period was 1979 - 1980, during the STATFJORD B platform construction. Maximum hardness was initially set at 280 $HV_{10}$, later increased to 300. Steel had been purchased for the bulk of the structure, a typical analysis was C 0,13 - Si 0,4 - Mn 1,45 - P 0,017 - S 0,002 - Cu 0,18 - Ni 0,07 - Cr 0,15 - Nb 0,039 - Al 0,046%.

Carbon equivalent was typically 0,41 - 0,43, calculated by the IIW formula

$$CE = C + \frac{Mn}{6} + \frac{Cr + Mo + V}{5} + \frac{Cu + Ni}{15}$$

For this type of steel, the formula is probably reflecting hardenability rather well, but note the ultralow sulphur content. This is believed to have the same effect as a 0,02 increase in carbon equivalent, when compared to a more conventional S = 0,03%. Anyway, the steel was purchased and thus no more a variable parameter, leaving only the the thermal influence as a working tool.

High preheat was applied, $150^{o}$ C being the maximum practical temperature. Stress relieving could only be used to a very limited extent, because of the size and shape of the structures. This left only the heat input during welding, and a number of procedures were qualified utilizing the temper bead technique to its full extent. This meant very accurate positioning and heat input control of the final capping beads. (Fig. 4) Fillet welds were a particular problem, as even small size fillets had to be positioned in 3 passes. Welders were specially trained, and a vast amount of grinding became necessary to combine this capping technique with weld surface acceptance criteria. The structures were finished and the quality OK, but welding manhours and cost far higher than for previous, similar type of work.

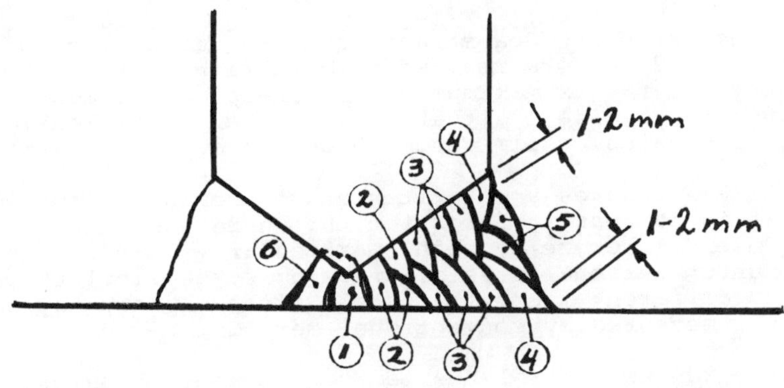


Fig 4. Weld bead sequence to avoid excessive hardness, position 2G.

| | | | | |
|---|---|---|---|---|
| 1) | Electrode | 3,25 mm, | 110-120 A, | ROL: 100-150 mm |
| 2) | " | 3,25 " | 120-140 A, | ROL: 160-200 " |
| 3) | " | 4 " | 160-190 A, | ROL: 200-250 " |
| 4) | " | 4 " | 160-180 A, | ROL: ca. 250 " |
| 5) | " | 4 " | 160-180 A, | ROL: 120-150 " |

## DEVELOPMENTS BASED ON LESSONS LEARNED

The STATFJORD B project had a number of contracts and subcontracts, giving a strong awareness among fabricators on the problem of HAZ hardness. Other projects during the same period had smaller, but similar difficulties. Everybody, including oil companies and their engineering consultants realized the necessity of a better balance between weld quality requirements and base material weldability.

Developments have been following two directions:

Firstly, specifications are moving towards the previously mentioned 350 $HV_5$ as a realistic maximum requirement. Some, however, stopping at 325 and some still hanging on to 300.

Secondly, steel specifications are now calling for lower carbon, lower carbon equivalents, forcing the steelworks into sophisticated thermomechanical treatment to obtain strength and toughness with a minimum of alloying elements. These are techniques previously used for high strength pipeline steels, now also for plate and pipe for structural purposes.

For recent projects, pearlite reduced steels with less than 0,10% C has been delivered, carbon equivalents typically 0,37 - 0,38. Fabricators no longer have serious problems meeting the hardness requirements, although preheats have been reduced and more natural, more economic welding procedures employed.

## CONCLUSIONS AND COMMENTS

Specifications for North Sea structures have during the last years been moving towards a more realistic, more fitness for purpose adapted philosophy, as far as maximum HAZ hardness requirements are concerned. However, the ideal situation, a master specification used by all the oil companies, still seems to be far away.

Hardness still dominates as a criterion to prevent hydrogen induced cracking. It is too easily forgotten that hardness in not uniquely related to microstructure as such, particularly in modern high strength, more unconventional steels. Cracking susceptibility can be very different in different microstructures, although they may have the same hardness measured by conventional methods.

Emphasis on hardness control has led to the use of pearlite reduced low carbon steels, apparently less crack sensitive. But the IIW carbon equivalent formula was never developed for this type of steel,and does not give a true indication of weldability. The Ito-Bessyo formula, still not commonly applied, might tell a more reliable story and ought to be introduced:

$$P_{cm} = C + \frac{Si}{30} + \frac{Mn + Cu + Cr}{20} + \frac{Ni}{60} + \frac{Mo}{15} + \frac{V}{10} + 5B$$

A word of warning must be sounded against going too far in steel development towards low HAZ hardness figures. Some of the new steels, although still fulfilling specification requirements, show dramatic reductions in HAZ notch toughness and must be checked by fracture mechanics methods before approval.

Parameters other than hardness influence cracking sensitivity. Microstructure has been mentioned, but still easier to control is hydrogen content. With some reservations it may be stated that the following parameter variations have the same influence on weldability:

| | |
|---|---|
| Carbon equivalent (IIW) | ± 0,05 |
| Hydrogen content | ± 3 ml/100 g weld |
| Preheat | ± 30° C |
| Weld heat input | ± 0,4 kJ/mm |

Increased preheating temperature is always expensive and should be avoided. But then, in contrast to the efforts to develop low CE steel material, what about

1. Welding consumables, manufactured, packed, stored and handles to give less than 5 ml of hydrogen instead of 10.

2. Welding consumables to be used at higher heat input, say 3 kJ/mm instead of 2, weld metal - fracture toughness fully preserved.

Consumables like these are available today, for manual and mechanized welding, but no classification system has included these particular properties in the standard designations. Far from being promoted as "the easy way out" they seem to present an interesting alternative to the very low carbon steels, the idea being to end up with the same resistance towards HAZ hydrogen cracking, in total.

## REFERENCES

Threadgill, P.L. (1981). Measurement of hardness in welds. Weld. Inst. Seminar Handbook, Bradford.

Gooch, T.G. (1982). Hardness and stress corrosion cracking of ferritic steel. Weld.Inst. Research Bulletin, 19, 241-246.

Still, J.R., George, M.J. and Terry, P. (1978). Economic welding procedures for demanding applications. Metal Construction, 10, 488 - 492.

Eide, G.H. and Valland, G. (1981). Practical experience with hardness control. D.n.V. document.

Brooks, T.I. and Hart, P.H.M. (1977). Do hardness measurements impress you? Weld. Inst. Research Bulletin 18, 69 - 72.

Nordbø, L. and Tandberg, S. (1981). Welding of steel structures for the STATFJORD B oil production platform. Grossen Schweisstechnischen Tagung, Essen -81

Cotton, H.C. (1983). Welding under water and in the splash zone. Portevin Lecture, IIW Annual Assembly, Trondheim, Norway.

# INTEGRITY OF NON-POST-WELD HEAT TREATED HEAVY SECTION WELD REPAIRS

J.W. Prince

*Ontario Hydro Research Division, 800 Kipling Avenue, Toronto, Ontario M8Z 5S4*

## ABSTRACT

Development work in the area of pressure vessel weld repairs has shown that satisfactory weld repairs can be made without the use of high temperature post-weld heat treatment provided special consideration is given to the welding procedure used. These efforts have led to the adoption of non-post-weld heat treated (PWHT) repair technique in the ASME Boiler and Pressure Vessel Code based on half-bead or temper-bead welding technology. At Ontario Hydro, temper-bead welding has been used successfully for repairs in nuclear and thermal steam raising systems as well as in wet sulphide (heavy water plant) service. Close control of heat input and deposition sequence will produce a fine-grained heat-affected zone with a low susceptibility to hydrogen cracking and good toughness properties. A major concern, however, has been the fact that residual stresses, that would otherwise be substantially reduced during PWHT, are not relieved. The potential significance of this factor regarding the integrity of temper-bead weld repairs is currently under investigation.

As part of the program, several integrity assessment methods were surveyed and compared based on their ability to deal with the effects of residual stress in an assessment. These include the techniques given in the ASME Boiler and Pressure Vessel code, PD 6493 from the British Standards Institution and the CEGB R6 method. Of those examined, the R6 method appears to be the most suitable assessment technique if residual stress effects are to be included in integrity evaluations.

KEYWORDS

Welding, Repair Welding, Power Plant, Pressure Vessel, Residual Stress, Integrity, Fracture Mechanics, Flaw Evaluation

## 1.0 INTRODUCTION

In recent years Ontario Hydro has qualified five temper-bead welding procedures for intended use in repair welding of existing thermal and nuclear plant components. Four of these are approved for use in nuclear plant applications which, by the nature of governing nuclear codes and regulations, are more closely administered and carefully executed. The Ontario Hydro temper-bead welding procedures were developed expressly to avoid the need to perform a conventional post-weld heat treatment on existing plant components that require repairs. Although the temper-bead procedure provides for hydrogen removal and refinement of the heat-affected zone microstructure, temper-bead welding does not allow for removal of weld residual stresses as does conventional high-temperature PWHT.

It becomes necessary, then, to provide a method that can assess the effect of these remaining residual stresses on a component's integrity when it is returned to service. Several engineering critical assessment or fracture mechanics evaluation techniques are available which can be used, to varying degrees of success, to include the effects of residual stress. The current approach used in Section XI of the ASME Boiler and Pressure Vessel Code, for example, assumes a linear elastic model for behaviour of vessel materials which implies that residual stresses should be added to service stresses for fracture mechanics assessments.

This approach can be unreasonably conservative if the material under consideration exhibits large amounts of plasticity or ductile tearing during fracture. To date, most steel pressure vessels in Ontario Hydro nuclear systems are fabricated using SA516 Grade 70 plate which, along with shielded metal arc weld (SMAW) deposits in these materials are known to have good toughness properties at operating temperatures. When carrying out an evaluation for a vessel, the allowable flaw size becomes diminishingly small if residual stresses of yield magnitude, which are not uncommon, are applied to component assessments done according to Section IX of the ASME code. If temper-bead welding is to be used for repairs, a more realistic approach of including welding residual stress effects in assessments must be adopted. This study outlines some observations regarding the influence of welding residual stresses on component fracture behaviour and suggestions are made on techniques for handling residual stresses in component integrity assessments.

## 2.0 BACKGROUND

### 2.1 Weld Repair Strategies

Once a vessel has been placed into service in a nuclear or thermal generating station, the task of major repair welding often becomes difficult. The vessel is invariably attached to a piping and support system and surrounded by process equipment and instrumentation which can present innumerable problems for preheating and preparation work not to mention the actual weld repair work. However, most of these difficulties may be overcome with careful planning and attention to thermal gradients ensuring that they are not too severe in the area around the repair. Upon completion of a conventional type weld repair, thick vessels designed according to ASME Section VIII or III must receive a high temperature PWHT at 650°C for approximately two hours. At these temperatures controlling thermal gradients becomes a formidable task particularly in regions of nozzles or other rigid connections.

Repairs in cylindrical areas of pressure vessels can be handled successfully if strategic placement of heating elements is used to produce a symmetric high temperature band during PWHT taking care to minimize temperature gradients. In some cases in-situ PWHT of large vessels would be impossible without damaging connections to the vessel or the vessel itself. When situations such as this arise, a welding procedure that does not require full high temperature PWHT is desirable. The temper-bead welding technique can provide this option by using a carefully controlled weld deposition sequence for auto-tempering of the heat affected zone (Lawson, 1979). This eliminates the need for PWHT and allows repairs to be made in areas that would otherwise be considered too difficult to repair using conventional methods.

## 2.2 Temper-Bead Weld Repairs

As the name implies the temper-bead welding process uses the controlled heat input from successive layers of weld metal to 'temper' the heat-affected zone (HAZ) of the first layer of weld metal deposited. Figure 1 illustrates schematically the mechanics of the temper-bead welding procedure. Temper-bead welding is a variation of the ASME Section XI half-bead weld repair technique in which half of the first layer of weld metal deposited is removed by grinding. Lawson (1979) showed that the use of successive layers of weld metal to control tempering/grain refinement in carbon steel vessel repairs is in fact justified. However, the use of a grinding step after the first layer of weld metal is questionable. Elimination of this grinding step and optimization of heat inputs between successive layers led to an improvement in the hardness and microstructure of the base material's HAZ. The practical difficulties in grinding only half the first layer were also of some concern, providing additional justification for eliminating this step from the procedure. At Ontario Hydro, temper bead welding procedures have been developed for several specific applications based on code requirements and service conditions. Welding parameters are listed in Table 1 for existing temper bead welding procedures in thermal, nuclear and heavy water (wet $H_2S$) plant.

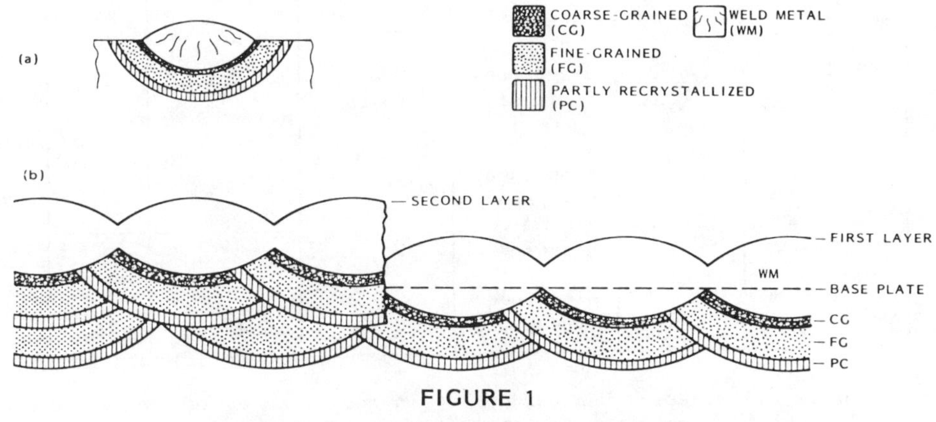


FIGURE 1

TEMPER-BEAD WELDING TECHNIQUE

a) HEAT AFFECTED AREAS OF A SINGLE WELD BEAD

b) CORRECTLY APPLIED TEMPER-BEADS SHOWING REFINEMENT OF ORIGINAL COARSE GRAINED HAZ

## 2.3 Avoiding Hydrogen Cracking

The greatest threat to the success of a weld repair in a heavy carbon steel vessel is hydrogen-induced cold cracking caused by the introduction of hydrogen into the steel during welding. Extensive studies have shown that hydrogen cold cracking is a problem only if a combination of events occur simultaneously after the weld has cooled to ambient temperatures. The weld metal or heat-affected zone must be relatively hard, there must be sufficient hydrogen present in the steel and there must be a tensile stress acting on the weld region. Conventional repair welding procedures are designed to avoid the occurrence of hydrogen cracking by incorporating a number of safeguards. The amount of hydrogen available during welding can be minimized by using iron powder low-hydrogen type electrodes that have been properly processed to remove excess moisture prior to use. A suitable preheat should be applied to the base plate during welding to slow down the cooling rate

TABLE 1

TEMPER-BEAD WELDING PROCEDURE
DATA FOR SA516 GR70 REPAIRS

<table>
<tr><th rowspan="2">Vessel Type</th><th rowspan="2">Preheat</th><th rowspan="2">Deposition Sequence</th><th rowspan="2">Post-Weld Bake</th><th colspan="2">Heat-Affected Zone Hardness (VHN 2.5 kg)</th></tr>
<tr><th>Range of Data Excluding Weld Toe</th><th>Peak Value Weld Toe</th></tr>
<tr><td>Thermal Plant ASME VIII</td><td>93°C</td><td rowspan="3">Layer 1 - 2.4 mm Electrodes Stringer Beads<br>Layer 2 - 3.2 mm Electrodes Stringer Beads<br>Subsequent Layers - 4.0 mm Electrodes<br>Electrode Type E7018</td><td rowspan="3">230°C<br>Two Hours</td><td>209-274</td><td>322</td></tr>
<tr><td>Nuclear Plant ASME III</td><td>177°C</td><td>170-150</td><td>296</td></tr>
<tr><td>Heavy Water Plant (Wet $H_2S$) ASME VIII</td><td>230°C</td><td>206-229</td><td>250</td></tr>
</table>

and avoid areas of excessive HAZ hardness. In addition, a high temperature post-weld heat treatment is used immediately after welding to temper any hard regions in the HAZ, reduce welding residual stresses and remove residual hydrogen from the weld metal and HAZ.

Procedural requirements of the temper-bead welding technique accomplish all of the above except for the removal of residual welding stresses, since a final high temperature PWHT is eliminated. As shown in Table I, preheat requirements are set with hardness control of the HAZ in mind. That is, thermal plant vessels designed under ASME VIII receive a minimum of 93°C preheat while procedures for wet $H_2S$ vessel repairs require a minimim of 230°C preheat. A post-weld bake out at 230°C for two hours is required in all cases to remove residual hydrogen.

The key to the metallurgical integrity of the temper-bead welding procedure is the accurate control of heat input and positioning of each weld pass. When administered correctly, the second layer of a temper bead weld repair will effectively temper the coarse grained region of the HAZ of the first weld pass. The tempering passes provide two benefits. A drop in hardness of the original HAZ suppresses the opportunity for hydrogen cracking to occur. In addition the softer, grain-refined, tempered microstructure has acquired improved fracture toughness properties.

Eliminating the full PWHT raises several questions concerning the integrity of temper bead weld repairs. Without PWHT, stress relief in the weld region is not effected, leaving residual stresses as high as yield magnitude in the repair area. In the next section typical residual stress magnitudes and distributions in heavy section weldments will be discussed. In addition, it will be shown that the influence of these residual stresses on component integrity can be illustrated using one of several engineering critical assessment methods.

## 3.0 POST-REPAIR INTEGRITY

### 3.1 Post-Weld Heat Treatment and Residual Stress

As discussed previously, it is often preferable to design in-situ weld repairs in vessels without the use of conventional post-weld heat treatment. High temperature PWHT, whether it is used for a complete vessel or for local treatment, can generate enormous problems for a structure which is tied into a system of piping and supports, due to thermal expansion and distortion during heating. If it can be shown that the structural integrity of a non-PWHT weld repair is not significantly different from a repair that receives full PWHT, then in many circumstances heavy section pressure vessel repairs can be performed with substantially decreased effort by eliminating PWHT.

Upon completion of a non-PWHT temper-bead repair, residual stresses remain unrelieved in and around the repair region. In some cases these residual stresses could influence the integrity of a component in the presence of a flaw or discontinuity. It then becomes a question of to what degree residual stresses influence material fracture behaviour. If non-PWHT repairs continue to be used in heavy section vessels whose integrity in the presence of a flaw can be shown by using an engineering critical assessment (ECA) or integrity assessment, then there must be a method available to include the effects of residual stress when an analysis must be done in the area of a weld repair.

In order to effectively assess a component's integrity under the influence of residual stress, it is important to have an understanding of the magnitude and distribution of residual stresses generated by a repair weld. Residual stress distributions in heavy section pressure vessel weldments have been studied in detail by a number of investigators. Although it is beyond the scope of this paper to examine residual stress distributions in great detail, there is some merit in reviewing the type of residual stress distributions expected in heavy section repairs.

Most published data pertaining to residual stress distributions in weldments are concerned with heavy section butt welds and repair welds. A number of these studies have been recently reviewed by Ruud and Dimascio (1981) in order to predict typical three-dimensional stress fields resulting from heavy section welds. The work concluded that the published residual stress distributions exhibited many similarities despite the diverse range of measuring techniques used to assess residual stress. Figure 2 shows schematically a typical transverse through-thickness residual stress distribution that would be expected for a full penetration butt weld or weld repair. For non-PWHT weldments, the distribution of residual stress typically varies in magnitude from yield level in tension at the plate surfaces to compressive stresses of similar magnitude in the centre of the plate. Results from stress-relieved specimens showed the same distribution of stresses, although substantially reduced in magnitude. Two other studies investigating half-bead repair-weld residual stresses produced distributions which were in close agreement with those shown in Figure 2 (Binkley, 1974; Rybicki, 1980).

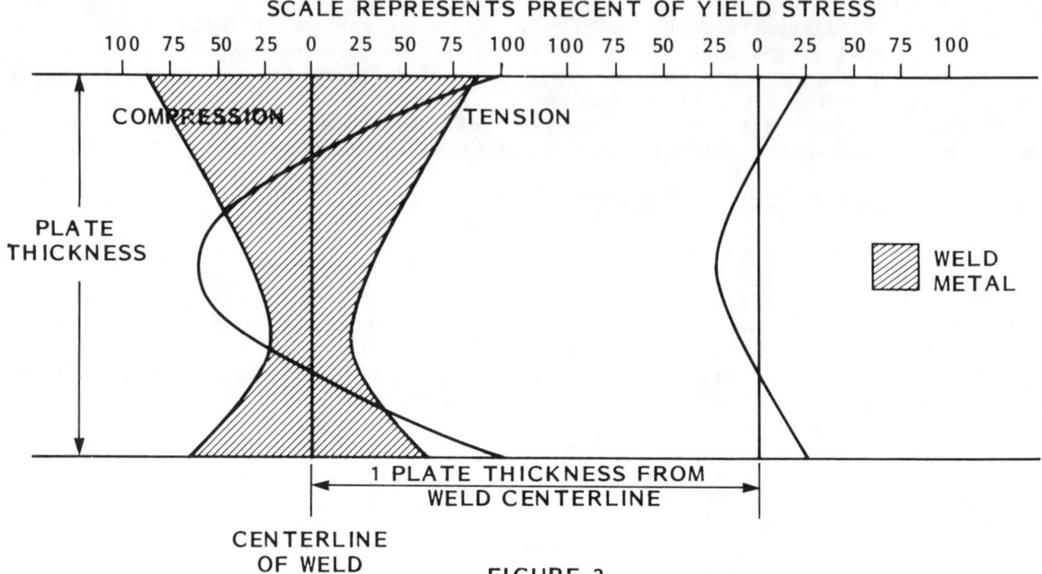


FIGURE 2

IDEALIZED THROUGH THICKNESS TRANSVERSE RESIDUAL STRESS DISTRIBUTIONS FOR FULL PENETRATION BUTT WELDS

In an ideal situation, full knowledge of residual stress distributions and magnitudes in a non-PWHT weld repair would be beneficial in assessing the integrity of the repair. This, however, is not realistic and, in most cases, published data must be relied upon to predict welding residual stresses if they are to be used in an assessment. In general, for non-PWHT weld repairs a conservative approach is to choose the material's tensile yield stress or some fraction of the yield stress as a value of residual stress if the true distribution or a reasonable estimate of the stress distribution is not known.

## 3.2 Integrity Assessment Methods

Several systematic assessment techniques are available for dealing with the integrity of structures or pressure vessels containing flaws. The ASME Boiler and Pressure Vessel Code uses a linear elastic fracture mechanics (LEFM) approach in its assessment methodology which tends toward excessive conservatism when dealing with materials which exhibit a high degree of plasticity. The British Standards Institution has published a document which provides guidance for determining acceptance levels for defects in welded joints. The document, PD6493, allows analysis using both linear elastic and general yielding fracture mechanics concepts based on the extent of loading a structure or vessel will see in service.

It is possible when using either of these techniques to include the residual stresses that would be present in a non-PWHT structure or vessel. However, if one assumes full yield level tensile residual stresses in regions under consideration, which are not unrealistic for surface regions adjacent to welds, then both of these assessment methods reduce significantly the margin of stress remaining for applied or primary stress. This difficulty arises from the inherent conservatism of the ASME code using LEFM even in situations when materials of substantial fracture toughness are used. The methods of PD6493 are not nearly so misleading since a material's ability to deform plastically is considered by using the crack tip opening displacement approach. However, PD6493 has safety margins included in the assessment method which are not explicitly stated. These margins are considered "safe" and although the margins are known to be conservative, the applied factors may create difficulties if specific safety factors are required.

Since the vast majority of pressure vessels and piping in power plants will exhibit ductile fracture at design temperatures in the presence of a flaw and sufficient stress, it seems reasonable that an assessment method used to analyze these vessels should be capable of dealing with ductile material behaviour as well as brittle fracture. In addition, to allow application to non-PWHT weld repairs, a more specific requirement is that the technique should use a rational approach when including residual stress in the assessment.

For materials possessing relatively high fracture toughness, stresses approaching the material flow stress can develop across the remaining ligament of a flawed region. A substantial amount of plastic deformation or significant ductile tearing could occur at the flawed region prior to failure. For this reason it is necessary to have an assessment method which will allow investigation of components over the full range of material behaviour from perfectly elastic to fully plastic.

### 3.3 The Failure Assessment Diagram

The Central Electricity Generating Board in the United Kingdom has developed an assessment technique known as the CEGB R6 method which satisfies the need for a reasonably simple method for assessing the integrity of structures containing defects over a wide range of material behaviour (Harrison, 1977). The technique has been designed to make the process of defect assessment more uniform by providing the assessor with an ordered sequence of operations. The R6 method differs from other assessment documents or codes in that safety factors are not included in the assessment procedure. Instead, a more accurate prediction of material behaviour is presented and it is up to the user to choose conservative factors to apply to a structure or vessel, based on experience or adopted code requirements.

The R6 method is based on a two criteria approach which states that there are two limits to the failure load of a structure containing flaws; fully plastic and linear elastic. Between these two limits Dowling and Townley (1975) suggest that elastic-plastic failure occurs when

$$K_r = S_r \left[\frac{8}{\pi^2} \ln(\sec(\frac{\pi}{2} S_r))\right]^{-1/2}$$

where

$S_r$ - stress ratio for fully plastic failure conditions

$$= \frac{\text{Loading Stress}}{\text{Material Flow Stress}}$$

$K_r$ - stress intensity ratio for linear elastic failure conditions

$$= \frac{\text{Stress Intensity}}{\text{Fracture Toughness}}$$

A plot of $K_r$ versus $S_r$, shown in Figure 3, produces a curve known as the Failure Assessment Diagram (FAD) which is the basis for the R6 method. When per-

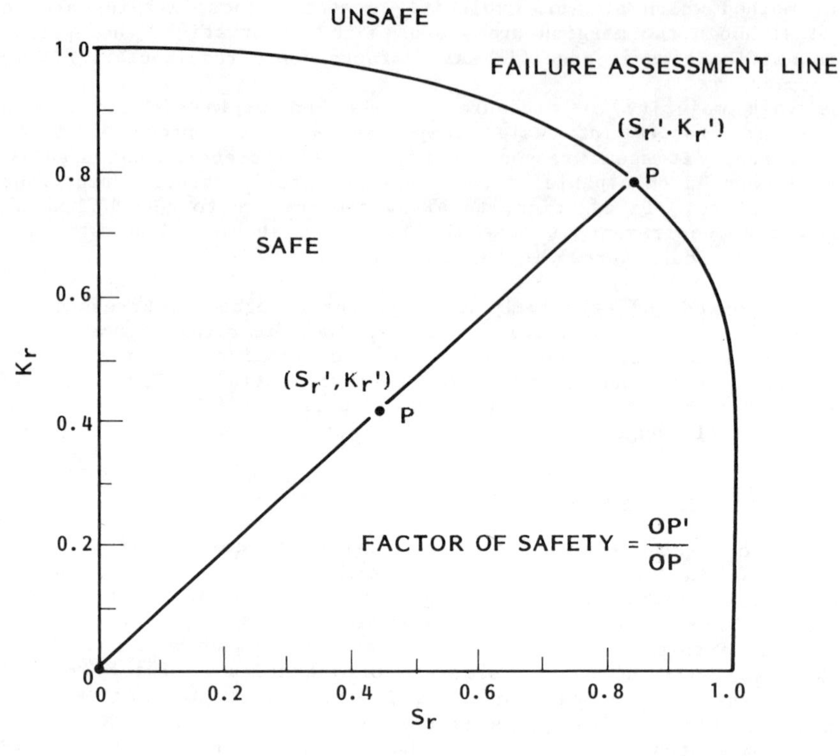


FIGURE 3

CEGB R-6 FAILURE ASSESSMENT DIAGRAM

forming an assessment, a point P with coordinates $S_r$, $K_r$ is plotted on the diagram. If this point lies on or outside the assessment line, failure is predicted; if it lies inside the line failure is avoided. Validation of the method using data from specimen and vessel test programs, confirms that the R6 FAD procedure provides an appropriate limit line for the avoidance of failures in ferritic structures (Harrison, 1979).

In the example shown in Figure 3, the margin of safety of the component can be easily determined by drawing a radial line from the origin through the point P and intersecting the failure assessment line at P'. The factor of safety is taken as the ratio OP'/OP. This line also represents the path of response or material behaviour as the load is increased on a structure. The curve given in Figure 3 is an idealized version of the failure assessment diagram which predicts failure at stress ratios of 1.0, when the loading stress reaches the flow stress or plastic collapse stress. The flow stress in this case is estimated to be the mean of the yield and ultimate tensile strengths:

$$\sigma_f = \frac{\sigma_y + \sigma_{UTS}}{2}$$

It was recognized that this limit did not constitute actual failure and that real material behaviour includes a substantial amount of plasticity in this range for the case of ductile materials. Tests of specimens with various crack lengths and geometries show clearly that $S_r$ can exceed a value of 1.0 without failure.

Figure 4 shows failure assessment curves for typical pressure vessel grade steels based on several crack length to plate thickness ratios. The details of the derivation of these curves is given in the appendix of the work by Bloom and Malik (1982).

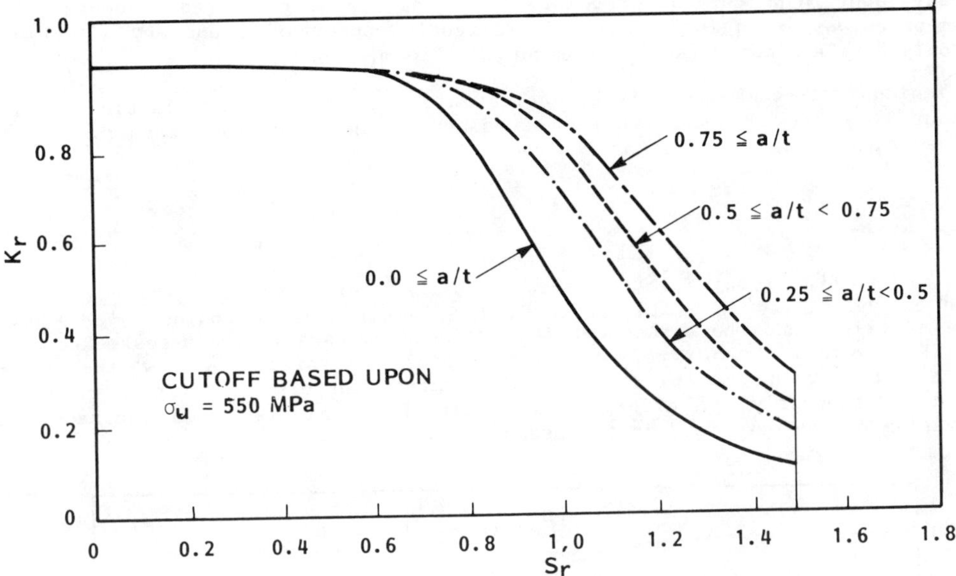


FIGURE 4

FAILURE ASSESSMENT CURVES FOR NUCLEAR GRADE PRESSURE VESSELS

### 3.4 Secondary Stress Effects

In the valid linear elastic regime of material behaviour, residual stresses can be treated as being equivalent to mechanical loading stresses. However, residual stresses do not influence the plastic collapse load of a structure and thus must be introduced into an integrity assessment such that this is taken into consideration. This was shown quite clearly in the Heavy Section Steel Technology test program at Oak Ridge National Laboratories (Bryan, 1981). Non-post-weld heat treated half-bead weld repairs were made in simulation test vessels, and artifical flaws were placed adjacent to the repair areas. The vessels were then pressurized in two separate tests, one at a low temperature to ensure the material was at its lower shelf toughness and one at a substantially higher temperature so that the material was in the upper transition region. Post-test analysis illustrated that for cases of low fracture toughness LEFM can be used to accurately predict the effect of combined pressure and residual stress. However, at values of greater fracture toughness when full section yielding can occur in the region of a flaw, residual stresses have little, if any, effect on fracture strength.

In terms of the failure assessment diagram, these results suggest that at $S_r = 0$, when LEFM applies, residual stresses will have a significant effect, but as $S_r$ increases in value, elastic-plastic behaviour predominates and as fully plastic failure is approached, residual stress is no longer important.

To include residual stress in an assessment using the failure assessment diagram, the vessel stresses must be considered independently. Mechanical loading or pressure loading generates primary stress ($\sigma^P$) while loading due to residual and/or thermal stress is considered as secondary stress ($\sigma^S$). The CEGB has shown that universal lower bound curves for combined loading can be generated. However, the assessment curves given in Figure 4 are slightly conservative and any adjustment would only have a minor effect on the curves (Bloom, 1982).

Since residual stresses are only effective in the case of linear elastic material behaviour they are included on the assessment diagram in that manner. $K_r$ is redefined as

$$K_r = K_r^P + K_r^S$$

while $S_r = S^P{}_r$ remains the same. Using this method, $K^S{}_r$ is first plotted on the FAD as shown in Figure 5 at $S_r = 0$. This new point, O', is used as the new origin and the assessment is continued as usual. The assessment point ($S^P{}_r$, $K^P{}_r$, + $K^S{}_r$ is plotted and the factor of safety is now the ratio O'A'/O'A. This value can be compared with safety factors in existing codes or assessed independently by performing a sensitivity analysis which includes the conservatism of the data used in the assessment.

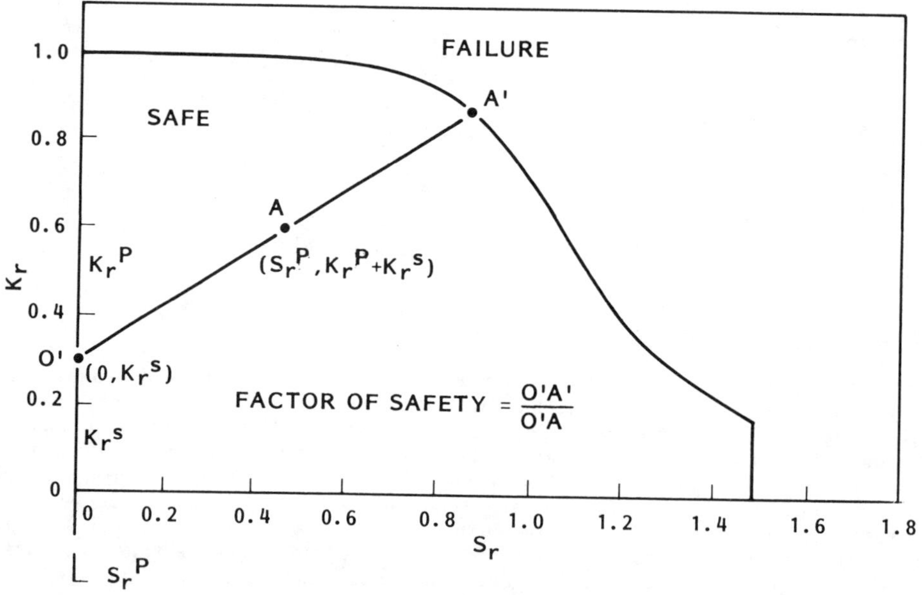


FIGURE 5

INCLUDING RESIDUAL STRESS AS A SECONDARY STRESS ($K_r^S$) ON THE FAILURE ASSESSMENT DIAGRAM

Using this method, it can be shown how improvements in fracture toughness or reduction in residual stress will influence a component's integrity in terms of resistance to fracture. The CEGB R6 approach provides a clear step-by-step method for integrity assessment which reduces many of the complexities of elastic-plastic analysis into a graphical procedure and allows for the incorporation of residual stresses in an effective and understandable manner.

Work is currently under way at Ontario Hydro in the area of component integrity assessment for non-PWHT repair welds. Test coupons containing temper-bead welds in heavy section pressure vessel steel plate have been prepared and residual stress magnitudes and distributions have been measured. Fracture toughness testing of the HAZ region is under way. The information generated in these studies is intended for use in integrity assessments for non-PWHT repairs to obtain a more accurate model of component behaviour.

## 4.0 CONCLUSIONS

Extensive experimental work and service performance have shown that the temper-bead weld repair method provides a successful means for repairing vessels without the use of conventional PWHT. However, complex residual stresses, often reaching tensile yield strength magnitude, are introduced when using the temper-bead repair method and remain unrelieved when the component is returned to service. If it becomes necessary to perform an integrity assessment on a vessel in the vicinity of a non-PWHT temper-bead repair, the CEGB R6 method provides a reasonably simple assessment procedure which can be used to include the effects of residual stress.

## REFERENCES


Binkley, N.C. and R.W. Herrmann (1974). An in-service reactor repair simulation. Welding in Nuclear Engineering, Deutscher Verlag Fur, Schweisstechnik, Dusseldorf.

Broom, J.M. and S.N. Malik (1982). A procedure for the assessment of the integrity of nuclear pressure vessels and piping containing defects. Electric Power Research Institute. Report NP2431, Research Project 1237-2.

Bryan, R.H., P.P. Holtz, S.K. Iskander, J.G. Merkle and G.D. Whitman (1981). Test of a thick vessel with a flaw in a residual stress field. J. Pressure Vessel Tech., 103, 85-93.

Dowling, A.R., and C.H.A. Townley (1975). The effect of defects on structural behaviour: A two criteria approach. Int. J. Pressure Vessels and Piping, 3, 77-107.

Harrison, R.P., K. Loosmore and I. Milne (1977). Assessment of the integrity of structures containing defects. CEGB Report R/H/R6-Rev 1.

Harrison, R.P., K. Loosmore and I. Milne (1979). Assessment of integrity of structures containing defects. Supplement 1 - Validation. CEGB Report R/H/R6 Supplement 1.

Lawson, W.H.S. (1979). Metallurgical Characteristics of temper-bead weld repairs. Ontario Hydro Research Division Report 79-129-K

Rudd, C.O. and P.S. Dimascio (1981). A prediction of residual stresses in heavy plate butt welds. J. Materials for Energy Syst., 3, 62-65.

Rybicki, E.F. and R.B. Stonesifer (1980). An analysis procedure for prediction of weld repair residual stresses in thick-walled vessels. J. Pressure Vessel Tech., 102, 323-331.

# INADEQUATE FIELD WELDING PROCEDURES LEAD TO MARINE BOILER SUPERHEATER HEADER FAILURES

R. Myllymaki
*Defence Research Establishment Pacific*
*FMO Victoria, B.C. V0S 1B0*

ABSTRACT

The paper describes the investigation of the causes of cracking in the superheater headers of several marine boilers. Metallurgical analysis showed that the cracks initiated from small hydrogen-induced welding cracks at the toes of fillet welds and propagated by corrosion fatigue over a period of years under complex thermal/mechanical stresses. Mechanical tests showed that the toughness of the header material was adequate to prevent brittle fracture in service, i.e. the header would leak before fracture.

A reliable ultrasonic inspection technique was devised to locate cracks and measure their length and depth of penetration.

KEYWORDS
Marine boilers, superheater headers, steel, hydrogen-induced cracking. ultrasonic inspection, corrosion fatigue.

## INTRODUCTION

A pre-refit inspection in October 1981, revealed that a crack extended into a Y100 superheater header [a 300 mm outside diameter, 32 mm thick, 2,700 mm long pressure vessel containing steam at 450°C and 3.8 MPa]. In attempting to remove the crack by grinding, it was found that the crack had extended nearly all the way through the header. This alarming observation resulted in all Y100 superheater headers being given a dye penetrant examination the results of suggested that the cracking problem was widespread. A rapid ultrasonics examination of another header discovered a crack about 19 mm deep; a serious crack indeed!

A number of basic questions were now asked:
1. What was the extent of the cracking and could it be determined reliably non-destructively?
2. Was the cracking restricted to or more common to one class of ship or were all classes involved?
3. Did the cracks represent an immediate safety hazard as well as making the boiler inoperable?
4. What was the cause of the cracking?
5. Could the headers be repaired or would the cracked headers have to be

replaced?

The investigation concentrated initially on the first question above and determined the size (length and depth) of the cracks in the headers. It was shown using this technique that while most of the cracks were shallow, less than 2mm, the cracks in one boiler inlet header were deep, extending up to 25 mm into the header. It was this header which was selected for the detailed metallurgical examination by DREP and Babcock and Wilcox, the boiler manufacturer, which is outlined below. Additional investigations were carried out at Defence Research Establishment Atlantic, in Halifax, and at Canmet in Ottawa.

## EXAMINATION AND TESTING OF HEADER

Both magnetic particle and radiography (X-ray) were used to locate the cracks in the header but ultrasonics could be used to measure the crack length and depth with good accuracy as shown when parts were sectioned for metallurgical examination.

A total of nine ships were inspected. In the total of 280 small (≃75 mm) fillet welds inspected, 39 crack indications were found, of which at least 15 could have been minor undercuts rather than cracks. The most serious cracks (up to 25 mm deep) were found in the bottom baffle of the inlet header which is the subject of this report.

All the cracks, except one, run circumferentially at the toes of the baffle/header fillet welds. The exception was a crack, 4 - 5 mm deep, and (≃38 mm long, running longitudinally near the bottom baffle of the port outlet header on another ship. This crack was situated between the fillet welds near the drain hole and was not connected to the welds at any point.

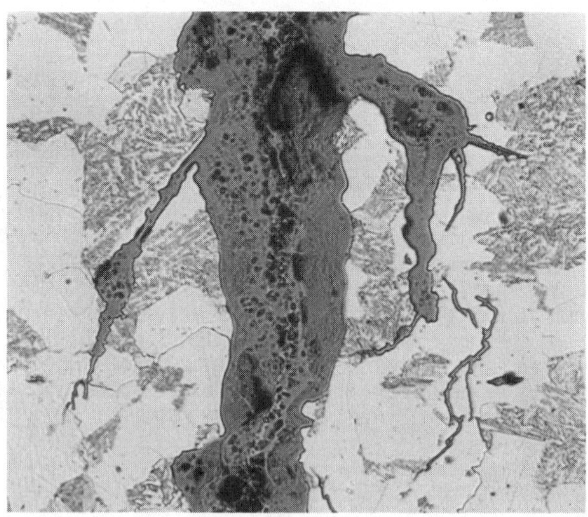

Fig. 1. Transgranular crack near surface of header. X250

It was noteworthy that only two small cracks (1 and 4 mm deep) were found in the 4 ships built after 1960. Numerous cracks were found in the headers on another ship, all of which were less than 4 mm deep except one which was as much as 13 mm deep in the bottom baffle of the port header

A chemical analysis was carried out on all the headers. The pre-1960 ships had headers forged from a nominal 0.5% Mo-containing steel with up to 0.27% C. The four ships built after 1960 had headers forged from nominal 1% Cr-0.5% Mo steel with up to 0.16% C. These steels are similar to ASME specifications SA 336, Class F1 and F12 respectively.

The cracks were filled with an oxide (Fig. 1) which was identified by X-ray diffraction as magnetite. No other oxides or materials were identified in the cracks. In some areas the oxide had a two-tone colour which suggests that two different forms of oxide were present.

Two tensile specimens were machine from the header material. The longitudinal and transverse to the tube axis results were respectively: UTS, (MPa) 543, and 537; YS (MPa) 364 and 334; % Elong., 28.5 and 21.0; % RA, 58.2 and 23.6. The longitudinal specimen showed fibrous fracture with considerable shear. The transverse specimen was about 75% crystalline flat fracture.

Charpy V notch specimens were taken from the inside surface of the header. Transverse and longitudinal specimens were tested in accordance with ASTM E23. The absorbed energy and fracture appearance are shown in Figure 2.

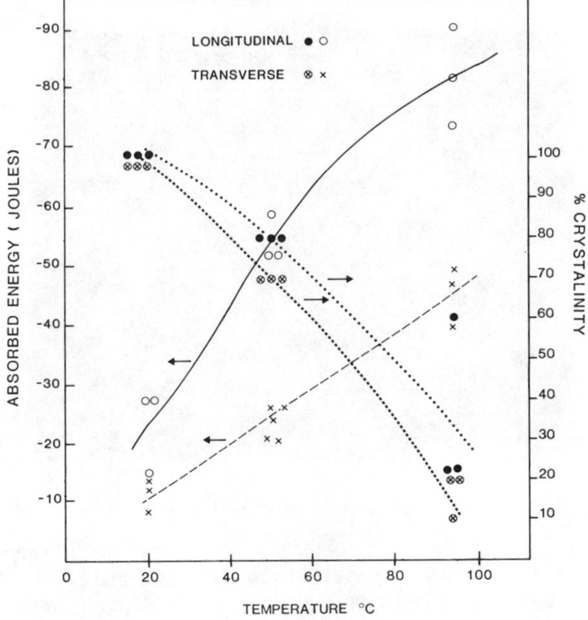


Fig. 2 Charpy V impact test results.

The P2 type drop weight specimens were machined from as close to the inside surface of the tube as possible. The crack starter weld was placed on this surface. The specimens were tested in a vertical 16,000 Joule capacity drop weight machine in accordance with ASTM E208. The nil-ductility temperature (NDT) is about 15°C.

A 50 Kip Materials Testing System (MTS) machine was used to determine the plane strain fracture toughness (KIc) in accordance with ASTM E399-78a. A C-shaped specimen (Fig. 5b in E399), with the following dimensions was machined from the header material: B = 36.3 mm W =32.8 mm X = 22.2 mm r1 = 120.6 mm r2 = 152.4 mm.

The specimen was fatigue pre-cracked to a total crack length of 14 mm at a load ratio, R = +0.14 and a test frequency of 5 Hz. The resulting value of KQ was found to be 63 MPa $\sqrt{m}$.

The value of KQ thus obtained is not a valid measure of KIc since the ratio of Pmax/PQ slightly exceeds the maximum value of 1.1. However, observation of the fracture indicated only plane strain behaviour in the fatigue region. Therefore the value of KQ thus obtained is representative of the fracture toughness of the material in its applied thickness and should closely approximate KIc.

Hardness measurements were made on polished cross sections of header material. The hardness (RB78) did not differ between the middle, outside, and inside surfaces. The micro-hardness (100 gm load) in the weld heat affected zone near the cracks shown in Fig. 3 was about 450 HV (Rc45). This, the hardest region, was within 1-2 grains of the fusion boundary.

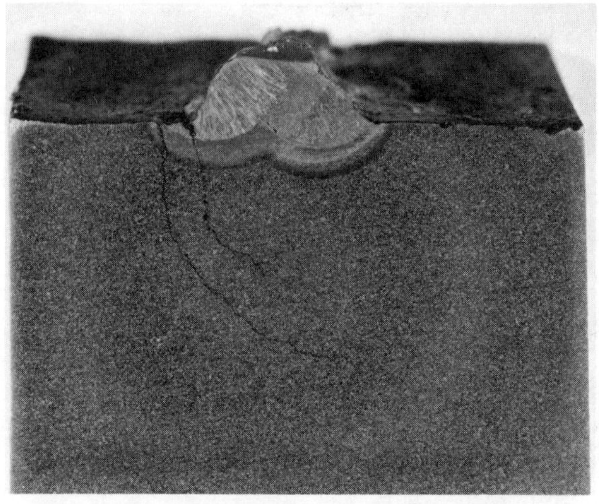

Fig. 3. Macro-section showing two cracks. X1

Numerous sections removed from the cracked regions for metallography, showed that most of these cracks travelled approximately perpendicular to the tube surface. However, two cracks changed direction, as shown in Fig. 3, with the tips running about parallel to the tube surface. The cracks tended not to extend to considerable depth without spreading laterally. It is also noteworthy that the cracks meandered about with no fixed direction.

The cracks in all the microsections appear quite similar. They are predominantly transgranular with a few intergranular excursions. Fig. 1 shows part of the main crack and a number of branch cracks all filled with oxide. The small branch cracks are transgranular and in some cases appear to follow slip planes near the end of the crack. Figure 4 shows the end of another crack with a blunted end and oxide-filled branch crack. There is no deformation at the end of the crack which extended deep into the header wall. There is also no sign of creep damage in any of the sections examined.

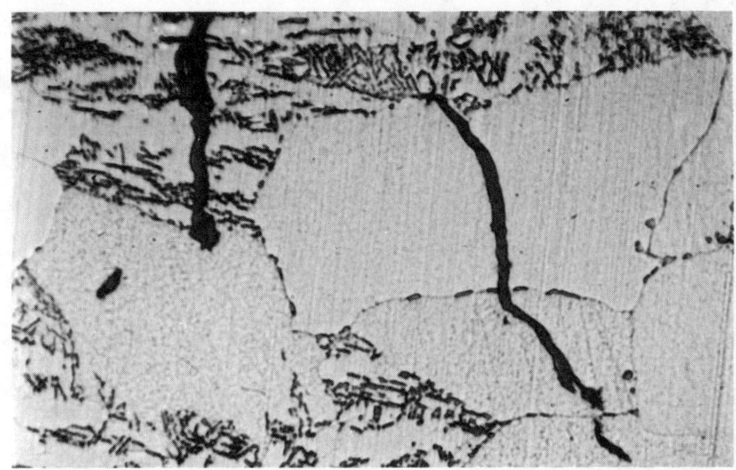

Fig. 4. Blunt end of oxide filled cracks. X720.

Sections through the cracks revealed a number of other interesting points. There was a crack similar to that shown in Fig. 3 which curved under the weld and ran about parallel to the header surface. The major cracks extended along the toe of the baffle/header fillet weld. Some other minor cracks were seen in this sample as shown in Fig. 5. There were several of these cracks, none of which travelled beyond the visible heat-affected zone (HAZ) of a weld which had been removed at some time. The overlapping heat affected zones suggest that at least 2 or 3 welds were made in this section at some previous time.

The header material microstructure consists of ferrite and pearlite grains. The grain size is mixed with some ferrite grains as small as ASTM 6 and some pearlite grains as large as ASTM 2. The ferrite is arranged in a Widmanstätten pattern within the pearlite grains.

Careful examination of the grain boundaries in the header material did not reveal any precipitation or intergranular cracks. There was no sign of hot cracking in the weld heat-affected zone nor was there any sign of graphitization. An adherent oxide formed on the inside surface (steam side) of the header. The oxide penetration did not show preference for ferrite or pearlite grains and only a few minor excursions were along the grain boundaries. There was no evidence of cracking other than that seen near the baffle welds.

A section of one crack was broken open and examined with the scanning electron microscope. The fracture face exhibited no easily recognisable fracture mechanism due to the extensive oxidation. The main fractographic feature that was still visible was the formation of a large number of secondary cracks.

The microhardness of the heat affected zone near the cracks shown in Fig. 3 was about 450HV. The effect of preheat 20°C, 95°C, and 150°C on the heat affected zone hardness of the header material was assessed by comparing three bead-on-plate welds. A 4 mm diameter E7018 electrode used at 150 amperes, 22 volts, and 140-150 mm per minute travel speed gave a heat input of about 1.2-14 KJ/mm.

This simulates the lowest heat input (which translates to the highest hardness) which would be used on the headers to make a 5-6 mm fillet weld in the overhead position. Microhardness measurements showed that the preheat had no effect on the maximum hardness of 450 HV.

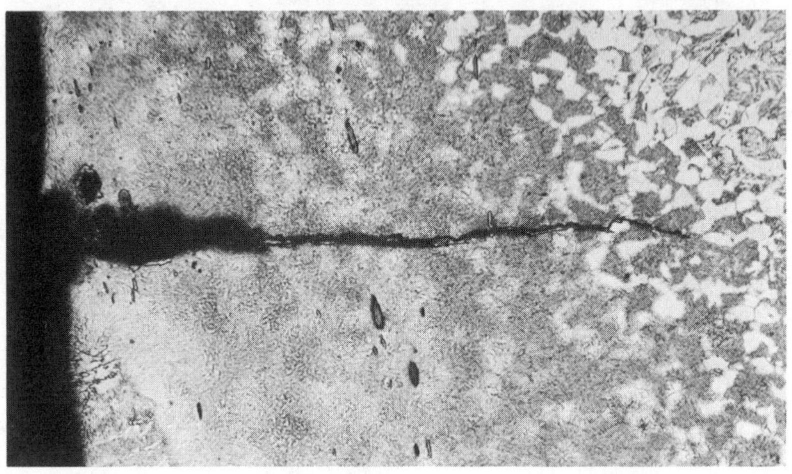

Fig. 5. Heat-affected zone crack. X45.

## DISCUSSION

### Extent of Cracking

A thorough ultrasonics inspection located and mapped 39 indications in a total of 280 fillet welds. Most of the indications were less than 5 mm deep and the smallest ones, 1.25 mm, are probably minor undercuts but this cannot be confirmed because of the limited access to the weld toes inside the headers. Two ships together account for 25 of the 39 indications. The lone longitudinal crack found was not located at the toe of the baffle/header weld and thus was quite different from the other indications.

Ultrasonics inspection, confirmed by sectioning, showed that the cracks running circumferentially in the header at the toes of the fillet welds were 19 mm to 25 mm deep. The two cracks departed from the circumferential direction and started to propagate more in the longitudinal direction. The crack propagation direction was influenced by the presence of the large inlet nozzle on this side of the header. Figure 3 shows that as the cracks got deeper they changed direction and started to follow a path nearly parallel to the header surface. There is some suggestion that the inlet nozzle baffle welds may be more susceptible to cracking than the other baffles, but the results are not conclusive.

Corrosion and oxidation have obliterated the point of initiation of the cracks. However, when one considers the ultrasonic inspection results and looks at the cross-section of the welds such as those shown in Figs. 3 and 5, an initiation mechanism can be postulated. All the cracks examined started from the toes of fillet welds. Figure 5 shows one of many heat affected zone cracks, typical of hydrogen-induced cracks found in steels that are susceptible to hydrogen-induced cracking. That the header steel with 0.27% carbon is susceptible to hydrogen-induced cracking is clear from the hard (≈450HV) martensitic microstructure in the heat affected zones. The critical hardness for hydrogen-induced cracking in the header steel is not known but 450 HV would be expected to crack under many conditions of restraint and weld metal hydrogen content (Graville, 1975; Brisson, 1968) The hardness of 450 HV approaches the maximum hardness expected for a steel containing 0.27% carbon (Krauss, 1978; Andrews, 1956; Boyer, 1977). Up to 50,000 hours operation at 450° C has not reduced the heat affected zone hardness significantly.

To avoid hydrogen-induced cracking, carefully controlled welding procedures must be used, including pre-heat (probably > 150°C and properly dried 400-450°C low-hydrogen welding electrodes. A preheat temperature up to 250°C had little effect on the heat-affected zone hardness. Therefore heat-affected zone hardness cannot be used to confirm that the preheat of 150°C which had been specified in previous welding procedures was in fact used. Considerably higher preheat temperature (perhaps > 250°C would be needed to appreciably lower the high hardness. This is not surprising because the martensite start (Ms) temperature for the header steel is about 400°C (Andrews, 1956; Boyer, 1977) and the 90% martensite transformation temperature is about 330°C. The preheat would have to be much higher to significantly slow the cooling rate at the Ms temperature and reduce the heat affected zone hardness in the 32 mm thick header.

There were very few ultrasonic indications in the 4 -"post 1960" ships. The header alloy in these ships is a chromium - molybdenum alloy with much lower carbon than the molybdenum alloy used for the headers on the other 5 ships. The lower carbon content will form a heat affected zone with a lower hardness and lower susceptibility to hydrogen-induced cracking. Further, it is suggested that these superheater header baffle plates were welded under controlled conditions in the fabrication shop and have not been rewelded in the difficult conditions aboard ship as in the case of many of the baffle plates aboard the other five ships.

It seems likely that subsequent to manufacture all of the baffles were replaced (in some cases more than once) in the "pre 1960" ships. As mentioned above, fully 25 of the total of 39 ultrasonics indications were found on 2 ships. This suggests that the welding procedure was not followed, or an inadequate procedure was used, or welding electrodes other than truly low-hydrogen were used, or perhaps a combination of all three. It is obvious that future repairs must use a proven welding procedure and be carefully supervised at all stages.

## Corrosion - Fatigue Mechanism of Cracking

The cracks are mainly transgranular (with a few intergranular excursions) and in some cases appear to follow slip lines. These cracks look very similar to the transgranular cracking in ferritic steels described as corrosion fatigue by Thielsch (1956). Corrosion fatigue cracks tend to follow a more direct path than that seen in the present case and normally more than one corrosion fatigue crack is seen in one location.

Corrosion fatigue suggests a mechanical-chemical combination in which the corrosion reduces the fatigue life or the dynamic stress increases the corrosion rate, or both. In the present case, wet corrosion as well as high temperature oxidation are possible. The oxide at the tips of the cracks is magnetite, a high temperature oxide which does not form at low temperatures from wet corrosion. The inner surface of the header is in good condition with no evidence of wet corrosion and no excess oxide present. The cracks are wider at the surface (Fig. 1) than at the tips (Fig. 4) but not very much wet corrosion or high temperature oxidation has taken place. There was very little evidence of pitting on the header surface. Thus the environmental effects are not very active.

The oxide which fills the cracks to the finest tips may plug the cracks, prevent the access of oxygen and stifle further cracking. Also the oxide may blunt the tips, reduce the stress concentration, and require more dynamic cycles to reactivate the cracking (Tomkins, 1979; Haugh, 1976; Wells, 1979). Oxide which blocks the crack can reduce the effective cyclic crack opening particularly when low stresses are involved (Tomkins, 1979).

The cracking mechanism is complex as it involves material/chemical/thermal/mechanical interactions which have undoubtedly changed with time. Fatigue accompanied by corrosion and high temperature oxidation is probably a reasonable description of what is happening. The wet corrosion likely played a larger role in the crack propagation early in the life of the header. As the cracks deepened the high temperature oxide filled the cracks, protected the surface, and prevented ingress of the wet environment to the crack tips. Periodic large stress cycles may cause crack advance or decohesion of the oxide from the metal surface. In either case, if oxygen is present, oxidation of the clean metal surface will be more rapid.

Other cracking mechanisms which have been considered during this investigation include: thermal fatigue, creep, stress-rupture, overheating, stress corrosion cracking. Simple thermal fatigue is recognised by a crazed pattern of crack which is not evident in the present headers. Creep or stress-rupture failures can be recognised by the presence of a multiplicity of intergranular cracks and in the case of stress-rupture by creep voids adjacent to the main fracture. No such evidence was found in the samples examined. Moreover one would not expect creep failure after a maximum of 50,000 hours service at 450°C and low service loads. There was no evidence of overheating in the microstructure of the headers i.e. no spheroidization of the pearlitic grains.

Stress corrosion cracking in low strength ferritic alloys is normally intergranular although there have been cases of transgranular cracking attributed to stress corrosion and intergranular to corrosion fatigue (Barsom, 1971; NTSB, 1970). Stress corrosion is normally much more branched than the cracks seen in these headers.

The oxide filled cracks would be expected to obstruct the movement of the specific crack-inducing substances to the crack tip. Thus, the other cracking mechanisms do not fit the evidence as well as corrosion fatigue. Hainsworth(1982) in the Babcock and Wilcox report on this failure also suggests that cracking had occurred in two modes, i.e. cold cracking at the toes of fillet welds followed by corrosion-assisted fatigue or stress enhanced corrosion.

## The Stresses

That the dynamic stress system near the inlet nozzle is complex is not surprising when one considers the geometry of a 150 mm diameter nozzle in a 300 mm diameter header as well as the thermal effects. Thermal gradients are caused by differences and fluctuations in temperatures between the inlet steam on the inside of the header and the outside surface on the fire-side and boiler room-side. Cooler steam and water introduced during boiler shut-downs cause periodic stress gradients near the inlet nozzle. Parts of the header away from the inlet nozzle experience fewer and lower magnitude thermal fluctuations. Superimposed on the thermal stress is the service stress due to 4 MPa steam pressure; about 17 MPa hoop stress and about 9 MPa longitudinal stress. In addition there is stress due to thermal expansion because the headers are constrained by the inlet and outlet nozzles. The inlet nozzle, hand holes, superheater tubes, and the baffle/header fillet welds all act as stress concentrators. Although the above points cause a complex stress system, the appearance of the wandering cracks both on a macro (Figs. 3) and the micro scale (Figs. 1 and 4) suggest a rather low principal stress. The branching near the ends of the cracks (Fig. 1) and the absence of any deformation or cracking or spalling of the oxide in the crack tips (Fig. 4) further suggest low stress levels.

## The Fracture Strength of the Header

The mechanical test results illustrate a number of interesting points. The yield strength is, perhaps, somewhat higher than expected at about 345 MPa. The rather low reduction of area in the transverse specimen is not entirely unexpected from a forged tube. The Charpy V notch results show that the longitudinal specimens have considerably better properties. Both the tensile and Charpy longitudinal specimens reflect the crack path in the header rather better than the transverse specimens.

The drop weight tests gave a nil-ductility temperature of about +15°C which corresponds to rather low Charpy values (Fig. 2). The fracture path in the drop weight specimens was oriented in the same direction to the cracks in the header.

Although a valid result was not obtained with the plane strain fracture toughness tests, it is felt that it is not far from being valid. The fracture appearance, both macro and micro, suggest that very little plasticity was present at the fracture. There was no evidence of a stretch zone between the fatigue pre-crack and the final fast fracture. The KIc value of $63\text{MPa}\sqrt{\text{m}}$ is somewhat encouraging. In addition, the crack in the C-shaped specimen was oriented parallel to the header axis, the most unfavourable direction for fracture toughness.

An approximate calculation for a hypothetical semi-elliptical surface crack extending 25 mm into the 32 mm wall and 50 mm long under an applied tensile stress of 24 MPa indicated that the stress intensity factor was $6\text{MPa}\sqrt{\text{m}}$). Note that this stress intensity is about on order of magnitude smaller than that required to cause rupture of the header.

All mechanical tests were conducted at less than 95°C. The superheater headers operate at about 450°C, a temperature significantly above the nil-ductility temperature of 15°C or the 50% fracture appearance transition temperature of about 75°C for the drop weight tear test and the Charpy V-notch impact test respectively. These results look even more favourable when one considers that the headers are slowly loaded in service. There may be some concern about hydrostatically proof testing the boilers at ambient temperatures, but heating the test water should overcome such apprehensions.

## CONCLUSIONS

The DREP investigation of the superheater header has shown the following:

1. The cracks initiated at the baffle/header fillet welds, likely from hydrogen-induced cracks at the toes of the welds. Inadequate field welding procedures are likely the cause of many of the large number of ultrasonic indications (25 of the total of 39) found on two ships.
2. The cracks propagated by a form of corrosion fatigue, in which the stress was low and the environment mildly corrosive. The blunt oxide filled crack tips suggest that during the later stages the cracking mechanism was fatigue and high temperature oxide formation rather than wet corrosion.

3. Cracking was not as bad in the "post 1960" headers probably because the baffles were installed under controlled welding conditions in the fabrication shop and were not re-welded in the difficult conditions aboard ship.
4. Ultrasonic inspection can be used with confidence to locate and determine the size of cracks in the superheater headers.
5. The cracks do not represent a safety hazard. The toughness of the header material is adequate to prevent a brittle fracture in service, i.e. the header will leak before fracture. All but the deepest cracks can be tolerated and monitored to estimate the rate of crack growth.
6. Weld repair in-situ is not recommended. Shallow cracks (and in many instances, cracks extending up to half way through the headers) can be removed by grinding.
7. When hydrostatic testing headers at ambient temperature, it should be mandatory to use hot water.
8. A carefully supervised proven welding procedure must be used to replace any baffle plates that have been removed.

## ACKNOWLEDGEMENT

The author acknowledges the contributions of many DREP and CFB Esquimalt personnel in this failure investigation; in particular, Mr B.W. Greenwood for the ultrasonic inspection procedures, Mr. P.D. Martin for the metallography and mechanical testing, and Mr K.I. McRae for the plain strain fracture toughness testing.

## REFERENCES

Andrews, K.W. (1956). "Empirical Formulae for the Calculation of Some Transformation Temperatures", Journal of the Iron and Steel Institute, Vol. 183, 349-359.

Barsom, J.M., "Mechanisms of Corrosion Fatigue Below KIscc", International Journal of Fracture Mechanics, Vol. 7, 163.

Boyer, H.E. (Ed.) (1977), Atlas of Isothermal Transformation and Cooling Transformation Diagrams, ASM, 120.

Brisson, J. (1968), "Study of Underbead Hardness in Carbon and Low Alloy Steels", Soudages et Tech. Connexes, 22 (11/12), 437-455.

Graville, B.A. (1975). "The Principles of Cold Cracking Control in Welds", Dominion Engineering Company, 1975.

Hainsworth, J. (1982), "Failure Analysis for Cracked Superheater Headers on B W Y-100 Marine Boilers", Babcock Wilcox, Reference No. 836-0591. September 30, 1982.

Haugh, J.R.,Skelton, R.P., Richards, C.E., (1976) "Oxidation - Assisted Crack Growth during High Cycle Fatigue of a 1% Cr-Mo-V Steel at 550°C", Materials Science and Engineering, Vol. 26, 167-174.

Krauss, G., (1978). "Martensitic Transformation, Structure and Properties in Hardenable Steels", in Hardenability Concepts with Applications to Steel, D.R. Daone and J.S. Kirkaldy (Eds), AIME, Warrendale, Pa. 229-248.

NTSB (1970) National Transportation Safety Board, Report No. HTSB-HAR-71-1, Washington.

Thielsch, H, (1965). "Defects and Failures in Pressure Vessels and Piping", Reinhold Publishing, 398.

Tomkins, B. (1979). "Elevated Temperature Fracture Mechanics", from Fracture Mechanics, Current Status, Future Prospects, Proceedings of a Conference, Cambridge 16 March 1979, Pergamon Press,

Wells, C.H. (1979). "High Temperature Fatigue", in Fatigue and Microstructures, ASM Materials Science Seminar, 14 October 1978, ASM, 307.

# SUBMERGED ARC STRIP OVERLAY WELDING OF 2.25-Cr-1Mo STEEL

R.S. Chandel and R.F. Orr
*Physical Metallurgy Research Laboratories*
*Energy, Mines and Resources Ottawa, Ont. Canada*

## INTRODUCTION

Owing to their excellent mechanical properties and high resistance to corrosion and hydrogen attack, austenitic stainless steels are the obvious choice for reactor vessels in the petrochemical industry. However, because of the high cost of this material, fabrication of a complete vessel from austenitic stainless steel is impractical. Therefore, in normal practice a composite of low-alloy ferritic steel and an austenitic stainless steel is used. Such a composite provides an excellent balance of strength, surface properties, and economy. The barrier thickness of austenitic stainless steel required to resist corrosion and hydrogen attack on base material is small compared to the total wall-thickness.

The only practical method of cladding for fabrications that involve large clad areas such as pressure vessels, or where the base plate exceeds 100 mm in thickness, is weld deposition (1,2). Various methods of cladding by welding have been used for this purpose (3). However, one of the most popular weld-deposition techniques is submerged-arc strip-cladding that has gained popularity in Europe and the U.K. This process deposits metal of excellent properties with limited dilution and at a high deposition rate (3). Furthermore, using a strip rather than a wire reduces the number of overlapped regions, thereby decreasing the numbers of sites where underclad cracking and other defects develop (4).

In principle, submerged-arc strip-cladding resembles submerged-arc welding except that in the former a strip (normally 60 mm x 0.5 mm) is substituted for the solid wire (5). The other difference is that in strip-cladding the arc burns at a number of points along the strip edge, so that a high arc-force at a single point is not developed, which in conjunction with the large weld-pool, means that the penetration into the base material is relatively low (6).

Whenever an austenitic stainless steel is deposited on a ferritic steel, the dilution of the weld metal by parent metal leads to a problem in meeting the desired compositions. In practice the effects of dilution have been overcome by depositing more than one layer of austenitic stainless steel. In petrochemical reactor vessels, the first or "buffer layer" normally consists of 309L stainless steel which contains about 23% Cr. The second layer of 308L or 347 stainless steel is deposited on top of the buffer layer.

To be acceptable, a stainless steel overlay is required to have the following characteristics (1,5,7):

1. A guaranteed minimum thickness of cladding.
2. A minimum thickness of cladding within the specification for the particular stainless steel.
3. A deposit of suitable microstructure and mechanical properties.
4. Complete fusion between parent metal and deposit.
5. A finely rippled bead surface, free from cracks, porosity, and slag inclusions.

The reasons for these requirements have been discussed elsewhere (3).

When an austenitic stainless steel is deposited on a ferritic base, a number of problems such as dilution from the base metal, formation of martensite, and diffusion of carbon from the base metal etc. are encountered which have an influence on the properties of the joints. The weld overlay process results in considerable residual stress; the Pressure Vessel Code requires varying post-weld heat treatments depending on wall thickness before service. Therefore, the aim of this work is to evaluate the metallurgical and mechanical characteristics of the weld overlays both in 'as-welded' and 'post-weld heat-treated' conditions.

## EXPERIMENTAL TECHNIQUES

### Material

The experimental base material was a 2.25 Cr - 1.0 Mo (SA-387 Grade 22 class 2) steel which is generally used for the fabrication of petrochemical reactor vessels because of its strength, toughness and creep resistance. The plate was 37.6 mm thick and was supplied in the normalized and tempered condition. The chemical composition of this base plate is given in Table 1 and its microstructure is shown in Fig. 1. The plate was cut into 305 x 406 mm pieces and the surfaces were cleaned to remove dirt and oxides.

The weld deposits were made by using 60 mm x 0.5 mm strips of 309L, 308L, and 347 stainless steels. The chemical compositions of these strips are given in Table 1. A matching neutral and non-compensating type of flux was used for welding. The strips and flux were manufactured by Sandvik.

### Welding

A Lincoln SA-800 submerged-arc unit was used in conjunction with the Sandvik welding head suitable for strip widths of 60 mm to 120 mm. The power source was a D.C. generator with variable potential.

The plates to be overlaid were tack welded to the support table on all four sides to prevent distortion. The flux was baked at 260°C for 2 h before use to remove moisture. Before welding, the strip was cut at an angle of 20° to facilitate starting the arc and was lowered to about 35 mm from the plate. Throughout the work the strip was held at 90° to the plate and the electrode was positive. The normal length of the deposit was 355 mm.

The first layer was deposited by using 309L stainless steel strip. In order to get a flat weld deposit and a smooth tie-in between beads, an overlap of 5 mm was used. A second layer of 308L or 347 stainless steel was deposited over the previously laid first layer of 309L stainless steel. The second layer was not placed directly over the first layer, instead it was offset such that its centre was 10 mm away from the centre of the first layer. The welding parameters were taken from the previous work (8) and are listed in Table 2.

## Post-Weld Heat Treatment

While some of the overlaid plates were retained for further investigation others were given a post-weld heat treatment in an argon atmosphere. The heat treatment cycle was in accordance with ASME Division I UCS-56 and consisted of heating to 677°C (1250°F) and holding at that temperature for 2 h. The heating and cooling rates were 56°C/h (100°F/h).

## Bend Tests

Side-bend test specimens were cut from both 'as welded' and 'stress relieved' plates. From each plate four specimens (two perpendicular to and two longitudinal to the direction of welding) measuring 9.5 x 50.8 x 152.4 mm were cut. The specimens were cut in such a way they contained the full thickness of the overlay. These specimens were then subjected to a 180° bending by using a 38 mm mandrel.

## Metallography

Metallographic techniques were used to reveal the microstructures in the parent metal, HAZ, fusion line, and clad layers. The microstructures in the base plate and HAZ were revealed by 2% nital etch. An electrolytic etch with a 2% chromic solution revealed the ferrite and carbides in the overlay. The ferrite contents of the overlays were determined by a Magne-Gage.

## Microhardness Survey

Microhardness measurements were carried out with a Tukon tester on the specimens cut perpendicular to the welding direction. Several hardness impressions were made perpendicular to the fusion line. The impression load was 500 g and the measurements were made in the base metal, HAZ, fusion line and overlays.

## Chemical Analysis

The chemical composition of the second layer of overlay was determined by conventional wet analysis as per QW 462.5 of ASME IX. To investigate the constitution of the fusion boundaries between the base plate and the first weld overlay and between the first and second layers of overlay a microprobe analyser was used.

# RESULTS AND DISCUSSION

## General Bead Characteristics

The deposited weld overlay beads were smooth with fine ripples. The slag was easily detached and left no impression on the surface. The "tieing-in" was such that the junction had a smooth bead contour. The total thickness of the overlay was 10-11 mm and both fusion lines were parallel.

The dilution$\left(= \frac{\text{fused base metal}}{\text{deposited metal + fused base metal}}\right)$

between base metal and 309L overlay was about 15%.

Extensive macroscopic examination of surfaces and various sections showed that the deposits were sound and free from cracks, porosity, and slag inclusions. When subjected to the guided bend test the specimens bent by 180° easily in both 'as welded' and 'stress relieved' conditions without cracking. In most cases some plastic deformation was observed in the form of wrinkling in the overlay portion of the specimen which indicates that the joints were fully ductile.

## Microstructure

Figure 2 shows the microstructure of HAZ, fusion line and overlay regions. The base material contains tempered bainite and ferrite. In the 'as welded' condition the HAZ microstructure consists of varying amounts of bainite and ferrite with an increasing amount of bainite closer to the fusion line. The increase in bainite is consistent with the fact that the peak temperature increases as the fusion line is approached. This process continues till the $A_3$ temperature is reached when an almost fully bainitic structure is observed. Near the fusion line where the temperature exceeds $A_3$, there are some signs of grain growth. The fusion line between base metal and 309L stainless steel is characterised by the presence of lightly etched martensite which is in the form of a layer or band. The overlay consists of austenite and ferrite where the ferrite contents of 309L, 308L and 347 stainless steel layers are 9.4%, 9.3% and 10.5%, respectively. The fusion boundaries between 309L, 308L, 309L, and 347 are only identifiable by the morphology of the ferrite grains.

In the unetched condition inclusions could clearly be seen uniformly distributed throughout the overlay. This is a problem inherent in submerged-arc welding, however, the inclusion size and distribution were similar to that reported in the literature (2,5,6,7). After stress relieving, considerable changes were observed in the microstructures of the HAZ and fusion line. In the HAZ the bainite experienced tempering as is evident from the presence of a number of carbide precipitates Fig. 4. The martensitic band at the fusion line became darker and wider making it easily visible. It has become darker and wider. Many workers have attributed the easy visibility and darkness of martensite to the precipitation of carbides, (5,6,9). During stress relieving the carbon from the base metal migrates towards the fusion boundary and forms chromium carbides. This accumulation of extra carbides in the fusion zone is probably responsible for the widening of the dark band. On the overlay side of the fusion line, precipitation and accumulation of carbides along the grain boundaries of austenite can also be seen. This seems to have reduced the ferrite content of this region. Recently it has been reported that these austenite grain boundaries where carbide precipitation takes place are potential sites for hydrogen disbonding of overlays (after prolonged exposure to the operating environment) during service (10). The post-weld heat treatment of 2 h seems to have had no noticable effect on the microstructure of base metal or overlays.

## Microhardness Survey

Figure 5 shows the microhardness values in the overlaid joint. For the 'as welded' condition the hardness curve can be divided into four distinct zones. Moving from the base metal to the overlay the flat portion of the curve represents the uniform hardness of the base metal. In the HAZ the hardness gradually increases to a peak value of 350 DPH and then falls to about 315 DPH. This increase in hardness is caused by the increase in the amount of bainite as shown by the microstructures in Fig. 1. The peak hardness can be attributed to a fully bainitic microstructure in the HAZ. The drop in hardness after reaching a peak value is from the presence of coarser bainite. In the third zone marked as 'fusion zone', the hardness suddenly increases to about 400 DPH and then drops dramatically to about 200 DPH. This increase in hardness is due to the presence of the 'martensitic band'. Thereafter a drop in hardness occurs on entering the overlay region. In the fourth zone the hardness remains relatively unchanged in the 309L, 308L, and 347 stainless steel overlay regions. All three overlays exhibited almost identical microstructures and, their hardnesses were similar.

After stress relieving the hardness dropped in every zone, however, the relative drop varied from region to region. The greatest drop in hardness occurred in the HAZ and fusion zone which can be attributed to the tempering of bainite and martensite. In the overlay and base metal the drop (in hardness) is very small as no significant changes occurred in the microstructures of these regions. It is possible that this small change in hardness is because of the release of residual stresses. A notable feature of the hardness curve after post-weld heat-treatment is that the HAZ exhibits uniform hardness, similar to that of the base metal.

### Chemical Composition

The analysis in weight % of Cr, Ni, Mn and Mo as determined by the microprobe analyser for the fusion and overlay regions of 'as welded' joints is shown in Fig. 6. It can be seen that on the overlay side the Cr, Ni, and Mn contents quickly rise over a distance of 100μm. At the same time the Mo content falls. These values are maintained for the rest of the overlay. Though some variations in the Cr, Ni and Mn contents are noticed at the fusion line between first layer and second layer, from a practical point of view these changes are insignificant. The narrow fusion zone and the uniformity of Cr, Ni, Mn and Mo contents thereafter indicates that the fused metal was completely homogenized. The chemical composition of the first layer also suggests that the requirements for Cr, Ni, and Mn can be met by a single layer overlay using 309L stainless-steel strip. However, the use of single layer overlay is restricted because of its higher carbon content. It would be very useful to know the carbon distribution of the first layer, however, it is not easily measured by microprobe analysis.

The complete analysis of 308L and 347 layers as determined by the wet method is given in Table 3 and indicates that they meet requirements of SFA 5.4.

## CONCLUSIONS

An attempt has been made to clad 2.25 Cr-1Mo steel plates with 308L and 347 stainless-steel by using a submerged-arc strip overlay process. A barrier layer of 309L stainless steel was used to counter the dilution effects from the base plate. The following conclusions can be drawn from the work reported here:

1. The submerged-arc strip welding process can be used successfully to clad 2.25 Cr-1Mo steel with 308L and 347 stainless steels, and defect free deposits can be obtained.

2. In the 'as welded' condition the microstructural constituents of HAZ, fusion line, and overlay are bainite and ferrite, martensite, and austenite plus ferrite, respectively.

3. During post-weld heat treatment tempering of martensite and bainite takes place which reduces their hardness.

4. Post-weld heat treatment of 2 h has no significant effect on the microstructures of base metal and overlays.

5. When subjected to guided bend testing, the overlay joints could easily be bent by 180° which indicates that they are ductile.

6. Composition requirements as specified in SFA 5.4 can be met in the 308L and 347 stainless steel overlays.

## ACKNOWLEDGEMENTS

The authors are thankful to Dr. J.T. McGrath for his interest and encouragement throughout the work. Our thanks are also due to Dr. R.H. Packwood and Mrs. V. Moore for carrying out microprobe analysis and Dr. M.J. Lavigne for analyzing the ferrite content of the overlays.

## REFERENCES

1-Marshall, A.B., M.F. Jordan, and J.L. Aston (1973). Welding and Metal Fabrication, 41, 8, 292-301.
2-Zenter, H. (1976). Ibid, 44, 4, 208-216.
3-Groach, T.G. (1978). Welding Research Abroad, 24, 6, 3-56.
4-Lavigne, M.J.A. "Possible designs and manufacturing sequences for a 10,000 barrel/day hydrogenation vessel for up-grading heavy oils and tar sands bitumen" (1979). Lab Report, MPR/PMRL 79-70(TR), CANMET, Energy, Mines and Resources Canada, Ottawa.
5-Bush, A.F. and P. Colvin (1969). Welding and Metal Fabrication, 37, 6, 234-241.
6-Almqvist, G. and N. Egman (1963). Ibid, 31, 7, 294-302.
7-Horsefield, A.M., G. Almqvist and C.H. Rosendahl (1966). British Welding Journal, 13, 5, 315-325.
8-Chandel, R.S. and R. Hoare, (1980). "Effect of process variables on the bead geometry of the submerged-arc strip-overlay welds", Lab Report MRP/PMRL 80-80(TR), CANMET, Energy, Mines and Resources, Canada, Ottawa.
9-Ohnishi, K. et.at. (1980). "Hydrogen induced disbonding of stainless steel overlay weld", a paper presented at the September 1980 PVRC meeting in New York.
10-Okada, H. et. al. (1982). "Solution to hydrogen problems in service: disbonding", a paper presented at the 1982 ASME conference in Washington.

TABLE 1 Chemical Composition of Materials wt %

| Material | C | Mn | Si | P | S | Ni | Cr | Cu | Mo |
|---|---|---|---|---|---|---|---|---|---|
| Base Metal (SA 387 Grade 22 Class 2) | .11 | .48 | .22 | .011 | .022 | - | 2.34 | - | .95 |
| Strip 309L SS | .016 | 1.77 | .42 | .010 | .003 | 13.15 | 23.55 | .037 | .03 |
| | (Ti - .07, Ta - .01, cb - .01) | | | | | | | | |
| Strip 308L SS | .014 | 1.78 | .25 | .007 | .010 | 10.2 | 19.8 | .03 | .08 |
| | (Ti - <.01, cb - <.01, Co - .02, N - .031, V - .015) | | | | | | | | |
| Strip 347 SS | .018 | 1.66 | .35 | .012 | .009 | 10.1 | 20.3 | .044 | .07 |
| | (Ta + cb - .61) | | | | | | | | |

TABLE 2 Welding Parameters

| | Current (A) | Voltage (V) | Travel speed (mm/sec) | E.S.O. (mm) | Polarity | Preheat temperature (°C) |
|---|---|---|---|---|---|---|
| 1st layer | 750 | 27 | 2.33 | 35 | Dc +ve | room temp. |
| 2nd layer | 750 | 27 | 2.33 | 35 | Dc +ve | room temp. |

TABLE 3 Chemical Composition of the Overlays at a Distance of 2.25 mm from the Surface (wt %)

| Material | C | Mn | Cr | Ni | P | S | Cu | Cb |
|---|---|---|---|---|---|---|---|---|
| 308L | .023 | 1.47 | 20.2 | 10.7 | .009 | .010 | .029 | - |
| 347 | .026 | 1.34 | 20.3 | 10.5 | .012 | .008 | .044 | .24 |

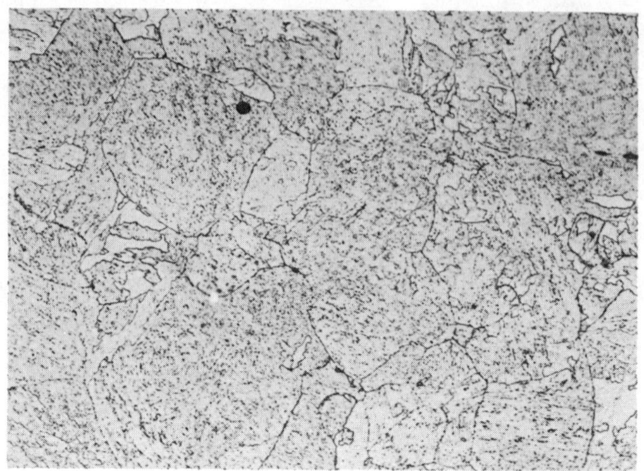

Fig. 1 - Microstructure of base metal. Etchant - 2% nital, X500.

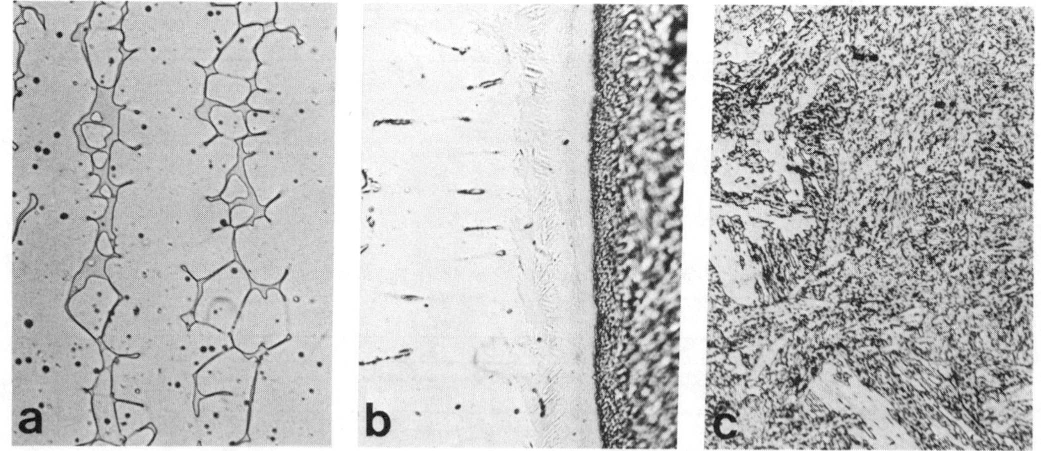


Fig. 2 - Microstructures in 'as welded' overlay. X500

(a) clad, etchant 2% chromic acid

(b) fusion zone, etchant 2% nital

(c) HAZ, etchant 2% nital.

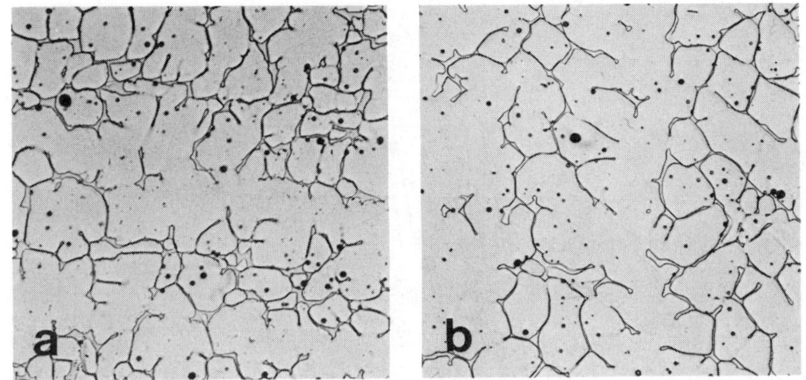


Fig. 3 - Microstructures 'as deposited' (a) 308L stainless-steel (b) 347 stainless-steel. X 500 etchant - 2% chromic acid.

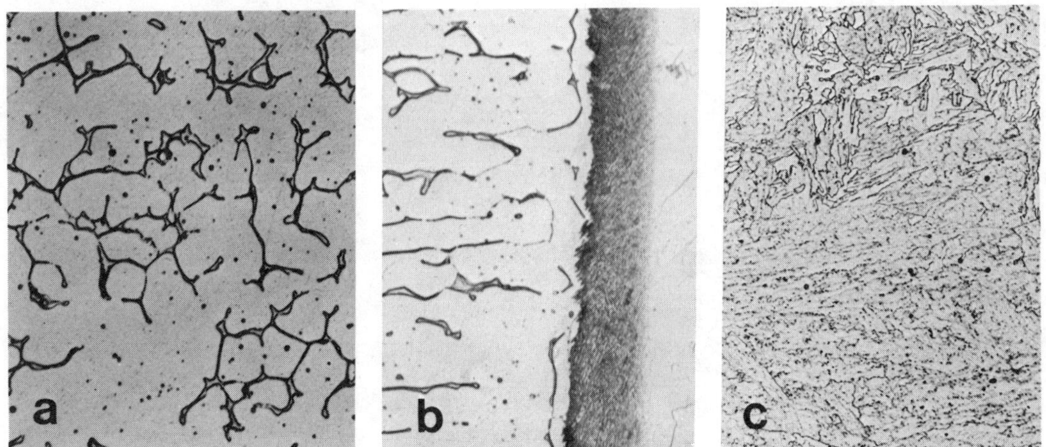


Fig. 4 - Microstructures of post weld heat treated overlay. X500.
(a) clad - (309L stainless-steel), etchant - 2% chromic acid
(b) fusion zone - etchant 2% nital
(c) HAZ - etchant 2% nital.

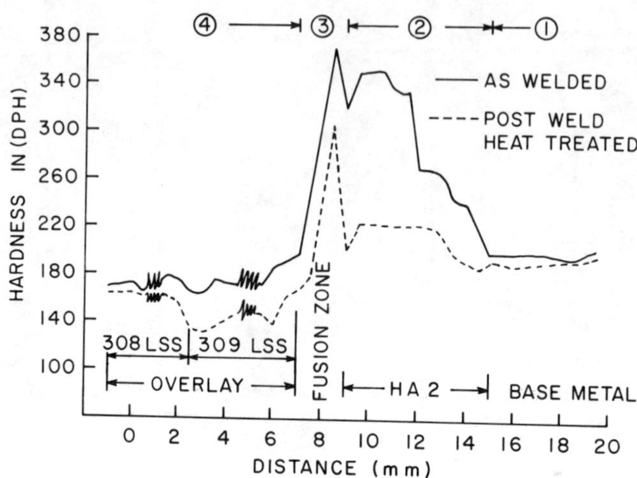


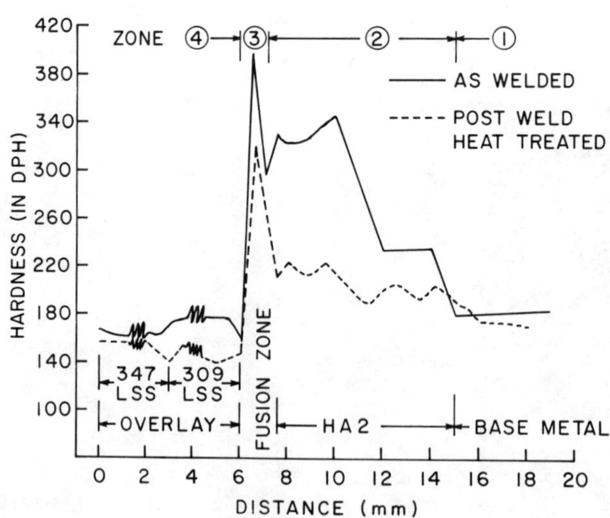


Fig. 5 - Microhardness distribution in a strip overlay joint.
(a) base metal 309L and 308L stainless-steel
(b) base metal 309L and 347 stainless-steel.

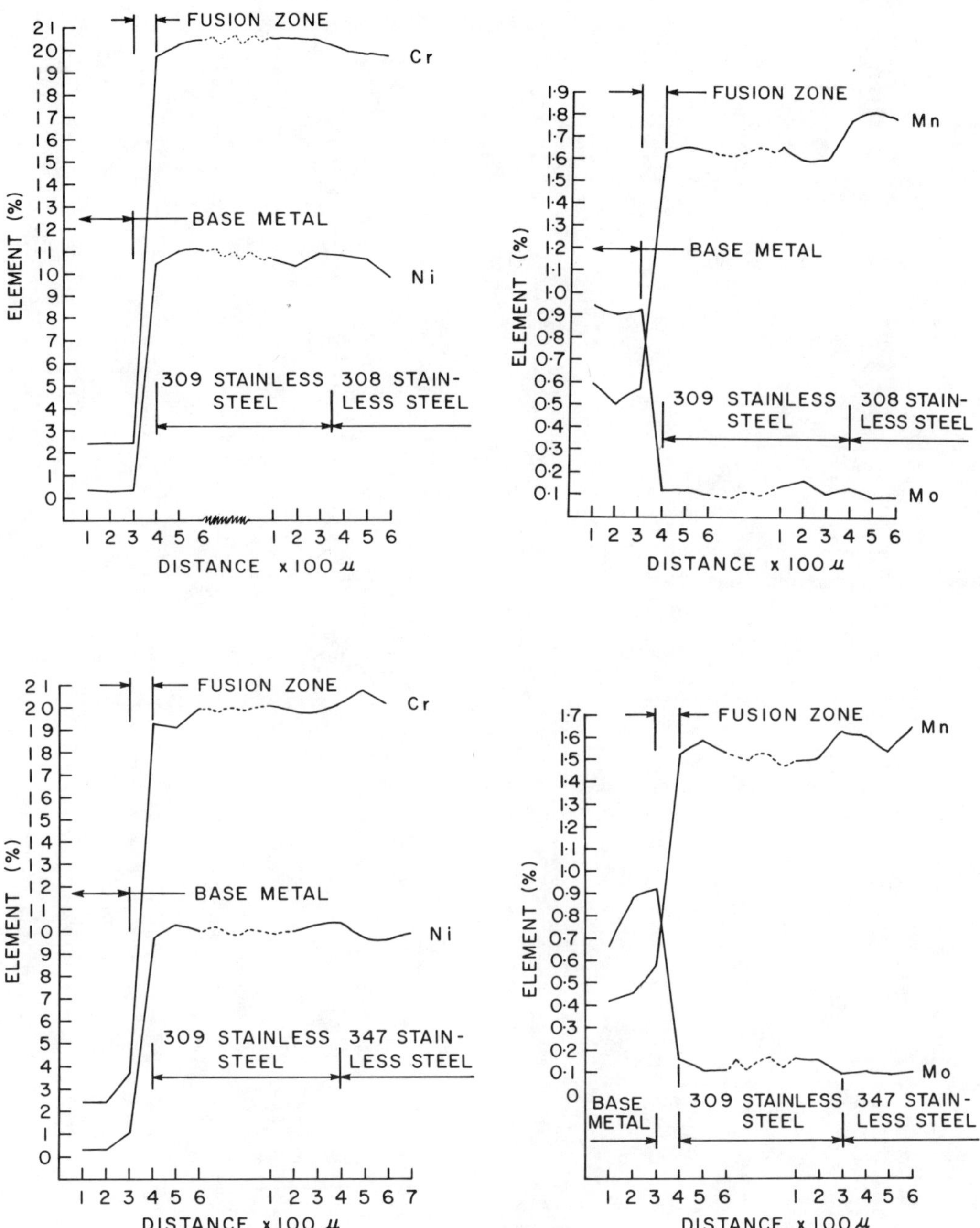


Fig. 6 - Distribution of Cr, Ni, Mn and Mo in a strip overlay joint.
(a) Cr and Ni in base metal, 309L and 308L stainless-steel
(b) Mn and Mo in base metal, 309L and 308L stainless-steel
(c) Cr and Ni in base metal, 309L and 347 stainless-steel
(d) Mi and Mo in base metal, 309L and 347 stainless-steel.

# QUALITY ASSURANCE IN THE FABRICATION OF OFFSHORE STRUCTURES

K.P. Bentley
*Quality Consultant,*
*172, Hills Road, Cambridge, United Kingdom*

## ABSTRACT

The author provides a set of activity guidelines to be used by owners or users and their quality assurance teams when embarking on major offshore construction projects and also by the fabricators and suppliers who will be required to build the structures.

These recommendations are based on the author's experience of actions and arrangements found to be particularly beneficial or undesirable and complement the usually accepted essential elements of quality management which are very well detailed in the British and Canadian standards.

Of particular importance to the user is the early use of QA to support procurement and the way in which technical specifications are written. For the fabricator emphasis is placed on the management of quality control and its effect on contractual relationships.

Several pitfalls and actions to be avoided are highlighted for all parties in the contractual chain and the possible benefits for them in using a good QA approach for major projects are reviewed. These ideas are introduced by a clarification of the standard definitions which are used throughout.

## KEYWORDS

Quality Assurance; Quality Control; Quality Losses; Quality Costs; Quality Attainment; Specification; Welding; Offshore; User/Owner

## BACKGROUND

The introduction of Quality Assurance to the offshore industry has not always occurred entirely happily largely due to some misconceptions and erroneous pre-conceived ideas as to what it really means and how it should be implemented.

Offshore construction projects are characterised by high investment and acute safety considerations and the need for Quality Attainment is clear. In recognising this, Quality Assurance has perhaps been assumed to be synonymous with Quality Attainment - the ultimate aim, whereas it should better be regarded as an important means to achieving that end.

We should therefore first consider what this overworked phrase Quality Assurance or QA really does mean especially in relation to other well worn phrases in the quality scene before considering in detail how to use QA for the attainment of quality in a large project.

## DEFINITIONS

Excellent formal definitions of the usual quality phrases can be found in standard text books. For this paper dealing specifically with offshore work, the following less formal definitions and associated comments are provided for clarification.

### Quality Assurance.

The entire range of managerial activities and arrangements, set up in recognition of possible foreseeable errors and designed to minimise problems arising during the project, the successful execution of which together with documentation can assure both the owner/user and the certifying authority that a satisfactory state of affairs exists.

- commonly referred to as QA
- its involvement clearly extends to any aspect of a project which may affect the end result such as design or procurement as well as construction work and inspection.

### Quality Control.

Specific and detailed measurements of the actual quality achieved for comparison with agreed standards or requirements so that acceptance decisions can be made.

- an important part of the overall quality assurance system.
- provides the necessary documented evidence on which formal acceptance by the appointed Certifying Authority depends.

### Quality Attainment.

Getting the construction technically correct by the required time.

- the ultimate aim of quality assurance and quality control.
- in practice it would probably be a more appropriate bearer of the abbreviation QA than quality assurance since this is really what many people have in mind when they say QA.

### Quality Losses.

The calculated losses to a manufacturing enterprise incurred as a result of mistakes or defects being revealed requiring further work to remedy or rectify.

- largely consist of materials, labour and re-inspection costs with overheads since the manufacturing resource as a whole is unavailable for other profitable activity during the re-work period.

### Quality Costs.

Expenditure made to reduce the likelihood of incurring quality losses and promoting quality attainment.

- largely comprise the costs of quality control staff and activity, i.e., inspection and non-destructive examination etc..

- may be regarded as an insurance premium against quality losses.

THE VALUE OF QA TO OWNERS AND FABRICATORS.

The prevention of things going wrong during a construction project or subsequent service life is the prime potential value of QA to the owner/user. He should therefore ensure that his management systems for giving assurance of quality reach every aspect of a project which may affect the end result i.e., he must recognise that any activity if not performed correctly or any material or component that may be unsatisfactory can lead to less-than-ideal situations which may effectively lower the technical integrity of a structure or impair its product performance and often with consequent penalties in time, cost or product capability.

The owners QA efforts should therefore not only promote the final technical quality but contribute to efficient project operation and progress by minimising unwelcome delays and extra costs.

Similarly fabricators require a smooth flow of work at site and in the construction shops to promote production efficiency and look for a ready acceptance of their products by owners and certifying authorities.

The economic advantages of getting it right first time cannot be over emphasised and a fundamental truth of all engineering and manufacturing enterprises is that whenever events turn out less than ideally someone always loses either in money or reputation and often both.

The significance of major structural or plant failures is easy to interpret in these terms as a direct loss of production time or equipment usage for the owner and everyone in the manufacturing chain whether main contractor, fabricator or material supplier may have back charges or a damaged reputation to stand.

It is equally true in the technical details of the fabrication shop where quality is actually created in the product. For instance the replacement of wrongly used material, rectification of a machining error or the repair of a defective weld represents a loss of manufacturing resource for which a sum of lost money, often surprisingly large, can be calculated. These are the losses which we collectively refer to as Quality Losses.

The avoidance of these potential losses in financial return, image, manufacturing resource or hard cash should be the principal aim of both owner and fabricator alike the owner through his policy and QA management and the fabricator through the efficient and effective Quality Control of his manufacturing activities.

Another specific advantage to the owner/user is that his QA system should systematically generate all the necessary documentation for achieving approvals and associated certification from Statutory Authorities covering the design, construction, installation offshore and final use of the platform or semi-submersible etc., so that the plant can be allowed to perform in production as intended.

Furthermore the system should provide detailed technical records relating to the materials used, the design, methods of construction, quality levels actually achieved including repairs or changes made and indicating any technical concessions or amendments to specification that were used. This enables both modifications to the structure or plant to be made in later years and routine maintenance during its service life to be undertaken on a sound technical basis.

## DO'S AND DONT'S FOR THE OWNER/USER.

If QA in offshore construction work has sometimes been less successful than it might have been it is probably because owners have not held sufficient conviction or commitment to the QA concept at the outset. They have often acted too late in the project for QA to be effective and possibly with the wrong kind of activity and effort.

Such a standpoint is in many ways to be expected since QA is too often portrayed as a virtuous neo-religious doctrine seemingly far removed from the reality of structural hardware and scepticism on the part of the practically experienced engineer is understandable.

Above all the QA we employ must be realistic, not idealistic, and genuinely relate to those human mistakes and oversights and technical errors etc., that we know from past manufacturing experience really might occur. If it is allowed to become ethereal and doctrinal then we shall probably be wasting effort and money and generating too much paper on the way.

The owner's QA department must never lose sight of the fact that the object of the exercise is to build a platform for profitable business.

Many of the comments in this paper are directed towards the user because it is he who ultimately determines the success and effectiveness of all quality assuring measures at all levels of the contractual chain by properly declaring at the outset his policy, strategy and detailed requirements regarding quality.

The following list of specific items or aspects have been found to be particularly important and are recommended to users embarking on a major offshore fabrication as follows:

1. Define the quality policy for the project - This should cater for statutory and safety requirements as well as structural engineering and process system functions.

2. Establish a QA team early in the project - Experience shows that the two items of project work from which most constructional problems emanate relate either to the technical specifications or to procurement. The writing of specifications and placing of orders and sub-contracts occur early in the project time span and the QA department must be established in time to contribute effectively to them both.

3. Ensure that the QA department has the authority necessary to fulfill the users policy - Declare this formally to all parties in the project, stating the QA department's terms of reference and defining the relationship of the QA department to other departments in the users organisation. This opportunity should also be taken for reminding the other departments of their inherent responsibility towards quality in their regular engineering or administrative work.

4. Ensure good liaison and effective communications between the QA and other departments - The QA department must be readily available to other departments, especially those involved in materials, design and construction methods etc., and have the means for rapid liaison with them in response to events. Proper links must therefore be established at all levels in the users project organisation.

The physical location of the QA department is important in this respect and should be situated close to the engineering or design teams preferably in the users central offices.

The appointment of a QA department to act for the owner as part of a broader management services package is seen as a formula for unwieldy communication and slow ineffectual QA activity since it effectively introduces a further layer in the chain of communications.

As a guiding principle owners should aim to make all contractual chains as simple and short as possible. In recent years the trend has been to complicate the management of major projects unnecessarily by introducing too many parties so that decisions and answers to questions arising can only be reached after considerable effort and delay. This situation encourages unilateral decision making at the point of action, under the time pressure of a major construction and can have serious quality implications.

5. Ensure that technical specifications accurately reflect the owners technical policy or requirements - Current trends amongst owners or their main engineering contractors are towards the writing of even tighter and more rigorous specifications for materials and construction practices etc..

Clearly if specifications are too lax then some contractual control is lost but it is extremely important that the levels of quality required are not overstated. Specifications should be as specific and not as general as possible in their application indicating as precisely as possible just how good a structure, product or weld really needs to be. They should be written with true regard to the likely freedom from weld defects, for instance, that is reasonably to be expected from the fabrication facilities available.

Specifications which pitch the required quality levels too high can lead to difficulties both technical and contractual during the fabrication stages. This may have any of three consequences:

(i) A rash of unnecessary repairs costing time and money and often with an overall technical deterioration, e.g., the introduction of further residual stresses and metallurgical damage often being overlooked in favour of a clear radiograph.

(ii) The temptation for staff in all parties to bend the rules to keep the work moving e.g., the designation of lack of fusion as a slag inclusion, sometimes difficult to distinguish, by an expedient technician when pressure vessel standards excluding all lack of weld fusion are applied to module structural steelwork. We can sympathise with but not condone such actions.

(iii) A host of concessions being given by the owner and sometimes variations to the ruling project specification having to be made, approved and distributed throughout the project organisation. Clauses in specifications cannot be waived at a nod from the owner or his representative without recourse to a formal, documented variation procedure because ultimately it is against the specification as written and formally amended that the Certifying Authority issues a Certificate of Compliance.

Furthermore, it is often forgotten that the status and authority of the fabricators quality manager, diligently striving to maintain standards of workmanship possibly in the face of conflicting production pressures may be seriously undermined if a stream of concessions is seen to come from the client.

The owners fabrication specification should therefore relate weld defect acceptance levels as far as is practicable to specific structural details classified to reflect the severity of engineering duty that applies, i.e., avoid simplifying the standards or specifications so that in the extreme, tubular node requirements are equally applicable to simple butt welds in secondary and minor bracing.

6. Adopt a flexible strategy for the application of specifications - The users strategy for implementing his specifications should allow some flexibility in their rigorous application so that defects which are truly insignificant on a localised engineering basis and to the integrity of the structure as a whole can be left alone. This must be declared to all departments in the users organisation, where emphasis should be to measure

weld defects, assess them correctly by sound engineering analysis, and only repair those which are truly unacceptable.

The guiding principle should be to find true justification for the acceptance of welds, materials or equipment put under question and not to take delight in demanding repairs or replacements on principle.

7. Do not impair the fabricators responsibilities by over-inspection - The fabricator has a clear contractual duty to control the quality of work produced by means of planned inspections and tests.

Surveillance by the owner to verify that this is done is usual and appropriate but if quality control breaks down or technical problems relating to workmanship arise at the construction site it is tempting for the user to get involved in the detailed testing of welds alongside the fabricator's inspection team.

The danger is that this may override and replace the fabricator's quality control, effectively taking from him the incentive to make the necessary improvements and incidentally relieving him of some contractual responsibilities. In short, over inspection can kill the fabricators initiative for action and must be resisted by the user; extra efforts should rather be focussed on the QC function and personnel rather than on the actual hardware.

8. For mid-project problems emphasise QC not QA. - Related to 7 above a spate of welding problems half way through a project should not be the signal for the user to demand a new QA system from the fabricator. Probing critically into matters like the records of internal audits and purchasing, calibration and stores etc., and demanding new manuals to be written will only help the next client. It is too late for that to have positive effect on the current job and corrective action must be directed to the quality control staff, procedures and inspection records on which the final stamp of approval for acceptance will be won.

## DO'S AND DONT'S FOR THE QA DEPARTMENT

A number of especially beneficial actions or activities that the users' QA department should undertake and some restraints which it should experience are suggested as follows:

1. Make an early contribution to the procurement of materials, equipment and services- By ensuring that all orders have the following:

(i) A statement concerning the inspection of items to be supplied. This must indicate clearly who will perform the various inspections, the standards or quality levels to be met and at what stages of production it will occur.

(ii) The name of the prime contact in the user's QA department with whom the contractor should liaise.

(iii) A statement to prevent uncontrolled subcontracting. The user must be informed and his approval sought before any fabricator concludes a sub-order.

(iv) A clear indication of any inspecting agency or certifying authority who must witness work and approve materials and results at various stages of production. The mechanism for calling in such a body must be clearly indicated.

2. Make assessments of potential suppliers as early as possible - Especially for long lead items and major fabricators and preferably before, or at the latest, during the period of specification writing, a parallel activity with which the QA department should maintain a close liaison.

If this is not done orders may be placed with unsuitable suppliers for the wrong reasons and the QA department may eventually find itself asked to retrieve situations by conducting assessments on suppliers who already hold an order, an impossible situation and a largely pointless exercise.

Technical assessments of vendors often fail to be fully effective or realistic because of the specialist technical knowledge and experience often required. Whilst QA departments are usually good at vetting inspection and NDE agencies they must not be afraid to enlist specialist technical assistance from other departments or from external sources. For instance, sub-contractors for welded fabrication must be assessed by a good welding engineer with managerial experience in a manufacturing company. Ideally he should be a permanent member of the users QA team for the entire project.

3. Report fully on the assessments made. - All assessments must aim to make a full and accurate appraisal of a potential suppliers true technical and managerial capabilities or lack of them as a solid basis for procurement selection and later contractual control purposes. They only have any value if properly reported and recorded and the conclusions should be stated clearly and without ambiguity including any data that was available. If any doubts exist recommend a re-assessment at a later date. No attempt should be made to use some kind of artificial scoring system by which the potential suppliers can be compared. Such methods are not suitable for major engineering projects and can give very misleading results.

Assessment reports should clearly indicate any non runners giving clear guidance to procurement in the selection of favoured suppliers and most important they should include judged estimates of the probable and possible extra costs and delays that should be expected if certain suppliers are used.

Reports should also propose the level and extent of inspection or surveillance that will be necessary for the user/owner to use both during fabrication and on completion before acceptance. These estimates can then be used by the QA department in drawing up its quality programmes and plans.

4. Clearly define a quality programme - In terms of procedures and methods to be used by all departments in the user organisation indicating how it is to be controlled and implemented by all parties in the project and at all levels.

For other departments this will involve the writing of documents with titles such as:

'Organisation of Objectives for Project QA'
'Documentation System for Jacket Construction'
'Guidance Procedure for Assessment of Suppliers QA/QC etc.'
"Project Information Recording Format' etc.

For the QA departments principal use the programme should also include sets of detailed inspection instructions and schedules with associated reporting procedures for use by the users surveillance staff resident at each site. Typically these documents would be entitled: 'Inspection Surveillance Procedure for ---' items such as:

Fabricated Pipework
Mechanical Equipment
Node Fabrication
Weld Procedure Qualification
Installation of Electrical Equipment
Module Steelwork Dimensional Control. etc.

5. Establish site surveillance teams - For the monitoring of contractors quality control performance including any sub contractors they may use.

This will involve teams of staff at each construction site having representative skills and expertise in welding engineering, visual inspection, non-destructive examination, materials control and documentation etc.

The manning levels of these teams should initially refer to the original assessment reports and should then be subject to adjustment during the project life to reflect the technical content of current work and the performance of the contractor.

6. Monitor problems and non-compliances - By maintaining documented records of all such events and ensuring that problems are dealt with and resolved by technical staff within the users organisation of appropriate technical skill and seniority.

In this connection the QA department must refrain from making technical decisions of acceptance - rework - reject significance. Exceptions to this may occur when the QA department has individuals with specific knowledge and experience but in general the QA department should not act as an internal technical consultancy because it is not in the direct line of technical responsibility.

7. Promote and stimulate communications - Between the users departments and with external bodies and between the various construction sites involved in a project to whom the QA department may represent a focal point.

This applies especially to the Certifying or Regulating Authorities involved to whom the QA department represents the user on all aspects of work acceptance and approval, the qualifications of welders and inspection staff and matters of amendment or variation to the specifications or other contractual requirements.

8. Control and collect all documented records of work and inspection - For presentation on a progressive routine basis to the Certifying Authority so that the necessary approvals can be formalised.

There is scope in this activity for computer storage of all records and reports so that rapid feed back of information relating to material usage, welding procedures, welders, inspection and NDE results, heat treatments, batches of welding consumables and where they were used and details of repairs etc., is possible.

## DO'S AND DONT'S FOR THE FABRICATOR.

The following recommendations are stated primarily for main or sub-contract fabricators but they apply also to suppliers of proprietary items such as mud pumps, cranes and process plant etc.

1. Maintain a representative quality manual - And have it readily available to potential clients especially for pre-contract assessments.

The fabricator should regard this as a means of demonstrating to a potential client how he controls his activities to achieve required results and it must therefore accurately reflect his true attitude towards quality, his organisation and where individual responsibilites lie and also outline the managerial systems used for controlling basic activities. Detailed working procedures for welding, heat treatment, and inspection etc., need not be included but reference to their existence and effective use should be made.

In addition to having a strong sales value for the fabricator, the quality manual is also a principal item of internal managerial control. It allows the workforce to feel the quality responsibility carried by the company.

2. Demonstrate that effective quality control activities are established - With facilities, staff and control systems suitable for the size and nature of the project involved. This must include reporting and the control of documents.

3. Employ directly the key quality control staff - Although inspection and NDE teams may be supplied by sub-contractors the fabricator must have the senior qualified staff such as an NDE engineer, materials controller or weld inspection engineer in his own employment to manage these functions, otherwise sub-contractual control is lost.

4. Be able to demonstrate the effectiveness of systems established - Through the quality manual or in detailed working procedures to control aspects such as material identification, welding, consumable storage, processing of purchase orders, the qualification of welding procedures, welder approvals and their deployment, heat treatment of welds, sub-contracting e.g. for painting etc., by making reference to documents such as work requests, purchase orders, inspection records, material test certificates, welder records, release notes and drawings etc.

5. Ensure that all contractual requirements and arrangements are fully understood - When accepting a contract. It should be clear to the fabricator which codes, standards or specifications etc., apply and the fabricator should possess the latest or intended edition of them. He should look carefully for and resolve any conflicting requirements or ambiguities and know all the points of liaison with the client, any nominated subcontractors, inspection or certifying agencies involved and be aware of all documents or reports that must be supplied to the client.

6. Do not reduce quality control effort because a client will be doing some inspections - The clients visiting or resident inspectors reports are his property and the fabricator probably will not see them and even less agree with them.

The fabricator's own quality control results safeguard his contractual position and are vitally important.

7. Use quality control results to raise efficiency and profitability - The information returned from quality control should be analysed to establish where quality losses occur and provide a basis for adjustment and corrective managerial action. Results can indicate for instance where training is required or where supervision needs to be strengthened, where equipment needs to be replaced or serviced and where material being used is unsuitable.

## SUMMARY COMMENT.

Although these guidelines and recommendations are intended primarily for offshore structures they can apply equally well to any large engineering construction project.

Such projects necessarily involve several contracted parties and contain concepts or aspects of a size and complexity which apparently dwarf the individual's detailed contribution. Human frailties and oversights are therefore more likely to prevail and quality assurance is therefore invoked by a contractual requirement to use quality management systems in accordance with the established national standards.

The advice offered in this paper complements these standards by recommending where emphasis should be laid for the greatest returns in providing safeguards and for ensuring that the best engineering opinions are not stifled but allowed to bear directly for the benefit of the project and the final construction.

# CODES, SPECIFICATIONS AND TRAINING FOR QUALITY ASSURANCE OF PIPING

L.E. Baxter
*Canadian Consultant, 920 Notre Dame Street, Lachine, Quebec H8S 2B9*

Designs of piping systems covering nuclear power plants, refineries, chemical plants, paper mills or transmission lines are all based on the same general requirements. Similar codes that control materials, design and fabrication procedures do not vary to any degree except in the frequency of verification and possibly the acceptance levels of quality control. The actual fabrication and inspection methods themselves are identical and consequently, the quality levels of each installation should be equal or at least similar.

In every case, there exists a quality level selected by the owners or consultants to which all suppliers and contractors must adhere. Economy is realized when the quality level requested by the owners has a proper and real relationship to the designed lifetime operation of the system.

Intelligent design of a system demands from engineers a complete analysis of all pertinent factors. The architect-engineer will design the systems, structures and components. In turn, the various manufacturers design and fabricate the components, that is, the equipment.

The design engineer must consider the interface relationship of systems and what quality level each system will require. He must also consider where non-critical systems require higher levels of quality control when they may be embedded or inaccessible following construction.

The design of a large power or processing plant includes administrative, quality and technical activities through standardized procedures and designs. Cost reductions and scheduling requirements have been instrumental in the development of standardization. The unfortunate aspect of standardization is that seldom do designers include provisions, or have the qualified staff, for inclusion of improved developments in fabrication methods or of components. The intent is that the general documents or "Higher Tier Documents" (1) provide the procedures by which the operating divisions assume the responsibilities delegated by Corporate policy documents, (2) establish additional detail necessary for divisional implementation (e.g. engineering, construction, quality assurance) and (3) provide direction for subordinate departments (e.g. the technical and support disciplines).

Lower tier documents consist of instructions, standards and guides which standardize the day to day implementation of departmental functions in the detail and manner required to assure overall program objectives. The intent of such documentation and planning is excellent but unfortunately, the realization is not always accomplished. In the quality assurance field, there is a tendency on the part of construction personnel to read nothing or zero into any document where the quality control requirements are not specific. Invariably, ASME or ANSI codes are quoted in a general preamble but there are usually no specific or detail requirements listed. In the case of some of the larger owners some detail on "Special Processes" such as welding and nondestructive testing have been developed but in such a general way that there is no selectivity for areas that may require a more stringent control.

Some standardized designs have been locked in for many years but one must realize that there are substantial developments in most disciplines over a span of twenty to thirty years and economy of both manufacture and fabrication demands evaluation of the current state of the trade.

As mentioned previously, most designs list few details of welding, heat treat or inspection requirements. They are usually of a general nature and simply referenced to a code. We believe that proper assurance verification will entail assurance planning which should assess every item and activity that must be controlled and verifies that the assurance measures applied are appropriate. Welding, nondestructive testing and piping specialists must have the opportunity for input and editing of the special process requirements before the plan is finalized and issued to suppliers.

When the plan has been approved or reviewed, then the next step is to prepare the detailed Technical Specification which will specify material type and sizes, fabrication details and restrictions, work to be done, work not included, schedules of shipments or progress, financial conditions and other pertinent information required by a supplier. At this stage, the welding engineer must review the welding specifications, including the requirements for submission of procedures, certification of welders, review of weld joints and methods to be used. The nondestructive testing engineer likewise, must use this opportunity to ensure that all code requirements are specified as well as any additional requirements requested by the discipline engineers or as agreed upon after discussion between design, project and special process engineers. Both welding and NDT engineers will have one more chance to rectify an unacceptable process when the contractor submits his welding and NDE procedures.

It is the nondestructive testing engineer's responsibility to ensure that code options and ambiguities be clarified. He must check to ensure that methods specified are practical in both technique and location.

A piping system requires that mechanical and project engineers re-check and clarify with the specialists, all areas of doubt so that the final bid document includes a technical specification that is clear, concise and specific.

The invitation to tender should include pipe flow lines, valve specifications, systems descriptions, material specifications, pipe hangers, guides, support and anchor locations, restraints of welded joints, joint configuration, material hardness requirements, material supplied as "Free Issue" and a complete and accurate description of the work to be done.

A list of applicable codes, specifications, laws and standards must be included.

In Canada, quality assurance levels to Canadian Standards Association Z299 series are usually listed and the fabricator must meet and be accepted to these standards previous to being invited to bid. Foreign suppliers are usually supplied with copies of the specific CSA standard and it is then the prerogative of the consultant or owner to evaluate, verify, accept or reject their Quality Assurance program to the specifications and standards of the project.

The bids are opened at a public meeting and then reviewed by the project engineer who again calls upon the specialists to review the tenders. One or more pre-award meetings with personnel from tenderers are usually held to clarify any areas that may be misunderstood or misinterpreted. The minutes of such meetings form part of the contractual requirements on both parties. As soon as a decision has been made, a telex of intent is forwarded to the successful bidder with instructions that he may proceed immediately. Comfirmation then follows, usually within a fortnight.

## CONTRACTS

There are many variations of contracts but two basic conceptions are most common. A contract may cover supply, fabricate and install or it may cover only one or two facets of procurement. Some consultants believe that separate contracts for fabrication and supply leave them better control of quality of materials and surveillance of production. On the other hand, contractors may be more critical of owner supplied materials, especially when free issue products are close to tolerances.

In the case of the Atikokan fossil generating station for Ontario Hydro, separate contracts were issued for material supply and for fabrication of spools and installation. Piping procurement contracts were again broken down into separate orders for high and low pressure piping, continuous cooling water piping and drainage piping. A number of companies supplied all the valves, fittings, pumps, strainers, hangers, supports and auxiliary equipment.

Successful pipe supply bidders have included companies in the United States, England and Canada. On receipt of purchase orders, the suppliers submitted, as required in the tender document and in CSA Z299, an "Inspection and Test Plan". Most I & T plans submitted are inadequate as they do not include sufficient detail. The piping specifications called up CSA Z299.1 and 2 quality levels and both of these specify that full details of manufacture and inspection must be submitted. We insisted on compliance to the standard as we believe that the I & T plan is one of the most important documents. When correctly prepared and submitted, it allows the consultants to evaluate the process of manufacture and ensures that inspection will be implemented at all critical points. The procedures that accompany the I & T plan specify the methods of inspection and it is here that the consultants must be wary and knowledgeable. Suppliers must understand that the submission and review of procedures implies the right of input by the consultant. A consultant may reject a procedure but unless an indication is given as to what is not acceptable, considerable time and effort may be wasted by a confused supplier. We will reject parts of procedures and write in references to codes or specifications as we interpret them. This allows a quick correction and return of an acceptable procedure.

The principal areas of inspection and witness points by the consultant's surveyor are as follows:

1 - Material specification and documentation
2 - Ultrasonic inspection of plate
3 - Rolled dimensions - diameter, wall thickness, length
4 - Welding
5 - Heat treat
6 - Pressure tests
7 - Radiography

8 - Descaling
9 - Magnetic particle inspection
10- Product analysis
11- Longitudinal tensile tests
12- Transverse tests
13- Flattening tests
14- Identification and marking
15- Release for shipment

Our normal practice on welded pipe would be to have a surveyor witness the NDE tests, check material certificates and spot check dimensions. He may also witness physical tests.

No matter how hard we try, there is always some event that alters the material or the dimensions to where a concession is requested by the supplier. We have had a sub-contractor roll the plate and tack the tubes and then to ensure that the ovality would not change, welded in spiders at both ends. When received at the contractor's plant, the removal required grinding to below the minimum wall and the necessity of"reviewed" repair procedures. We have also had weld material classed as A-No.2 under ASME Section IX QW.422 where the analysis only reached 0.35 MO. instead of 0.40 min.

As pipe from the producer is to be shipped to the contractor, as free issue material, it is necessary to carefully evaluate the results of out of tolerance material. The spool fabricator must not exceed hardness maximums so if he has difficulty with hardness, he will more than likely, and with some justification, blame it on out of tolerance materials supplied by the owners. Other types of pipe supplied such as seamless or spiral would all follow the same general procedure varying only as per demands of their specific characteristics.

Inspection procedures are important in the production of all types of pipe. Designers must decide whether code requirements are sufficient or whether additional inspection is to be done. Procedures must be carefully reviewed to ensure that the techniques submitted are satisfactory. Ultrasonic procedures must indicate scan techniques, angle propagation, acceptance levels and details regarding calibration.

Certification of operators to an acceptable standard is important and must be shown in the procedure.

Radiography procedures must include technique sheets which detail exposure data, source type and size, source film distances, film type and developing techniques.

Many codes, in our opinion, incorrectly call for magnetic particle or penetrant inspection. It is incorrect, we feel, because while both methods are regarded as surface NDE methods, magnetic particle inspection is much more sensitive than penetrant inspection when used on magnetic materials. Penetrant, although one of the oldest methods going way back to kerosene and chalk, has been refined into a product of at least ten different sensitivities or seeabilities and was originally developed for use on non-magnetic materials. Since the introduction of the pressure can, red dye penetrant has been questionably used as a substitute for other more sensitive inspection methods. With the exception of the aircraft and space industries, few designers or manufacturers even consider yet specify the use of more appropriate penetrants for inspection of critical components. A definitive statement in the Technical Specification of tenders could easily clarify the ambiguity of some codes. It may be noted that paragraph 8.3 of ANSI B16.34-1977 specifies that the magnetic particle method should be used on all ferritic steel and liquid penetrant will be used only on austenitic steel and other non-magnetic materials. The author is aware that ANSI B16.34 is a valve specification, but logic would indicate that the same reasoning should apply for all types and shapes of magnetic and non-magnetic materials.

Some suppliers feel that the option as listed in some codes indicates a supplier choice. This paper strongly disagrees and maintains that the option should be dependent on the material and if not then on the consultant or owner. It has been necessary to reject many procedures and insist on what is regarded as an applicable method.

Inspection methods should include the most efficient process available as the acceptance levels will then determine whether a discontinuity is rejectable or not. The idea of using an inferior method that will not show all discontinuities is repugnant as a slight error in technique may well eliminate an indication of a critical defect. There is not universal concurrence in the belief that proper inspection will use the most efficient method and then evaluate the results against practical acceptance levels.

Radiography is a common modern test method used to evaluate pipe welds both in manufacture and for field welds. Plant methods are pretty well established and most fabricators follow codes and specifications to the letter. Here again, we recommend a close review of plant procedures to ensure that procurement and supply standards are the same.

Radiography in the field does present other problems. We have found in Canada a growing tendency to use isotopes in contradiction to the code and to use source film distances that are inadequate and which will provide misleading results. Section V of ASME states in paragraph T.242 that Ir.192 may be used on a minimum thickness of 19 mm. It then states that the minimum thickness may be reduced when the radiographic techniques used demonstrate that the required radiographic sensitivity has been obtained. The latter part of this paragraph has been the origin of much dissension and disagreement. Operators of many companies, but especially those of commercial test facilities who are most often involved in field installations, regard T.242 and T.243 as licence to use Ir.192 in almost all cases of field welds. Such interpretation should not be permitted and consultants should insist that X-ray machines be used wherever the combined thickness of pipe wall plus the weld re-inforcement is under 19mm. The physical limitations of X-ray machines are recognized for inaccessable areas but one should not give blanket approval to any sub-contractor. Site Quality Assurance representatives must authorize any deviation from "Reviewed" R. T. procedures and those procedures should not be "Reviewed" unless they conform to the consultant's interpretation of the code.

We are convinced by experiment and field experience that knife edge defects such as cracks and lack of fusion may easily be missed when high energy sources such as Ir.192 are used on thin materials. The inherent unsharpness caused by lateral movement of secondary electrons in film causes a diffusion of the indication. It is most difficult to correctly interpret or even see tight discontinuities under such circumstances. There has also been an unfortunate move to decrease exposure time by reducing source film distances. Piping is especially susceptible to this trend as many operators find it easier to use a panoramic shot from the I.D. or a contact shot from the O.D. We have recently experienced certified senior operators who have proposed panoramic shots on four inch diameter pipe with a wall thickness of 25.4mm and over. The relevant deflection of the rays and the geometry of the source to film configuration would cause a penetrameter (ASME) on the inside of the pipe to be blown up and distorted to over twice its size. The operator erroneously assures himself that he is achieving proper sensitivity by placing the penetrameter on the film side of the material where distortion cannot take place. Good radiography should restrict geometric unsharpness to half that of the inherent unsharpness which would then give us a total of only 10% more than the minimum possible unsharpness. The ASNT Nondestructive Testing Handbook recommends a minimum SFD ratio of 35 to 1 when using a 50 curie source of Iridium

192.

The author agrees fully with the results published some time ago by R. Halmshaw in the British Journal of NDT in which he reviewed the parameters used to determine a good radiographic technique. Summarization of his paper confirms a minimum source film distance based on the following equation:

$$a = T\left(\frac{S}{U1} + 1\right)$$

Where a = the source film distance
Where T = the specimen thickness
Where S = the effective source diameter
Where U1 = the inherent unsharpness because of electron scatter in the film emulsion.

Consultants should accept the wording of T.242 and T.243 of Section V, ASME, provided a proper test technique is developed and witnessed by a surveyor in accordance with Para. 6 of ASTM-E142 "Controlling Quality of Radiographic Testing". To date, we have not had any contractor present us with a request to prove out a technique that does not conform with our interpretation of the code. We believe we are now achieving proper radiographic inspection that is both in compliance with the code and is good technical practice. The results have more than justified the small increase of exposure time.

A review of current techniques used on many pipe installations might reveal a number of questionable practices and techniques.

Section III para. NB-5130 of the ASME code permits laminar type of discontinuities on weld preparations if they do not exceed 25.4mm in length. I will not comment on the advisability of welding a joint that contains a 25.4mm laminar discontinuity but if we refer to Section VIII you will not find any paragraph that will permit a similar discontinuity on a weld preparation. The Canadian Structural code CSA STD. W. 59 under para. 5.3.3 requires that a laminar discontinuity 1" or less be removed if the edges are to be welded. It would therefore appear that nuclear Section III of ASME is less severe in this instance than either Section VIII or CSA W. 59.

Many suppliers believe that if they write up their procedures on Section III they will be accepted for all other divisions. That belief is not factual as we do not want to have joints welded that contain up to 25mm laminations. We therefore insist on compliance to Section VIII plus the additional requirements in technical specifications.

There are also differences in radiographic film density requirements and acceptance criteria in the two sections. I recommend that suppliers read the "Tender Documents", especially the technical specifications that are contractual requirements.

There are some requirements which I like to see on field or shop isometrics. We appreciate the inclusion of "Reviewed" welding procedure numbers, and details of nondestructive testing. We must have agreement as to the interpretation of percentage inspection. If our technical specification calls up 5% or 10% R.T. on non-critical systems, we want to know how the percentage is determined and we establish that our surveyor will do the selection. Is the 5% based on total welds of each system, of each spool; or on the complete installation? Have we clearly specified and is our interpretation acceptable to the contractor? Do you include individual welder qualification welds as part of the percentage or are those welds additional to the requirements? Are the number of spools on each isometric drawing limited or not? Where is the verification plan to ensure that all welds are examined as per the technical specifications and satisfactory repairs completed as required? A piping system, whether plant or pipe line, includes valves, fittings and other appurtenances. There are often problems of alignment and these must be overcome if sound welds are to be realized.

In addition to ANSI B31.1 and B16-34 which cover pressure-temperature ratings, dimensions, tolerances, material, nondestructive examination requirements, testing and marking for cast, forged and fabricated steel flanged and butt welding and valves, we include other specifications which indicate acceptance parameters and also reference all other pertinent standards and specifications. These may include standards from Canadian Standards Association, American National Standards Institute, American Society of Mechanical Engineers, Canadian Government Specifications Board, Manufacturers Standardization Society of the Valve and Fittings Industry, American Society for Testing and Materials and the American Iron and Steel Institute and a number of others as applicable. The owner specifications are in addition and supplementary to ASME or ANSI codes and are contractual requirements.

It therefore behooves the supplier to review and understand the requirements before submitting a tender.

One area of supply where problems may develop and where care must be exercised is in the preparation and dimensions of buttweld ends of pipe, valves, welding web flanges and pipe fittings. There have been occasions where radiography of welded mismatched joints have resulted in difficulty of film interpretation. If the incident angle of exposure is such that the mismatch of the near side of the weld to source is projected over the area on the opposite or film side of the pipe being examined, the film reader has little way of knowing whether he is viewing a true discontinuity or an acceptable mismatch. If he determines that it is an acceptable mismatch, then he cannot be certain that the recorded higher density area is not screening a true defect. If he is to be certain, he must re-shoot at a different incident angle that will project the mismatch over another area. He can then calculate the location of the mismatch and view the total area under examination. It is usually much more economical and safer to provide good fit-up.

Most fabricators, to avoid problems of fit-up, will bore fittings, valves and pipe to ensure an identical I. D. of the interface. The concern here is to ensure that the wall thickness is not reduced to below 87.5% of nominal. Care must be used on joining pipes such as A106, Sch.120, where one pipe is of minn. O. D. of 272.5mm with nominal thickness of 21.4mm which would result in an I. D. of 251 mm, and the other with a max. O. D. of 275.4mm and a minimum wall of 18.8 mm. The I. D. difference would be 5.6mm which equals a mismatch of 2.7mm. As the I. D. mismatch should not exceed 1.6mm then the main O. D. and nominal wall thickness of 272.5mm and 21.4mm would require removal of 2.7mm to provide zero mismatch. Removal of 2.7mm would result in an unacceptable thickness of 18.7mm so the fabricator would endeavor to remove only 2.7mm or less to maintain a minimum thickness of 18.7mm.

It is easy to imagine the care that must be exercised in a spool fabrication shop to avoid excessive removal and possible rejection or repair of the pipe.

Valves will often present problems as rough welding ends do not always contain enough material to allow machining to match wall thickness of all pipe schedules. As specified in ANSI B.16.25, special internal machining of valves should be a matter of agreement between the manufacturer and the consultants or owner while maintaining the general contour of the welding bevel as specified in ANSI B16.25. Again it is the responsibility of the design engineer to ensure that purchase requisitions are specific enough to eliminate major problems of fit-up and excessive machining. Careful consideration must be given to the fact that the valve manufacturer has the option of providing his choice of contour profile provided it is of a shape avoiding sharp re-entrant angles and abrupt changes of slope that do not exceed a taper of 1 : 3.

Time does not permit a full discussion re fit-up, weld preparations and non-destructive testing. These subjects are constantly under review, but it would help the end fabricators and substantially reduce costs of installation if the tolerances were tighter on manufacture of pipe, valves and fittings. The consultants also have the option of tightening their specifications which may result in additional costs on supply but an overall saving when fabrication and installation costs are considered.

Most of the problems of radiography have originated in carelessness or incorrect procedures which have produced films of poor density, incorrect alignment and variations that should be unacceptable when the work is repetitive, as on a pipeline. Knowledgeable surveillance is the only way to ensure a minimum quality level.

There is one aspect of inspection of pipeline welds that we feel should be re-evaluated. As many of you know, heavy wall pipe is inspected by placing an isotope of Ir.192 or cobalt 60 inside the pipe through an access hole drilled in a boss or through the pipe. The source is centered under the weld and a 360 degree exposure is completed. We must doubt the reliability of the I. D. exposure technique because of the small source film distance and thick wall. The penetrameter is placed on the film side of the specimen so the operator will feel comfortable about the shot when in reality, the sensitivity of such an exposure is almost negligible. It will not locate the critical weld defects of linear cracks, lack of fusion or lack of penetration and consequently, this type of inspection should be questioned.

Internal diameter radiography to code requirements (?) has been supplemented by ultrasonic inspection on some projects where we encounter thick wall pipe. If the contractor is able to establish realistic procedures using proper equipment then we will accept the radiographic method without the U. T. supplement. The availability of high activity sources now permits techniques acceptable in regard to sensitivity and reasonable in regard to exposure time.

The use of back-up rings and weldable inserts occasionally caused ultrasonic inspection problems of interpretation for lack of penetration but now as we do not permit backup rings and seldom use inserts, the objections to U. T. are no longer justified. Our critical systems root passes are all tungsten arc welded (GTAW). Any other welds regarded as difficult or critical are also GTAW root welded. This usually ensures a sound root pass and an easy root for ultrasonic examination.

The Canadian Standards Association's Z299 series of quality assurance standards are fast being recognized internationally as one of the most thorough systems in the world. Mr. L. E.Stebbing, a Fellow of the Institute of Quality Assurance in the United Kingdom, visited several countries in Europe, North America and Asia and reported the following:

> "My overseas experience shows the Canadians to have the right approach to Quality Assurance, one only has to study the Canadian Standards Association's Z299 series to verify this. Their definition of Quality Assurance reads as follows: Quality Assurance means a planned systematic pattern of all means and actions designed to provide adequate confidence that items or services meet contract and jurisdictional requirements and will perform satisfactorily in service."

Our national standards, when called up by procurement authorities will do much to ensure a quality product, but the system can only work when it is called upon. Fortunately, more and more of our construction consultants and also general industry purchasers are insisting on observance of quality standards. The CSA standards and those of the Canadian Welding Bureau, the Canadian Government Specifications Board, American Society of Mechanical Engineers, ANSI, ASTM and many others are all pertinent and applicable to quality. We often hear of complaints about the amount of paper generated by Quality Assurance but we must recognize that quality verification cannot be achieved without documentation.

High management must decide if they wish to travel the quality assurance route. If that decision is made, then authority for its implementation must go with the responsibility. Production responsibilities must be divorced from quality assurance functions and the administrator of the latter must not be subjected to pressures or schedules. It is doubtful wisdom to think that all well designed systems complete with quality assurance safeguards may be circumnavigated by production requirements. If such a premise is true, then the quality assurance requirements are redundant. If the quality assurance requirements are necessary, then the producer faces uncalculated risks if they do not conform. No one can afford to have field engineers, surveyors, or inspectors altering design requirements without review and approval of design personnel.

## TRAINING

All training programs are beneficial and should be supported by individual companies, professional technical societies and government. Our technical colleges and schools of technology are doing an excellent job but the demands are great and the curriculum broad. The Welding Institute of Canada and the Canadian Society and Foundation for NDT are training personnel for certification by the Canadian Welding Bureau, pressure vessel authorities and by the CGSB. They are doing a yeoman's job but there are some weaknesses and they must be addressed or at least, recognized.

1 - Operators should realize that when certified, they have achieved the minimum capability necessary to pass the examination.

2 - Technical education should be a continuous process encouraged by company and government sponsors.

3 - Some degree of coordination should be developed between schools of technology and universities to allow graduate students of the former to progress through the universities to a degree. Credit should be given to students for equivalent courses studied in the lower institutions.

4 - Welding and NDE operators are taught specialized courses by the respective societies. There is an urgent need for expansion of subject matter to include a wider range. Society programs would benefit from courses on metal processing, machine shop, foundry practices and report writing. NDE operators may be certified to radiograph material but often their interpretation on welds consists of slag and lack of penetration. They simply do not know, nor seem to have been taught to any degree about "The source or origin of Defects". Education is expensive today. The student is asked to pay $400.00 to $600.00 for a week course, plus living and travelling expenses for a specific course that will allow him to sit for the certification examination. The examination where practical tests are involved, costs almost as much. At such cost, few individuals can afford the certification, so companies must support and subsidize the programs. It is little wonder that an operator who has passed his examination after paying out almost $1,000.00, feels he now knows everything there is to know about radiography.

Today, education is a business and a source of revenue for societies. The policy of some societies that insists on participation in their training program before being permitted to write the certification examination is understandable only when considered as a source of revenue for the society. A nominal charge for the costs of the examination would stimulate independent home study and bring on stream a larger number of trained workers.

In summary, education is very expensive, it suffers from the lack of trained and experienced teachers, it is narrow in concept, and specialized to possibly a dangerous degree. Incentives are recommended for capable technicians and operators to progress to higher levels of education so as to broaden their capabilities. Quality Assurance personnel in the metal working fields should have broad and specific experience in welding, nondestructive testing, machine shop practice, foundry practice, drafting or blueprint interpretation, finishing and materials handling. Such personnel are valuable and should be developed to engineering technicians and engineers for design and layout consultation.

## CONCLUSION

The following recommendations should be considered to achieve a quality piping system:

1 - Procurement specifications be specific and precise in their requirements for material, welding and inspection procedures. NDE requirements must be called up in relation to the needs of the job and not because of overenthusiasm of an NDE specialist or to grudgingly comply with a code.

Special Process personnel should review and provide input to technical specifications.

2 - Close surveillance of pipe production procedures and review of production procedures are mandatory if quality pipe is required. Confirmation of inspection personnel qualifications and the independence of Quality Assurance to production are essential.

3 - Field installations must include proper procedures, reviewed by the consultants or owners, proper equipment and power sources, certified personnel and again, surveillance.

Quality control of procurement, pipe production and field operations will do much to ensure a safe and economical pipe system. Without such control, we live in hope but without any basis for confidence.

## REFERENCES

Crocker and King. (1967). Piping Handbook 5th Edition. McGraw Hill.
Halmshaw, R. (1971). Industrial Radiology Techniques, Wykeham Publications, Ltd.
McMaster, R. C. Nondestructive Testing Handbook.
Stebbing, L. E. March 1980. Quality Assurance. Volume 6, No. 1.
Willenbrock, Jack H. and Thomas, H. Randolph. Planning, Engineering & Construction of Electric Power Generation Facilities.

# "...BECAUSE OF QA"

G.J. Melnyk, P.Eng.
*Ontario Hydro, Pickering G.S. "B", Box 1000, Pickering, Ontario Canada L1V 2R8*

ABSTRACT

In this presentation, the distinction and relationship between QA requirements and other requirements is illustrated using typical construction activities in order to reduce any confusion that may exist as a result of the use of the expression "...because of QA".

KEYWORDS

Project control, management controls, QA program, Audit program, Procedures, Records, Nuclear Construction.

INTRODUCTION

The title of my talk is "...because of QA". How often have you seen someone roll back their eyes when told to do something "...because of QA"? How often have you heard someone using "...because of QA" as an excuse? Blaming everything on QA provides an easy but often inappropriate out.

When faced with the task of implementing or working within a QA program, the 18 criteria which read so nicely seem to create nothing but confusion and misunderstanding.

My objective is twofold. One is to reduce the confusion by showing the relationship between QA requirements and other requirements so that each can be seen in the proper perspective. The second is to show the relationship between all these requirements and a QA program with audits.

I will illustrate these relationships by using actual schedule activities performed during construction at Pickering G.S. "B" and overlaying various requirements on top of them. I will identify the source of requirements for records and procedures when work proceeds as planned and for records and procedures which come into play when work does not go as planned.

## RECORD AND PROCEDURE REQUIREMENTS

### When Construction Activities Proceed as Planned

The first slide (Fig. 1) is a schedule which illustrates typical activities carried out during construction. There are two activities which are prerequisite to the whole sequence of events for the installation of any equipment or system. They are the receipt of information and material. Once we have received the drawings (our kit instructions) and the material (our kit parts) then we are ready to proceed with the actual construction activities. This is so, for both the nuclear and conventional portions of the plant.

While carrying out these activities, we are required to generate records and the requirement for these records may come from designers, regulatory authorities, or from various codes and standards. In addition to generating records in response to these requirements, we must perform some activities in accordance with specific procedures. For example, we are required to follow approved procedures when assembling reactor components. When welding metals, we have available a whole spectrum of approved and registered welding procedures for use on various materials, sizes and configurations and for various services. Also, we have a range of procedures for the installation of various categories of piping and tubing systems.

Up to this point, I have been talking about requirements which apply to our construction activities and these are illustrated on this slide. So far, QA has not been mentioned.

Now I would like to add to this schedule, the QA activities that must be addressed because of the implementation of a Quality Assurance Program. These activities are additional to the normal construction activities. They are shown as dashed lines.

As a result of QA or I should say "...because of QA", there are specific activities required (Fig. 2) and they are, to plan construction work prior to the start of installation, maintain a calibration program for calibrators used to set up permanent plant instrumentation and controls and maintain a records storage facility.

In the next few slides, I will superimpose the record and procedure requirements resulting from QA. Records are required for two of the activities. One of them is the planning activity and the other the calibration activity. This next slide highlights activities where QA procedures were implemented primarily out of respect for Murphy's Laws - if anything can go wrong, it will! Where there are significant consequences of things going wrong we have procedures to provide additional management controls and to increase confidence that the activities are carried out as required.

### When Construction Activities Do Not Proceed as Planned

So far we have seen where and why records and procedures apply to construction activities when work goes as planned.

If work does not go as planned, there are three processes available to provide the necessary management controls. I will show this with the next few slides.

The boxed activity (Fig. 3) is one that contains the requirements in the form of instructions and specifications that must be followed in carrying out downstream activities. When something cannot be or is not installed in accordance with the requirements specified on the drawings, we have a problem. When this occurs, say for the activity "Piping and Tubing" (something is upside down instead of right side up), we rely on problem solving techniques to resolve the problem. That is, we define the problem, flag the item to prevent inadvertent use and develop and evaluate alternatives to resolve the differences between the specified and installed condition. We pick the most appropriate alternative, obtain agreement and approval from those who issued the original drawings and record the agreement and its results. The same techniques for problem solving apply to activities on this next slide and are highlighted accordingly.

Our QA Manual contains a number of procedures which can be used selectively to solve problems and to control activities. These procedures allow us to obtain concessions from designers or cause drawings revisions, cause corrective action with suppliers, control changes to installed systems and quarantine items as required. We use these procedures to control what we must and when we must. The records these procedures generate are simply forms where the necessary information is recorded and which follow the progress of the problem and the item. Some of these records are important for future use and are retained for periods of time and some have no use whatsoever once a problem has been resolved. In the latter case they were only a mechanism to cause certain things to happen. For example, an approved concession request becomes an integral part of the specified requirements and must be kept until the drawing is revised. The other records do not have much value once the problem has been resolved. That is, it may be of no interest that changes were made to correct a deviation as long as it now is in accordance with drawings. Also it may be of no interest that an item once was quarantined. These records are not kept long and are disposed of in a controlled manner when convenient.

The next process (Fig. 4) that is available relates to the controls that are applicable to the supply of material prior to its receipt on site. If we received all material through the Designer/Supply and Procurement Division route, there would be no need to concern ourselves with these controls. However, there are many instances where we fabricate or initiate procurement of these items on site. When we make something or buy something, we must ensure that the specifications and controls that would have been provided to a supplier are equally in place.

The controls we apply when purchasing material include specifying requirements, preparing requisitions, evaluating tenders and placing the business. For example, for the "Piping and Tubing" activity, pipe and fittings may be purchased from site and pipe hanger assemblies fabricated on site. Other activities where these controls may apply are shown on the next slide.

The procedures we have are few and are used when necessary. They allow us to buy the items on behalf of the Designer and to initiate the requisitioning and purchasing processes so that our Supply and Procurement Division becomes involved during the manufacturing phase. They also allow for the provision of drawings and other requirements such as materials, inspections and tests, code classifications and records for items manufactured at site.

To minimize the chances of Murphy's Laws being demonstrated, a third process (Fig. 5) within our QA Program provides for verification of requirements while activities are being carried out. For the activity "Site Handling" for example, we may rely on inspection records, check lists and test records to verify the work being performed by our trades or contractors. Items that are sensitive to the elements or have special handling requirements are controlled in this way. There are other activities that may receive special attention because of management's interest to minimize the risks of problems occurring. These are highlighted as well.

What procedures do we have? There are a few contained in our QA Manual. They deal with such things as the requirement to confirm that all prerequisite activities, records and procedures have been satisfactorily completed prior to pressure testing, the requirement to verify the completeness and correctness of a system prior to turnover from Construction to Commissioning and the requirement to verify the work of site contractors performed on behalf of the owner.

## RECORD AND PROCEDURE SUMMARY

I would like to summarize the presentation to this point (Fig. 6) with the following slides. Here we have a typical schedule of activities which includes Quality Assurance activities. I have shown where procedures and records are required based on QA, Design and Regulatory Authority requirements. We then looked at three processes available to control problems and deviations, the purchase and manufacture of items on site and the verification of work where Murphy's Laws might hurt us. I have attempted to show the sources of requirements and relationship among them.

## PICKERING'S QA PROGRAM

When anyone chooses to follow a QA program, and addresses the 18 criteria they are simply implementing a management system to control activities so that the necessary requirements are met and a high level of assurance is provided. Pickering G.S. "B" was one of the first Canadian nuclear projects to implement a formal Construction and Installation QA Program and I would like to briefly review two aspects of it, auditing and self-review.

As far as the audit program is concerned, what is required? Well, there is a requirement to have planned and scheduled audits of all activities performed by our staff and trades as well as those activities performed by contractors. The purpose of the audits is to assess adherence to all of the requirements discussed. The auditors have no direct responsibility for the work being audited, they must follow written procedures and then cause action to be taken on audit findings. What do we do at Pickering? Well we have a set of procedures that deals with the scheduling of audits, preparing for audits, writing reports describing what took place during the audits and the closing out of audit findings.

As far as the self-review, what is required? There is a requirement to perform an annual review of the administration of the QA Program and of its effectiveness in meeting objectives. It is performed by the Resident Engineer and a written report is issued to the Construction Manager, Manager of Engineering and the Project Manager. What do we do relative to these requirements at Pickering? We do just what's required. We review how we administer our program and the significance of audit findings. We decide if and what recommendations are appropriate for the following year. We then write and issue a report entitled "Program Review" for distribution to the Managers.

That concludes my presentation, and I hope that it has cleared up some of the confusion - not added to it.

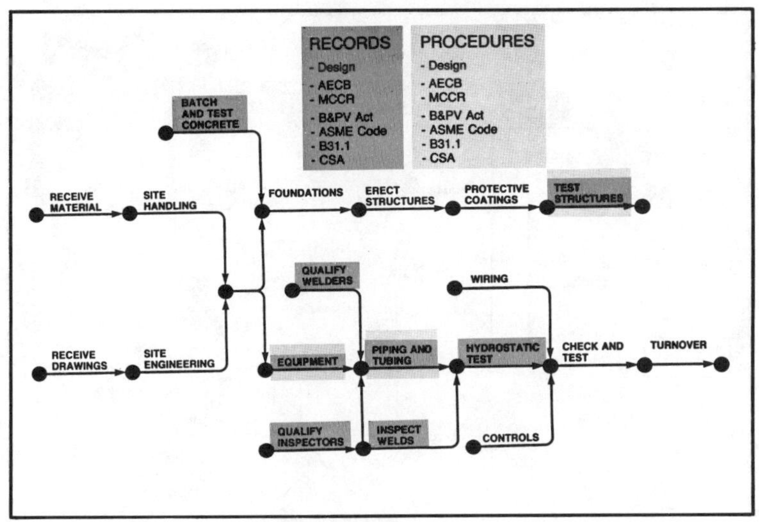


Fig. 1. Slide composite #1 from the section RECORDS AND PROCEDURE REQUIREMENTS

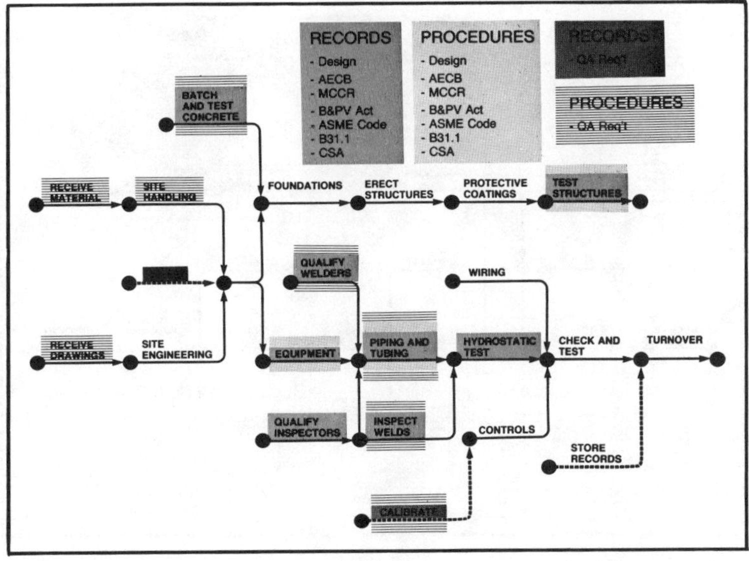


Fig. 2. Slide composite #2 from the section RECORDS AND PROCEDURE REQUIREMENTS

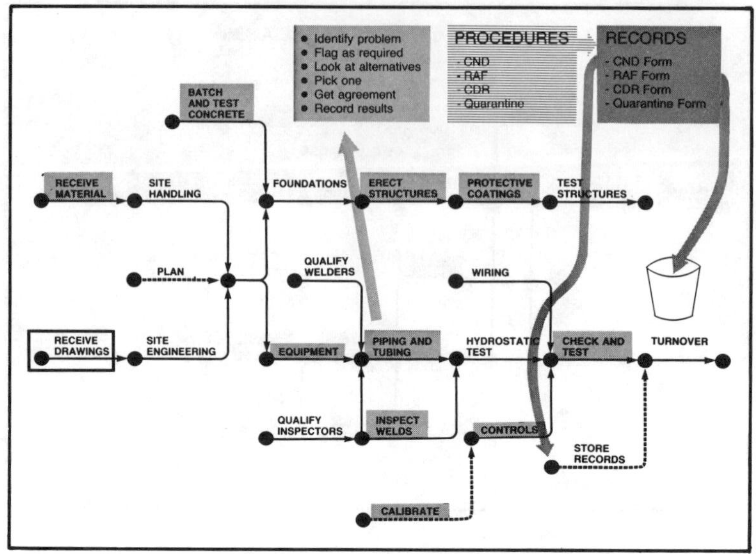


Fig. 3. Slide composite #3 from the section RECORDS AND PROCEDURE REQUIREMENTS

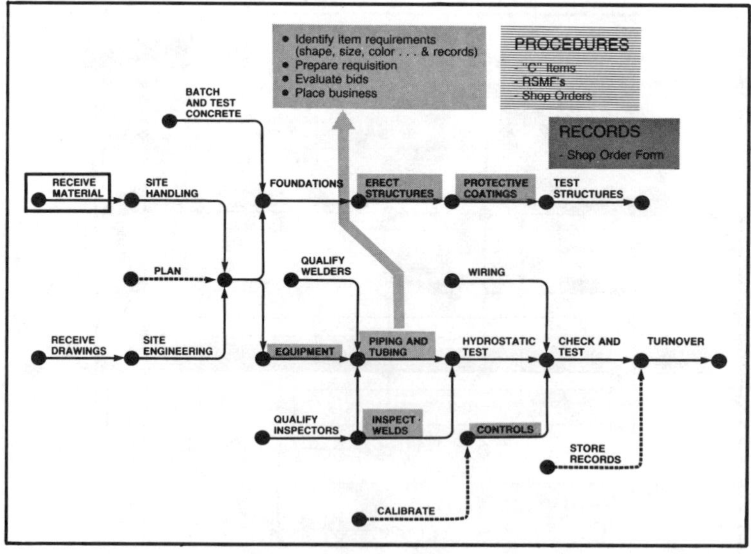


Fig. 4. Slide composite #4 from the section RECORDS AND PROCEDURE REQUIREMENTS

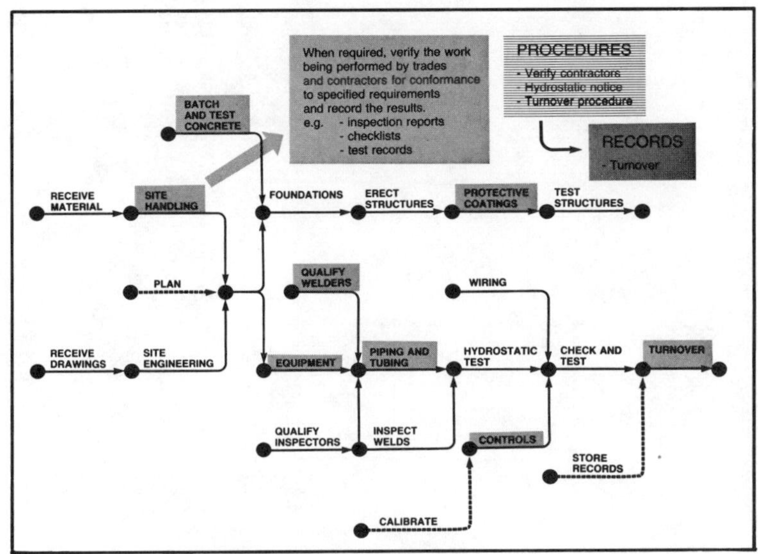


Fig. 5. Slide composite #5 from the section RECORDS AND PROCEDURE REQUIREMENTS

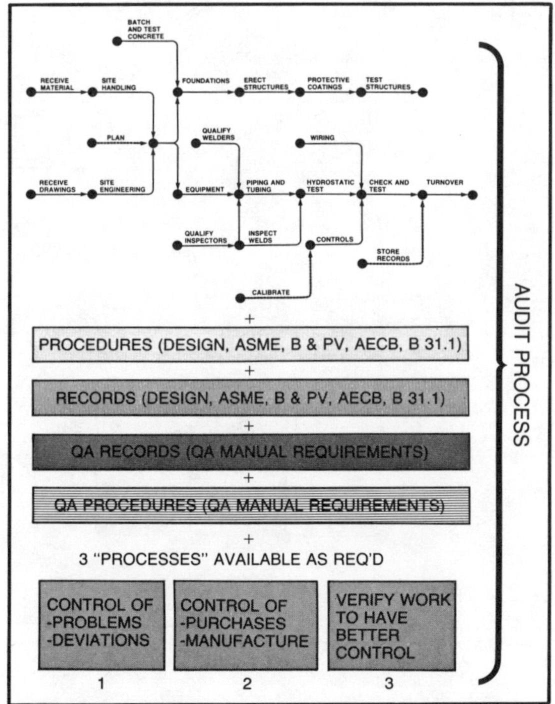


Fig. 6. Slide composite #6 from the section RECORD AND PROCEDURE SUMMARY

# ULTRASONIC DEBRIS SCANNER IN ENERGY PLANT PIPING

by
M. Macecek
*Techno Scientific Inc., Toronto, Downsview M3J 2C6*

ABSTRACT

Heavy wall piping is frequently used in the construction and operation of an energy plant. For a variety of reasons a considerable amount of debris can be left in that piping.

While the general presence of the debris can be detected from the flow disturbance or general process control, the actual location, its size and its detrimental effects on the integrity and the operation of the plant are difficult to assess.

The ultrasonic scanner is capable of finding any debris equivalent to a reflector 50 mm (2") in diameter and 1 mm (1/16") thick in the 500 - 1000 mm (20" - 40") O.D. pipes. A wall thickness of up to 100 mm (4") can be used. The unit uses eight contact transducers covering over 90° of the circumference at the bottom of the pipe. Debris equal to or larger in size than the test reflector will trigger an alarm.

The scanner is a highly portable device which, in spite of its complexity, can be operated by unskilled personnel.

A principle of operation and the system description are given and possible problem areas are discussed.

Application to the nuclear industry and possible applications to other energy related projects are described.

KEYWORDS

Ultrasonics, Debris, Transducer Array, Gating, Multiplexing, Go-No Go Display.

INTRODUCTION

Heavy wall piping is the backbone of the operations of an energy or petrochemical plant. The debris left in the piping during construction or resulting from a component failure somewhere in the system can lead to a plant shutdown or an accident. While the general location of the culprit can be traced from the flow charts or the response of the control system, the exact location is very difficult to specify. X-ray or Gamma radiography can be used to locate the debris in some instances but access from both sides of the pipe is required. If at all possible the process is slow, expensive and constitutes a health hazard. Ultrasonics may be considered as an attractive alternative.

## APPROACH TO THE PROBLEM

The ultrasonic approach to debris monitoring is reasonably straightforward and can be understood with the help of figure 1. An ultrasonic beam passes from a transducer which is contact, gap, or immersion coupled and penetrates the vessel wall. The vessel wall acts as a lens which forms the passing ultrasonic beam (1). It also generates a number of multiples which are reflected back to the probe. The divergent beam continues through the liquid until it encounters the far wall of the vessel. The beam is again divided; part of it is reflected and part of it enters the wall. For debris monitoring the reflected beam only is of interest. Since the beam impinges at normal incidences the signal levels are high. The A-Scan presentation of the case can be seen in figure 2.

If the debris is present its near edge acts as a reflector to the beam. An echo is received before the far wall echo (see figure 2). One can gate out the far wall echo while opening the gate within the wall of the vessel. Any signal within the gate is a debris signal. Due to the restrictions of the geometry and the properties of the ultrasonic beams full coverage of the inside of the vessel cannot be achieved. This aspect will be discussed later on.

## DESCRIPTION OF THE SYSTEM

While the principle of operation is simple, several critical conditions have to be met for the system to perform satisfactorily. These are the choice of transducer, its beam shape, sensitivity, and waveform shape. Some other requirements are, the sufficient gain of the system, isolation of the different channels, sufficient speed of multiplexing, and simplicity of the display. The scanner detailed below satisfies these conditions. The scanner is capable of finding any debris equivalent to a reflector 2" (50 mm) in diameter and 1/16" (about 1 mm) thick in the 19 - 26" O.D. (about 482-660 mm) header pipes. The unit uses transducers covering over 90° of the circumference at the bottom of the pipe. Debris equal to or larger in size than the test reflector will trigger an alarm.

The UT pipe scanner consists of four main parts:

- an adjustable band housing the transducers
- control electronics unit
- power supply
- an 80 ft (25 m) long umbilical cord connecting the control unit to the power supply.

The overall view of the system is shown in Figure 3. The system was designed as a go-no go system which is very easy to interpret. The block diagram is shown in figure 4. The system uses 8 individual transducers which are individually pulsed, an analog multiplexer and two gates: one which monitors the acoustical coupling and the other which is the debris gate. A resetable peak detector signal is demultiplexed for display on LED's.

## TRANSDUCER BAND

The transducer band is the backbone of the scanner. It is highly portable and moves easily along the pipe on four rollers with ball bearings. The curvature of the band is designed to accommodate headers with diameters of: 19-26" (482-660 mm) and wall thickness of 1.75- 3" (66-76 mm).

This is accomplished by having the band made in three segments, each approximately 8" (200 mm) in length. Two large diameter spacers can free the segments allowing for the curvature adjustments . Each transducer is held in position by a plastic holder which is spring loaded and allows for the setting of the unloaded height of the transducer. Thus, the transducer face will be in contact with the pipe surface for all curvatures. The adjustment of curvature is simply accomplished by tightening two knobs. A BNC connector joins the coaxical cable from each transducer to a control unit housing the electronics. The choice of transducer is critical since it has to insure the coverage between neighbouring elements, gain uniformity, and sufficient sensitivity to trigger the alarm. Transducers one MHz, .5" (12.5 mm) diameter wiht low damping were chosen after some experiments. The beam from each transducer is divergent enough to insure overlap between neighbouring sensors.

## ELECTRONICS

The control unit contains the pulsers, the receiver and the evaluation logic, as well as the control and setting knobs for the operator.

The system consists of:

- eight individual pulsers (one for each transducer)
- one common receiver connected to any transducer through the analog multiplexer
- gating circuitry for both the acoustical coupling and the debris signals
- level comparator and a digital multiplexer
- LED displays for the alarm and for lack of coupling signals
- master clock and address freeze-up circuitry.

The front panel of the box contains all the displays and controls (see fig.3), viz:

- adjustment of the debris gate delay (acoustical coupling gate is preset internally)

- address freeze-up which allows the operator to lock on one transducer (for debris sizing, lack of coupling and laminations checks).

- gain adjustment for different pipe diameters.

- gate delay button for back wall signal monitoring.

POWER SUPPLY

The power supply to run the equipment is contained in a separate box joined by an umbilical cord with two connectors. The length of the cord is 80 ft ( 25 m) . The power supply is fused and requires one 110V receptacle. A battery pack can be used for remote field applications.

OPERATIONS

After assembly the band is placed on the pipe and the power turned on. The pipe should be well covered with a UT couplant, such as Hamikleer. The high gain required for the debris signal results in the amplification of wall multiples (see figure 5). Typically up to 10 multiples are easily seen resulting in a period of 200-250 us (eg. 5-10" or 125-250 mm from the top of the header) where no debris can be detected by receiving the echoes. This dead zone is not detrimental to the operation of the unit. Adjustment of the gain is followed by the adjustment of the gate. First the gate is opened to include the back wall signal. The gate is then moved forward until the back wall signal is out of the gate. Any signal in the gate now will be a debris signal. Due to the out -of-round tolerances ( up to 1/8" or 3 mm fig. 6) it may be necessary to move the gate forward to the same extent.

The presence of debris can also be verified by observation of the signal from the adjacent transducers. If the debris is large enough to cause reflection in several channels the LED display indicates this. On this basis a single indication of the display may be neglected. This is applicable only when very thin debris is sought. For thicker debris the out of round tolerancies can be made irrelevant by moving the gate farther away. The push button switch arrangement makes it easy for the operator to verify the back wall signals by extending the gate to include the back wall.

DEFECT SIZING

Sizing or outlining the boundaries of the debris is simple. Once the debris signal is detected the operator can lock on one channel. Using the equivalent of a dB drop method he outlines the defect by simply marking the edge where indication on the proper LED'S are switched on or off. Although the unit has a total gain of 110 dB only 70-90 dB are required. Too much gain results in spurious signals being generated.

POSSIBLE PROBLEM AREAS

Since the operator only sees the LED display and hears the buzzer for a lack of couplant signal, he should be aware of the problems he might encounter during the inspection:

These are:

- signals from the nozzle's hole,
- laminations,
- a lack of the debris signal combined with a lack of back wall signal.

The nozzle hole can be recognized by a lack of back wall signal.

In the case of laminations in the pipe material, no debris will be detected. The lack of coupling alarm may also sound depending on the depth of the lamination.

In case of large reflectors (large size debris), there may be times when debris signals do not appear due to unfavorable orientation of the reflector. There will also be a lack of back wall signal, to allow for a decision whether this is debris or a nozzle hole. In general the back wall signal or lack of it is not a reliable measure of reflectors (debris) in the pipe. More commonly, the multiple reflections in the debris will extend into a back wall position. Common sense by the operator is required to judge the marginal cases.

The transducers are chosen to have a gain response within + 2 dB. The signal response as a consequence is reasonably uniform across the circumference of the pipe. The variations and the signal gain can be compensated for by an adjustment in the pulsers to reflect the sensitivity of transducers. The speed of inspection when compared with a single probe manual testing is up to 10 times faster with additional advantages brought about by the multiplicity of probes.

## CONCLUSIONS

The scanner is a highly portable device which offers great ease of operation. It substantially reduces the man-rem expenditures during the inspection in nuclear applications. Other applications like off-shore, petrochemical and pressure vessel testing are possible. A follow-up data acquisition system might also be added to provide the link with a control microcomputer thus giving a permanent record of the data. But the device is adequate for the intended applications without any additional equipment. The operations are simple enough so that semi skilled-labour may be used. The system is rugged and compact yet portable for field applications.

## REFERENCES


1. J. Krautkramer, H. Krautkramer., Ultrasonic Testing of Materials, Springer Verlag, N.Y., 1969, pp 251-257.

2. I. Bradael, Characterization of Ultrasonic Transducers, Research Techniques in Non-destructive Testing, Volume III, Academic Press, New York, 1977 pp. 175-215.

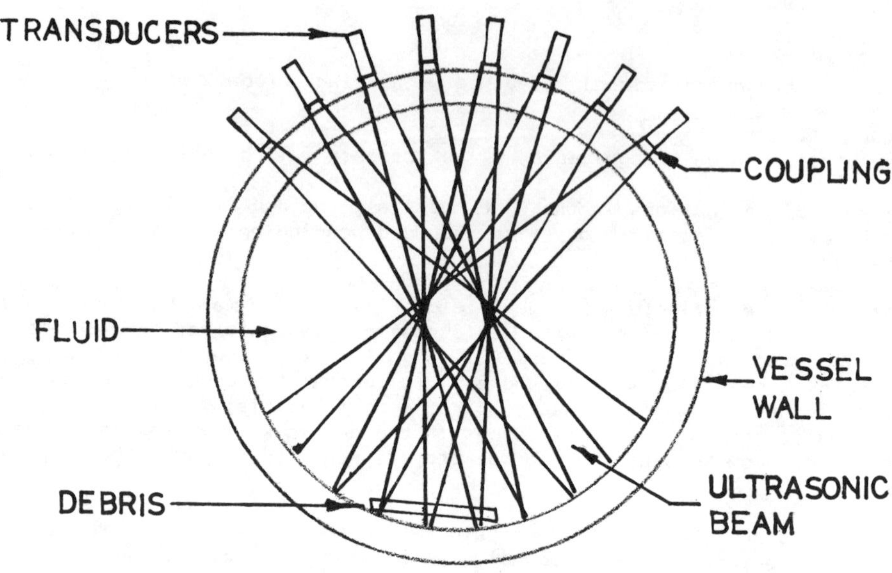


Fig. 1 Principle of operation of the ultrasonic debris scanner.

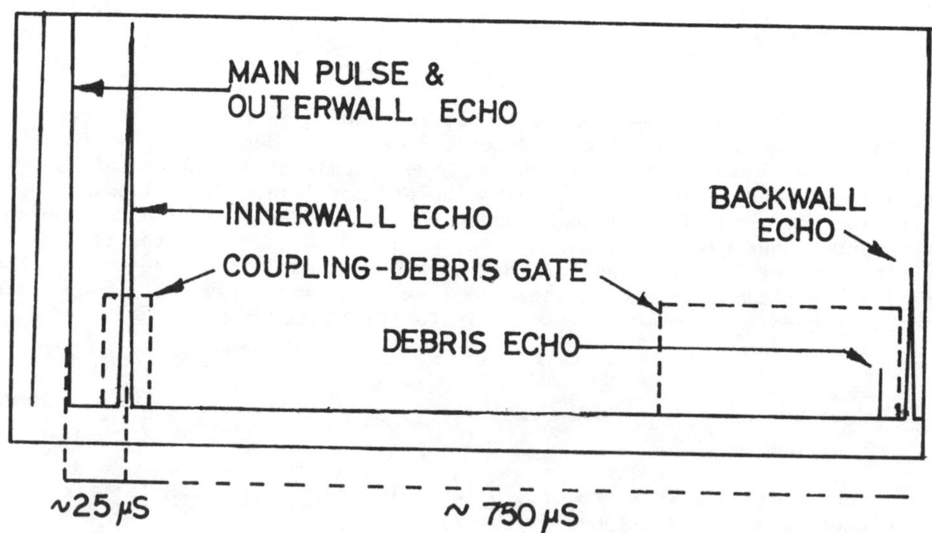


Fig.2 A-scan presentation of the debris scanner signals.

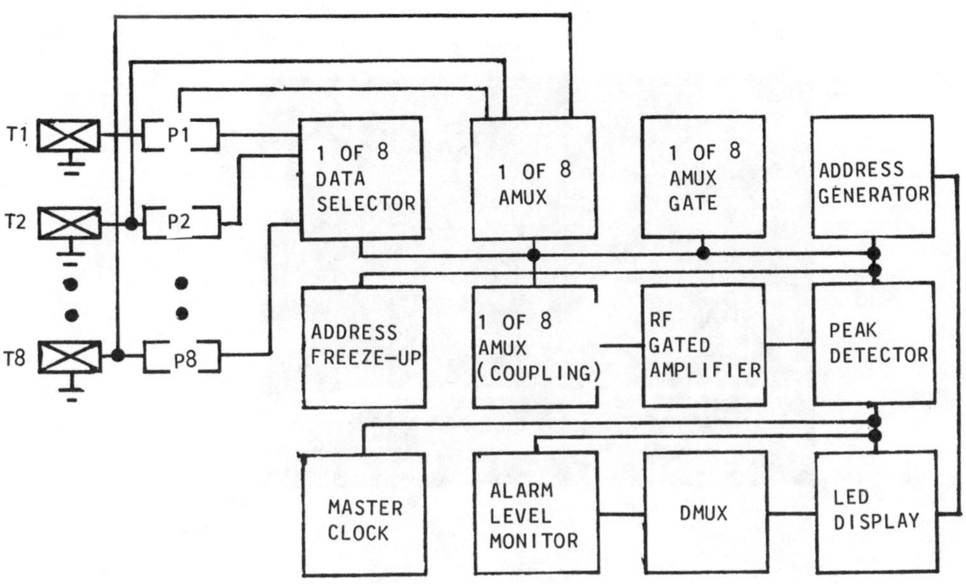


Fig. 3 A block diagram of the scanner

Fig.4 Overall view of the system mounted on a 19" pipe.

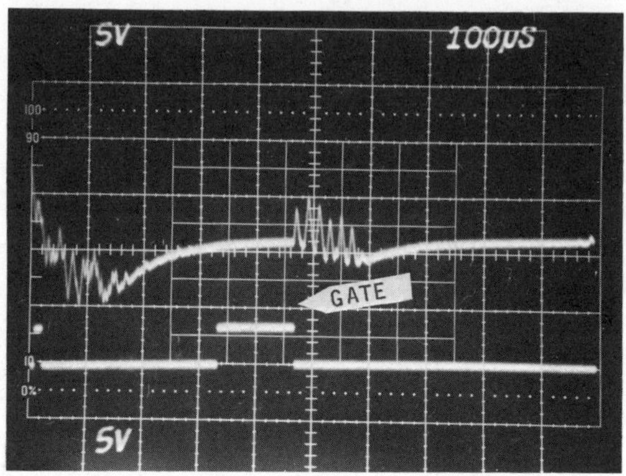


Fig. 5 Photograph of the signals showing the multiples in the wall.

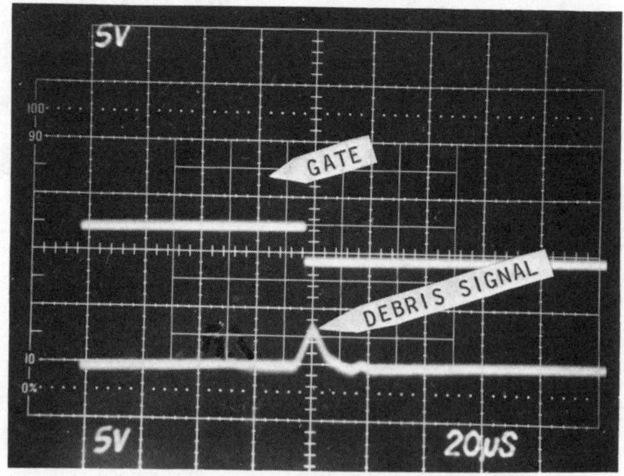


Fig.6 The actual appearance of the debris signal

# THE PROPERTIES OF PIPELINE GIRTH WELDS PRODUCED BY ARC WELDING PROCESSES

P.R. Kirkwood, K. Prosser and P.J. Boothby

*British Gas Engineering Research Station*
*Killingworth, Newcastle upon Tyne, England*

ABSTRACT

This paper reviews the mechanical properties of pipeline girth welds produced by a number of arc welding processes which have been studied by British Gas. Consideration is also given to process characteristics, such as welding speed and ease of use, since these have an important bearing on the acceptability of the process. Traditional stovepipe welding techniques, using cellulosic electrodes, normally produce welds with adequate toughness for land pipeline applications. However, increasingly severe requirements for offshore pipelines and the dangers of weld metal as well as heat affected zone cracking, have led to greater interest in low hydrogen welding techniques. Vertical down low hydrogen manual electrodes are shown to be capable of producing welds with good toughness, but may require further development to improve operating characteristics. The semi-automatic gas metal arc welding process has suffered from a reputation for poor reliability and inconsistent welding quality which has limited its use. The introduction of self-shielded flux-cored filler wires has led to renewed interest in semi-automatic welding. Fast welding speeds are possible, but there is a problem in achieving consistent levels of toughness with present carbon-manganese flux-cored consumables. The properties of welds produced using mechanised gas metal arc welding systems are discussed. It is shown that further development of commercially available filler wires is desirable in order to increase the ease with which requirements for weld toughness demanded by the application of 'fitness for purpose' standards can be met. Finally, the results of early trials with a mechanised gas-shielded flux-cored welding system are given.

KEY WORDS

Weld Metal Toughness
Mechanical Properties
Pipeline Girth Welds
Carbon Manganese Steels
SMA Welding
GMA Welding
FCA Welding
Hydrogen Cracking

## INTRODUCTION

Traditionally, field welding of cross-country pipelines has been carried out using the 'stovepipe' technique with cellulose-coated manual metal arc electrodes. When using these electrodes large quantities of hydrogen are introduced into the weld and its heat affected zone (HAZ) and avoidance of hydrogen cracking, particularly in the HAZ, has been a major consideration. Welding procedures which seek to minimise the occurrence of hydrogen cracking have been developed and these have largely been successful. Many thousands of kilometres of pipeline have been constructed worldwide using cellulosic electrodes, normally with HAZ cracking frequencies of less than 1% of the total welds made during construction. Developments in linepipe steel metallurgy have kept pace with requirements for increased pipe strength and diameter so that HAZ hydrogen cracking is not a major problem even in X70 grade pipelines, although experience beyond this strength level is still fairly limited (Ref 1,2). However, high strength pipeline fittings, such as bends, tees, flanges, etc have not benefited from the same improvements in weldability as linepipe and can still present HAZ cracking problems if welded directly in the field with cellulosic electrodes. In addition concern has been expressed over the dangers of weld metal, rather than HAZ, hydrogen cracking when using cellulosic electrodes designed to match high strength linepipe and fittings (Ref 3).

British Gas has experienced both types of problem in recent years and this has led to a greater interest in low hydrogen welding processes for site welding of pipelines. Even using the best radiographic practice, it is not possible to guarantee detection of all types of hydrogen cracking in pipeline girth welds (Ref 4). This is especially true for laybarge construction where radiographic conditions are seldom ideal and the additional insurance of magnetic crack detection of the root bead normally carried out during field welding of land pipelines is not feasible. British Gas has, therefore, taken the step of insisting on low hydrogen welding processes for the construction of its offshore pipelines. To date this has normally involved the use of mechanised arc welding systems, although one short length of pipeline has also been constructed with low hydrogen manual metal arc electrodes.
For land pipelines British Gas continues to use cellulosic electrodes, although for large projects increasing use has again been made of mechanised arc welding systems.

In addition to the need to specify welding procedures which ensure freedom from cracking, increasing importance is now being placed on the achievement of good levels of weld fracture toughness. Traditional pipeline design criteria have led to fairly modest Charpy toughness requirements for land pipelines (our own requirement being 27J average at $0^{o}C$). These levels are readily attained by most traditional welding processes. However the use of defect acceptance levels based on fitness-for-purpose considerations, and more stringent requirements for offshore pipelines, have led to the specification of minimum crack tip opening displacement (CTOD) toughness levels, as well as higher levels of Charpy toughness. The values specified, particularly those for CTOD, have created some problems especially with the mechanised arc welding processes.

This paper presents a review of the properties of pipeline girth welds produced by arc welding processes, with particular reference to weld metal toughness. Special consideration will also be given to the operating characteristics of those low hydrogen welding techniques which are possible replacements for pipeline welding with cellulosic electrodes.

## MANUAL METAL ARC PIPE WELDING TECHNIQUES

### Cellulosic Electrodes

Early pipeline girth welding was carried out successfully using low strength cellulosic electrodes of the E6010 type. With the development of higher strength linepipe, the demand arose for electrodes with improved tensile properties. However, use of these to make the complete girth weld, often led to problems in the root due to the poorer ductility of the high strength electrode deposits. Consequently, for many years British Gas has adopted welding procedures specifying E6010 or E6011 electrodes for the root bead and occasionally hot pass. This is in order to deposit weld metal with sufficient ductility and notch toughness to accommodate stresses arising from the pipeline front end 'lift and lower' operation, without undue stressing of the critical HAZ region or failure from root bead discontinuities such as slag intrusions. For filling and capping passes E7010 or E8010 electrodes are used, depending on pipe strength level, to enable the completed weld to overmatch the pipe for strength. This ensures that any workmanship defects in the weld metal are not subjected to high strains during subsequent stressing of the pipeline. Table 1 shows typical properties obtained for different grades of cellulosic electrodes.

TABLE 1 Comparison of Tensile and CTOD Properties for Cellulosic Electrodes (All-Weld Tests)

| Electrode Type[1] | 0.2% RP N/mm$^2$ | RM N/mm$^2$ | Elongation, % | Reduction in Area, % | CTOD[2,3] at 20$^\circ$C, mm |
|---|---|---|---|---|---|
| E 6010/6011 | 382-396 | 468-483 | 23-26 | 60-68 | 0.3-0.4 |
| E 7010 Al, G | 451-617 | 568-694 | 16-28 | 60-67 | 0.1-0.3 |
| E 8010-G | 474-658 | 562-750 | 15-16 | 42-56 | 0.07-0.2 |

Notes [1]Several electrodes of each type tested.

[2]CTOD specimens had 0.15 mm machined notch (not fatigue pre-cracked).

[3]Nine specimens tested for each electrode type.

In order to produce cellulosic electrodes meeting E8010 and E9010 strength classifications manufacturers have increased electrode carbon contents giving weld deposits with levels as high as 0.18%C in some cases. These carbon levels are considerably higher than those present in modern linepipe steels (typically 0.1%C) and may mean that avoidance of weld metal hydrogen cracking becomes the dominant factor governing the choice of preheat temperature in the welding procedure. British Gas has experienced such cracking, although in that particular case the weld metal carbon content was also augmented by dilution of a high chemistry fitting material into the weld pool (Ref 5).

In addition to a higher risk of cracking, there is a danger that the high carbon weld deposits may be less tolerant to procedure variation with regard to toughness properties although British Gas has not, to date, experienced any toughness problems with cellulosic electrodes. The toughness data generated has been for the major part by Charpy testing and a typical transition curve for an E7010-G electrode is given in Figure 1. Only a limited amount of CTOD testing has been performed on actual girth welds and the results obtained, as for the Charpy testing, have been good, with values of 0.2 mm CTOD at temperatures down to minus $10^{o}$C. Where it has been necessary to apply retrospective fitness-for-purpose analyses to land and offshore pipelines, welded with cellulosic electrodes, CTOD levels of this order have been sufficient to provide realistic acceptable defect sizes (Ref 6).

## Low Hydrogen Electrodes

Cellulosic electrodes remain in favour, despite the limitations referred to above, because of their rapid welding speeds and their ability to cope with the variations in weld joint preparation which inevitably arise on site. Therefore it is desirable for low hydrogen pipewelding electrodes to compete with cellulosic electrodes adequately in these areas. Since vertical down welding is the quickest manual girth welding technique, new low hydrogen electrodes for pipeline welding have been designed for this type of use (the LHVD electrodes). British Gas has been using similar LHVD electrodes for many years for fillet welding applications on live gas pipelines but it is only recently that LHVD electrodes suitable for butt welding have been available commercially. Typical welding times recorded for a LHVD welding procedure and a comparable cellulosic 'stovepipe' procedure are given in Table 2. It is evident that the total welding and cleaning time for the LHVD electrode is in this case 20% shorter than that for the cellulosic electrode. This is largely the result of the greater deposition efficiency of the iron-powder containing LHVD electrode against the non iron-powder cellulosic electrode. Root bead welding times for the LHVD electrode however are up to 60% longer than those for the cellulosic electrodes, for two reasons. Firstly, because it operates with a less penetrating arc, the LHVD electrode requires a larger root gap and must be used with a smaller electrode size than the cellulosic electrode. Consequently run out lengths for the LHVD electrode in the root are shorter, welding speeds are slower, and stop/starts more frequent. Secondly, in order to ensure satisfactory fusion at stop/starts the previous weld crater must be ground before striking each new electrode, adding considerably to the total root bead welding time. Since the time for root bead welding is the critical factor controlling the overall rate of progress on a pipeline welding spread, the LHVD electrode is at a disadvantage in this respect unless additional root bead welders can be used.

It has been claimed that root bead welding speeds equivalent to those occurring with cellulosic electrodes have been obtained with a LHVD electrode on a 450 mm diameter pipeline in Australia (Ref 7). However, our own trials on 600 mm diameter pipe have not reproduced this behaviour and it is possible that the small diameter pipe gives an unduly optimistic comparison because of the reduced number of stop/starts involved.

TABLE 2 Comparison of Welding Times for Cellulosic and Low Hydrogen Vertical Down Welding

a) Low Hydrogen

| Run No | Run Type | Electrode Size mm | Welding Time | | Cleaning Time | |
|---|---|---|---|---|---|---|
| | | | Mins | Secs | Mins | Secs |
| 1 | Root | 3.25 | 7 | 50* | 2 | 25 |
| 2 | Filler | 4.00 | 4 | 10 | 1 | 40 |
| 3 | Filler | 4.00 | 6 | 10 | 1 | 05 |
| 4 | Filler | 4.00 | 8 | 20 | 2 | 05 |
| 5 | Filler | 4.00 | 6 | 45 | 1 | 30 |
| 6 | Side Stripper | 4.00 | 2 | 30 | – | 35 |
| 7 | Cap | 4.00 | 10 | 10 | 1 | 25 |
| | TOTAL TIMES | | 45 | 55 | 10 | 45 |

Electrode - E 8018-G, Deposition efficiency 116%
Pipe Size - 609 mm OD x 16 mm W.T. (two welders used)
Joint Preparation - 30° bevel, 1.5 mm root face, 2.5 mm root gap

* Includes grinding stop/starts.

b) Cellulosic

| Run No | Run Type | Electrode Size mm | Welding Time | | Cleaning Time | |
|---|---|---|---|---|---|---|
| | | | Mins | Secs | Mins | Secs |
| 1 | Root | 4.00 | 4 | 48 | 3 | 18 |
| 2 | Hot Pass | 4.00 | 4 | 54 | 1 | 42 |
| 3 | Hot Fill | 5.00 | 4 | 12 | 2 | 08 |
| 4 | Fill | 5.00 | 6 | 10 | 3 | 16 |
| 5 | Fill | 5.00 | 8 | 10 | 2 | 04 |
| 6 | Fill | 5.00 | 8 | 00 | 2 | 26 |
| 7 | Fill | 5.00 | 7 | 30 | 2 | 21 |
| 8 | Stripper | 5.00 | 2 | 45 | 1 | 18 |
| 9 | Cap | 5.00 | 8 | 20 | 1 | 32 |
| | TOTAL TIMES | | 57 | 49 | 20 | 05 |

Electrode - 4.00 - E 6011, Deposition efficiency 72%
- 5.00 - E 7010-G, Deposition efficiency 83%
Pipe Size - 609 mm OD x 16 mm W.T. (two welders used)
Joint Preparation - 30° bevel, 1.5 mm root face, 1.5 mm root gap

When using the cellulosic electrode to weld pipeline tie-ins and fittings a vertical up technique is normally used, since this is better able to cope with the less than perfect fit-up normally present. It is anticipated that the low hydrogen electrode may also have to be used with a vertical up technique in these circumstances for similar reasons. Although the new LHVD electrodes will operate in a vertical up mode, traditional E7016 electrodes of the type used for offshore constructions are normally superior in this respect and would be expected to be favoured for these situations.

Recent procedure qualification trials for a 450 mm diameter 22 mm wall thickness offshore pipeline were carried out using a LHVD electrode of the E8018-G type. Typical test results are given in Table 3 and Figure 1. As expected, weld toughness is superior to that obtained using cellulosic electrodes, although weld metal oxygen and nitrogen levels are higher than one would normally expect from a conventional low hydrogen electrode. The need to keep ahead of the advancing slag discourages excessively slow welding speeds and therefore welds produced normally consist of thin layers with a high degree of refinement. However, difficulties of slag removal and associated increased repair times, both more prevalent at increased pipe wall thicknesses, can occur when using LHVD electrodes and the contractor eventually opted to construct the pipeline in question using a conventional vertical-up E7016 low hydrogen electrode. This illustrates the importance of ensuring all-round performance, including for example, ease of use and welder appeal, in addition to good mechanical properties when introducing any new manual welding process. Nevertheless, the LHVD electrodes may be seen as an important step forward in pipeline electrode development. Nickel containing versions of the LHVD electrode are currently in use on a 450 mm diameter process pipeline and have proved capable of meeting low temperature Charpy requirements which were not consistently achievable with cellulosic electrodes (Ref 8).

TABLE 3 Properties of Girth Weld Produced with LHVD Electrode (E 8018-G)

(450 mm dia x 22 mm WT X65 Pipe)

Tensile Properties

| | |
|---|---|
| 0.2% RP ($N/mm^2$) | 536-573 |
| RM ($N/mm^2$) | 646-689 |
| Elongation % | 26 |
| Reduction in Area % | 69-70 |

Toughness ($0^oC$)

| | |
|---|---|
| Charpy V | 73,122,103J |
| CTOD | Range: 0.40-0.66 mm<br>Mean (of 6 tests): 0.53 mm |

| ELEMENT | C | Si | Mn | S | P | Al | V | Nb | Ti | Ni | Cr | Mo | Cu | O ppm | N ppm |
|---|---|---|---|---|---|---|---|---|---|---|---|---|---|---|---|
| % WT | .08 | .49 | 1.38 | .013 | .013 | .003 | .02 | .002 | .014 | .02 | .03 | .005 | .02 | 443 | 204 |

## SEMI-AUTOMATIC PIPE WELDING TECHNIQUES

### Gas Metal Arc Welding (GMAW)

In view of the present interest in low hydrogen welding techniques for pipeline use it is perhaps somewhat surprising that more use has not been made of the semi-automatic GMAW process. This process was first used for UK pipeline construction in the mid 60's when $CO_2$ shielding gas was employed (Ref 9). It offers the advantage of a low hydrogen weld deposit with the flexibility of a manual welding process and is suitable for front end, tie-in and pipe to fitting welds. Welding times are again faster than with the cellulosic electrode because of fewer stop/starts and the reduced necessity for slag removal. The early trials with the process were plagued by difficulties caused by poor equipment reliability and the widespread occurrence of lack of fusion defects in the weld. Although it is claimed that both problems can now be overcome, the process has been very slow to recover from these initial setbacks. In particular disquiet has been expressed over the avoidance of lack of fusion defects and their reliable detection if present. Because filling runs are normally deposited using a vertical up technique and the weld pool freezes rapidly because of the dip metal transfer, there is a temptation to 'block weld' thereby promoting lack of fusion (Ref 10).

Although these difficulties, combined with a shortage of suitably trained operators, have limited its use in the UK the process has seen something of a revival recently in Europe (Ref 11). For this particular project weld radiographic examination was supplemented by manual ultrasonic examination in order to encourage freedom from lack of fusion defects.

The consumables used for semi-automatic GMAW of pipelines have not changed significantly since its early application when a C-Mn-Si + ½% Mo wire was used in conjunction with $CO_2$ shielding gas. Although very little development work has been carried out since those early trials, a recent review of commercially available filler wires (Ref 12) suggests that the ½% molybdenum wire gives the best combination of as-welded properties at intermediate heat input levels (Table 4). These trials were, however, carried out with Argon-20% $CO_2$ shielding gas and toughness properties may be somewhat better than those obtained with the 100% $CO_2$ gas shield normally used for field welding of pipelines.

TABLE 4 Toughness Properties for GMAW with ½% Mo Filler Wire (From Ref 12)

CONDITIONS: 1.2 mm dia wire, 280A, 28-30V, Heat Input 2KJ/mm
Shielding gas, Argon, +20% $CO_2$

CHARPY V: 27J Transition temperature $> -75^{\circ}C$
41J Transition temperature $-65^{\circ}C$

CTOD ($-10^{\circ}C$) 0.62, 0.55 mm

CHEMICAL ANALYSIS - WELD METAL

| ELEMENT | C | Si | Mn | S | P | Al | Nb | Ti | Ni | Cr | Mo | Cu | O ppm | N ppm |
|---|---|---|---|---|---|---|---|---|---|---|---|---|---|---|
| % WT | .13 | .31 | 1.59 | .010 | .015 | .006 | - | - | .18 | .09 | .43 | .27 | 440 | 102 |

## Flux Cored Arc Welding (FCAW)

Although gas-shielded flux cored wires suitable for girth welding are available, most recent interest has centred on the self-shielded types since they have an obvious advantage for site use. The present self-shielded wires for pipeline welding belong to the same family as those well established for offshore constructions, the major change being their use in the vertical down direction. This facility, combined with the relatively fierce arc, gives the wires a 'feel' not dissimilar to the cellulosic 'stovepipe' welding electrode. Weld deposition rates are high and the penetrating arc is claimed to reduce the lack of fusion problems associated with the bare wire process. Typical welding times for a 609 mm OD x 14 mm WT pipe are given in Table 5. Comparison with Table 2 (albeit for slightly thinner wall pipe) shows that root bead welding speeds are comparable with those of the cellulosic electrode, while overall welding times are likely to be quicker than both the LHVD and cellulosic electrodes.

TABLE 5 Welding Times for Vertical Down Self Shielded FCA Welding

| Run No | Run Type | Electrode Size mm | Welding Time Mins Secs | Cleaning Time Mins Secs |
|---|---|---|---|---|
| 1 | Root | 1.7 | 5.55 | 1.15 |
| 2 | Filler | 2.0 | 5.07 | 2.10 |
| 3 | Filler | 2.0 | 5.00 | 3.25 |
| 4 | Filler | 2.0 | 5.27 | 1.30 |
| 5 | Side Stripper | 2.0 | 2.45 | 1.06 |
| 6 | Cap | 2.0 | 8.40 | 1.35 |
| TOTAL TIMES: | | | 32.54 | 11.01 |

Electrodes - Root - E70T - G
Fill and Cap - E70T-G

Pipe Size - 609 mm OD x 14 mm WT (one welder used
- welding times are averaged half butt times)

Joint Preparation - $30^{\circ}$ bevel, 1.5 mm root face, 2.0 mm root gap

The mechanical properties of welds made using two E70T-G wires in combination, one for the root and one for the filling passes are given in Table 6. Considering Charpy V notch impact behaviour, upper shelf energy values are superior to those obtained in welds made with cellulosic electrodes but the ductile-brittle transition temperature is disappointingly high (Figure 1). The good upper shelf performance of these welds has been attributed to their low oxygen and sulphur levels (Ref 13). The poor transition behaviour, which is also reflected in a large scatter in CTOD results at $0^{\circ}$C, may be a result of the high proportion of coarse as-deposited microstructure in this particular weld (Fig 2). A previous weld, made using a stringer bead technique, gave considerably better toughness behaviour, indicating that the plain carbon-manganese wires may be very procedure sensitive.

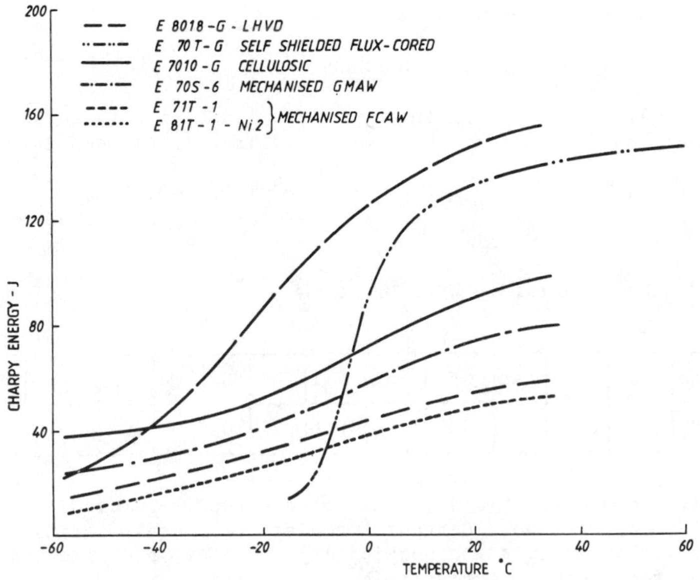


*FIG 1:* CHARPY V-NOTCH TRANSITION BEHAVIOUR FOR VARIOUS PIPELINE GIRTH WELDS

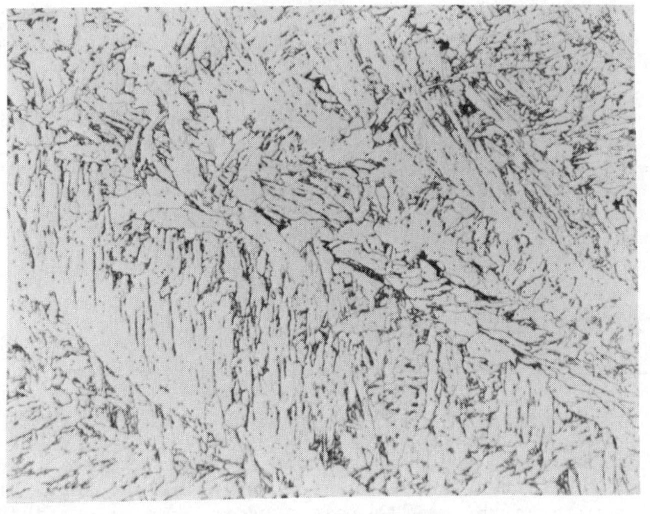

X500

Fig 2 Weld made with self shielded flux cored wire-as-deposited structure

TABLE 6 Properties of Girth Weld produced with Self-Shielded FCAW

(609 mm OD x 16 mm WT X52 Pipe)

TENSILE PROPERTIES

| | |
|---|---|
| 0.2% RP ($N/mm^2$) | 449-472 |
| RM ($N/mm^2$) | 592-596 |
| Elongation % | 29-30 |
| Reduction in Area % | 72-74 |

TOUGHNESS

| | |
|---|---|
| Charpy V | 27J Transition temperature -10°C |
| | 41J Transition temperature - 7°C |
| CTOD | (0°C) 0.04-0.92 mm |
| | (Mean of 6 tests 0.35 mm) |

CHEMICAL ANALYSIS - WELD METAL

| ELEMENT | C | Si | Mn | S | P | Al | V | Nb | Ti | Ni | Cr | Mo | Cu | O ppm | N ppm |
|---|---|---|---|---|---|---|---|---|---|---|---|---|---|---|---|
| % WT | 0.12 | .29 | .87 | .006 | .008 | .81 | .002 | .006 | .016 | .51 | .02 | .03 | .01 | 83 | 223 |

A current problem with some self shielded flux cored wires is related to the nature of the flux formulation. Current formulations contain barium compounds which, besides presenting the possibility of safety problems if present in welding fume, can give rise to barium containing slag in the weld. Since barium has a similar linear coefficient of absorption to steel for X-rays in the range 200 - 300 KeV this slag may not readily be detected using normal radiographic procedures (Ref 14). It may, therefore, be necessary to change to low energy X-rays or, alternatively, $\gamma$-radiography in order to ensure detection of slag inclusions. Both these techniques have disadvantages, low energy X-rays mean longer exposure times and conventional $\gamma$- radiography is inferior to X-radiography for detecting fine cracks.

The reluctance of UK contractors to adopt the process may be related to other considerations. Firstly special power sources are needed to supply the low operating voltages used with some self-shielded wires, in addition to the usual wire feed equipment. Also some difficulty has been found in achieving consistent root bead penetration around the pipe circumference and trials to date in the UK have been unsuccessful because of this problem.

## MECHANISED PIPEWELDING TECHNIQUES

### Gas Metal Arc Welding (GMAW)

British Gas has made extensive use of mechanised GMAW pipewelding systems both for land and offshore pipelines (Ref 15, 16). These systems give a low hydrogen weld deposit and overall welding speeds are normally at least 20% faster than those for stovepipe welding. It is well known that a major problem with this process is a susceptibility to lack of fusion defects at the sides of the narrow joint preparation. However, compared with semi-automatic GMAW, these defects are orientated to allow a greater chance of detection by conventional radiography.

Since the defects often fall outside the length limits imposed by traditional codes, in which acceptance criteria are based on good workmanship standards, repairs may be called for which are not strictly necessary on a 'fitness-for-purpose' basis. While the execution of such repairs may not be too much of a hardship on a land pipeline, costly delays may result when laying offshore pipelines, since repairs may entail halting the laybarge. As a result, British Gas has made use of the engineering critical assessment analysis permitted in British Standard BS 4515 to incorporate revised defect acceptance levels into its specification for mechanised girth welding of offshore pipelines. In order to use this approach, it is first necessary to demonstrate adequate weld toughness at the minimum design temperature, as measured by the CTOD test, to ensure that naturally occurring defects do not exceed the critical size for failure. For mechanised GMAW systems lack of sidewall fusion defects may be of considerable length but are normally limited to one weld run in depth (approximately 3 mm). Using this information, a minimum required CTOD level for a particular pipeline and laybarge configuration may be calculated, a level of about 0.15 mm being typical for current projects (Ref 17).

A review of CTOD test results obtained by British Gas, both from equipment evaluation and procedure qualification trials carried out using various mechanised welding systems, indicates that this level of minimum CTOD is not always readily achievable when using the combination of consumables (filler wires) and commercial equipment currently available. The most accessible method for improving weld metal toughness in mechanised GMA welding is through changes in filler wire type, since the possibility of varying welding parameters is limited due to the constraints imposed by the GMA process, coupled with the requirement to maximise production rates (i.e. number of butt welds per day).

In trials, with three commercial filler wires using a number of welding systems, a C-Mn-Si filler wire with small titanium additions was found to give the best overall performance in terms of deposition characteristics and CTOD toughness (see Figure 3b). A traditional double deoxidixed C-Mn-Si filler gave the poorest toughness (see Figure 3a), whilst a wire containing 1% Ni gave intermediate toughness (see Figure 3c) and inferior deposition characteristics than the other filler wires. The mediocre performance of the 1% Ni wire is somewhat surprising in view of previously reported results (Ref 18). However, the observed trend in toughness behaviour can be explained in terms of the weld metal microstructure developed. The weld microstructure produced using the C-Mn-Si wire contained a large proportion of grain boundary ferrite and ferrite with aligned MAC[1] (see Fig 4a), both microconstituents having low cleavage resistance due to their large effective grain size.

The C-Mn-Si-Ti filler wire gave a weld metal microstructure which, although containing a moderate amount of grain boundary ferrite, had a reduced proportion of ferrite with aligned MAC, this latter constituent having been replaced to a large degree by intragranular acicular ferrite (Figure 4b). The microstructure of this weld would be expected to give improved toughness, although the continued presence of grain boundary ferrite and some side plate structure indicates further room for improvement.

[1]MAC - Martensite, Austenite or Carbide

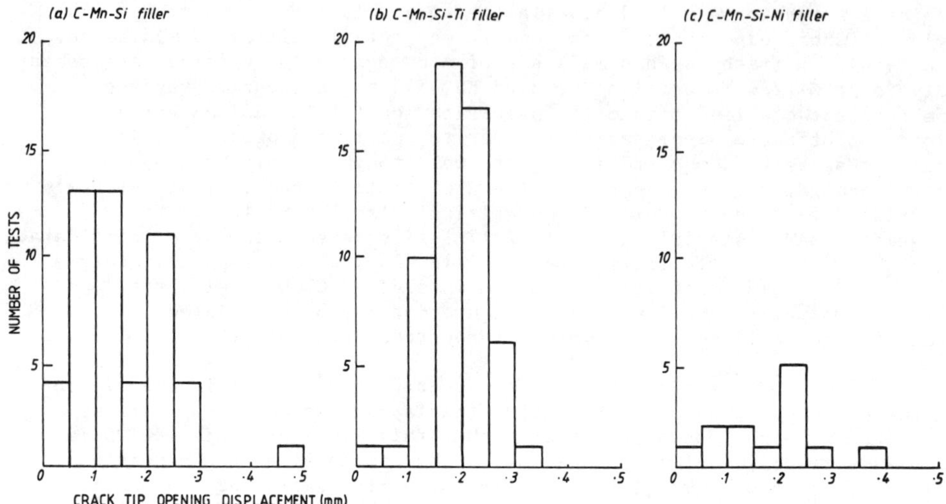


*FIG 3* : COMPARISON OF C T O D RESULTS FOR MECHANISED G M A WELDS MADE WITH THREE DIFFERENT FILLER WIRES

The C-Mn-Si-Ni wire deposit had a finer microstructure than that produced by the other two filler wires (Figure 4c). Although the amount of grain boundary ferrite and coarse ferrite with aligned MAC was seen to be significantly less than in the welds described above, the acicular ferrite was of a poor quality, i.e. not fully developed and interlocking. Instead the acicular ferrite transformation appears to have given way to the formation of lower temperature (bainitic) transformation products (see light etching areas of Fig 4c). Thus it is felt that whilst the toughness of the C-Mn-Si-Ti wire deposit would have benefitted from a slightly increased hardenability (to reduce the amount of grain boundary ferrite and ferrite with aligned MAC), in the case of the welds made using the C-Mn-Si-Ni filler wire, the hardenability of the weld metal was too high, resulting in a high yield strength deposit and consequently toughness below the optimum. This hypothesis is supported by the results of all weld tensile tests. The yield strength of the weld produced by the C-Mn-Si and C-Mn-Si-Ti wires were around 90 ksi whereas for the Ni bearing deposit a value in excess of 100 ksi was obtained.

The strength of all these deposits considerably overmatch existing grades of linepipe and are a consequence of the rapid cooling associated with the low heat input weld procedure ($\Delta$t 800 - 500$^{o}$C cooling times of 3 - 4 seconds being typical) and the high levels of silicon necessary in the filler wire for deoxidation purposes. It would be advantageous to 'trade-off' some of this high strength for improved toughness and this suggests that more subtle use of alloying additions such as nickel may be necessary than is seen in current wire formulations.

Limited trials have been carried out using a ½% molybdenum filler wire of the type referred to previously. Microstructural observations and high recorded tensile strengths suggest that this wire produces weld metal with too much hardenability at the very low heat inputs involved to possess good toughness. Similarly, trials with an experimental titanium-boron containing filler wire in an argon-5% oxygen shielding gas were disappointing, although in this case further trials are to be undertaken using pure argon shielding gas in order to aid recovery of the titanium and boron.

Variations in the level of residual elements are also believed to influence the toughness of mechanised GMA welds, and this is the subject of current investigations. Although this behaviour was not studied specifically in the work reported here, differing nitrogen levels in the three filler wires and corresponding weld deposits correlate reasonably well within the toughness ranking of the welds, indicating that some effect may be present (see Table 7).

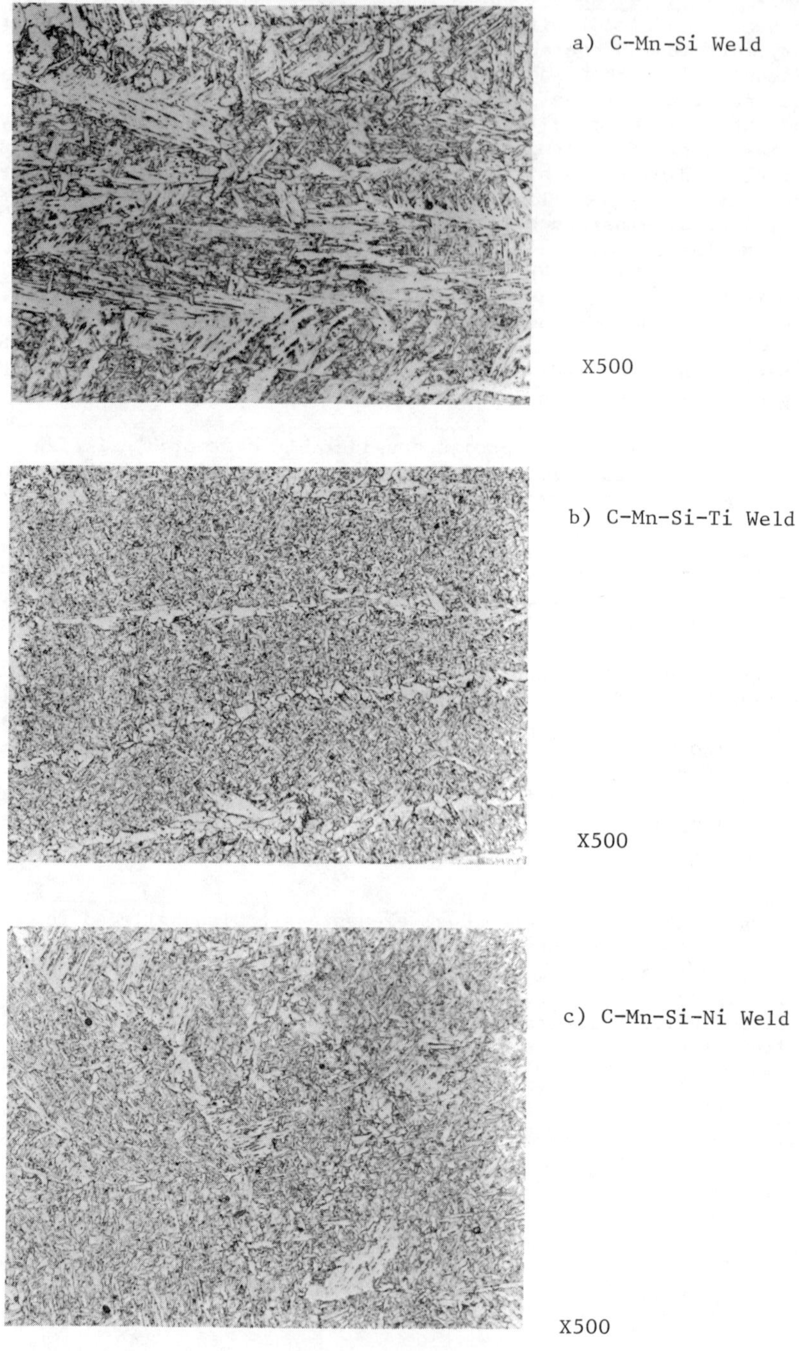


Fig 4 Microstructure of mechanised GMA welds (as deposited region of filling passes)

TABLE 7 Properties of Girth Welds Produced with Mechanised GMAW
(914 mm OD x 16 mm WT X65 Pipe)

WELD METAL TENSILE AND CHARPY PROPERTIES

| Filler Wire Type | 0.2% RP ($N/mm^2$) | RM ($N/mm^2$) | Elongation (%) | Reduction in Area (%) | Charpy V at $0^oC$ (J) |
|---|---|---|---|---|---|
| C-Mn-Si | 599-608 | 702-735 | 21-23 | 55-62 | 35 (ave) |
| C-Mn-Si-Ti | 615-680 | 719-748 | 23-24 | 64-65 | 69 (ave) |
| C-Mn-Si-Ni | 730-735 | 802-815 | 23 | 62-66 | 79 (ave) |

CHEMICAL ANALYSES - PARENT PIPE, FILLER WIRES AND WELD METALS

| ELEMENT | C | Si | Mn | S | Al | V | Nb | Ti | Ni | Cr | Mo | Cu | O ppm | N ppm |
|---|---|---|---|---|---|---|---|---|---|---|---|---|---|---|
| Pipe | .11 | .31 | 1.53 | .010 | .032 | .069 | .047 | .005 | .03 | .053 | .011 | .027 | 50 | 100 |
| C-Mn-Si Wire | .07 | .85 | 1.51 | .011 | .005 | .010 | .002 | .003 | .01 | .030 | .010 | .150 | 100 | 110 |
| C-Mn-Si Ti Wire | .09 | .76 | 1.48 | .013 | .004 | .010 | .002 | .059 | .02 | .040 | .010 | .170 | 150 | 90 |
| C-Mn-Si Ni Wire | .08 | .69 | 1.35 | .006 | .029 | .020 | .002 | .123 | 1.28 | .050 | .010 | .118 | 180 | 120 |
| C-Mn-Si Weld | .10 | .66 | 1.27 | .017 | .009 | .020 | .006 | .005 | .03 | .056 | .011 | .136 | 480 | 150 |
| C-Mn-Si Ti Weld | .09 | .57 | 1.30 | .014 | .009 | .018 | .005 | .011 | .04 | .060 | .012 | .151 | 380 | 90 |
| C-Mn-Si Ni Weld | .08 | .56 | 1.23 | .008 | .015 | .029 | .014 | .044 | 1.06 | .039 | .018 | .180 | 470 | 110 |

## Flux-Cored Arc Welding (FCAW)

A recent development has been the introduction of a mechanised welding system designed to operate with gas-shielded flux-cored wires. The system is currently used in the vertical-up direction and possible advantages for the system over existing GMAW systems are its suitability for welding thick wall pipe, a reduced tendency for lack of fusion defects, and reduced heat affected zone hardness levels as a result of the higher heat input. The equipment is not, at the present time, capable of making a root bead in an open root gap, and therefore trials have been undertaken using it for filling and capping runs only. Welds were made in 1050 mm diameter x 29 mm wall thickness X65 pipe using two flux-cored consumables, one a carbon manganese type of E71-T1 classification and the other a nickel bearing wire of E81-T1-Ni2 classification, both using argon-20% $CO_2$ shielding gas. The root run was deposited from inside the pipe using semi-automatic GMAW, and the joint filled from outside in approximately 11 passes. Both flux-cored wires produced acceptable weld profiles, although slag removal was extremely difficult with the E81-T1-Ni2 wire. Mechanical properties are given in Table 8, and Figure 1, and it is apparent that the toughness results are poor, even for the nickel bearing weld, when compared with mechanised solid wire GMA welds and previously published results for vertical down FCAW (Ref 12).

TABLE 8 Properties of Girth Weld Produced with Mechanised FCAW System (1050 mm dia x 29 mm WT X65 Pipe)

TENSILE PROPERTIES

| | 0.2% RP ($N/mm^2$) | RM ($N/mm^2$) | Elongation % | Reduction in Area (%) |
|---|---|---|---|---|
| E71-T1 weld | 659-665 | 735-738 | 23 | 64 |
| E81-T1-Ni 2 weld | 701-719 | 763-775 | 11-15 | 31-36 |

CHARPY AND CTOD PROPERTIES

| | Charpy V at $0^oC$ (J) | CTOD at $0^oC$ (mm) |
|---|---|---|
| E71-T1 weld | 55 (ave) | 0.1-0.16 (mean 0.13) |
| E81-Ti-Ni 2 weld | 69 (ave) | 0.06-0.08 (mean 0.07) |

CHEMICAL ANALYSES - PARENT PIPE AND WELD METALS

| ELEMENT | C | Si | Mn | S | P | Al | V | Nb | Ti | Ni | Cr | Mo | Cu | O ppm | N ppm |
|---|---|---|---|---|---|---|---|---|---|---|---|---|---|---|---|
| Pipe | .09 | .31 | 1.36 | .010 | .009 | .043 | .09 | .04 | .010 | .06 | .17 | .02 | .02 | 7 | 54 |
| E71-T1 | .09 | .75 | 1.83 | .019 | .012 | .005 | .03 | .02 | .035 | .006 | .04 | .01 | .04 | 831 | 39 |
| E81-Ti-Ni 2 | .08 | .51 | 1.19 | .019 | .017 | .005 | .04 | .02 | .033 | 2.57 | .05 | .01 | .03 | 748 | 175 |

## SUMMARY

The test results presented in this paper are from a series of trials conducted by British Gas, some experimental, some part of procedural tests for specific projects. As such, a range of pipe dimensions and strength levels have been used, although most welds have been produced in pipe manufactured from conventional Nb or Nb-V control rolled plate of a type commonly used in the British Gas transmission system. It is believed, therefore, that general comparisons between welding processes may be drawn despite individual differences in degree of dilution, number of weld passes etc. It is apparent from the results that the mechanical property requirement which is most likely to cause concern is that of notch toughness. While most processes are capable of meeting the moderate Charpy toughness levels specified for onshore natural gas pipelines in the UK, some processes may have difficulty meeting the more onerous requirements demanded for some offshore projects and for pipelines with design temperatures below $0^{o}C$. Increased use of the 'fitness-for-purpose' approach to weld defect acceptance has highlighted the need for good CTOD properties.

Although adequate toughness can generally be achieved using cellulosic electrodes, there is an increasing trend towards the use of low hydrogen processes to ensure freedom from hydrogen cracking. Of the low hydrogen processes available, manual welding with LHVD electrodes has given, in our experience, the best CTOD toughness properties. Semi-automatic welding with self-shielded carbon manganese flux-cored consumables appears to be procedure sensitive and acceptable levels of CTOD toughness cannot consistently be achieved. Both of these welding processes suffer from operational difficulties although these should be capable of solution given further attention.

Mechanised GMAW systems offer the best all round performance at the present time where the size of project justifies the additional logistical difficulties involved in their use. However, there is potential for improvement in the level of weld metal toughness produced by this process and work in this area is currently underway at the British Gas Engineering Research Station and other research establishments. Mechanised FCAW appears in need of considerably more development work.

The gas tungsten arc system has not been considered here, but the ability to weld in pure argon shielding gas should, in theory, enable production of clean, high quality welds. In addition separate control of heat input and deposition rate via hot or cold filler wire additions should aid positional weld pool control. Finally, increased use of newly developed solid state pulsed power sources will benefit the application of most arc processes, particularly the mechanised systems.

## CONCLUSIONS

Cellulosic electrodes are being used successfully at the X70 strength level, but our own experience shows care must be taken to avoid weld metal and HAZ hydrogen cracking, particularly if these electrodes are used to weld pipeline fittings.

Low hydrogen vertical-down electrodes give good mechanical properties and are likely to be preferred to semi-automatic gas shielded metal arc welding if a manually operated low hydrogen process is required. Further work, and a greater emphasis on welder training, is necessary to overcome operational difficulties with this, otherwise promising, technique.

The semi-automatic flux-cored arc welding process is capable of rapid weld speeds when used vertical down with self shielded wires. However, there is a problem in achieving consistent levels of toughness with present carbon-manganese filler wires.

Mechanised systems using gas shielded metal arc welding give good overall performance provided a fitness for purpose approach to defect acceptance is taken. Although the commonly applied CTOD level of 0.15 mm used in this approach can normally be met with existing consumables, improved consumables would allow a greater latitude for procedural variation on site.

## ACKNOWLEDGEMENTS

The authors would like to thank the British Gas Corporation for permission to publish this paper. Thanks are also due to the staff of the Mechanical Testing Section of the British Gas Engineering Research Station, particularly A Clyne, for carrying out the test work.

## REFERENCES

1. North, T H, Rothwell, A B, Glover, A G and Pick, R J. 1982
Weldability of high strength linepipe steels.
Welding Journal, 61(8), 243s-256s.

2. Marzoli, I and Scopesci, L. 1982
The use of API 5LX 70 steel in the TransTunisian Pipeline.
I.I.W Commission XI E document 3/82

3. Cotton, H C
Discussion session 1979. 2nd International Conference on Pipewelding. Welding Institute, London

4. Lumb, R F and Fearnehough, G D. 1975
Toward better standards for field welding of gas pipelines.
Welding Journal, 55(2), 62s-71s

5. Prosser, K and Cassie, B A. 1981.
Field welding and service experience with gas transmission pipelines.
Metals Society Conference 'Steels for linepipe and pipeline fittings' London

6. Harrison, J D. 1977
The Welding Institute studies the significance of Alyeska Pipeline defects.
Welding Institute Research Bulletin, April 1977, 93-95

7. Rumble, C B and Kalb, J G. 1982
Low-hydrogen vertical down pipeline welding and mechanical properties of electrodes.
Europipe '82 Conference. Basel, Switzerland.

8. Turner, J., Murphy Pipelines Ltd. 1983
Private communication.

9. Breeze, H and Chetcuti, J. 1969
Experiences with $CO_2$ welding of pipelines.
Proceedings of pipewelding conference. Welding Institute, London

10. Doherty, T. 1970
Practical aspects of manual $CO_2$ welding of pipe butt joints.
Metal Construction, August 1970, 323-326

11. Anon. 1980
Saipem sets new pipelaying record.
Pipeline Industry, July 1980, 39-42

12. George, M J, Still, J R, Terry, P. 1981
Gas metal arc welds for high toughness applications - microstructural and other factors.
Metal Construction 13(12), 730-737

13. Graville, B A. 1982
Flux cored wire welding.
Report PR-140-137 by Welding Institute of Canada for American Gas Association

14. Redmayne, I. 1979
Interpretation of weld radiographs.
British Journal of NDT, September 1979, 275-276

15. Cassie, B A and Avery, T B. 1981.
Laying a 42 inch diameter pipeline across the Firth of Forth.
Metals Society Conference 'Steels for Linepipe and Pipeline Fittings'

16. Anon. 1981.
Automatic welders speed Nacaps Scotland project.
Pipeline Industry 54(3), 27-29

17. Hopkins, P, Jones, D G and Fearnehough, G D. 1983
Defect tolerance in pipeline girth welds.
Symposium on behaviour of circumferential cracks in pressure vessels and piping ASME, Oregon, USA

18. Archer, G L, Hart, P H M and Stalker, A Q. 1977
An assessment of pipeline girth welds.
WRC Monograph. 'Welding Linepipe Steels', 147-175.

# THE TOUGHNESS OF MECHANIZED GAS METAL ARC WELDS

D.V. Dorling and A.B. Rothwell

*NOVA, AN ALBERTA CORPORATION*

ABSTRACT

The mechanized gas metal arc welding (GMAW) process has been used extensively by NOVA and its affiliate company, Foothills Pipe Lines, on recent large diameter pipeline projects. The welding consumables selected were required to meet certain minimum weld metal mechanical properties as well as Charpy V-notch impact values specified by the Companies. In addition, fracture toughness, assessed by the CTOD test, was measured as part of the welding procedure qualification; these data were required as input parameters for engineering critical assessment calculations. The paper describes an evaluation of available GMAW consumables: the relationship between weld metal chemistry, microstructure and toughness for low heat input GMA welds is discussed in detail.

KEY WORDS

Gas metal arc welding, mechanized pipe-welding, weld-metal toughness.

## INTRODUCTION

The mechanized gas metal arc welding (GMAW) process has been used extensively by NOVA and Foothills Pipe Lines on large diameter pipeline projects for the last three years. The initial selection of welding wires to be used with the CRC system revealed a number of commercially-available consumables capable of meeting the strength and Charpy V-notch impact values then required for the welding of the CSA Z245.2 Grade 483 pipe material (yield strength 483 MPa; impact energy 27 J at -5$^{\circ}$C). However, application of the engineering critical assessment methods approved by the Northern Pipeline Agency for use on the prebuild section of the Alaska Highway Gas Pipeline Project necessitated an increased impact energy requirement (40 J at -5$^{\circ}$C) and, more importantly, the crack tip opening displacement (CTOD) testing of welding procedure qualification coupons. The unexpectedly low CTOD toughness values obtained from a limited survey of commercially available consumables demonstrated the need for a more detailed investigation into the factors that affect toughness in these low heat input gas metal arc welds.

Although much has been done to define the often complex relationship between weld metal chemistry, microstructure and mechanical properties and, further, to control the microstructure in flux-shielded processes such as shielded metal arc welding (SMAW) and submerged-arc welding (SAW), there is little comparable data on gas metal arc welds. Widgery (1974) studied the effects of deoxidation elements on the toughness, as measured by the CTOD test, of $CO_2$-shielded, mild steel, bead-in-groove deposits. George (1981) conducted a survey of the toughness of multipass gas metal arc welds in 50 mm structural steel to BS 4360 50D for offshore applications. However, both investigations were directed at GMAW in the spray mode of metal transfer at relatively high heat inputs (1.7 - 2.0 kJ/mm), whereas, for pipeline construction, mechanized GMAW operates in the short circuiting mode at a calculated heat input of between 0.3 and 1.0 kJ/mm, depending on location in the joint.

For this reason, a detailed metallurgical evaluation of six commercially-available consumables, together with one manufacturer's experimental consumable, has been carried out. The effects of deoxidation/alloying elements on micro-structure and toughness were studied using a single mechanized welding procedure in a typical high-strength pipeline steel.

## MATERIALS AND EQUIPMENT

Three wires were evaluated using sections taken from procedure qualification testpieces produced for NOVA by CRC Automatic Welding in Houston. However, extensive CTOD testing on mechanized gas metal arc welds by the Welding Institute of Canada for the American Gas Association (Glover,1981) had revealed little effect of test location around the circumference and subsequent test welds were produced in the flat position, on pipe steel in plate form, and using a simple laboratory set-up to simulate the CRC mechanized welding procedure.

### Parent Material

Girth welds were prepared from a single length of 1067 mm O.D., 12 mm W.T., Gr. 483 pipe to CSA Standard Z245.2. Laboratory-scale test coupons were prepared from a single plate of 12 mm thick Gr. 483 MPa pipe material. The chemical compositions are given in Table 1.

Table 1 Chemical Compositions of Pipe and Plate

| | C | Mn | P | S | Si | Ni | Cr | Mo | Cu | Al | V | Nb |
|---|---|---|---|---|---|---|---|---|---|---|---|---|
| Pipe | .07 | 1.41 | .008 | <.005 | .28 | .02 | .20 | .11 | .01 | .031 | .040 | .041 |
| Plate | .08 | 1.42 | .010 | .005 | .28 | .02 | .22 | .12 | .02 | .041 | .040 | .042 |

### Welding Consumables

The 0.9 mm diameter, copper-coated GMAW wires were obtained from three manufacturers and included carbon-manganese, 1.2 % nickel, 1.2% nickel-titanium, 2.5% nickel, molybdenum-titanium-boron, manganese-molybdenum and manganese-molybdenum-titanium chemistries.

### Welding Equipment

1067 mm O.D. girth welds were completed using the CRC mechanized welding system, which consists of an automatic, internal welding machine for the root bead with mechanized, external welding units for the hot, fill and cap passes.

The 400 mm long by 330 mm wide test plates were welded using the equipment illustrated in Figure 1. A combination travel carriage and oscillator provided the range of travel speeds along with the mode, frequency and width of oscillation required to simulate the CRC welding process. A direct current, constant potential power supply and a constant wire feed system completed the arrangement.

Fig. 1 Welding equipment used for laboratory simulation of CRC welding process

## EXPERIMENTAL PROCEDURE

### Welding Procedure

All welds were completed using the procedure detailed in Table 2. The only exception was one weld with the manganese-molybdenum wire where 90%A/10%$CO_2$ was used as the shielding gas. Completed welds were radiographed to ensure conformance to Clause 5.2.11 of CSA Standard Z184-M1983 prior to test coupon extraction.

Table 2 Mechanized GMAW Procedure - Summary of Welding Parameters

| Pass | CTWD, mm | Gas flow rate $m^3/h$ | WFS m/min | Voltage | Current, A | Travel Speed, mm/min |
|---|---|---|---|---|---|---|
| 1(internal) | 6.4 | 2.25 | 10.7 | 19 | 200 | 760 |
| 2 | 8.9 | 2.85 | 12.7 | 25 | 260 | 1140 |
| 3 | 12.7 | 2.85 | 10.9 | 23.5 | 200 | 380 |
| 4 | 12.7 | 2.85 | 10.9 | 22.5 | 200 | 380 |
| 5 | 7.9 | 1.40 | 9.7 | 18.5 | 210 | 330 |

All passes made with 0.9 mm wire, DC reverse polarity
Passes 1 and 5 75A/25$CO_2$: all others 100% $CO_2$

### Mechanical Testing

Standard Charpy V-notch specimens were prepared with the notch positioned at the centre of the weld and the notch plane parallel to the weld direction and perpendicular to the plate surface. These were broken at various temperatures to obtain transition curves for each wire evaluated.

Two all-weld-metal tensile testpieces of 25 mm gauge length and 6.3 mm diameter were extracted and tested in accordance with ASTM A370, 1981.

Three t x 2t CTOD specimens were machined from each weld with the notch at the weld centreline and in the through-thickness direction. Three t x t CTOD specimens, surface notched through the root at the weld centreline, were extracted from selected welds. Testing was according to BS 5762 at a temperature of $-5^{\circ}C$.

Finally, a hardness survey was carried out on polished and etched transverse sections at a load of 0.5 kgf.

### Chemical Analysis

Broken all-weld-metal tensile specimens were used to supply weld metal for chemical analysis. Non-gaseous elements were analyzed spectrographically and gaseous elements by the high-temperature, inert gas fusion technique.

### Metallographic Examination

Transverse sections were prepared by mounting in bakelite, wet grinding on graded silicon carbide papers and polishing on rotating cloth pads, using 6 micron and 1 micron diamond dust pastes. The sections were then etched in 2% nital for identification of the nature and relative quantities of the microstructural features present.

## RESULTS

### Chemical Composition

Table 3 shows the results of all-weld-metal chemical analyses on the eight experimental welds. Levels of vanadium and niobium appear to indicate some variability in dilution level, which would not be anticipated, but recovery of these relatively easily-oxidised elements may be affected by other deoxidants present. All oxygen contents for the welds made with 100% $CO_2$ for fill passes are in the range 0.05 - 0.075%: the weld made with a shielding gas of much lower oxygen potential (90%A/10%$CO_2$) had an oxygen content of 0.03%.

Table 3 Chemical Composition of Experimental Weld Metals

| | C | Mn | Si | Ni | Cr | Mo | Cu | Ti | Al | V | Nb | B | O |
|---|---|---|---|---|---|---|---|---|---|---|---|---|---|
| C-Mn | .09 | 1.18 | .29 | .03 | .12 | .15 | .19 | <.005 | .046 | .007 | .007 | - | .050 |
| 1.2 Ni | .11 | 1.37 | .55 | 1.02 | .13 | .04 | .08 | <.005 | - | .016 | .015 | - | .063 |
| 1.2 Ni+Ti | .08 | 1.31 | .59 | 1.01 | .11 | .05 | .19 | .052 | - | .015 | .013 | - | .064 |
| 2.5 Ni | .08 | 1.18 | .47 | 1.60 | .16 | .06 | .15 | .009 | - | .018 | .018 | - | .074 |
| Mo-Ti-B | .10 | 1.56 | .15 | .05 | .09 | .47 | .22 | .15 | - | .006 | .007 | .0013 | .049 |
| Mn-Mo | .09 | 1.78 | .53 | .02 | .12 | .37 | .37 | <.005 | - | .007 | .011 | - | .048 |
| Mn-Mo+Ti | .08 | 1.78 | .51 | <.01 | .11 | .15 | .15 | .074 | .018 | .020 | .019 | - | ND |
| Mn-Mo (90A/10$CO_2$) | .11 | 1.84 | .54 | .05 | .09 | .41 | .30 | <.005 | - | .006 | .011 | - | .030 |

## Tensile Properties

Table 4 shows tensile test results for the experimental welds. With the exception of the molybdenum-titanium-boron wire, all deposits showed yield strengths in the range 610-800 MPa; these values are somewhat higher than would be considered ideal, from the point of view of fracture toughness, and result from the extremely low heat input associated with the mechanized GMAW process (0.8 kJ/mm for fill passes). The yield strength of the molybdenum-titanium-boron deposit is higher, at 960 MPa, indicating the effectiveness of the molybdenum-boron addition in increasing the hardenability of the weld metal.

Table 4 All Weld Metal Tensile Properties

| | YS MPa | UTS MPa | Elong. % | RA % |
|---|---|---|---|---|
| C-Mn | 613 | 680 | - | 54 |
| 1.2 Ni | 658 | 746 | - | - |
| 1.2 Ni + Ti | 691 | 773 | 25 | 53 |
| 2.5 Ni | 645 | 746 | 25 | 50 |
| Mo-Ti-B | 960 | 1003 | 18 | 42 |
| Mn-Mo | 735 | 843 | 25 | 60 |
| Mn-Mo + Ti | 619 | 745 | 24 | 62 |
| Mn-Mo (90A/10$CO_2$) | 801 | 858 | 23 | 50 |

## Microstructure

Fig. 2 shows representative microstructures, taken in the unreheated fill pass region, of the eight weld deposits; hardnesses measured in the same region are also shown.

The carbon-manganese weld shows relatively large amounts of grain-boundary ferrite and ferrite side-plates, together with acicular ferrite in the primary grain interior (Fig. 2a). The microstructure is consistent with the fact that this wire showed the lowest hardness and tensile properties of all eight deposits. The addition of 1.2% nickel led to a considerable reduction in the amount of ferrite side-plates; the majority of the microstructure consists of fairly fine acicular ferrite, with fine veins of grain-boundary ferrite (Fig. 2b). Little visible change was brought about by the addition of titanium to this composition (Fig. 2c), though side-plates were virtually eliminated; there was also a modest reduction in hardness.

An increase in nickel content to 2.5% produced a microstructure which appears less desirable than at 1.2% nickel; some side-plates are visible, and the grain-interior microstructure tends towards fine bainite (Fig. 2d); this is reflected in the somewhat higher hardness of this deposit.

An increase in manganese and molybdenum, relative to the carbon-manganese wire, resulted in the virtual elimination of grain-boundary and side-plate ferrite; the entire microstructure consists of very fine acicular ferrite (Fig. 2e). Again, however, this deposit appears to have been slightly over-alloyed, as indicated by the elevated hardness level. No discernible change in microstructure occurred when the shielding gas composition was changed to 90%A/10%$CO_2$ (Fig. 2f), but there was a further increase in hardness. The reduction of the molybdenum content to give 0.15% in the deposit (Table 3), together with the addition of titanium, led to a slight coarsening of the acicular ferrite and the reappearance of some grain-boundary ferrite (Fig. 2g), while the hardness fell some thirty points.

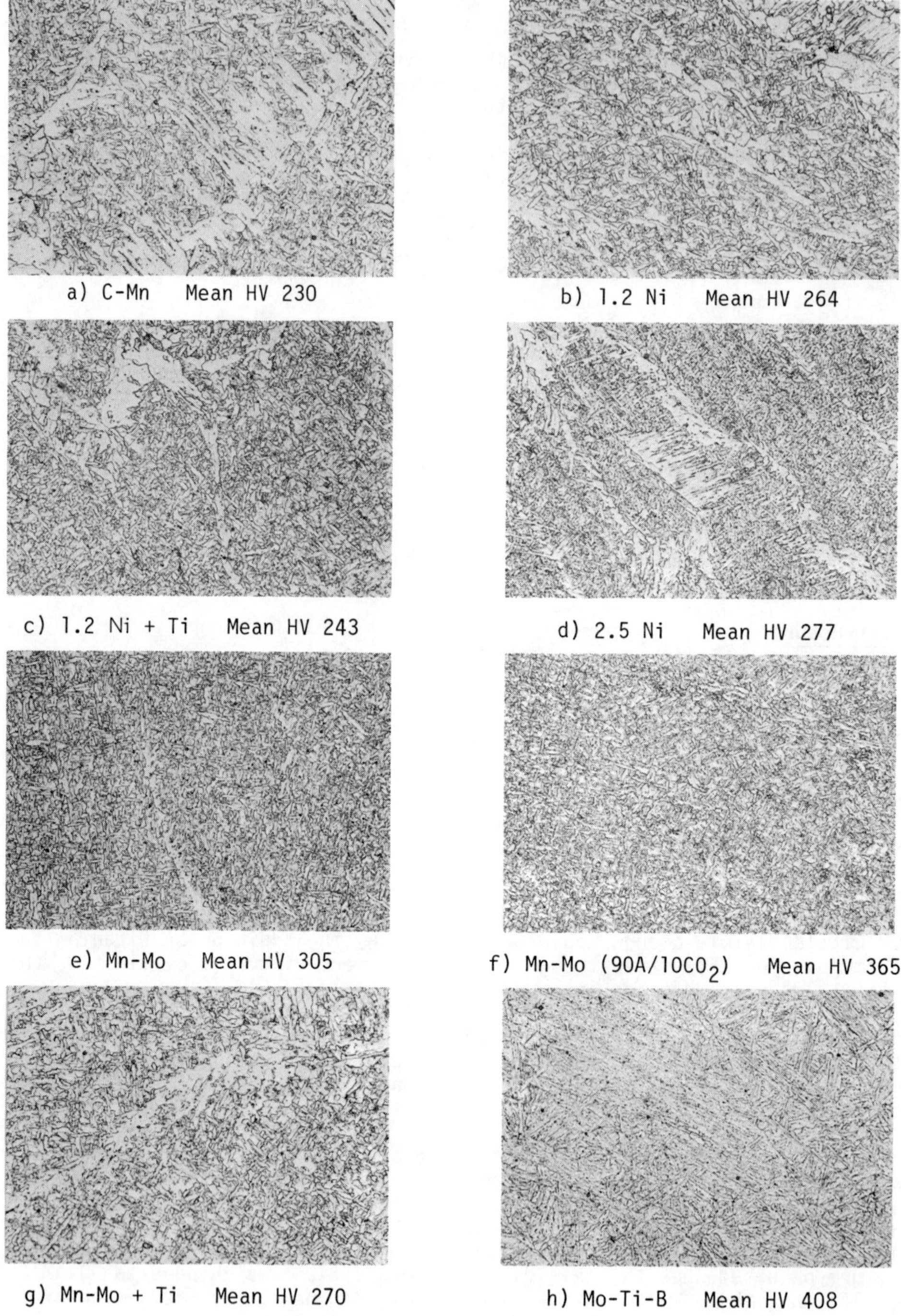


Fig. 2 Typical fill-pass microstructures of experimental welds (x400)

The molybdenum-titanium-boron wire gave a deposit consisting exclusively of martensite and bainite, (Fig. 2h) with a hardness of over 400 HV; the comparison of these results with those observed in other molybdenum-bearing deposits confirms the effectiveness of boron in increasing hardenability, despite the relatively high oxygen potential of the arc environment.

## Charpy Toughness

Fig. 3 shows impact transition curves for the eight experimental welds. The results for the carbon manganese wire are relatively poor, as regards both shelf energy and transition temperature, and company requirements for buried service (40 J at - 5°C) were barely met. All the alloyed wires, with the exception of the molybdenum-titanium-boron wire, gave superior results, the best performance being obtained with the titanium-alloyed wires; molybdenum-containing deposits generally had somewhat better toughness than nickel-containing deposits. The one weld made with 90%A/10%$CO_2$ shielding gas gave the best toughness of all.

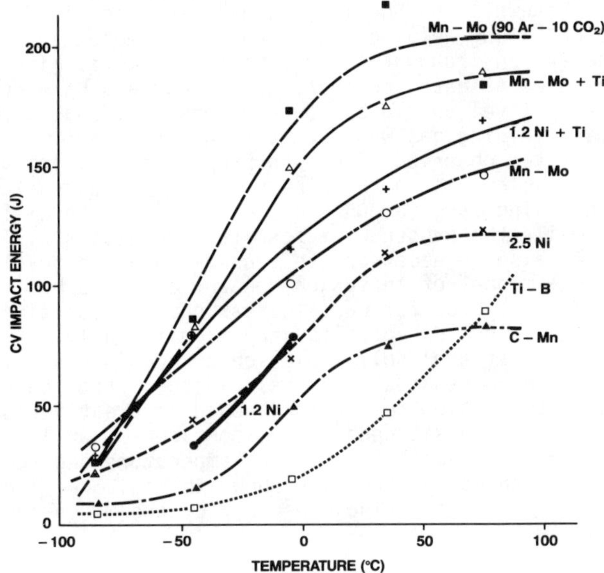


Fig. 3 Impact transition curves for experimental welds

## CTOD Testing

Table 5 shows the results of CTOD tests on the eight welds; in most cases, tests were carried out on both t x 2t testpieces, with a through-thickness notch, and on t x t testpieces notched from the inside surface. All testpieces failed in a stable, ductile tearing mode, with the exception of the 1.2% nickel and the molybdenum-titanium-boron welds. These failed unstably after initial stable crack extension, and CTOD values shown are $\delta_u$ . Results for t x t testpieces were lower than for t x 2t testpieces; this is expected when failure occurs by stable tearing, and results from the reduced ligament rather than from any local embrittlement. With minor exceptions, the CTOD results parallel the Charpy values presented above.

Table 5 CTOD Values at -5°C, mm

| | t x 2t | t x t |
|---|---|---|
| C-Mn | 0.13, 0.18, 0.21 | |
| 1.2 Ni | 0.05*, 0.05*, 0.12* | 0.04*, 0.07, 0.06* |
| 1.2 Ni + Ti | 0.22, 0.21, 0.22 | 0.16, 0.14 |
| 2.5 Ni | 0.16, 0.16, 0.19 | 0.13, 0.12, 0.12 |
| Mo-Ti-B | 0.08*, 0.06*, 0.09* | |
| Mn-Mo | 0.18, 0.21, 0.19 | 0.13, 0.13. 0.14 |
| Mn-Mo + Ti | 0.35, 0.37, 0.35 | 0.15, 0.14, 0.14 |
| Mn-Mo ($90A/10CO_2$) | 0.28, 0.35, 0.27 | 0.14, 0.11, 0.11 |

* $\delta_u$ - remainder $\delta_m$

## DISCUSSION

Factors affecting weld metal toughness have been reviewed in some detail in a recent publication (Rothwell and Dorling, 1983). In general, very complex interactions can occur among welding conditions, base metal and consumable chemistries and the arc environment to determine weld metal microstructure and properties. In the present case, however, heat-input and base composition are constant, the dilution level should be low and constant, and, with one exception, the same shielding gas was used throughout. Thus, it should be possible to interpret the observed toughness properties primarily in terms of the effect of alloying elements in the wire on basic microstructural components; microstructures containing large amounts of acicular ferrite will give the best overall toughness, while the presence of ferrite side plates will generally be detrimental. It will also be necessary to consider any changes in oxygen level, which is the main determinant of inclusion content and thus has a strong effect on Charpy shelf energy and CTOD for ductile tearing. In addition, though the total nitrogen content is not expected to vary significantly, it may be present either primarily in interstitial solid solution or combined as nitrides (in particular, in the titanium-bearing deposits); in the latter case, its deleterious effect on transition temperature would be greatly reduced. Finally, the strength level of the deposit needs to be considered; an increase in yield strength may adversely affect both transition temperature and resistance to ductile tearing. The changes in toughness shown in Fig. 3 will now be considered in the light of these principles, and of the microstructural variations shown in Fig. 2.

The low-hardenability, carbon-manganese wire gave relatively poor toughness properties, which is consistent with the large proportion of ferrite side-plates and the correspondingly restricted amount of fine, acicular ferrite. The addition of 1.2% nickel to the wire brought about some improvement in toughness, though additional testing showed that the shelf energy was only around 90 J. This improvement may be related to the decrease in side-plate ferrite; in addition, there may be some intrinsic effect of nickel in solid solution. An increase in nickel content to 2.5% (1.6% in the weld) did not produce any significant improvement in transition temperature, but the shelf energy did increase somewhat: in all probability, much of the beneficial effect of nickel was offset by the transition to a less-desirable, harder microstructure (Fig. 2d). The addition of titanium to the 1.2% nickel wire resulted in a very significant improvement in toughness, though there appeared to be only limited changes in microstructure (Fig. 2c). The modest reduction in hardness could account for some of this improvement, but the effect of titanium appears to be a complex one. Several authors have proposed that titanium compounds, either nitrides (Ito and Nakanishi, 1976) or oxides (Watanabe, 1980) may promote the nucleation of acicular ferrite. In addition, the formation of titanium nitride will improve toughness by the removal of nitrogen from solid solution (Bernard, 1978). The titanium compound formed will depend on the oxygen and nitrogen potential of the arc atmosphere and weld pool, and may also differ with welding conditions. The reduction in hardness related to the titanium addition is consistent with nitride formation, but it should be noted that there is also a small decrease in carbon content between these two welds.

An increase in manganese and molybdenum, relative to the carbon-manganese wire, resulted in a significant improvement in toughness. This change occurred despite a very considerable increase in hardness, and can be related to the establishment of a microstructure consisting predominantly of very fine acicular ferrite, with only a few, narrow bands of grain-boundary ferrite persisting (Fig. 2e). Any change in inclusion population brought about by a change to a less-oxidising gas, using the same manganese-molybdenum wire, was clearly not detrimental as far as ferrite nucleation was concerned (Fig. 2f), and was decidedly beneficial as regards shelf energy. This weld gave the best Charpy properties of all those examined; while the operating characteristics would not have been satisfactory for field use, as a result of the less-stable arcing conditions and consequent increase in spatter and frequency of fusion defects, these results do indicate the potential benefits which could be achieved by optimizing shielding gas composition. Current work on pulsed power sources, which will allow the spray mode of metal transfer to be extended to all-position welds, is expected to make an important contribution in this area.

The reduction in (deposit) molybdenum content to 0.15%, together with the addition of titanium, resulted in an improvement in toughness relative to that achieved with the manganese-molybdenum wire without titanium: this weld had the highest toughness of all those made under 100% $CO_2$. The microstructure, though consisting largely of acicular ferrite, does not appear as fine as those of the higher molybdenum welds discussed previously, and somewhat more grain-boundary ferrite is present (Fig. 2g). The deposit is considerably softer, however, and this, together with any effect of titanium on interstitial nitrogen content, as discussed above, must account for the observed improvement in toughness.

The molybdenum-titanium-boron weld, at the low heat input characteristic of the mechanized GMAW process, developed an excessively hard microstructure, which resulted in a very poor impact transition temperature. A similar effect has been noted with these wires, which were designed for high heat inputs, even in submerged-arc welds at ~2.0 kJ/mm (Rothwell and Dorling, 1983).

There is a relatively good correlation, in most cases, between the CTOD results (Table 5) and the Charpy results discussed above. It may be more appropriate to compare CTOD results at -5°C with Charpy shelf energy, rather than energy at the same test temperature since, when the testpiece thickness is little more than 10 mm, the CTOD transition should shift to significantly lower temperatures, in materials with normal strain-rate sensitivity (Dolby, 1981). On this basis, few anomalies exist; the low CTOD value of the molybdenum-titanium-boron weld may be associated with its high yield strength, which will lead to lower strain-rate sensitivity and a much more limited transition shift from Charpy to CTOD. In addition, even at full ductile tearing, a lower CTOD will be observed for a given Charpy energy since, under these conditions, energy absorption is expected to be proportional to the product of flow stress and CTOD (Rice, 1968). Thus, it is not surprising that the manganese-molybdenum-titanium weld, with a yield strength of 619 MPa, shows a higher CTOD than the low-oxygen manganese-molybdenum weld, with a yield strength of 801 MPa. Similarly, the high value of CTOD observed in the carbon-manganese deposit, relative to its Charpy shelf energy, can probably be partly accounted for by its low yield strength. The only result which appears to be anomalous is that for the 1.2% nickel wire; there is no ready explanation for the unstable fracture, and associated low CTOD values, in this weld.

## CONCLUSIONS

The toughness of welds deposited by the mechanized GMAW process, using wires of a wide variety of chemical compositions, has been assessed by both Charpy and CTOD testing, and related to the predominant microstructural features in the fill pass region. The following conclusions can be drawn.

(i) Excellent toughness properties can be achieved using both manganese-molybdenum- and nickel-alloyed systems.

(ii) Titanium additions caused a significant improvement in toughness in both alloy systems; while this cannot be explained in detail, it may be accounted for by the ferrite nucleation effect of titanium nitride or titanium-containing oxides and also, possibly, by a reduction in interstitial nitrogen content.

(iii) With the exception of the effect of titanium, changes in toughness were well accounted for, on a qualitative basis, by the microstructural observations.

(iv) Molybdenum-titanium-boron wires designed for high heat-input welding gave poor toughness, as a result of the excessive hardness of the deposit.

(v) A shielding gas of low oxygen potential, while unacceptable from an arc stability point of view, gave indications of the excellent properties which may be achieved by optimizing the gas composition. In the future, the use of pulsed power-sources may permit much greater flexibility in the choice of shielding gas.

(vi) CTOD values, with limited exceptions, showed a good correlation with Charpy shelf energy. When yield strength variations were taken into consideration, only one anomalous result persisted.

## REFERENCES

Bernard, G., F. Faure and P. Maitrepierre, 1978. Properties of welds in submerged arc welding. In A. B. Rothwell and J. M. Gray (Ed.), Welding of HSLA (Microalloyed) Structural Steels. ASM, Metals Park, OH. pp. 187-211.

Dolby, R. E. (1981). Charpy-V and COD. Met. Cons., 13, 43-51.

George, M. J., J. R. Still and P. Terry (1981). Gas metal arc welds for high toughness applications. Met. Cons., 13, 730-737.

Glover, A. G. (1981). Fracture testing of pipe line girth welds. Publication No. L51409, Pipeline Research Committee of the American Gas Association, Arlington, VA.

Ito, Y. and M. Nakanishi (1976). Study of Charpy impact properties of weld metal. Sumitomo Search, 15, 42-51.

Rice, J. R. (1968). A path-independent integral and the approximate analysis of strain concentration by notches and cracks. J. Appl. Mech, Series E, Trans ASME 35(2), 379-386.

Rothwell, A. B. and D. V. Dorling (1983), The toughness properties of girth welds in modern pipeline steels. ASM International Conference on Technology and Applications of HSLA Steels. (In press).

Watanabe, I., M. Suzuki and T. Kojima (1980). Applicability of large-current MIG welding technique. Int. Conf. on Welding Research in the 1980s, Osaka, Japan. Paper B8.

Widgery, D. J. (1974). Deoxidation practice and toughness of mild steel weld metal. Weld. Res. Intl., 4, 54-80.

# THE SELECTION OF WELDING CONSUMABLES AND THE DEVELOPMENT OF WELDING PROCEDURES FOR STEELS FOR OFFSHORE CONSTRUCTIONS

G.W. Dawson B.Sc., C. Eng., F. Weld 1

*Senior Welding Engineer*
*Murex Welding Products*
*Waltham Cross, Hertfordshire, England*

## ABSTRACT

This paper considers some of the important factors affecting as welded joint toughness. Consumable selection in terms of alloy and coating type is discussed. The welding procedural variables covered are essentially fill procedure and root treatment, including choice of preparation, effect of restraint and back-gouging. The relationship between moisture and weld metal hydrogen is also considered. Typical weld procedures used by North Sea fabricators are illustrated.

## KEY WORDS

Toughness, charpy, COD, weld procedure, restraint, configuration, backgouging, current selection, hydrogen.

## INTRODUCTION

Offshore platforms operating in the North Sea are designed to withstand some of the worst weather conditions encountered anywhere in the world. Wave heights of 100 ft., gusts of winds to 150 mph and conditions below freezing demand the welded joints in the critical areas of these huge very complex structures to have the maximum integrity, freedom from defects and in particular enhanced notch ductility.

The aim of the paper is to provide guidelines on the selection and use of consumables and welding procedures to maximise the performance of welded joints particularly in regard to notch toughness. The results and comments are based on a long history of liaison with fabricators of large North Sea offshore structures backed by test data obtained from the laboratories of BOC Murex.

## CONSUMABLES SELECTION

### Manual Electrodes

Composition and coating. The early 1970's saw C-1%Mn electrodes such as Fortrex 35A, which had an excellent track record for achieving tough welded joints in pressure vessels and nuclear reactors, a natural choice for the offshore fabricator.

However, toughness performance, while acceptable when assessed by charpy testing down to -20°C, proved inadequate in the as-welded condition when evaluated by full through thickness COD testing at -10°C. Fabricators had to resort to costly stress relieving of critical joints. Even the performance of 2.5Ni (E8018-C1) electrodes used by many fabricators in the mid '70s to avoid stress relief never gave entirely satisfactory COD performance when evaluated under the BS5762 requirements.

The subsequent years have seen a remarkable advance in consumable design, both as regards welding performance and toughness properties. Specification requirements have also progressively increased in severity particularly regarding weld toughness. They now usually call for as-welded cap and root weld toughness at -40°C and at intervals into the HAZ. A minimum COD performance is also specified for both weld and HAZ. These requirements can now be satisfied by C-Mn electrodes of the E7016/E7018 type such as Ferex 7016. This has been achieved by concentrating development effort on high basicity/high gas shielded coatings, innovations in deoxidation balance, and the selection of low-impurity core wire and coating constituents.

To promote the formation of a high proportion of tough acicular ferrite in the structure of the as-deposited areas of the weld, this electrode is designed to deposit an undiluted manganese level of about 1.6%. This means that electrodes of this type do not strictly conform to the upper manganese limits of AWS A5.1 - 81.

To equip the electrode such as Ferex 7016 with a high proportion of basic type materials the coating has been restricted to an EXX16 type. Additionally the coating is kept thin to provide easier access for narrow "difficult to get at" joint configurations.

Nevertheless the present generation of C-Mn electrodes still require strict attention to weld procedure control if the required properties are to be obtained in the weld. Figure 1 illustrates this.

A restriction on the maximum electrode size may also be placed on the fabricator for positional welding. This is commonly 4.0mm and is a disadvantage economically since this position is often used to approve other positions where a larger size would give acceptable results.

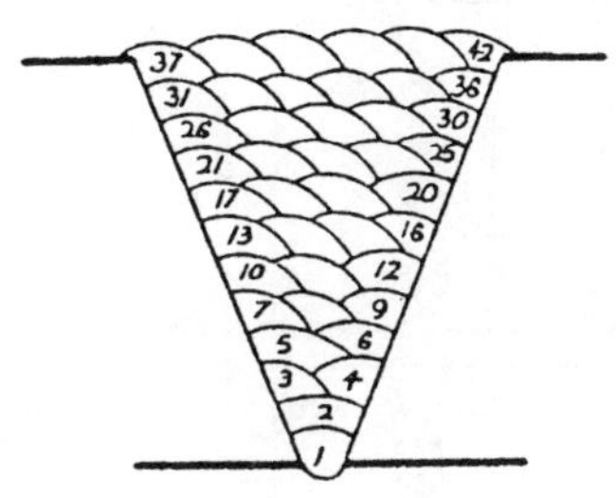


Material: BS 4360 : 50D
Thickness: 50mm
Position: ASME IX 3G
Preheat: 150°C min.
Interpass: 250°C max.
Runs:

1-4 3.25mm at 90-110 amps
5-42 4.0mm at 140 amps
Run out length: 180:200mm
Heat Input: 1.8-2.1 kJ/mm

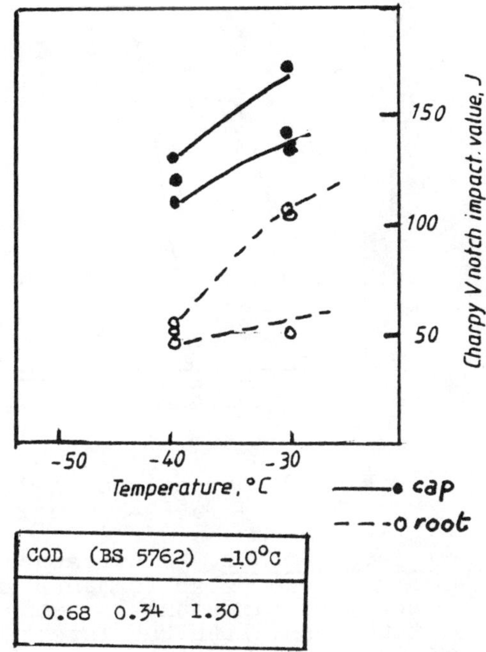


| COD (BS 5762) -10°C | | |
|---|---|---|
| 0.68 | 0.34 | 1.30 |

Fig. 1 Typical V joint COD weld procedure

The advances in coating formulation have however also provided the basis for nickel alloying to provide better, more consistent levels of charpy and COD toughness at the limits of C-Mn systems. This is demonstrated in Fig. 2 which compares COD transition and charpy data for Ferex 7016 and Hi-trex 7016-C1L, its 2.5Ni counterpart.

2.5 Ni electrodes of this type also offer the fabricator the attraction of using larger sizes for positional welds while maintaining good toughness.

Additionally by careful adjustment of manganese and carbon, 2.5Ni consumables can be designed to produce lower levels of tensile strength to reduce the risk of S.C.C. and micro-fissuring. The lower strength also achieves additional improvements in toughness over conventional E8018-C1 types.

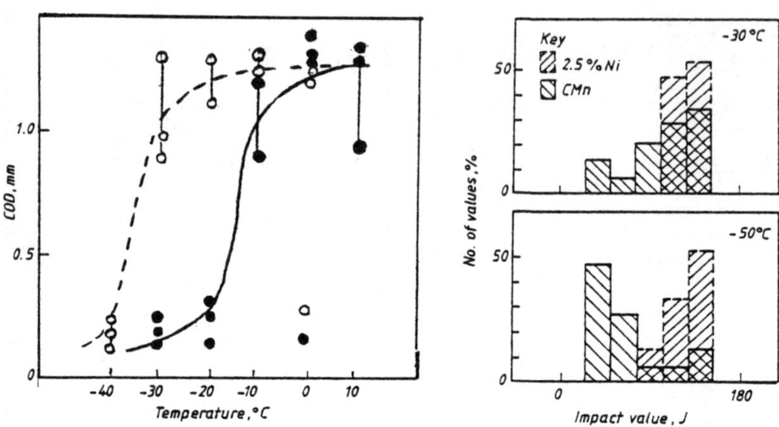


Fig. 2 Comparison of COD and charpy performance for C-1.6Mn and 2.5Ni welds

Recent developments. A collaborative research programme with British Research Establishments and Universities has undertaken a further more fundamental examination of factors affecting weld metal toughness, with the intention of offering the fabricator the opportunity of achieving more consistent weld metal performance at higher deposition rates.

This work, initial aspects of which have been reported by Judson and McKeown (1) identified inclusion size rather than type as exerting a major influence on promoting cleavage fracture in the weaker pro-eutectoid ferrite of the as-deposited weld metal structure. The findings have now been incorporated into the development of a new consumable, Ferex STF. This electrode uses a C-1.6%Mn system similar in composition to Ferex 7016 but further advances in formulation have ensured that the inclusion size is within the size limits identified to enable the weld metal to develop a higher upper shelf energy and delayed onset of cleavage fracture.

The potential of this electrode is demonstrated by Fig. 3 which shows the test results achieved for a V joint welded and tested in the U.K. by the Welding Institute.

These results illustrate that although two consumables might have a similar weld metal composition this does not necessarily mean they will give the same properties. When making an initial selection of potential candidates, detailed back-up data should therefore be obtained from the manufacturer including results on actual joints.

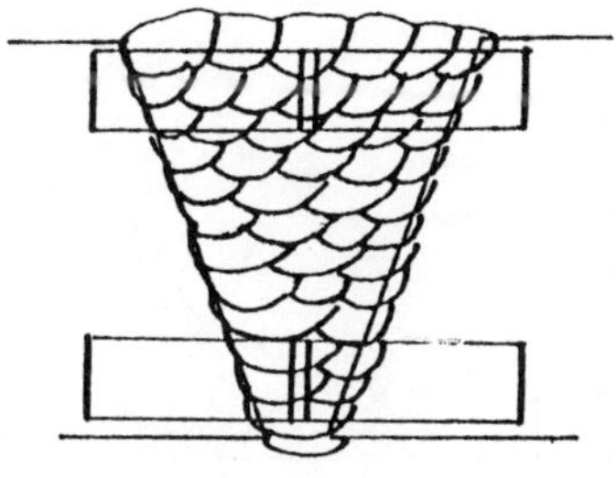

Material: BS4360:50D
Thickness: 50mm
Position: ASME 1X - 3G
Preheat: none
Interpass: 250°C max.
Runs:

1-4 3.25 at 65-105 amps
5-fill 4.0 at 135 amps
run-out length: 100-180mm
heat input: 3.1 kJ/mm

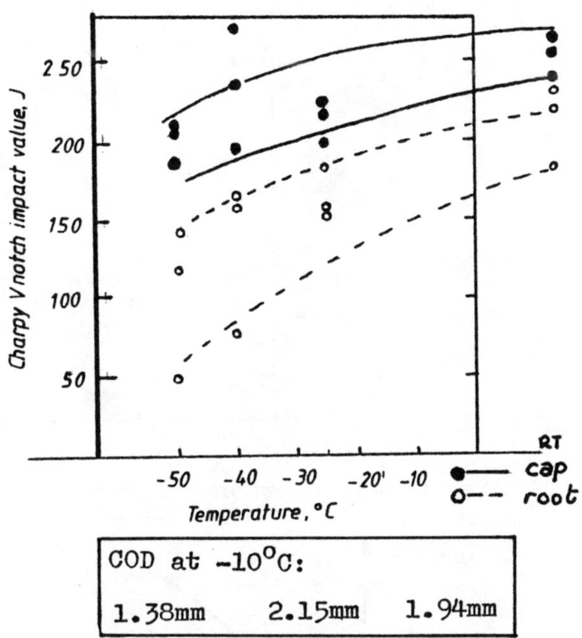


Fig. 3 COD and charpy weld procedure results for Ferex STF

Moisture and hydrogen. Moisture in the coating of a basic low hydrogen electrode is the greatest potential source of hydrogen in the deposited weld metal. It may be chemically combined with the coating constituents, in particular silicates used as binding agents. Additionally, depending on the hygroscopic nature of the coating, moisture may be physically absorbed from the atmosphere after baking. Absorbed moisture is more loosely bound and therefore generally more easily removed by baking, but it can become chemically associated.

Work in the U.K. (2) has shown there is no relationship between moisture content and hydrogen which can be applied for all electrodes in a generalised manner as currently provided in AWS 5.1-81. This is demonstrated by Fig. 4 which shows coating moisture contents and corresponding diffusible hydrogen levels for a number of different electrodes. The particular correlation must be established for a given electrode, Fig. 5 before moisture can be used as an albeit conservative, control test for diffusible hydrogen.

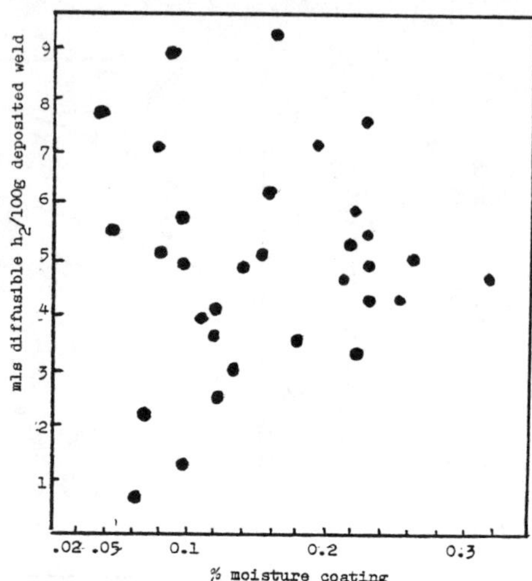


Fig. 4 Correlation of moisture and diffusible hydrogen

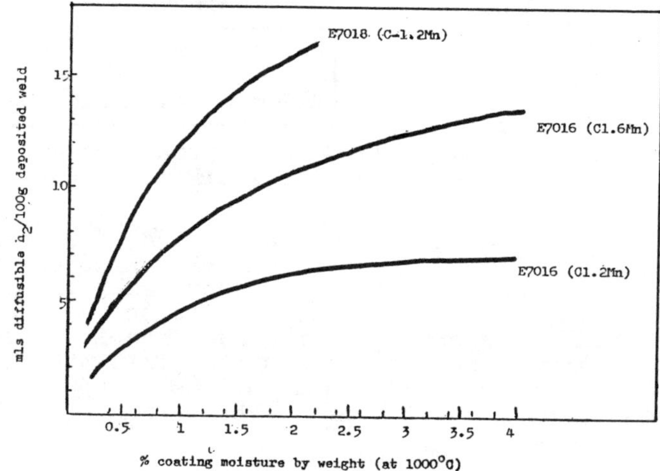


Fig. 5 Moisture and diffusible hydrogen for specific consumables

While basic low hydrogen electrodes must be stored in a dry place preferably at least 2°C above external air temperature and an RH of 0-60% storage is not the most important factor. It is the ease with which a particular electrode gives up its moisture on subsequent baking and the ease of conversion of moisture to weld metal hydrogen that is critical. Table 1 shows data for Ferex 7016 electrodes after exposure to humid conditions followed by baking at various temperatures.

| | Condition | mls $H_2$/100g. weld Ferex 7016 (C-1.6Mn) |
|---|---|---|
| | As manufactured | 3.56 |
| a) | 8 days at 25°C 90% RH | 41.18 |
| b) | As a) re-dry 250°C 1 hr | 6.43 |
| c) | As a) re-dry 300°C 2 hrs | 4.71 |
| d) | As a) re-dry 350°C 2 hrs | 4.33 |
| e) | As a) re-dry 400°C 1 hr | 2.93 |

Table 1 Effect of Baking after atmospheric exposure

Work by the author's company and others (3) has highlighted the dangers of assuming that the electrode temperature is the same as the oven recordings. The time lag between the oven indicating temperature and the electrodes in the middle of a closely packed stack can be 8 hours.

A baking temperature of up to 500°C does not affect the chemistry of the coating but can cause coating fragility and for this reason the number of re-bakes is often limited to three.

Submerged arc and semi automatic consumables

Advances in submerged arc consumable development has followed a broadly similar route to manual consumables. For example, fluxes used initially were of the acidic or semi-basic type and deposited weld metal high in oxygen. While adequate charpy toughness was achieved to -20°C, acceptable COD performance was confined to the stress relieved condition.

The development of all mineral basic fluxes usually of the agglomerated type, which give lower oxygen levels, have achieved significant improvements in toughness properties in the as-welded condition and consumables such as Satinarc BX400/Bostrand WB3, (a 1.5% Mn wire) can readily achieve root and cap charpy performance at -40°C and COD properties at -10°C in multi run welds. Subsequent development has concentrated on maintaining toughness properties at higher deposition rates. Multi power fluxes such as Satinarc BX400 using 2% Mn wires which operate at heat inputs of 5kJ/mm with deposition rates of 16Kg/hour are now available. Data contrasting single and multi-power techniques are illustrated in Fig. 6.

The addition of iron-powder with a similar composition to the filler wire during welding is employed by some fabricators to obtain further deposition improvements without exceeding the 5kJ/mm maximum heat input limit specified by offshore fabrication codes.

Solid wire Mig consumables were quickly discarded because of an inability to give acceptable toughness and the prevalence for lack of fusion defects in positional welds. However gas shielded cored wires have a greater potential. It is possible to adjust alloying to achieve a more desirable manganese level and manganese silicon ratio than with currently available solid Mig wires. Cored wires with nickel additions designed for use with argon rich shielding gas to combine excellent deposition rates

Single wire weld

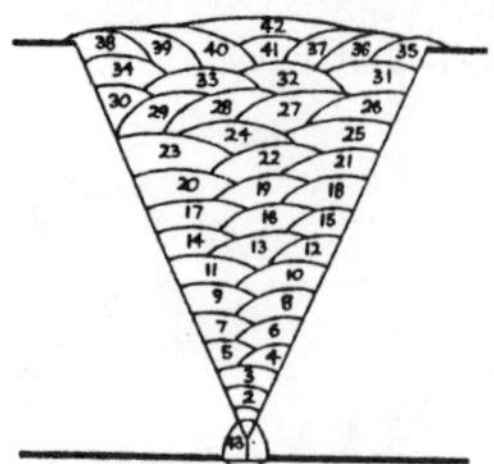

| | |
|---|---|
| Material: | BS 4360:50D |
| Thickness: | 63mm |
| Position: | ASME IX 1G |
| Treatment: | Preheat 120°C min. |
| | Interpass 200°C max. |
| Polarity: | DC positive |

Procedure:
Run:
1 MMA or Mig
2–5 4.0mm at 450–550 amps 27 volts
450mm/min.
(2.1 kJ/mm)
6–42 4.0mm at 600–650 amps 32 volts
400–450 mm/min.
(2.4–3.1 kJ/mm)
43 4.0mm at 700 amps 33 volts
400 mm/min.
(3.5 kJ/mm)

Charpy V notch impact value, J

Temperature, °C

| COD | (BS 5762) | -10°C | |
|---|---|---|---|
| mm | 0.35 | >2.0 | 0.70 |

Multi power weld

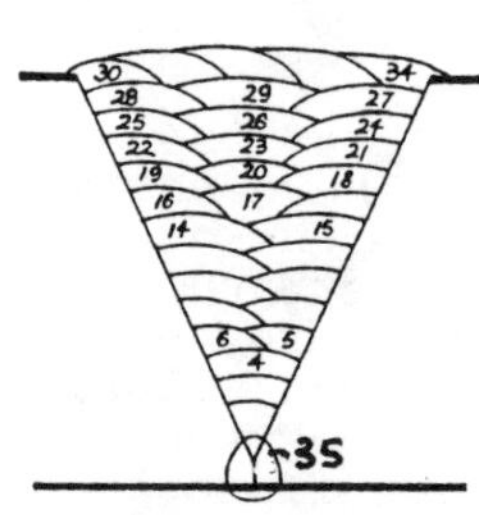

| | |
|---|---|
| Material: | BS 4360:50D |
| Thickness: | 63mm |
| Position: | ASME 1X 1G |
| Preheat: | 120°C min. |
| Interpass: | 250°C max. |
| Polarity: | DC lead |
| | AC trail |

Run:
1 MMA or Mig
2–3 4.0mm at 500–600 amps 27–30 volts
450–500 mm/min.
4–34 4.0mm at Lead 600 amps 32 volts
Trail 650 amps 39 volts
600mm/min.
(4.5kJ/mm)
35 4.0mm at 700 amps 33 volts
400 mm/min.
(3.5kJ/mm)

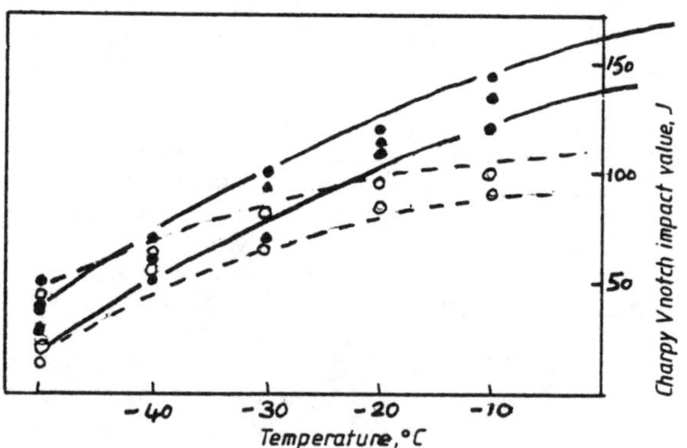


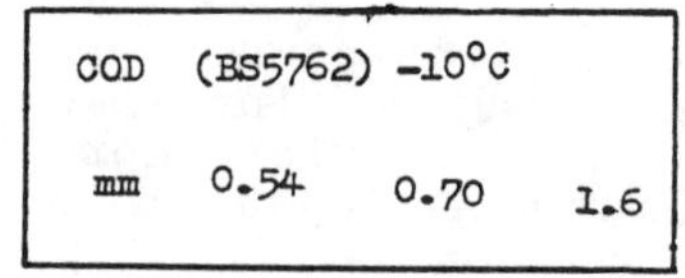

| COD | (BS5762) | -10°C | |
|---|---|---|---|
| mm | 0.54 | 0.70 | 1.6 |

Fig. 6 Single and Multi-power submerged arc procedures

with good as welded COD and charpy toughness properties but with comparable fume generation to solid wire Mig wire welding are available. The smaller 1.2mm size has reawakened the possibility of all positional welding.

## WELD PROCEDURE VARIABLES

### Fill Sequence

The most important factors in a fill procedure which influence joint toughness are the thickness of the individual bead and the consistency of the sequence.

Refined weld metal is generally the most tough and as cast material the least tough. Thin beads are more likely to cool faster and therefore develop a greater proportion of tougher acicular type ferrite. They are also more likely to be refined to a greater depth by the subsequent weld beads.

Toughness has very little to do with calculated heat input alone, which takes into account only travel speed and not weave width. This is illustrated by Fig. 7 showing the toughness performance for several different procedures. While the wide weave procedures are deposited at a very high calculated heat input, the individual layers are thin and the toughness correspondingly high. The danger with this procedure is the possibility of introducing a thicker bead into an otherwise predominately refined area. This could act as an initiation point for early fracture.

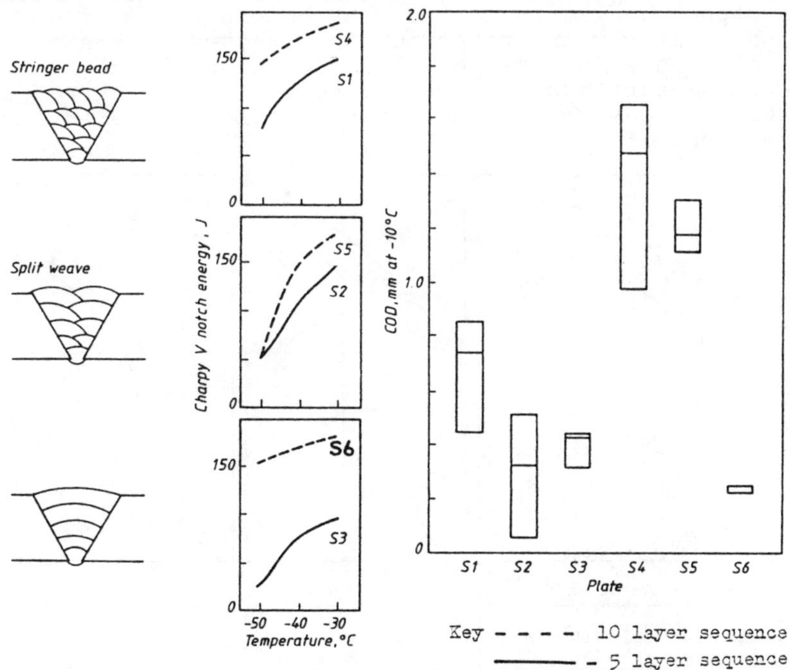


Fig. 7 Effect of bead sequence on toughness

Figure 8 indicates that sequences such as split weaving, chosen to achieve the greatest proportion of refined weld metal at the test notch do not necessarily give high toughness elsewhere. The stringer bead sequence, providing it deposits sufficiently thin beads is considered the best for overall consistency in performance.

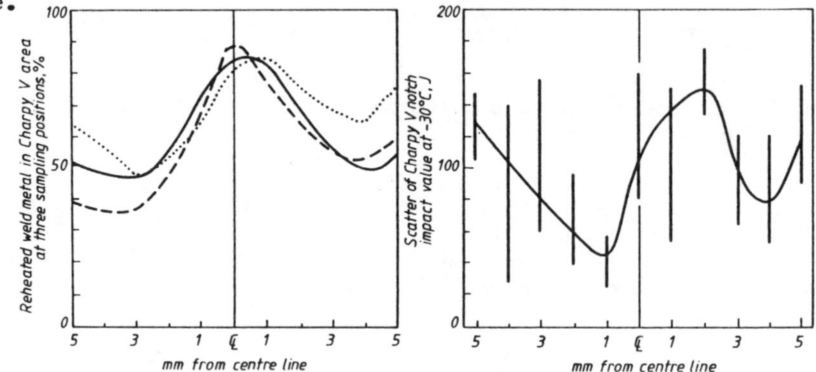


Fig. 8 Variation of charpy values with location of refined weld metal

Effect of current. Welding in the vertical position has generally been acknowledged as the most stringent and most specifications accept that this position qualifies for joints made in other positions. However joints made in the downhand position often thought to be the least onerous may give very poor results particularly in production test plates. This may be because welds made in this position are deposited at significantly higher welding currents than recommended. As demonstrated in Fig. 9 this can have a very detrimental effect on alloy recovery and properties.

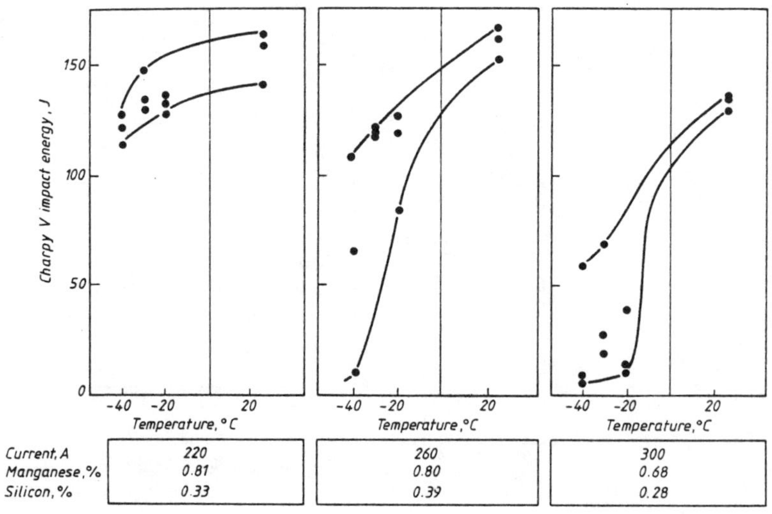


Fig. 9 Effect of welding current on charpy toughness

## Root Treatment

The root of the weld is usually the least tough. In addition to the adverse effect of dilution from the parent material, this area is subjected to strain aging which, as demonstrated by Berkhout and Van der Brink (4) has a disastrous effect on properties. The options open to the fabricator in minimising the adverse effect of the root are; joint configuration, choice of electrode size, restraint and backgouging.

Choice of joint configuration. As suggested by John (5) having the root in the centre of the joint as in a double V preparation, corresponds with the position of maximum triaxiality of stress. This is the worst place for achieving satisfactory COD performance.

This can be avoided by using single V preparations which as demonstrated by Fig. 10 generally give higher levels of COD properties, but not necessarily root charpy levels. A school of thought suggests that it is more serious to have a poor toughness area on the surface of the joint where fatigue cracks are more likely rather than surround it with tougher material in the centre of the weld.

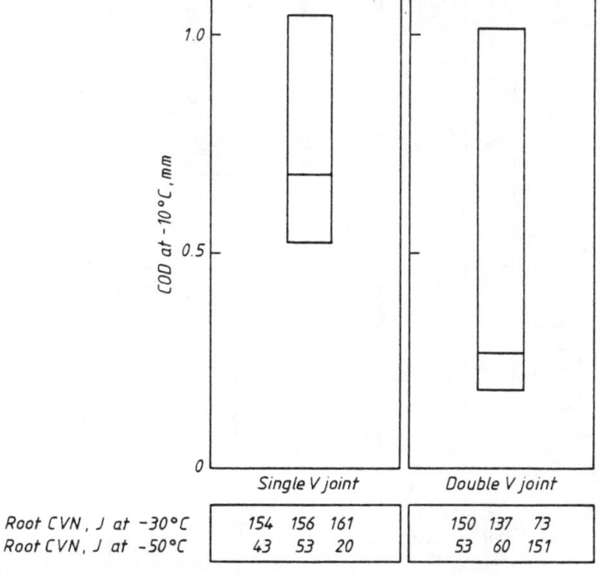


Fig. 10 Single and double V joint toughness

Attention to the initial layers in the root area is particularly important. Narrower 45° preparations now commonly adopted for single Vs should be filled with thin layers using smaller 3.25mm electrodes, changing to a larger size only when the welder can comfortably achieve an initial first split of thin beads. Wider 60° preparations can usually be welded earlier with a larger size.

Restraint. Strain aging is accentuated by plastic flow of the weld. Thus if the plates are allowed to move during welding, the properties of the root show even greater deterioration. For this reason it is crucial that procedure test plates are firmly anchored by the use of restraining bars to prevent both angular distortion (butterflying) and lateral contraction. Figure 11 compares the effect on toughness for two similar joints welded under varying degrees of restraint. While the need for restraint is commonly acknowledged the degree required for it to be effective is rarely applied in practice. Strong backs are often of inadequate size to resist movement and the effectiveness may be reduced by large mouseholes. Tacking is insufficient, strong backs should be welded on both sides across the full face of the test plate.

On double sided joints, strong backs must not be removed from one side until those on the other side have been welded in place.

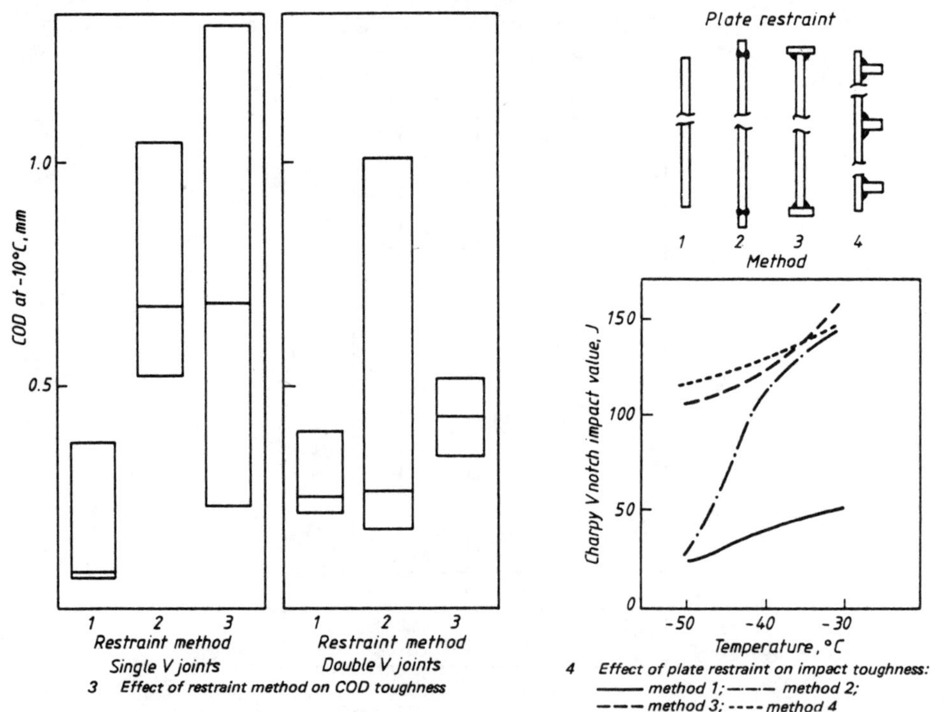


Fig. 11 Effect of restraint on toughness

A method of achieving effective restraint is illustrated schematically in Fig. 12.

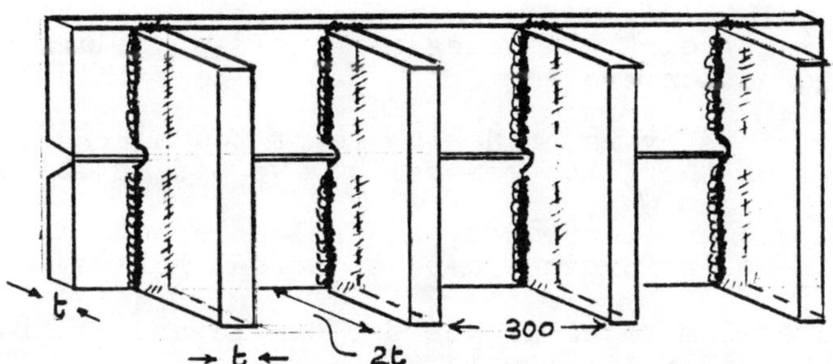


Fig. 12 Technique illustrating effective restraint

Backgouging. Backgouging to completely remove the root area is an obvious if costly remedy to the problem of root deterioration but for it to be even partially effective, this must be done properly and the strain damaged, heavily diluted area removed completely. The backgroove should also be wide enough to permit the welder to easily fill it with thin beads without blocking. Figure 13 illustrates a recommended technique for achieving adequate backgouging.

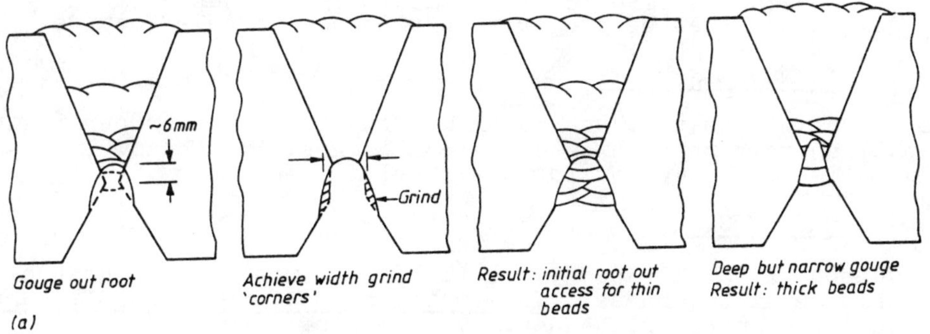


Fig. 13 Recommended technique for Effective backgouging

It has become common practice in submerged arc welding to disperse the root of single V joints by a "punch through" run at high currents using alloy wire. While acceptable for single V joints this should not be carried out on double V joints since the confinement of the preparation can result in a very undesirable depth to width ratio and subsequently, cracking at the centre line.

## SUMMARY

Based on the points discussed in this paper, guidelines to achieve satisfactory weld joint toughness of a weld procedure and on subsequent production are summarised as follows :

1. Single V joints by removing the least tough area away from the centre of the joint generally give better COD performance.

2. Use low hydrogen electrodes, and select specific types for which back-up performance data on actual joints is available.

3. Use basic agglomerated fluxes for single and multi-power welding on multi-positional welds.

4. Use gas shielded flux cored wires of the 2.5 Ni type. The 1.2mm size can be used for all positional applications.

5. The maximum restraint must be applied to prevent angular distortion and lateral contraction during welding. Tacking is insufficient.

6. A stringer bead sequence which deposits thin beads generates the best toughness.

7. High currents may be detrimental to toughness.

8. Backgouging is beneficial particularly on double sided joints but must be properly carried out to be effective.

## ACKNOWLEDGEMENTS

The author wishes to thank Murex Welding Products for permission to publish this paper and to his colleagues, Dr. McKeown, D.J. Ellis and P. Judson for their helpful advice.

## REFERENCES

1. Judson, P. and McKeown, D., (1982) Advances in the control of weld metal toughness. International Conference on Offshore Welded Structures, London. The Welding Institute.

2. Chew, B. (1981) Hydrogen control of basic coated MMA electrodes, part III - relationship between coating moisture and weld hydrogen. CEGB Research Division Report Marchwood Engineering Laboratories.

3. Boniszewski, T. and Paveley, D.A. (1978) Timing at temperature when drying welding electrodes before use. Metal Construction. The Welding Institute.

4. Berkhout, C.F. and Van der Brink, S.H. (1978) How strain aging influences the toughness of some weld metals. Welding & Metal Fabrication. The Welding Institute.

5. John, R. (1980) Factors affecting COD properties in MMA welds. Seminar on COD - Fact or Fiction? The Welding Institute.

# NON IRON POWDER NI-BEARING ELECTRODES FOR LOW TEMPERATURE ENVIRONMENTS

J. Koivula* and J.H. Rogerson**

*Technical Research Centre of Finland, Metals Laboratory, Finland

**Cranfield Institute of Technology, School of Industrial Science, England

## ABSTRACT

In site welding and repair and maintenance welding in low temperature environments - arctic regions - manual metal arc welding is likely to be preferred. The question to be answered is whether or not the electrodes which are being developed have a sufficiently wide tolerance to welding variables to avoid hydrogen cracking and to achieve a satisfactory toughness.

Weld microstructures of commercial electrodes 7016 CIL and 8016 G and of a new development, 10016 G were examined with a range of heat inputs and run temperatures. For 8016 and 10016 mainly fine grained acicular ferrite was obtained with the whole range of variables used. For 7016 a mixed ferrite structure was obtained. Acicular ferrite content decreasing from 52% to 30% within heat input range 1.5-4.5 kJ/mm.

Tekken tests showed that the root pass in a butt weld with double V-preparation in normal restraint conditions could be welded without preheat with 7016 electrodes. For 8016 a minimum preheat of +50$^{o}$C is needed and at least +100$^{o}$C for 10016 electrodes.

For toughness testing only Charpy-V test was used. High upper shelf energies were obtained - a level of over 200J for 7016 down to ∿ 150J for 10016 with increasing acicular ferrite content. Impact toughnesses, $C_v$ > 40J were obtained down to -85$^{o}$C for 10016, to -75$^{o}$C for 8016 and to -80$^{o}$C for 7016 with the lowest heat input, Q = 1.5 kJ/mm, with slightly decreasing toughness with increasing welding energy. A more detrimental effect on impact toughness was found by PWHT. For 10016 and 8016 electrodes $C_v$ = 40J was as high as -20$^{o}$C and -15$^{o}$C before and after PWHT, respectively.

## KEYWORDS

MMAW; weld metal mechanical properties; weld metal toughness; weld metal microstructures; non iron powder consumables; effect of PWHT.

## INTRODUCTION

For various types of fabrication to operate in arctic regions, there is a require ment to achieve weldments with satisfactory properties in steel of yield strengths in the range 400-700 N/mm$^2$.

In the long term gas shielded welding techniques may provide the most satisfactory technical answer but in the short and medium term, and even in the long term for site welding and repair and maintenance welding, manual metal arc welding is likely to be preferred for such applications. The problems in developing welding electrodes relate to the difficulty in achieving a balance between toughness and freedom from hydrogen cracking at high strength levels. The hardenable compositions which are needed to give a satisfactory strength/toughness combination are undesirable in that they lead to risks of hydrogen induced cracking. The experience is that the higher the strength level that is required the closer the control which is needed on welding procedures (including moisture control of the electrodes).

The question to be answered is whether or not the electrodes which are being developed have a sufficiently wide tolerance to welding variables for them to be reliably used in practice.

The work concentrated on the low hydrogen non iron powder electrodes of the AWS 7016, 8016 and 10016 types. The electrodes are all position, Ni bearing variants which are likely to meet the property and weldability requirements. Parent plate materials used were BS 4360 grade 50D, plate thickness 50mm for 7016 and RQT 700, plate thickness 30mm for 8016 and 10016 electrodes. The former was normalised and the latter was in quenched and tempered condition.

Three aspects were considered in this work: 1. Weld microstructures determined by electrode type, solidification conditions and post solidification cooling rate. 2. The evaluation of weld metal cracking tendencies. 3. The determination of weld metal mechanical properties.

## EQUIPMENT AND MATERIALS

Manual metal arc welding was used throughout the tests. The consumables selected for this work formed a series of increasing tensile strength levels 500, 600, 700 $N/mm^2$. All the electrodes were basic non iron powder type with a Ni addition of 0.9-2.9 wt%. AWS specifications being 7016 CIL, 8016 G and 10016 G. Table 1 gives typical all weld metal chemical compositions.

TABLE 1 Typical all Weld Metal Chemical Composition of the Consumables

| Electrode | C | Mn | Si | S | P | Ni | Cr | Mo | Nb | V | Al | Ti | Cu |
|---|---|---|---|---|---|---|---|---|---|---|---|---|---|
| 7016 CIL | .045 | .6 | .30 | .010 | .015 | 2.4 | | | | | | | |
| 8016G | .055 | 1.7 | .40 | .010 | .015 | .9 | .03 | .10 | .01 | .01 | - | .04 | .02 |
| 10016G | .045 | 1.9 | .30 | .010 | .015 | 2.9 | .05 | .30 | | | | | |

Parent plate materials used were BS 4360 grade 50D, plate thickness 50mm for 7016 and RQT 700, plate thickness 30mm for 8016 and 10016 electrodes. The former was normalised and the latter was in the quenched and tempered condition. Chemical analyses of the plates are shown in Table 2.

TABLE 2 Chemical Compositions of Parent Materials

| | C | Mn | Si | S | P | Cr | Ni | Mo | Al | Cu | Nb | V | Ti | B |
|---|---|---|---|---|---|---|---|---|---|---|---|---|---|---|
| 4360,50D | .16 | 1.34 | .04 | .001 | .006 | .12 | .07 | .01 | .04 | .03 | .02 | .05 | .003 | - |
| RQT 700 | .16 | 1.23 | .34 | .018 | .012 | .08 | .12 | .17 | .16 | .19 | .04 | - | .028 | .0019 |

## EXPERIMENTAL PROCEDURES

The work was divided into three parts: 1. test welds for thermal and microstructural analysis, 2. weld metal cracking tests, where Tekken test was employed, 3. butt welded tests for determination of weld mechanical properties.

### Microstructural analysis

Test welds for microstructural analysis were laid on V-grooves as shown in Fig. 1. The kind of preparation was chosen for simplicity but representing the same dilution and weld thermal cycles found in welding Tekken weld cracking tests and the root run in double-V preparations.

Welding was carried out both in the flat and vertical positions with large variations in the range of heat input. Working temperature was also varied to simulate weld thermal cycles in arctic conditions, and in normal conditions with preheat. Interpass temperature was approximately 100°C.

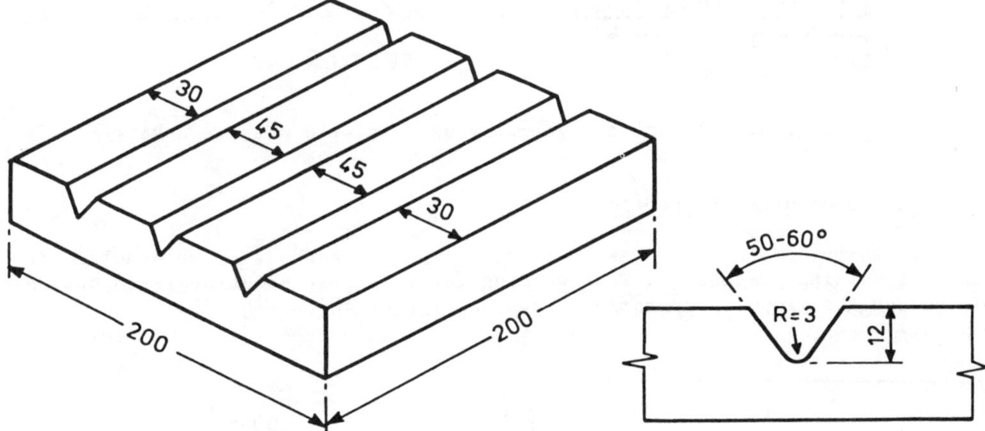


Fig. 1. Test plate used in the test welds for thermal and microstructural analysis.

Weld cooling rate was measured with a Pt/Pt-Rh- thermocouple immersed in the weld pool. From the cooling data, cooling times between 800-500°C and 1400-200°C were calculated, and contributed to the study of austenitic transformation products and hydrogen diffusion conditions, respectively.

For better distinction of the austenite decomposition temperatures, which are between 800 and 400°C for C-Mn steels, a derivative of the weld temperature versus time curve was also determined (Alcantara, 1982)

The weld metal microstructures were identified from optical micrographs using the terminology described by Abson and Dolby (1980). A point counting method was used to determine the volume fraction of the constituents present. The transformation temperature from the thermal analysis helped to confirm the identification.

### Tekken tests

The Tekken test has proved to be the most reliable among the self restraint types of cracking tests to assess cold cracking in the weld metal (41), provided that a symmetrical preparation is used. Figure 2 shows the test piece geometry used in this work.

The main variables in the tests were the weld hydrogen content and the working temperature.

Electrodes were prepared to three different moisture contents with the intention of approximately reaching the weld hydrogen contents ($H_{DIIW}$) 3, 6 and 9 ml/100g deposited filler material.

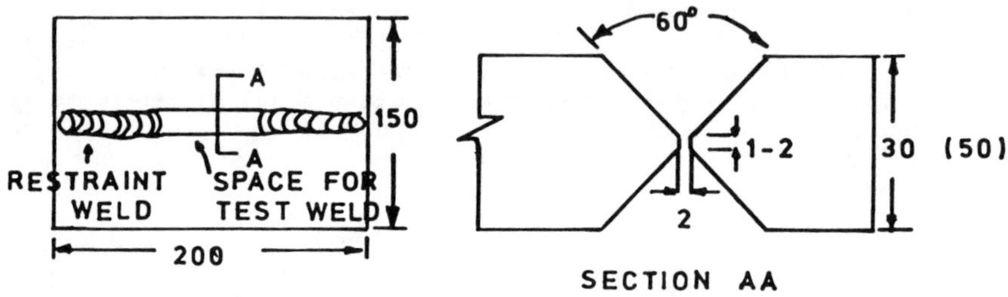


Fig. 2. Tekken test geometry used in weld metal cracking tests.

Weld metal mechanical properties

The main object of the work was to measure impact toughness of weld metal with a range of welding procedures. Test welding for this purpose was carried out on plates and the groove preparation shown in Fig. 3.

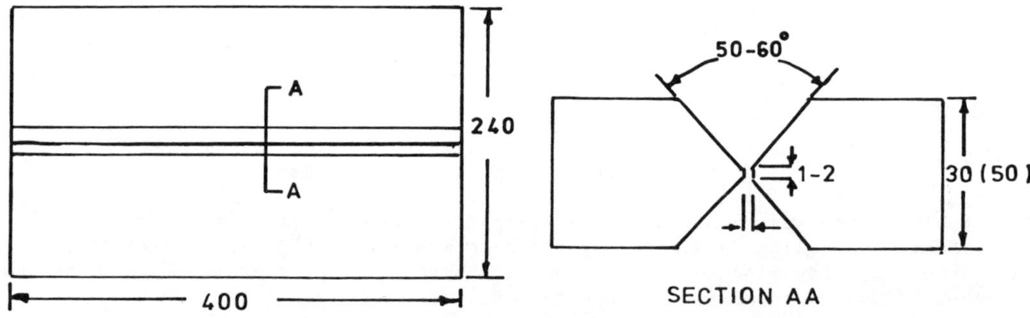


Fig. 3. Test plate used for butt test welds.

Tests were welded in both flat (1G) and vertical (3G) positions. Welding energy was varied from 1.5-4.0 kJ/mm. Welding current was selected according to manufacturer's recommendations. The plate was allowed to cool down to about 80°C before the next pass was laid. For 10016 electrodes a higher interpass temperature, 100-110°C was used to be on the safe side against hydrogen cracking.

Charpy-V impact toughness transition curves were determined both in the root and subsurface areas. Fracture surfaces of the Charpy-V specimens were studied under a microscope to determine the type of fracture and the temperature with 50% of brittle fracture. The reverse of Charpy-V specimens were surveyed to estimate the proportions of columnar, coarse grained and fine grained weld metal microstructures.

## RESULTS AND DISCUSSION

### Weld metal microstructures

A quantitative description of weld metal microstructures is given in Table 3 and examples of optical micrographs in Fig.4

Thermal analysis showed that the austenite transformation start temperature is reduced in the following sequence: 7016, 8016, 10016. This is accompanied by a clear refinement of microstructure as can be seen in Fig.4

TABLE 3 Quantitative Description of Weld Metal Microstructures

| Electrode | Specimen No. | Heat Input/ Welding Pos./ Run Temp. | Grain Boundary Ferrite (+ polyg. ferrite) | Acicular Ferrite | Ferrite with aligned M-A-C | Ferrite-Carbide Aggregate |
|---|---|---|---|---|---|---|
| 7016 | M 1 | 1.5 | 25 | 52 | 23 | - |
| | 2 | 3.0 | 22 | 43 | 33 | 2 |
| | 3 | 4.5 | 36 | 30 | 35 | 0 |
| | 4 | 3.0/3G/ | 27 | 48 | 25 | 0 |
| | 5 | 4.5/3G/ | 31 | 41 | 27 | 1 |
| | 6 | 3.0//+100 | 32 | 38 | 30 | 0 |
| | 7 | 3.0//-50 | 21 | 60 | 19 | 0 |
| 8016 | 8 | 1.5 | 9 | 86 | 5 | - |
| | 9 | 3.0 | 8 | 77 | 15 | - |
| | 10 | 4.5 | 11 | 83 | 6 | - |
| 10016 | 15 | 1.5 | 6 | 91 | 3 | - |
| | 16 | 3.0 | 2 | 97 | 1 | - |
| | 17 | 4.5 | 4 | 94 | 2 | - |

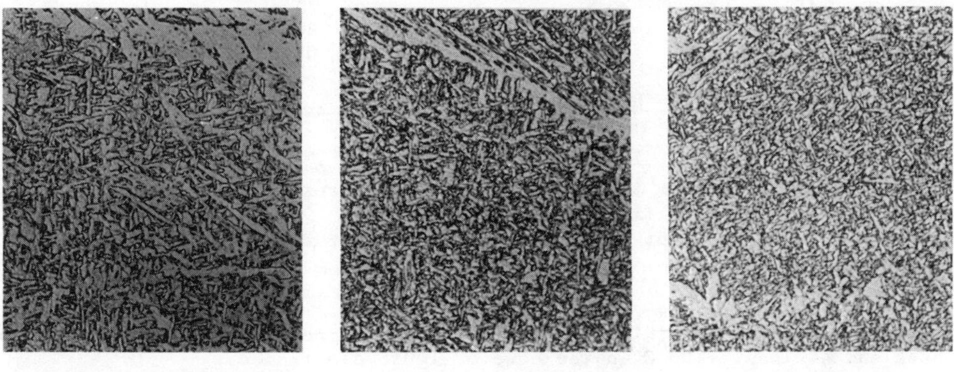

a. 7016 b. 8016 c. 10016

Fig. 4 Weld metal microstructure, Q = 1.5 kJ/mm (200x).

The bulk of the microstructure of 8016 and 10016 is acicular ferrite almost entirely independent of heat input between 1.5 and 4.5 kJ/mm. The microstructure of 7016 instead seems to vary more with welding energy. Acicular ferrite content decreases with increasing heat input. This is replaced by both polygonal ferrite and coarse lamellar products. The same tendency is found when run temperature is varied from -50°C, through room temperature, up to 100°C. The volume fraction of ferrite with aligned carbides is clearly reduced by decreasing working temperature and/or welding energy, that is when increasing cooling rate.

The weld metal microstructure of 8016 is mainly acicular ferrite with some grain boundary and polygonal ferrite. With intermediate heat inputs some ferrite with aligned austenite or carbides is formed. The width of ferrite veins is relatively large as was the case with the ferrite found in 7016 weld metal.

The microstructure of 10016 is almost entirely very fine grained acicular ferrite, the amount of grain boundary ferrite tending to increase with decreasing heat input. With intermediate heat inputs (3.0 kJ/mm), sharp dark grain boundaries were found indicating carbide precipitation. This precipitation will gradually disappear when heat input is further increased.

## Weld metal hydrogen cracking

Tekken tests with a symmetrical joint preparation showed that the root pass in a butt weld with double V preparation in normal restraint conditions can be welded without preheating using 7016 CIL electrodes provided that the electrodes are well baked giving not more than $H_{DIIW}$ = 6 ml/100g as weld hydrogen content (Fig.5).

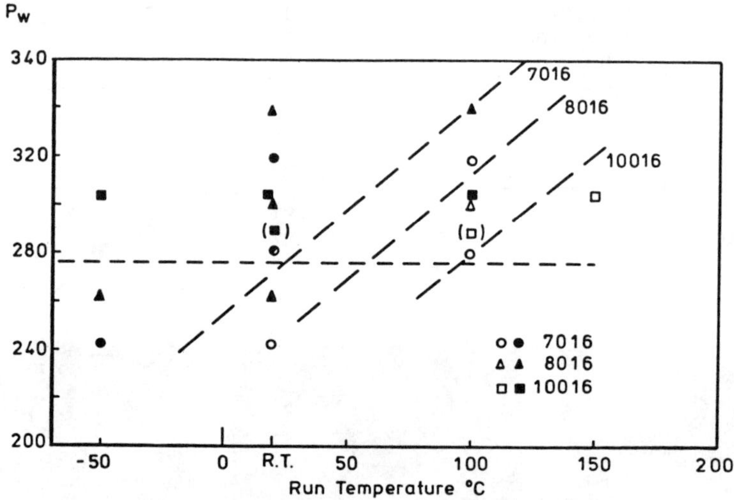


Fig. 5 Tekken test results, clear figures - no cracking, black figures - cracked.
Saw cut edges (x = 300mm) used for the tests in brackets.

In the same conditions a minimum preheat of +50°C is needed if 8016 G electrodes are used and at least +100°C in the case of 10016 electrodes.

The weld metal hydrogen cracking risk is more severe, for 7016, 8016 and 10016 electrodes than predicted by Yurioka's formula (Yurioka and co-workers, 1981).

$$(t_{100})_{cr} = 10500\,(P_w - 0.276)^2 \text{ where}$$
$$P_w = P_{cm} + 0.075 \log H_{JS} + 0.15 \log (0.017K_t\,\sigma w)$$

It is therefore proposed that formulas which employ $P_{cm}$ are not suitable for the kind of filler metals obtained here with low carbon, Ni-bearing consumables.

## Weld metal impact toughness

Charpy-V transition curves for the test welds are given in Figs. 6, 7 and 8. Table 4 gives the $C_v$ energy values at -60°C and the temperatures where 50% of the impact test fracture surfaces were identified as "brittle" (FATT).

Impact toughness results can be discussed in terms of welding technology factors and microstructure, the latter as far as optical micrographs and occasional scanning electron micrographs on fracture surfaces can explain the observed behaviour.

TABLE 4 $C_v$-Energies at -60°C and the FATT Temperatures

| Consumable | Welding Procedure | $C_v$-energy at -60°C/J root | subsurface | FATT Temp/°C root | subsurface |
|---|---|---|---|---|---|
| 7016 | low Q. 1G | 58 | 165 | -38 | -53 |
| | 3G | 28 | 132 | -29 | -54 |
| | high Q.1G | 40 | 110 | -20 | -48 |
| | high Q.1G | 25 | 95 | -30 | -33 |
| 8016 | low Q. 1G | 53 | 95 | -42 | -60 |
| | 3G | 41 | 70 | -20 | -52 |
| | high Q.1G | 35 | 80 | -30 | -50 |
| | high Q.1G | 21 | 65 | 0 | -36 |
| 10016 | low Q. 1G | 65 | 72 | -60 | -57 |
| | 3G | 54 | 40 | -52 | -50 |
| | high Q.1G | 58 | 47 | -45 | -45 |
| | high Q.1G + PWHT | 18 | 35 | 0 | - 1 |

### 7016 CIL

Increasing welding energy decreases the volume fraction of acicular ferrite so that the amount of M-A-C ferrite increases instead. That means a reduction in the amount of high angle boundaries at the expense of low angle boundaries, less energy is then needed for fracture. Transition temperature is also increased with increasing welding energy. It can thus be concluded that increasing acicular ferrite content in 7016 weld metal will lower the transition temperature.

The big difference in transition temperature or toughness value at a given temperature (Fig. 5) can be explained by the usual arguments of high dilution percentage and severe strain ageing conditions at the root compared to subsurface passes.

Although Table 4 shows an increase in lamellar products in the microstructure with increasing heat input it should be noted that no clear side plate structures can be seen in any of the specimens under optical microscopy.

The best impact toughness and lowest transition temperature were obtained with the lowest heat input. Similar microstructures were achieved when welding was carried out at -50°C (Table 4) with Q = 3 kJ/mm. Very low heat input was unfortunately not employed in low temperature welding.

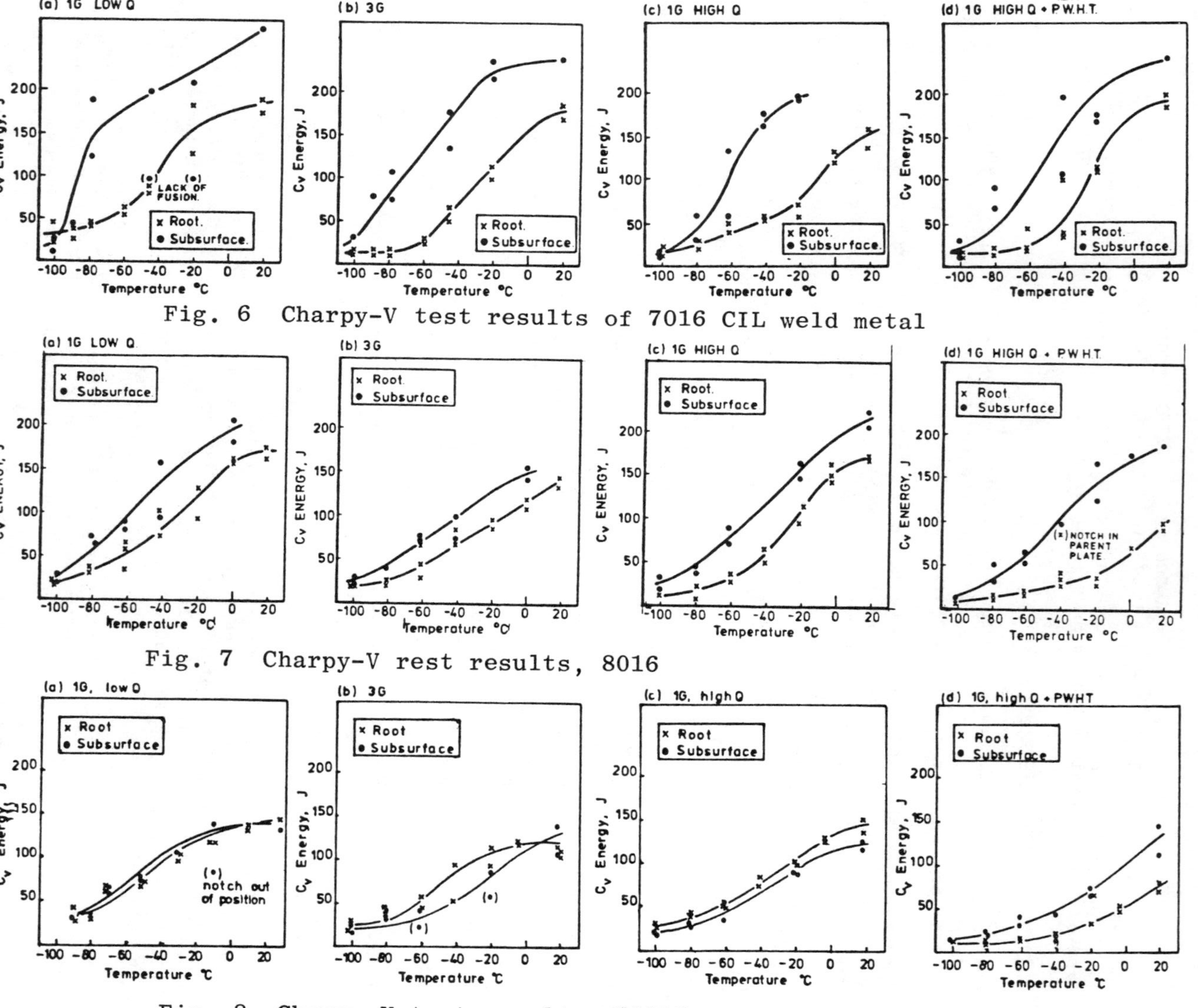


Fig. 6 Charpy-V test results of 7016 CIL weld metal

Fig. 7 Charpy-V rest results, 8016

Fig. 8 Charpy-V test results, 10016

8016 G

The weld metal microstructure did not change much with a wide range of heat inputs employed. Ferrite with aligned martensite-austenite and/or carbides does occur with intermediate welding energy (Table 3). This is reflected in the poorer impact toughness results (Fig. 4b and Table 4).

The increased upper shelf energy and slightly higher transition temperature after PWHT cannot be explained without more detailed microstructural analysis. It can be seen that the transition temperature fell only in the root area. This gives rise to a conclusion that some recovery from strain ageing occurred under stress relief.

The transition temperatures (FATT) are nearly the same as for 7016 (Table 4), although the acicular ferrite content is considerably higher and the structure inside the grains is finer (Fig. 4). The difference between root and subsurface toughness is not very high, the post weld heat treated test being an exception.

Looking at the chemical composition (Table 1) there are deoxidation agents, nitride and carbonitride formers, Ti, V and precipitation strengthening agents, e.g. V. It is thus expected that the filler metal would not be prone to strain ageing. The small difference between root and subsurface toughness supports this proposal.

Some of the carbide precipitates (possibly $Mo_2C$) coarsen and embrittle the structure after stress relief. Embrittling elements possibly come from the parent plate, RQT 700, because the region with higher dilution becomes more embrittled.

10016 G

Strain ageing will normally be more severe in the root area than in the bulk of the weld and dilution is higher, too. These are the most common reasons for poor root toughness. In this case, however, weld metal toughness in the root is the same, sometimes even higher than the subsurface Charpy-V properties.

The as welded microstructure, which is almost entirely fine acicular ferrite, (Fig. 4) seems to be superior in toughness compared to a tempered structure which undergoes possible $Mo_2C$ precipitation coarsening which is one possible reason for embrittlement after stress relief.

Comparison between 7016, 8016 and 10016 electrodes

In the root area Charpy-V energy values are slightly higher for 8016 than for 7016. Marginally higher values still are measured with 10016. This can be explained by the increasing acicular ferrite content. The same does not, however, hold true for toughness in the sursurface area (Table 4). A considerably higher upper shelf toughness is obtained for weld metal of 7016 than for 8016. Charpy-V test results are even lower for weld metal of 10016 remaining even beneath those of the root area. The reason for this is obviously the low Mn content of 7016 compared to the other consumables. The same effect was previously proposed by Ito and co-workers (1978).

The microstructure of 7016 weld metal is clearly more sensitive to variations in heat input than 8016 and 10016 weld metals. The subgrain size is finer going from 7016 to 8016 and to 10016. Both showing the opposite to the measured differences in subsurface toughness between the weld metals of different consumables. Again, to explain the observed behaviour, a very detailed microstructural investigation would be needed.

## CONCLUSIONS

### Weld metal hydrogen cracking

Tekken tests with a symmetrical joint preparation showed that the root pass in a butt weld with a double V preparation in normal restraint conditions can be welded without preheating using 7016 CIL electrodes provided that the electrodes are well baked giving not more than $H_{DIIW}$ = 6 ml/100g as weld hydrogen content.

In the same conditions a minimum preheat of +50°C is needed if 8016 G electrodes are used and at least +100°C in the case of 10016 electrodes.

The weld metal hydrogen cracking risk is more severe for 7016, 8016 and 10016 electrodes than is predicted by the Japanese formulas e.g.

$$(t_{100})_{cr} = 10500\ (P_w - 0.276)^2 \text{ where}$$
$$P_w = P_{cm} + 0.075 \log H_{JS} + 0.15 \log (0.017K_t\ \sigma w)$$

### Weld metal microstructures and toughness

Weld metal microstructure obtained by 8016 and 10016 was mainly acicular ferrite with fine subgrain size. 7016 was more sensitive to changes in welding energy, the acicular ferrite content decreasing from 52% to 30% when heat input was increased from 1.5 kJ/mm to 4.5 kJ/mm.

Charpy-V transition temperature, measured as the temperature (FATT) where 50% of the fracture surface is interpreted as ductile, is lowered in the root area with increasing acicular ferrite content.

Increasing welding energy increases the transition temperature in the root area even in the cases (8016, 10016) where the microstructure was not much affected by heat input. The transition temperature was little affected by welding energy and nearly the same values (-52 $\pm$ 2°C in 3G position) were obtained for all the consumables (7016, 8016 and 10016) in the subsurface area.

The upper shelf toughness at -60°C was considerably higher for 7016 weld metal than for the others (8016, 10016), low Mn content of 7016 being a possible reason.

For 7016 weld metal the root area toughness is much lower than the toughness in the bulk of the weld, tentatively because of higher dilution which will increase specially the weld Mn and Cr contents. The effect of strain ageing might be affected in the same manner.

For 8016 and 10016 weld metals the difference between the root and subsurface toughness is nearly erased, possibly because of the lower toughness of the weld at low temperatures.

The Charpy-V energy at -60°C decreased after PWHT for 7016 and the subsurface transition temperature increased, but a slightly better root toughness was obtained, because of the ageing effect of PWHT.

For 8016 the transition temperature increased about 30°C after PWHT. 10016 weld metal also suffered a considerable loss in toughness when stress relieved. A rise of 45°C in transition temperature was measured. In both cases there is a possibility of $Mo_2C$ carbide precipitation formation.

ACKNOWLEDGEMENTS

This work was carried out at Cranfield Institute of Technology. Financial support was given by CBI (Confederation of British Industry), VTT (Technical Research Centre of Finland) and W. Ahlström Foundation, Finland.

REFERENCES

Abson, D. J. and R. E. Dolby (1980). A scheme for the quantitative description of ferritic weld metal microstructures. Welding Institute, IIW Doc. IXJ-29-80, 12.

Alcantara, N. G. (1982). Weld metal hydrogen cold cracking. Ph.D. Thesis, School of Industrial Science, Cranfield Institute of Technology.

Ito, Y., Nakanishi, M., Katsumoto, N. and Nojima, K (1978). Submerged arc welding of 3.5% Ni steel. Metal Construction, Vol. 10, No. 5, 253-256

Yurioka, N., S. Ohsita, and H. Tamehiro (1981). Study on carbon equivalent to assess cold cracking tendency and hardness in steel welding. Symposium on pipeline welding in the 80's, Australian Welding Research Association, Melbourne, 15.

# FRACTURE TOUGHNESS OF WELD HEAT AFFECTED ZONES (HAZs) IN STEELS USED IN CONSTRUCTING OFFSHORE PLATFORMS

H.G. Pisarski* and R.J. Pargeter**

*Senior Research Engineer**
*Senior Research Metallurgist***
*The Welding Institute, Abington, Cambridge. UK*

**ABSTRACT**

This paper describes the metallurgy of weld heat affected zones (HAZs) in carbon-manganese microalloyed steels and methods by which their resistance to fracture initiation can be evaluated. The significance of brittle regions within the HAZ are discussed primarily with regard to ensuring structural integrity by avoiding fracture initiation.

**KEYWORDS**

Weld HAZs; carbon-manganese microalloyed steels; microstructure; fracture toughness testing; crack tip opening displacement (CTOD) tests; crack arrest.

**INTRODUCTION**

Assessments of fracture resistance in welded steel structures normally involve the measurement of toughness in at least three regions. These are the parent material, weld HAZ and weld metal. The weld HAZ presents difficulties in toughness measurement owing to its narrow width; for example, in carbon-manganese steels welded at arc energies typical of those used in the quality fabrication industry the transformed HAZ will usually be no more than 3 to 4mm wide. Within this distance there is a very wide range of microstructures and properties so that the region of importance may only be about 1mm wide. Very few procedure test specifications are written so as to ensure that such small regions are sampled. Despite their small size, brittle regions within the HAZ can have a significant influence on the integrity of a structure with respect to failure by brittle fracture. A number of catastrophic brittle fractures of engineering structures, including pressure vessels, storage tanks and bridges, in which the fracture started in the HAZ, testify to this (Harrison, 1983).

It has been argued by some that the narrow width of the HAZ and the higher hardness often associated with it compared with the parent plate tends to reduce its significance, even though it may contain locally brittle regions. However, as fabrication or service induced defects often lie in HAZs and also, as welds are often in regions of high local stress concentration and tensile residual stresses, the conditions for brittle fracture initiation are favoured, especially if the edge of the defect lies in an embrittled zone. Whether a crack, once initiated, stops or continues to propagate depends on the arrest toughness of the surrounding material and the energy available for fracture. In most structural steel, the arrest toughness is significantly below the initiation toughness, and consequently arrest is unlikely unless the crack driving force decreases rapidly as the crack extends. Although it is possible to design structures for crack arrest, the techniques available at present are either too empirical or too theoretical to have general application. Furthermore, with the majority of thick section carbon-manganese structural steels currently available it is difficult to achieve the necessary toughness to ensure arrest in many service conditions (of high stress and low temperature). Consequently, it is usual to design to prevent fracture initiation. This aspect forms the main theme of this paper.

## THE METALLURGY OF HAZs IN CARBON-MANGANESE AND MICROALLOYED STEELS

The HAZ comprises all that parent material which has been affected in any way by heat generated by the welding process. Since heat diffuses away through the plate, the thermal cycle becomes less severe (principally lower peak temperature) with increasing distance from the fusion boundary. The microstructural and other changes brought about by the thermal cycle are therefore also a function of distance from the fusion boundary.

In carbon-manganese steels, a number of metallurgical effects take place as the temperature increases. At temperatures of only 100 to 200°C interstitial solutes can interact with the strain fields of dislocations and give rise to accelerated strain ageing embrittlement (Baird, 1963). This will not be visible optically or probably even in the electron microscope, but nevertheless can give rise to marked reductions in toughness if there is sufficient free interstitial solute present (Baird, 1971) (nitrogen or carbon). There is usually sufficient strain associated with welding to produce the plastic deformation and hence dislocations with which the solute may interact. The strain will be increased if the weldment is free to move during welding instead of being restrained by strongbacks or the rest of the structure. At temperatures of above around 400°C iron carbides will tend to spheroidise and coarsen. This will generally have little noticeable effect on toughness, although thicker carbides do facilitate cleavage fracture initiation (Oates, 1968). It is also possible for alloy carbides to form a very fine stable precipitate above around 600°C. This is most likely to occur in controlled rolled microalloyed steels, which may have some microalloying (niobium and/or vanadium) remaining in solution particularly with higher microalloy contents and lower carbon contents. Recent work at The Welding Institute has indicated that such precipitation can cause serious embrittlement. All this region, where the peak temperature remains below the ferrite:austenite transformation temperature, is known as the sub-critical HAZ.

Above the $AC_1$ temperature (about 700°C), ferrite begins to transform to austenite but this transformation will not go to completion below the $AC_3$ temperature (about 800°C). The region exposed to temperatures between the $AC_1$ and $AC_3$ is known as the intercritical HAZ. The ferrite grain size will be unaltered but higher carbon regions will have retransformed, usually to finer pearlite. At low magnification in the optical microscope the boundary between this and spheroidised regions of sub-critical HAZ may be difficult to distinguish.

Where peak temperatures exceed the $AC_3$ the HAZ is termed supercritical, but contains two very different regions. Where the $AC_3$ temperature is only just exceeded there is little opportunity for grain growth, whereas as the peak temperature and time above the transformation temperature increase (i.e. towards the fusion boundary) the austenite grain size becomes progressively coarser. When the fine grained austenite cools and retransforms, multiple ferrite nucleation takes place (assisted by the high density of grain boundary nucleation sites) and a finer grain size ferrite/pearlite (or bainite) structure is formed. This, the grain refined HAZ, is invariably the toughest part of the HAZ and often has a higher toughness than the parent plate. In coarser grained regions, however, ferrite nucleation is delayed and lower temperature transformation products such as bainite and martensite may form. The colony size of those constituents (which is the effective grain size for cleavage) will increase with the increase in austenite grain size towards the fusion boundary, and the lowest toughness in an HAZ is often found in this grain coarsened region, close to the fusion boundary.

In a multipass weld part of the HAZ produced by one pass will be reheated and constitute part of the HAZ produced by a subsequent pass (Fig. 1). The microstructure of the previous HAZ has little effect on those regions of the subsequent pass HAZ which are fully reaustenitised, but the microstructure produced in the lower temperature regons will depend on the initial microstructure. Thus, when martensite is present in the previous HAZ this will be tempered, generally having a beneficial effect on toughness. In bainitic regions carbide will form from pockets of retained austenite or martensite between the ferrite laths, and spheroidisation and coarsening of carbides may take place. The effect on toughness will depend on the nature of the retained phases which are destroyed and the extent of carbide coarsening. Precipitation of fine alloy carbides or nitrides may also take place in regions close to the previous fusion boundary where the temperature had been high enough to take the original alloy carbides back into solution. This will generally produce a further reduction in the toughness of this region, which may already have been the least tough part of the HAZ. In the grain refined region of the HAZ some spheroidisation of the pearlite may take place, and strain ageing must be possible, but with no significant microstructural change, the good toughness of this region is unlikely to be seriously impaired.

From the above brief outline it should be clear that the HAZ is not a homogeneous region. Minimum toughness is most commonly found in the grain coarsened HAZ close to the fusion boundary, although in some cases in the as-welded condition minimum values have been measured just beyond the visible HAZ (Dolby, 1979). It should further be recognised that the total extent of the visible HAZ (which just extends into the sub-critical region) even from a fairly large weld (4kJ/mm in 25mm plate) would be about 3mm of which only 1mm would be grain coarsened bainitic structure. This is bounded on one side by weld metal and on the other by grain refined HAZ, both of which commonly have very good toughness.

With regard to the levels of toughness in the two regions where the lowest toughness is usually found (the coarse grained and the sub-critical HAZ) two aspects should be considered, namely the composition and thermal history of the parent material, and the welding procedure.

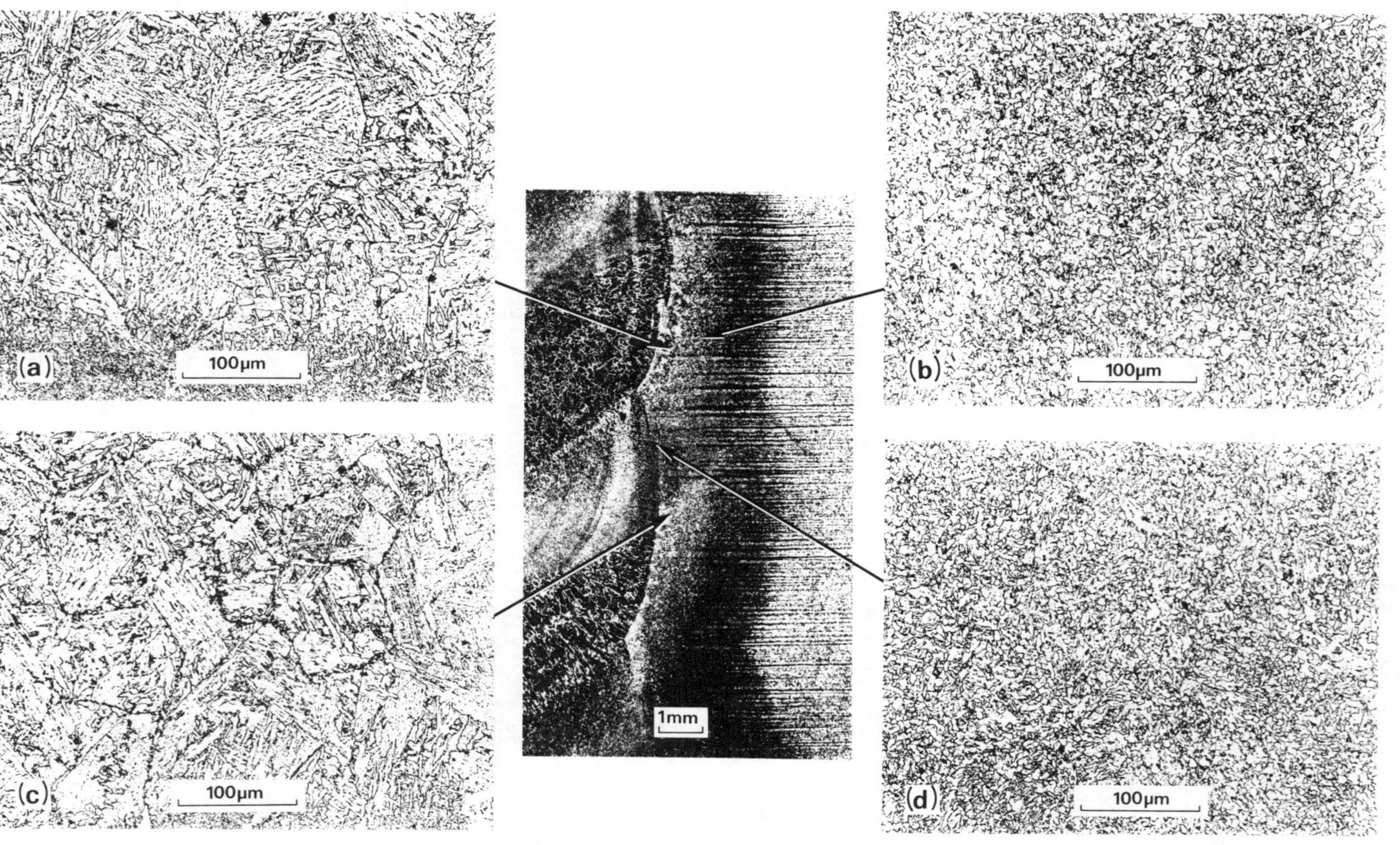


Fig. 1. HAZ regions in a multipass weld.
a. Grain coarsened HAZ, as deposited.
b. Fine grained HAZ, previously parent plate.
c. Intercritically reheated grain coarsened HAZ.
d. Fine grain HAZ previously grain coarsened HAZ.

The effects of composition are complex and not fully understood. Occasional low CTOD values have been found with carbon-manganese-silicon-aluminium-niobium steels, regardless of carbon, manganese or silicon level, particularly in thick section material welded with arc energies of around 3kJ/mm or greater. Non-microalloyed steels tend to give better toughness in the coarse grained region, but fabricators are unwilling to accept the higher carbon equivalent values necessary to maintain strength. Although there have been suggestions that low levels of nickel (approximately 0.5%) could be beneficial for HAZ toughness, this has not yet been demonstrated. Grain growth, and hence the width of the coarse grained HAZ is restricted to a certain extent in aluminium treated steels. This effect is exploited to a much greater degree in titanium treated 'high energy input resistant' steels (Threadgill, 1981), although these are only available in restricted thickness and composition ranges at present. The toughness of the sub-critical HAZ may not be adequate if the aluminium-nitrogen ratio is not sufficient (≤2:1 soluble aluminium:nitrogen) to prevent strain ageing due to free nitrogen[1]. Furthermore, low sub-critical HAZ toughness may occur in some controlled rolled microalloyed steels due to precipitation hardening effects. This is due to the presence of non-equilibrium levels of uncombined microalloying elements, and the problem is exacerbated by a low carbon content. It should be recognised that very much more highly alloyed regions can exist near the centre line of certain plates, in particular some made by continuous casting, and where the HAZ intersects these regions, toughness may be reduced.

Usually post weld heat treatment (PWHT) is carried out with the principal aim of reducing residual stresses, and as a result improvements in defect tolerance can be obtained even if toughness is unaltered or slightly reduced. However, in the sub-critical HAZ, strain ageing effects are reversed by PWHT and fine precipitation may become overaged, thereby improving the toughness of this region. In the case of the coarse grained HAZ in microalloyed steels very little improvement in toughness is commonly found however, and furthermore it is possible, although not proven, that very low carbon steels (<0.1%) are more resistant to tempering than similar higher carbon steels. In steels with niobium and vanadium micro-alloying additions together, significant reductions in toughness in the coarse grained regions have been observed as a result of PWHT (Rothwell, 1979).

The effects of welding procedure on HAZ toughness are complex. The nature and extent of the primary HAZ is affected, principally by the arc energy and interpass temperature. Furthermore, it is modified by subsequent passes, this being controlled by the detailed sequence and placement of subsequent beads, their arc energy and the relevant interpass temperature.

The grain size of grain coarsened regions is coarser in high arc energy welds (Davey, 1981), giving lower cleavage resistance, and details of the welding procedure (current, voltage and travel speed) are secondary to the actual level of arc energy. It should be recognised, however, that very low heat inputs can lead to the formation of more brittle constituents (martensite/lower bainite) in some steels which is of particular importance if the weld is to remain untempered. The risk of hydrogen cracking is also increased, and very low arc energies

---

1 It should, however, be recognised that high aluminium contents can cause a degradation of weld metal toughness in some high dulution welds (Terashima, 1982).

may be economically undesirable. It has been reported that strain ageing in the sub-critical HAZ is more pronounced at lower heat inputs (Dolby, 1979). This must be due to greater strain in the tested region as a result of the lower heat input and care should be taken when applying these results to practice, where geometry and restraint conditions may be different.

As indicated above, the refined HAZ often has better cleavage resistance than even the parent plate. In a multipass weld, therefore, it is desirable to transform as much grain coarsened into grain refined HAZ as possible by subsequent passes. The extent of the grain coarsened HAZ is principally a function of arc energy, but the extent of the refined HAZ also increases with interpass temperature (Evans, 1978). The amount of primary HAZ which is reheated can be further increased by controlling the degree of overlap through the angle of attack (Dolby, 1979).

## MEASUREMENT OF HAZ TOUGHNESS

### Charpy Tests

The Charpy V test is probably the most common test used to assess HAZ toughness. Typically in weld procedure tests the specimens are extracted at specified locations through the thickness of the plate and the notches placed at specified distances from the weld fusion boundary. For example, a typical procedure in offshore construction is to take specimens from three HAZ locations, namely; the fusion boundary (f.b.), f.b. +2mm and f.b. +5mm. Although a certain amount of information on the variation in HAZ toughness can be obtained by such tests, their primary aim is to ensure that a certain quality workmanship is maintained; the nature of the tests, especially with respect to the often arbitrary position from which the specimens are taken, prevents them from being much more. For example, since most specimens are taken from a weld containing an inclined bevel the f.b. is not parallel to the notch, so that along the length of the notch a mixture of different microstructures will be sampled. In welds deposited at low arc energies the HAZ will be narrow and the specimens sampling regions other than the f.b. will sample predominantly parent plate or weld metal; so that the toughness values may be unrepresentative of the HAZ. Since the Charpy test measures the energy required for crack initiation and propagation over the whole crack front, the presence of any localised brittle zones may be masked by the surrounding high toughness material. When occasional low toughness values are measured, re-tests are often specified and if these tests meet the requirements then the procedure is deemed satisfactory. Nevertheless, for reasons which have been described above, the occasional low value may be a genuine indication of a local brittle zone which has been missed by the other specimens. In general, however, it is not possible with the Charpy test to assess the signifiance of the toughness values measured with respect to the brittle fracture resistance of a structure. The tests may provide a qualitative assessment of toughness if correlations with large scale tests or structural behaviour have been established, but even so, this is an indirect measure of fracture resistance. Fracture mechanics tests on the other hand can provide a quantitative measure of toughness which may be used to calculate the fracture resistance of a structures. These tests are discussed in detail below.

Fracture Mechanics Tests

Where the fracture design philosophy is to prevent brittle fracture initiation, the regions of potential low toughness within a weldment must be identified and the toughness of these regions established by a fracture mechanics tests, such as those based on $K_{Ic}$/CTOD tests. Unlike the Charpy V test the results from fracture mechanics tests provide a fracture characterising parameter which enables a structure's fracture resistance to be assessed. Consequently, the design of the test specimen and how it is extracted from a weld requires careful consideration if the result is to be of practical use in design.

For material or welding procedure selection purposes, however, it is also necessary to use a test which is convenient to carry out and gives reproducible results. Furthermore, often when such assessments are required the detailed design of the structure to be built may not have been finalised, particularly in the offshore industry, and consequently the toughness requirements necessary and possible location of fabrication/service defects are unknown. In these circumstances it is often convenient to establish a lower bound estimate of the overall HAZ fracture toughness.

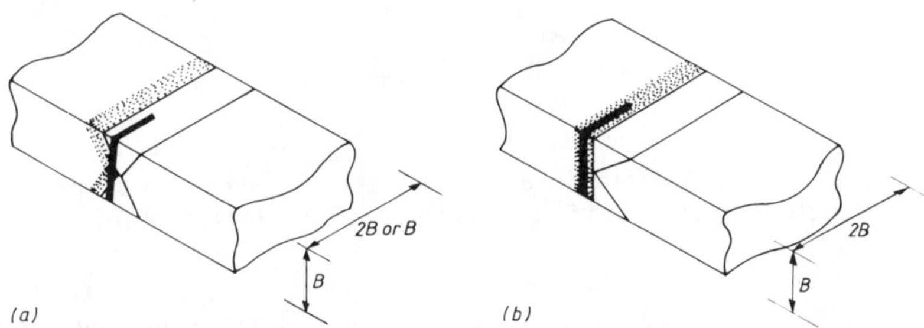


Fig. 2. Through thickness notches in preferred geometry CTOD testpieces.

A lower bound estimate of toughness can be obtained by using a specimen test geometry which maximises constraint, such as the preferred CTOD testpiece to BS 5762 (BSI, 1979). Where full component thickness, e.g. plate thickness, specimens are extracted (as is required by BS 5762) then the notches must be placed through the thickness of the weld, as shown in Fig. 2(a). Unfortunately, sampling HAZs in typical double V or single V joints in this way means that only part of the crack front actually samples the HAZ and misleading results can be obtained. This is considered by the authors to be more of a problem with testing thick section material compared with thin. In the former only a small proportion of the fatigue crack front will sample HAZ. The region of low toughness may be missed completely and even if it is not, recent work indicates that the CTOD value will be influenced by the proportion of low toughness constituent sampled. This problem can be overcome by utilising a joint preparation with a straight edge preparation, such as a K or half K, as shown in Fig. 2(b). Thus the whole of the fatigue crack can be maintained in the HAZ and parallel to the fusion boundary.

A drawback with using a special joint preparation however, is that it may necessitate a different welding procedure from that which the tests are supposed to qualify. Since HAZ toughness depends not only on the plate chemistry but also welding procedure, the HAZ toughness in the K preparation may be different to that in a more conventional preparation. The important variables affecting toughness are thought to be the angle of attack between the electrode and the edge preparation, and the degree of weld bead overlap. To some extent these difficulties can be minimised by tilting the plate to be welded so that the angle of attack and degree of bead overlap are the same on the perpendicular side of the preparation as in the angle bevel, as shown in Fig. 3 for a weld intended to be made in the flat position.

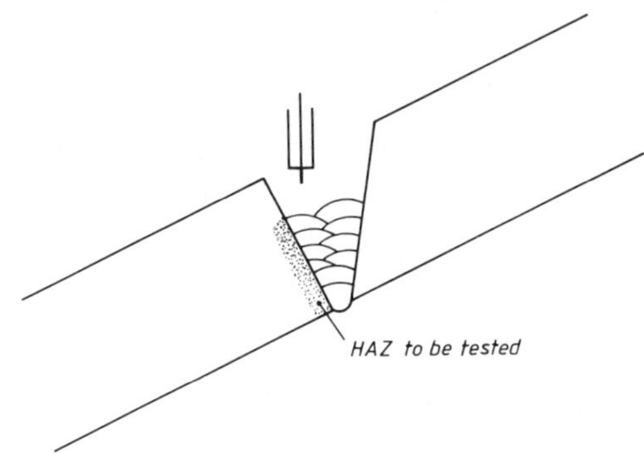


Fig. 3. Tilting the plate in a half K joint in order to simulate the angle of attack and bead overlap of an angled bevel.

The results obtained from preferred CTOD testpieces notched into the vertical side of a K weld may provide too pessimistic an estimate of toughness for the assessment of specific defects. For example, it may be possible through non-destructive examination (NDE) to ensure that defects above a certain size are unlikely in a weld, and therefore that such high levels of constraint are not experienced. Furthermore, the preferred geometry specimen tends to emphasise the properties of the HAZ towards the centre of plate thickness, as this is the region subject to the highest constraint in the test. If the region where defects are most likely is close to the weld surface, then the toughness values obtained may not be relevant.

Estimates of HAZ toughness more appropriate to the structure can be obtained by attempting to model in the test specimen

i. the orientation of the crack likely in the structure;

ii. the constraint that crack will be subject to.

The orientation of the defect can be modelled by placing the notch in the HAZ from the original plate surface, as shown in Fig. 4(a). The experimental difficulties in locating the crack tip in the HAZ can be minimised by using either a K or half K preparation. Alternatively, the test specimens can be extracted from the joint preparation which will be used in the structural weld (Figs 4(b) and 4(c)). A very careful specimen preparation technique is necessary in this case, however in order to ensure that a substantial part of the fatigue crack front is located in the HAZ and not in the weld metal or parent plate.

The precise location of the fatigue crack tip is of special importance in surface

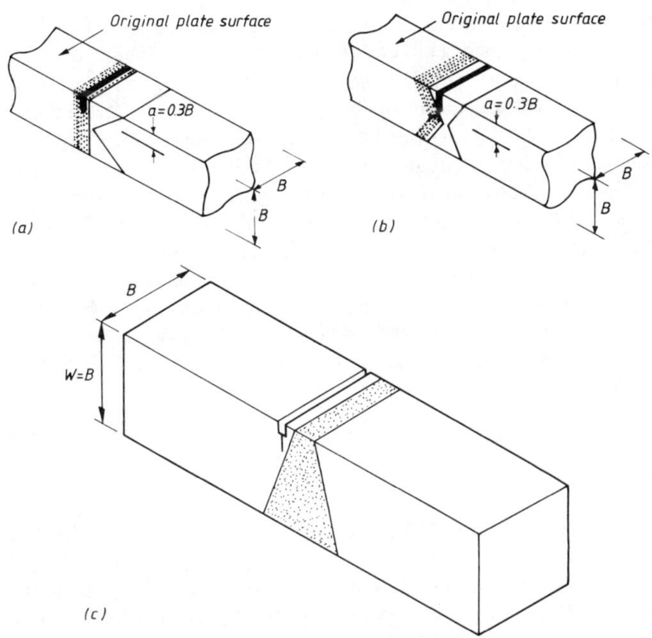


Fig. 4. Surface notches in subsidiary CTOD testpieces.

notched test specimens. With through thickness specimens both high and low toughness regions will be sampled by the notch front, but the fracture will tend to initiate from the lowest toughness region. On the other hand, generally only one specific region will be sampled by the crack tip in a surface notched specimen. If the fatigue crack tip samples fine grained HAZ a higher toughness will be recorded than if the crack tip position is moved only slightly into a grain coarsened region. Thus, with surface notched specimens small changes in crack tip depth can have a significant influence on the toughness value measured. This factor must be appreciated by the experimenter/designer when assessing the results from such tests. The importance of sectioning each specimen after testing to resolve these issues is discussed in the next section.

The problem of modelling the constraint that a defect is subject to in a test specimen is difficult to resolve. Constraint depends on a number of factors, including crack depth, mode of loading and crack shape. It is difficult to quantify but usual practice, based on empirical observations, is to attempt to match the constraint in two ways. First, by ensuring that the thickness of the specimen is the same as the structural section of interest, and secondly by ensuring that the notch depth is equal to or greater than the maximum expected in the structure. The use of shallow notches however may preclude the use of the standad CTOD formula given in BS 5762. This formula is only valid for crack depth to specimen width (a/W) ratios in the range 0.15 to 0.7. Where a/W is less than 0.15 calibration tests will be necessary to calculate CTOD. Alternatively, a double clip gauge technique may be employed so that by extrapolation the CTOD at the original fatigue crack tip can be estimated (Willoughby, 1981). However, a drawback of such procedures is that they are non-standard and are therefore only appropriate in special circumstances.

Shallow notched specimens have been used to evaluate the susceptibility of material to strain ageing embrittlement to which a number of failures have been attributed, especially where repair welding has taken place. One method used to simulate the possible degradation in toughness has been to deposit a single weld bead onto the surface of a plate or into a shallow groove (which has been machined onto the plate). The bead is then notched and a fatigue crack extended into the desired HAZ region. Typically this would be the sub-critical HAZ of the original weld. However, trials may be necessary to establish the most susceptible position. Subsequently, a second bead is deposited over the first to subject the crack tip to a thermo-mechanical cycle. It has been found that welding across the notch rather than parallel to it results in greater embrittlement, and also greater effects have been observed with relatively blunt notches than with fatigue cracks. After re-notching to pick up the original crack tip the specimen is tested conventionally.

## Sectioning of Specimens

Scatter of results observed when testing HAZs can be caused in two ways. Firstly, scatter will occur as a result of inherent material variability within the specific HAZ region sampled, and where tests are conducted at a temperature within the transition regime, scatter can be especially high. Secondly, scatter will occur as a result of experimental errors brought about by the difficulty of placing the crack tip in the same HAZ region in each test. For example, very different toughness values will be measured if in one specimen the tip samples the grain coarsened HAZ, whilst in another it samples the fine grained region. By sectioning

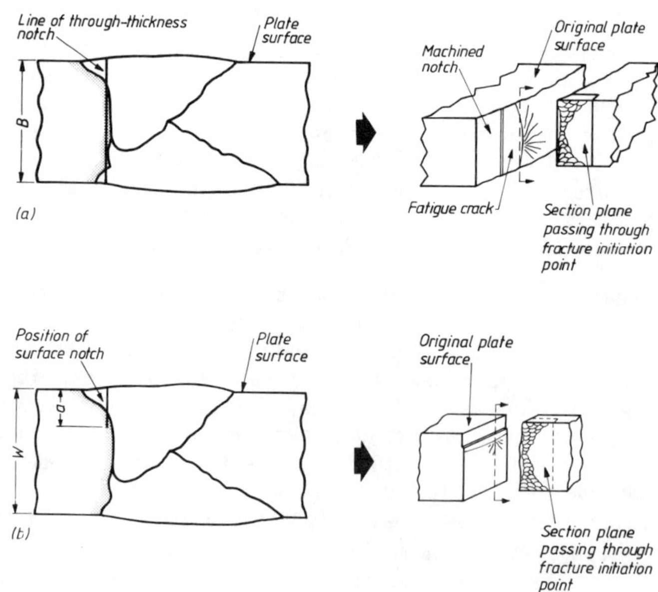


Fig. 5. Examples of sectioning techniques for:
(a) through thickness notched, and
(b) surface notched specimens.

the specimens after the test to identify the microstructures sampled by the crack tip and in particular at the fracture initiation point, the sources of scatter can be isolated. The authors consider that it is vital that all HAZ specimens are sectioned after testing and that sectioning is considered as an integral part of the HAZ fracture toughness test, otherwise judgements may be made on the basis of unrepresentative data.

Examples of sectioning techniques applied to through thickness and surface notched specimens are shown in Fig. 5. At the end of the test the fracture faces are first visually examined to identify the primary initiation point. The weld metal side of each fracture face pair is then sectioned on a plane perpendicular to both the fracture and original plate surfaces. After polishing and etching, the section can be examined to establish the HAZ microstructure sampled. This plane of section which is a standard transverse weld section, is advised as slight errors in locating the plane will only have a slight effect on the microstructure observed. Examples of specimens prepared in this way are shown in Figs 6 and 7 for through thickness and surface notched specimens respectively. The figures show crack tips correctly sampling the region of least toughness, and specimens where the notch tip missed this region and unrepresentative toughness values were obtained. Figure 7(b) shows the notch tip ending in weld metal. During testing the crack first extended by tearing and then jumped sideways through a grain coarsened HAZ and arrested in the fine grain HAZ. Although a CTOD of 0.63mm was recorded, the toughness of the grain coarsened HAZ (measured on another specimen) was only 0.02mm, and the value of 0.67mm merely confirmed that the weld metal toughness was high. Clearly if sectioning had not been carried out misleading estimates of HAZ topughness would have been obtained. This example also illustrates that when testing the HAZ the presence of tearing is not necessarily indicative of high toughness. Tearing may propagate the crack through tough material until it reaches the brittle zone and fracture occurs. The toughness value obtained from such a test is clearly at best an upper bound estimate of toughness. It should be recognised that if sectioning reveals that the region

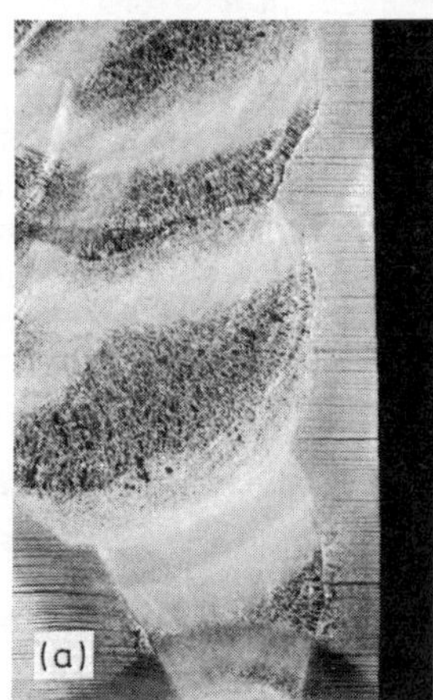


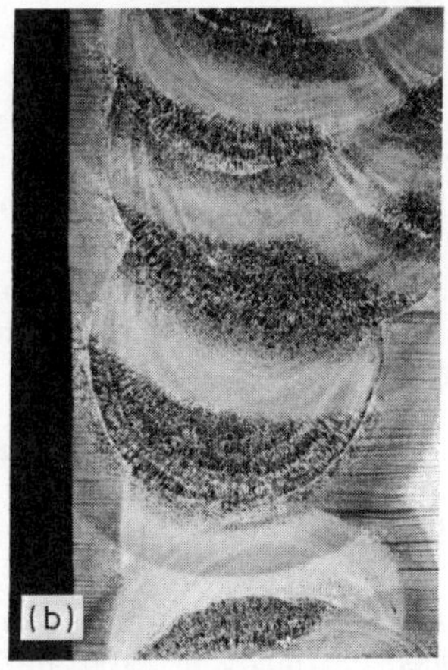


Fig. 6. Sections from through thickness notched CTOD specimens,
(a) $\delta_u$ = 0.55mm,
(b) $\delta_c$ = 0.07mm. (Mag. x3)

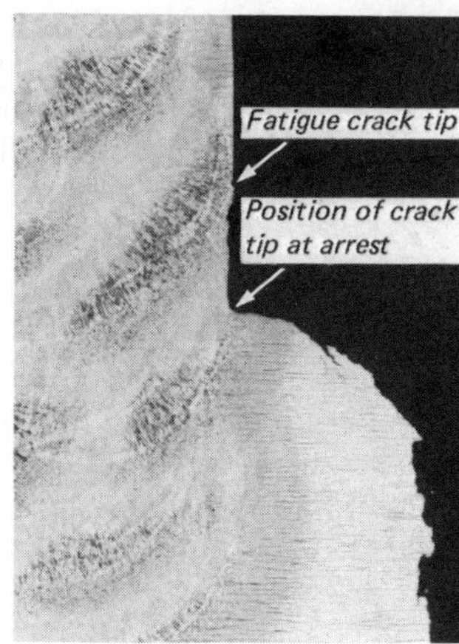

Fig. 7. Sections from surface notched CTOD specimens,
(a) $\delta_c$ = 0.02mm. (Mag. x3).
(b) $\delta_u$ = 0.67mm. (Mag. x8).

of suspected low toughness has not been sampled, further tests should be carried out, and procedure test welds should be large enough to allow for this possibility.

## SIGNIFICANCE OF HAZ TOUGHNESS

In a recent review, Harrison (Harrison, 1983) identified some instances of catastrophic brittle fracture in large engineering components. In all the examples discussed, the fracture initiated from HAZs and in most cases from comparatively shallow surface defects. These were often located in regions of geometric stress concentration. Although the parent material often had excellent initiation toughness (under essentially static loading), once a fracture had initiated in a brittle HAZ the parent material was not able to arrest the running crack. Thus, in general, reliance cannot be placed on maintaining structural integrity by achieving a high initiation toughness in the parent material as this will not necessarily ensure good arrest toughness, and steps must be taken to ensure that the toughness of the HAZ is adequate to prevent initiation there. The same is also true of the weld metal, though this is not the subject of the present discussion.

In order to assess the significance of the fracture toughness values measured in the test specimens described above, fracture assessment procedures such as those described in PD 6493 (BSI, 1980) can be used. However, these provide relatively crude guidance of the assessment of HAZs when located in regions of geometric stress concentration. Improved methods for taking account of the stress gradients present at welds when using the CTOD design curve have been proposed by Kamath (Kamath, 1981). Recently, Smith and others (Smith, 1982) have used a method which combines elastic finite element analyses and the CTOD design curve to assess the CTOD toughness requirements for an offshore platform. Finite elements were used to establish the variation in stress intensity factor with crack length for the toes of butt welds and attachment welds. By using these techniques, realistic estimates of the HAZ toughness requirements necessary to avoid brittle fracture initiation can be established. Where it is not possible to be confident that the toughness values calculated will be achieved, the consequence of brittle fracture initiation must be assessed carefully. The risk may be accepted but if not, changes to the welding procedure and/or material may be necessary to improve toughness. However, it is by no means certain that once a crack has initiated in the HAZ it will move out of it. The direction of crack propagation will depend on many factors, such as, the joint preparation, the relative toughness of the HAZ and adjacent regions and the direction of the principal stress. Assuming that the crack does turn out of the HAZ, then experience indicates that arrest may occur if the operating temperature relative to the nil-ductility transition temperature for the material is above the crack arrest temperature curve (after appropriate correction for plate thickness) in the Pellini Fracture Analysis Diagram (Pellini, 1976). Where these operating temperatures cannot be achieved then more sophisticated fracture mechanics based methods are necessary to demonstrate arrest. These techniques involve the measurement of the arrest toughness $K_{Ia}/K_{IA}$ for the material and careful fracture mechanics assessments.

Nevertheless, even when an arrest capability is demonstrated, the designer must satisfy himself that, in the longer term, structural integrity is not compromised. For instance, extension of the arrested crack may occur during service by fatigue or subsequent brittle fractures during peak loading conditions. Thus the significance of additional crack extension on integrity must be evaluated. It must also be recognised that the presence of the arrested crack may violate limit states other than brittle fracture, e.g. leakage, buckling, plastic collapse. For the reasons discussed above it is prudent to design structures such that the conditions for fracture initiation are avoided. However, to cater for the remote chance of initiation occurring in some local brittle zone such as within an HAZ, an arrest capability in the surrounding material is desirable. Thus structural integrity is ensured by combining both approaches to fracture control.

## CONCLUDING REMARKS

It is essential that a good assessment of HAZ fracture initiation resistance is obtained prior to fabrication, since, structural integrity cannot generally be guaranteed on the basis of parent material crack arrest capabilities alone. Careful consideration should be given to the type of fracture toughness specimen which is to be employed, in order to strike a balance between accurate modelling of the service situation and the provision of standard reproducible results. In view of the small size and inhomogeneous nature of the HAZ, assessment of the results should always include examination of the test specimens by metallographic sectioning to identify which regions of the HAZ have been tested.

## REFERENCES

BAIRD, J. D. (1963). Strain ageing of steel, a critical review. Iron and Steel, May and June, 180-192 and 326-334.

BAIRD, J. D. (1971). The effects of strain ageing due to interstitial solutes on the mechanical properties of metals. Metals and Materials, 5 (2). Review.

BSI (1979). BS 5762. Crack opening displacement (COD) testing. British Standards Institution.

BSI (1980). BS PD 6493. Guidance on some methods for the derivation of acceptance levels for defects in fusion welded joints. British Standards Institution.

DAVEY, T. G. (1981). The effect of welding parameter changes on the heat affected zone microstructure and toughness of single pass welds in 25mm thick C-Mn steels. Welding Institute Confidential Members Research Report 156/1981.

DOLBY, R. E. (1979). HAZ toughness of structural and pressure vessel steels - improvement and prediction. Welding Journal Research Supplement 58 (8), 225s-238s.

EVANS, G. M. (1978). Effect of interpass temperature on the microstructure and properties of C-Mn all weld metal deposits. IIW Doc. No. IIA 460-78.

HARRISON, J. D. (1983). Why does low toughness in the HAZ matter? The Welding Institute Seminar, Coventry, England.

KAMATH, M. S. (1981). The crack tip opening displacement (CTOD) design curve: some proposals for incorporating stress gradient effects. Fitness for Purpose Validation of Welded Constructions, The Welding Institute Conference, London, Paper 23.

OATES, G. (1968). Effect of hydrostatic stress on cleavage fracture in a mild steel and a low carbon manganese steel. JISI 206.2 (8), 930-935.

PELLINI, W. S. (1976). Principals of structural integrity technology. Office of Naval Research, Arlington, VA, USA.

ROTHWELL, A. B. and others. (1979). Heat affected zone toughness of welded joints in micro-alloy steels. Canmet Report 79-6.

SMITH, I. J., H. G. PISARSKI, H. ELLIS, N. J. PRESCOTT (1982). Assessment of fracture toughness requirements for the deck structure of the Hutton Tension Leg Platform using finite elements and the CTOD design curves. 14th Annual OTC, Houston, USA, Paper OTC 4430.

TERASHIMA, H. and P. H. M. HART (1982). Effect of aluminium in C-Mn steels on microstructure and toughness of submerged arc weld metal - progress report. Welding Institute Confidential Members Report 186/1982.

THREADGILL, P. L. ((1981). Titanium treated steels for high heat input welding. Welding Institute Research Bulletin, 22, (7), 189-196.

# RESIDUAL STRESSES AT GIRTH WELDS IN PIPES

R.H. Leggatt
*The Welding Institute, Cambridge, U.K.*

ABSTRACT

The results of previous experimental and theoretical studies of residual stresses at girth welds are drawn together to produce a unified theory for calculating the reaction components of the residual stresses at a multipass girth weld in terms of the weld shrinkage behaviour. The theory is illustrated with reference to experimental measurements in a 15.5 × 610 mm pipe, and is used to conduct a parametric investigation of axial bending stresses at the weld.

The theoretical study confirms the observation based on the review of published measurements that the axial stresses at the weld are primarily a function of wall thickness, and show a clear trend from tensile inner surface stresses in thin walled pipes (below 16 mm) to compressive stresses in thick walled pipes (above 25 mm).

KEYWORDS

Girth welds; residual stresses; wall thickness; tendon force; angular distortion; pipes; thin cylinder theory; bending stresses.

## INTRODUCTION

When making an assessment of the significance of defects in welded joints, it is necessary to make allowance for the residual stresses caused by the weld. In the absence of any better information, the residual stresses in a weld which has not been heat treated are generally assumed to be tensile and of yield magnitude (PD 6493, 1980). However, in cases where the defect lies in a region where the residual stresses are less than yield, this introduces an unnecessary degree of conservatism. One example of particular interest is the root region of circumferential welds in pipes, which may actually be in compression under certain conditions, as will be shown in this article. In general, the accuracy of a fracture assessment can be improved if the residual stress distribution is known. The linear elastic stress intensity factor due to an arbitrary residual stress distribution may be calculated using published solutions for certain simple defect geometries (Tada, 1973; Terada, 1976) or, more generally, using finite element methods. Calculated values of the stress intensity due to residual stresses are used in the CEGB R6 procedure (Harrison, Loosemore and Milne, 1980) and in suggested modifications to the CTOD design curve (Kamath, 1981), and thus allow residual stresses to be included in post-elastic fracture assessment. Hence, accurate knowledge of the residual stress distribution at a weld permits the degree of conservatism in the calculated allowable defect size to be reduced.

## PREVIOUS RESEARCH

The stresses at hoop welds in pipes have been investigated by many previous workers. Rybicki and co-workers (1977) and Makhnenko, Shekera and Izbenko (1970) have carried out transient elastic-plastic thermal analyses of the stresses developed during welding. The latter authors presented some interesting two-dimensional stress contour plots, which illustrate the sensitivity of the stress distribution produced by a given welding procedure to the pipe geometry. Fujita, Nomoto and Hasegawa (1980) used a somewhat simpler elastic-plastic analysis to produce design curves for the stresses generated by single pass welds in pipes of 5 or 10 mm wall thickness. Guan and Liu (1979) and Vaidyanathan, Todaro and Finnie (1973) describe elastic analyses in which the tensile hoop stresses in the weld are treated as an equivalent external radial loading, distributed over the relevant section of the pipe. Then, using classical elastic cylinder theory, the distribution of residual stresses outside the tensile zone is calculated as a function of the pipe geometry and the hoop shrinkage of the weld.

Vaidyanathan and co-workers (1973) also described a simpler approach, applicable when the pipe diameter is large compared with the width of the tensile zone, in which the applied radial load is concentrated at the weld centreline. White and Dwight (1978) use the same concept, and have coined the term "tendon force" to describe the imaginary applied line load which would produce the observed overall shrinkage in the direction of welding. White (1977) had previously shown that the tendon force in flat plates was broadly independent of the plate geometry. Scaramangas and Porter Goff (1981) extended this work to show that the flat plate tendon force, calculated as a function of the welding conditions, was also applicable to hoop welds in cylinders. White and Dwight (1978) introduced another important variable into the analysis: the angular distortion about the weld axis. As will be shown in this article, the angular distortion has a significant (and sometimes dominant) effect on the residual stress distribution in full penetration multipass welds.

Experimental and theoretical values of the axial stress on the weld centreline on the inside of the pipe are plotted in Fig. 1. The data were obtained from Burdekin (1963), Makhnenko (1970), Brust (1980), Leggatt (1982) and various unpublished measurements made at The Welding Institute and Cambridge University Engineering Department. A distinction has been drawn between the actual value of the axial stress on the inner surface ($\sigma_i$) and the value at the inner surface of the bending component of the axial stresses ($\sigma_b$). The theoretical curves on the figure will be discussed in a later section. There is a clear trend in the results from tensile inner surface stresses for thicknesses of 11 mm or less, to compressive stresses for thicknesses greater than 25 mm.

## THEORY

In the present paper, the results of previous research are drawn together to produce a unified theory for calculating the reaction components of the residual stresses at a multipass hoop weld in terms of the tendon force and angular distortion at the weld. The theory is illustrated with reference to experimental measurements in a 15.5 mm thick 610 mm outside diameter pipe, and is then used to conduct a parametric investigation of the axial bending stresses in the root region.

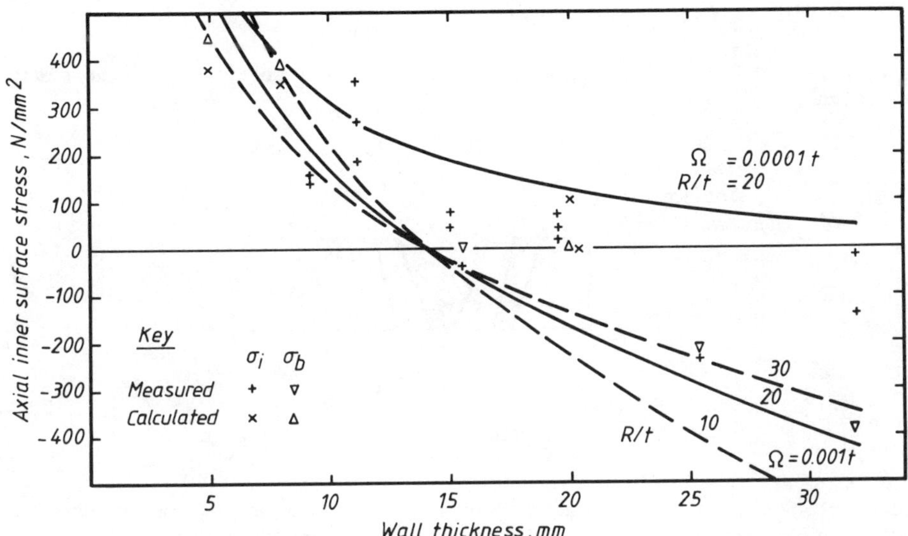


Fig. 1. Variation of inside surface axial residual stress with wall thickness.

Hoop Tendon Force

The tendon force locked into a weld is defined as the imaginary line load which would produce the observed shrinkage parallel to the weld, and is approximately independent of restraint. It may therefore be estimated on the basis of the assumption of infinite restraint against any hoopwise expansion of the heated material during welding. Then, the residual hoop stress at any point is a function of the maximum temperature reached, and the hoop tendon force is obtained by integrating the stress distribution.

For a surface weld bead, the maximum temperature as a function of the distance r from the weld centreline is given approximately by:

$$T_{max} = \frac{2}{e\pi} \cdot \frac{\eta q}{vr^2} \cdot \frac{1}{\rho c} \qquad [1]$$

where $\eta$ is the arc efficiency, q the weld power, v the weld travel speed, and $\rho c$ the volumetric specific heat.

Inside the region where the thermal strain $\alpha T_{max}$ is greater than twice the yield strain $\sigma_Y/E$, yield occurs in compression during heating, and then again in tension during cooling. For thermal strains between $2\sigma_Y/E$ and $\sigma_Y/E$, the material yields in compression during heating, but unloads to a sub-yield tensile strain during cooling. Material whose maximum thermal strain is less than $\sigma_Y/E$ does not yield, and ends at zero stress for the idealised conditions of infinite hoopwise restraint.

The width of the zone containing yield magnitude tensile stresses (Fig. 2) can be found by substituting $T_{max} = 2\sigma_Y/E\alpha$ in equation 1:

$$r_{1,2} = \sqrt{\frac{1}{e\pi} \cdot \frac{E\alpha}{\sigma_Y \rho c} \cdot \left(\frac{\eta q}{v}\right)_{1,2}} \qquad [2]$$

where $(\eta q/v)_1$ and $(\eta q/v)_2$ are the heat inputs for the surface beads on sides 1 and 2 respectively.

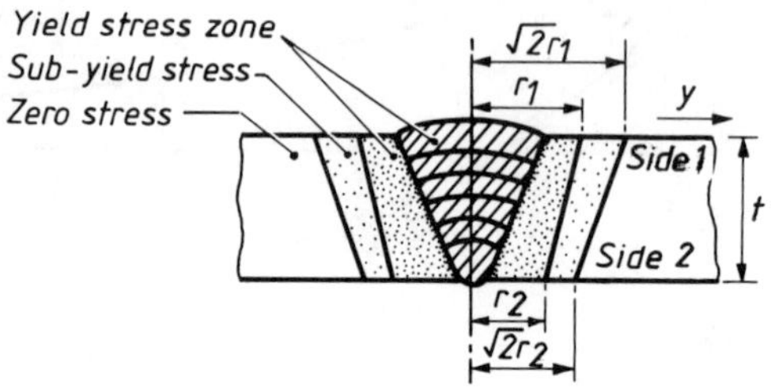


Fig. 2. Hoop tensile zone at girth weld under infinite hoop restraint.

If the envelope of the yield zone is assumed to form a trapezium, as shown in Fig. 2, then the total force locked into the tensile zone is given by:

$$F = A_W \sigma_{YW} + \{1.17t(r_1 + r_2) - A_W\}\sigma_Y \qquad [3]$$

where $A_W$ and $\sigma_{YW}$ are the area and yield strength of the weld, and the factor 1.17 allows for the sub-yield tensile stresses between r and $\sqrt{2}\,r$.

## Angular Distortion

Empirical or semi-empirical formulae for angular distortion due to single or multi-pass butt welds in unrestrained flat plates have been published by several workers (for review, see Leggatt, 1980), and the effect of restraint on angular distortion at single pass welds has also been investigated by Leggatt (1980) and Watanabe and Satoh (1961). However, the scatter of experimental results and the inconsistencies between different investigators' formulae are such that it would be unrealistic to expect to achieve an accuracy of better than 50 per cent in predicting restrained angular distortion at single pass welds. The present author is not aware of any published work in which angular distortion at restrained multipass welds has been systematically investigated, nor of any wide ranging study of the angular distortion at hoop welds in pipes, where there is the additional factor of bending moments about the weld caused by circumferential shrinkages.

The other possible route to distortion prediction is via a full scale elastic-plastic transient thermal model of the stresses and strains developed during welding, such as those currently in use at The Welding Institute and other research centres. However, this is a most complex field of study, still in its infancy, and the computed results cannot be regarded as reliable without extensive experimental verification under relevant welding conditions.

It therefore seems likely that, for the foreseeable future, angular distortion values for use in pipe residual stress calculations will be determined experimentally.

Elastic Thin Cylinder Theory

Timoshenko and Woinowski-Krieger (1959) give the inwards radial deflection due to axisymmetric shear force $Q_o$ and moment $M_o$ per unit length applied to a cylinder (see Fig. 3b) as:

$$w = - \frac{e^{-\beta x}}{2\beta^3 D} \left[\beta M_o(\cos\beta x - \sin\beta x) + Q_o \cos\beta x\right] \qquad [4]$$

and hence:

$$\frac{dw}{dx} = \frac{e^{-\beta x}}{2\beta^2 D} \left[2\beta M_o \cos\beta x + Q_o(\cos\beta x + \sin\beta x)\right] \qquad [5]$$

$$\frac{d^2w}{dx^2} = - \frac{e^{-\beta x}}{2\beta D} \left[ 2\beta M_o(\cos\beta x + \sin\beta x) + 2Q_o \sin\beta x\right] \qquad [6]$$

where $\beta = \left[\frac{3(1-\nu^2)}{R^2t^2}\right]^{\frac{1}{4}}$, and $D = \frac{Et^3}{12(1-\nu^2)}$

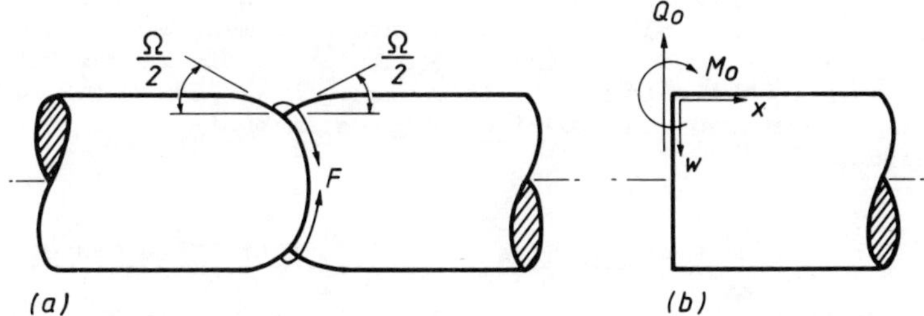


Fig. 3. (a) Shrinkage parameters Ω and F.
(b) Corresponding loads at weld centreline.

In order to model the deflections due to a hoop weld centred at x = 0, the applied forces $Q_o$ and $M_o$ are related to the angular distortion Ω and tendon force F which characterise the shrinkages of the weld.

The radial shear force is related to the tendon force F by considering radial equilibrium at x = 0:

$$Q_o = \frac{-F}{2R} \qquad [7]$$

The bending moment $M_o$ is found by setting $\frac{dw}{dx} = -\frac{\Omega}{2}$ at x = 0 in [5]. Hence:

$$M_o = \frac{-1}{2\beta}(\beta^2 D\Omega + Q_o) \qquad [8]$$

Substitution for $M_o$ and $Q_o$ in [4], [5] and [6] gives the radial deflection and its derivatives at any distance from the weld. Outside the tensile zone, the membrane component of the hoop stress is proportional to the radial deflection:

$$\sigma_{\theta m} = - \frac{Ew}{R} \qquad [9]$$

The axial bending stresses on the inner and outer surfaces at any section inside or outside the tensile zone are a function of the curvature:

$$\sigma_{xb} = \mp \frac{Et}{2(1 - \nu^2)} \cdot \frac{d^2w}{dx^2} \qquad [10]$$

The hoop curvature $\frac{1}{R} \cdot \frac{d^2w}{d\theta^2}$ is zero because of axial symmetry; hence the hoop stresses have a Poisson bending component:

$$\sigma_{\theta b} = \nu \sigma_{xb} \qquad [11]$$

## EXPERIMENTAL PROGRAMME

Measurements were made on a girth welded pipe specimen designated W65. The pipe dimensions were 610 mm outside diameter, 15.5 mm wall thickness and 400 mm length. The girth weld was at mid length. The pipe material was API-5L-X65, with $\sigma_Y$ = 510 N/mm$^2$, E = 207 000 N/mm$^2$, $\nu$ = 0.3. The weld was made using the Saipem/Arcos Passo Automatic Pipeline Welding System, with two welding heads which moved simultaneously round opposite sides of the pipe, starting at the twelve o'clock position and meeting at six o'clock. The abutting ends of the pipe were prepared with a single vee preparation, and were brought into alignment with an internal jacking system. The weld was deposited from the outside in six passes with an average heat input of 800 J/mm. The weld metal yield strength was 540 N/mm$^2$. Fuller details of these welding procedures and experimental programme are given in a previous report (Leggatt, 1982).

### Measurement of Surface Residual Stresses using Centre Hole Gauges

Residual stresses on the surfaces of the specimens were measured using centre hole rosette gauges. The centre holes were formed using a rotating air-abrasive jet (Beaney, 1976).

Measurements were made of the residual stresses on the inner and outer surface of specimen W65 at various distances from the weld centreline at the four o'clock position. This position was chosen to be as far as possible away from any axial asymmetry caused by the weld stop position or the axial seams of the abutting pipes. The measured stresses will be discussed later in conjunction with measurements made by the relaxation method.

### Measurement of Linear Components of Residual Stresses by Relaxation Method

The linear component of the residual stresses is the sum of the membrane and bending components of the through-thickness distribution of in-plane residual stresses at any section. The linear components of the stresses were measured between the three and four o'clock positions on specimen W65 using the relaxation method. This involved sawing out small strain gauged blocks from the pipe. The relaxed strain in the hoop and axial directions were measured on separate blocks at eight different positions along the length of the pipe.

Residual stresses measured by relaxation and centre hole methods are plotted on Fig. 4, along with a theoretical stress distribution which will be discussed later. In the elastic region, i.e. at distances more than 30 mm from the weld, the two measurement methods generally agreed to within 50 N/mm$^2$. The relaxation measurements appeared to give a smoother distribution, indicating that the surface stresses were subject to some scatter, possibly due to surface damage.

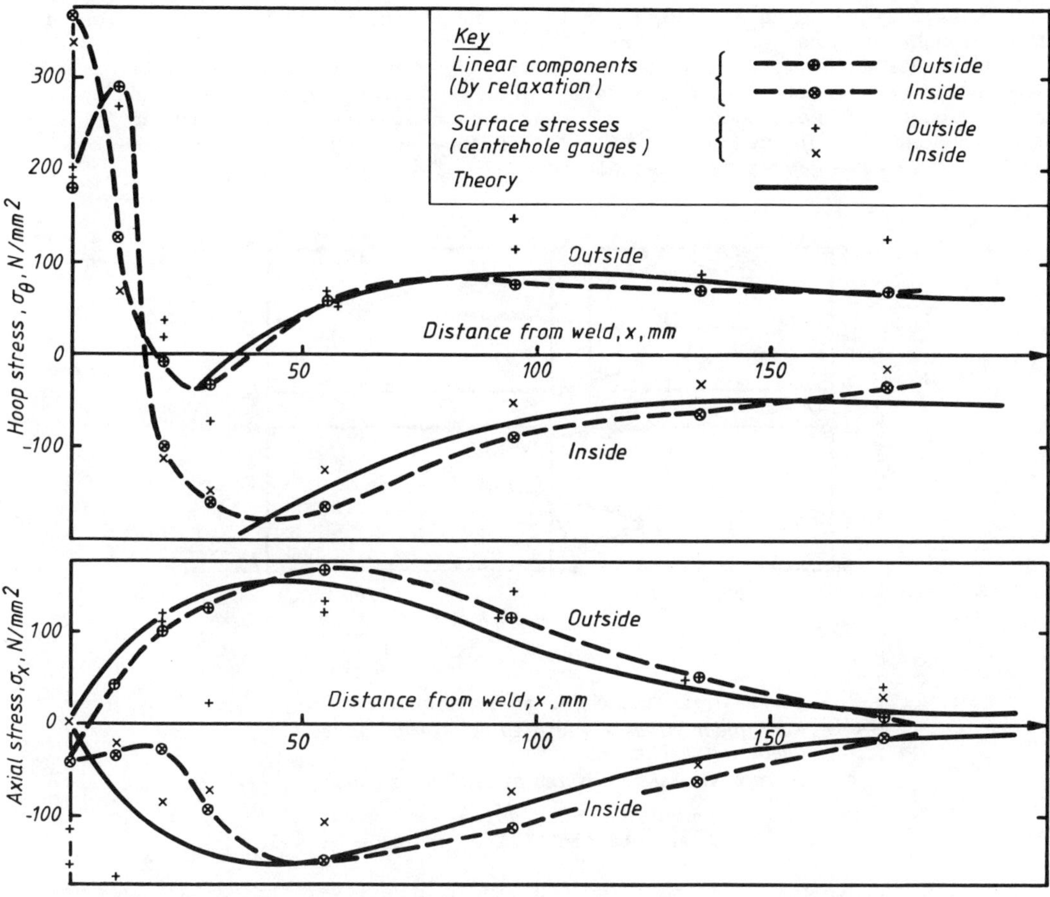


Fig. 4. Axial variation of hoop and axial stresses.

The hoop stresses were tensile in the weld and adjacent parent metal, going into compression at about x = 18 mm and then blending into the background elastic distribution at about x = 35 mm. At positions remote from the weld, the stresses did not go to zero, but tended towards a steady value of about ± 60 N/mm$^2$ which is assumed to have been caused by hoop bending stresses locked in during fabrication.

The axial stresses were predominantly tensile on the outside and compressive on the inside, and tended to zero at the weld centreline and at x = 175 mm. However, the axial stresses on the outer surface were significantly different from the linear components in the weld, which suggested that the through-thickness distribution was strongly non-linear near the centreline. The occurrence of large compressive stresses on the outer surface directly contradicts the conventional assumption of yield magnitude tensile stresses, and was further investigated by layering of the axial relaxation block.

## Measurement of Through-Thickness Variation of Axial Stresses by the Layering Method

The gauges attached to the inner surface of the axial relaxation block were used to monitor the strain changes when layers 1 mm thick were successively removed from the outer surface using a shaping machine. The theory for the reconstruction of the through-thickness stress distribution from the relaxed strains is described by Leggatt and Kamath (1981), and is a simplified version of the method originally described by Rosenthal and Norton (1945). The non-linear stresses remaining in the relaxation blocks were added to the linear components previously measured by relaxation, and are plotted on Fig. 5. The surface stresses previously measured using centre hole gauges are also plotted.

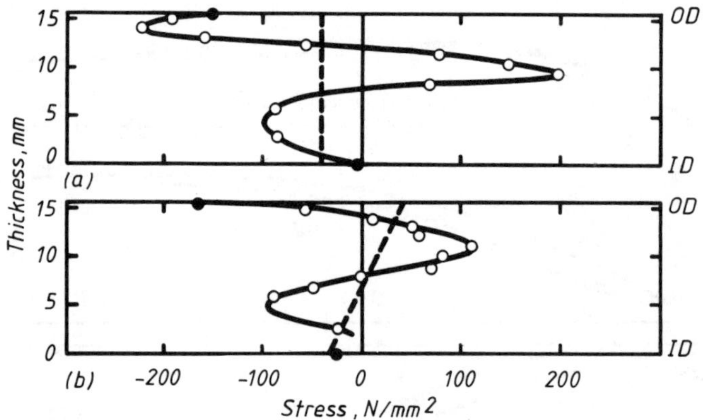


Key

- - - - Linear stress components measured by relaxation
—o— Non-linear stresses measured by layering added to linear components
● Surface stresses measured by centrehole gauge

Fig. 5. Through-thickness variation of axial stresses. (a) On weld centreline (b) Edge of weld.

The graphs show that there were large non-linear stress variations through the thickness in the weld metal, with a range of about ± 200 $N/mm^2$ on the centreline, diminishing to ± 100 $N/mm^2$ at the edge of the weld. At 20 mm and more remote locations there were no significant non-linear stresses. The surface stresses were generally in line with the trend of the underlying values, though perfect agreement could not be expected as the two sets of readings were made at locations separated by 80 mm in the hoop direction. The stresses just below the outer surface of the weld metal were compressive, which explains the apparent discrepancy between the relaxation and surface measurements at this point.

## Measurement of Angular Distortion

The deflection profile of the pipes was measured at one position on specimen W65 and at five positions on a similar specimen by scanning along the pipe with a dial gauge sliding on a straight rigid slideway. Measurements were made at 10 mm intervals over a span of 300 mm. The angular distortion was deduced by fitting tangents to the measured deflection profiles at the weld. The results are given in Table 1, and were similar at all positions except twelve o'clock, the weld start position, where the distortion was about 50 per cent larger.

TABLE 1 Angular Distortion

| Specimen | Position, o'clock | Angular Distortion, degrees |
|---|---|---|
| W65 | 3 | 0.72 |
| W74 | 3 | 0.71 |
| W74 | 4 | 0.60 |
| W74 | 6 | 0.60 |
| W74 | 9 | 0.62 |
| W74 | 12 | 0.95 |
| Average (excluding 12 o'clock) | | 0.65 |

## APPLICATION OF THEORY

### Present Results

The theory described earlier in this article has been used to calculate the membrane and bending stresses in the pipe specimen W65 as a function of pipe geometry, material properties and weld shrinkage parameters $\Omega$ and F using equations [4] to [11]. The tendon force calculated assuming $(q/v)_1$ = 1000 J/mm, $(q/v)_2$ = 800 J/mm, $\eta$ = 0.8, $A_W$ = 148 $mm^2$ and $\alpha/\rho c$ = 0.0034 $mm^3$/J, was F = 200400 N/$mm^2$. The measured angular distortion was $\Omega$ = 0.01256 radians. The derived values of the centreline loads were $Q_o$ = -343 N/mm and $M_o$ = 661 N. An assumed hoop bending stress of 60 N/$mm^2$ and a corresponding Poisson axial bending stress of 18 N/$mm^2$ were added to those given by [10] and [11] to allow for locked-in bending stresses caused by the fabrication of the axial seam. The hoop and bending components were summed to give the surface stresses and are plotted on Fig. 4.

It should be noted that the theory applies only to the reaction components of the stresses and does not attempt to predict the hoop tensile stresses or the non-linear component of the through-thickness variation of axial stresses in the immediate vicinity of the weld. The agreement between theory and experiment is good. The overall shape of the distributions (harmonic and exponential decay) confirms the applicability of elastic thin cylinder theory. The correct prediction of the magnitude and position of the maximum stresses validates the use of the weld shrinkage parameters $\Omega$ and F to characterise the stress distribution.

### Parametric Survey of Axial Bending Stresses

The axial bending stress at the weld centreline can be obtained by combining [6], [7], [8] and [10] and setting x = 0. Hence, the bending stress on the inside surface is given by:

$$\sigma_{xo} = \frac{3}{\beta t^2}\left[-\beta^2 D\Omega + \frac{F}{2R}\right] \qquad [12]$$

This formula has been used to calculate $\sigma_{xo}$ for the following range of input parameters:

| | |
|---|---|
| Thickness | : $6 < t < 32$ mm |
| Radius | : $R/t = 10$, 20 or 30 |
| Tendon force | : $F = t \times 12500$ N |
| Angular distortion | : $\Omega = 0.0001t$ or $0.001t$ radians |

The selected values of t and R cover the range of dimensions for which experimental results are available. The tendon force F is theoretically proportional to t and $r_1$ (the radius of the yield zone, see [3]) but, since $r_1$ is controlled by the weld heat input and is independent of the pipe dimensions, F is in practice largely dependent on t. As stated previously, there are no consistent data available for the prediction of angular distortion at girth welds. However, there is some evidence that the angular distortion at an unrestrained multipass butt weld is roughly proportional to the number of passes (Hansen, 1973), which in turn is roughly proportional to the wall thickness. Values of $\Omega = 0.0001t$ or $0.001t$ radians have been adopted to give an indication of the likely range of bending stresses for the cases of low and high angular distortion.

The calculated values of the inner surface axial bending stress are plotted in Fig. 1 and compared with the published values described previously. In view of the approximate nature of the assumed values of F and $\Omega$, the theoretical curves are reasonably successful in providing bounds to the available data. Many of the data points lie close to the lower bound curves corresponding to $\Omega = 0.001t$ radians. However, more tensile stresses were found in a number of specimens, indicating either a low angular distortion or a high tendon force. Such conditions are likely to occur when the weld is laid using a high deposition rate process.

The curves allow certain general conclusions about the nature of axial stresses at hoop welds to be drawn. The stresses are primarily a function of thickness, and are relatively insensitive to the pipe radius. For thicknesses less than 16 mm, the inner surface stresses are always tensile, and are approaching tensile yield for thicknesses less than 10 mm. For thicknesses greater than 25mm, all reported values of the inner surface stress are compressive, and there would therefore appear to be a good case for abandoning the conventional assumption of yield magnitude residual stresses for root defects under these conditions. In the intermediate thickness range $16 < t < 25$ mm, the inner surface stresses may be of either sign. In critical fracture assessments, accurate knowledge of the through-thickness distribution of axial stresses will almost certainly show that the maximum residual stress is less than tensile yield, and will therefore lead to a significant increase in the tolerable defect size.

An interesting corollary is that the application of a stress relief heat treatment will lower the compressive stresses in the root in thick walled pipes, and may therefore decrease the fracture strength or fatigue endurance of the structure. Reduction of fatigue life in stress relieved specimens with root defects has been observed in tests at The Welding Institute (unpublished).

## SUMMARY AND CONCLUSIONS

Measurements have been made of the residual stresses and angular distortion at a six pass circumferential butt weld in a 610 mm × 15.5 mm pipe.

1. The hoop stresses were found to be tensile in the weld and adjacent parent metal, and then decreased rapidly with distance from the weld before

stabilising to a steady, non-zero value associated with hoop bending effects assumed to have been locked in during fabrication of the seam welds.

2. The axial stresses were predominantly bending stresses, tensile on the outer surface and compressive inside. The stresses were at a maximum of ± 130 N/mm$^2$ at about 50 mm from the weld, and tended to zero at the weld centreline and at large distances from it.

3. Measurements of the local stresses on the surface using centre hole gauges generally agreed well with measurements of the linear components of stress using relaxation methods, except in the case of inner surface axial stresses in the weld.

4. Measurement of the through-thickness distribution of axial stresses in the weld by the layering method showed that the stresses were compressive on the outer surface, had a peak tensile value of 200 N/mm$^2$ 6 mm below the surface, and were zero at the inner surface.

A theory was presented which relates the axial distribution of residual stresses to pipe geometry, hoop tendon force and angular distortion. The tendon force can be calculated from the welding conditions, but the angular distortion must be determined experimentally. The theory agreed well with the experimental results. It was used to conduct a parametric survey of axial inner surface bending stresses, and explained the observed trend in published results from tensile stresses for thin walled pipes ($t < 16$ mm) to compressive stresses in thick walled pipes ($t > 25$ mm).

## ACKNOWLEDGEMENTS

This investigation was funded jointly by the Engineering Materials Requirements Board of the Department of Industry and Research Members of The Welding Institute. The author is grateful to Mr. A. Scaramangas of Cambridge University Engineering Department for the provision of residual stress data which will be included in his doctoral thesis.

## REFERENCES

Beaney, E. M. (1976). Accurate measurement of residual stresses on any steel using the centre hole method. Strain, 2 (3), 99-106.

Brust, F. W., Kanninen, M. F. and Kuznetsov, E. N. (1980). Methods for controlling weld-induced residual stresses. In Workshop Conference on Computations and Measurements of Residual Stresses and Crack Growth in Welded Structures, Battelle Columbus Labs., 62-68.

Burdekin, F. M. (1963). Local stress relief of circumferential butt welds in cylinders. British Welding Journal, 10 (9), 483-490.

Fujita, Y., Nomoto, T. and Hasegawa, H. (1980). Deformations and residual stresses in butt-welded pipes and spheres. IIW Doc. X-963-80.

Guan, Q. and Liu, J. D. (1979). Residual stresses and distortion in cylindrical shells caused by a single pass circumferential butt weld. IIW Doc. X-929-79.

Hansen, B. (1973). Formulae for residual welding stresses and distortions, Svejsecentralen, Copenhagen.

Harrison, R. P., Loosemore, K. and Milne, I. (1980). Assessment of the integrity of structures containing defects. CEGB Report R/H/R6 - Rev. 2.

Kamath, M. S. (1981). The crack tip opening displacement (CTOD) design curve: some simple proposals for incorporating stress gradient effects. Weld. Inst. Research Report 147/1981.

Leggatt, R. H. (1980). Distortion in Welded Steel Plates. Ph.D. Thesis, Cambridge University.

Leggatt, R. H. and Kamath, M. S. (1981). Residual stresses in 25 mm thick weld metal COD specimens in the as-welded and locally compressed states. Weld. Inst. Research Report 145/1981.

Leggatt, R. H. (1982). Residual stresses at circumferential welds in pipes. Weld. Inst. Res. Bulletin, 23, 181-188.

Makhnenko, V. I., Shekera, V. M. and Izbenko, L. A. (1970). Special features of the distribution of residual stresses and strains caused by making circumferential welds in cylindrical shells. Automatic Welding, 23 (12), 43-48.

PD 6493 (1980). Guidance on some methods for the derivation of acceptance levels for defects in fusion welded joints. British Standards Institution.

Rosenthal, D. and Norton, J. T. (1945). A method of measuring triaxial residual stresses in plates. Weld. Jnl., 25 (5), 295-307.

Rybicki, E. F. and others (1977). A finite element model for residual stress in girth-butt welded pipes. In Numerical Modelling of Manufacturing Processes, ASME Winter Meeting, Atlanta, 131-142.

Scaramangas, A. and Porter Goff, R. F. D. (1981). Residual stress and deformation in welded pipe joints. In Joints in Structural Steelwork, Proc. Int. Conf., Middlesbrough, UK.

Tada, H., Paris, P. and Irwin, G. (1973). The Stress Analysis of Cracks Handbook. Del Research Corporation.

Terada, H. (1976). An analysis of the stress intensity factor of a crack perpendicular to the welding bead. Eng. Frac. Mechanics, 8, 441-444.

Timoshenko and Woinowski-Krieger (1959). Theory of Plates and Shells. McGraw Hill.

Vaidyanathan, S., Todaro, A. F. and Finne, I. (1973). Residual stresses due to circumferential welds. Trans. ASME, J. Eng. Materials & Tech., 95 (4), 233-237.

Watanabe, M. and Satoh, K. (1961). Effect of welding conditions on the shrinkage distortion in welded structures. Weld. Jnl., 40 (8), 377s-384s.

White, J. D. (1977). Longitudinal shrinkage of a single pass weld. Cambridge Univ. Eng. Dept. Tech. Report No. CUED/C-Struct/TR57.

White, J. D. and Dwight, J. B. (1978). Residual stresses in large stiffened tubulars. Cambridge Univ. Eng. Dept. Tech. Report No. CUED/C-Struct/TR67.

Note: Welding Institute Reports may be restricted to Research Members of The Welding Institute.

# STRAIN ANALYSIS OF HOT TAP BRANCH CONNECTIONS UNDER EXTERNAL LOADINGS

Prepared by B. Gross*, D.J. Warman* and Dr. A.G. Glover**
*TransCanada PipeLines Ltd.
**Welding Institute of Canada

ABSTRACT

Branch connections made onto operating pipelines, known as hot taps, were evaluated under bending loads. A full encirclement reinforcing saddle, which is incorporated into the hot tap assembly, was examined in terms of its ability to reduce the stress concentration at the branch to mainline junction. External loads were applied to an ex-service hot tap and the resulting stress network was measured. The strain distribution across the joint was established and the effectiveness of the reinforcing saddle was ascertained by applying further loadings after the reinforcement was removed. The results have shown that the incorporation of a reinforcing saddle produces a significant reduction in the stress concentration at the mainline to branch junction.

KEYWORDS

Hot tap, strain analysis, bending stresses, reinforcement.

## INTRODUCTION

A hot tap is a method of attaching a branch pipe to an existing gas transmission line without interruption to the flow of gas in that line. The method consists of attaching a branch pipe directly to the mainline using a full penetration tee joint groove weld. This branch, and the mainline pipe are then reinforced in the immediate vicinity by a full encirclement sleeve.

Gas is subsequently admitted into the branch by attaching a full opening valve through which a pressurized internal boring machine drills out the material contained within the branch to mainline joint allowing gas to enter the branch. The boring machine is then retracted back through the valve which is then closed thus allowing the boring machine to be removed. A branch line can then be attached via the flange to the valve completing the hot tap network.

Hot taps have generally been regarded as a temporary or semi-permanent fixture to be replaced by a purpose-made forging when conditions permit. The high cost of replacing such hot taps has prompted an investigation into their integrity with the intent of making them a permanent fixture.

Branch connections used in gas transmission must be capable of not only withstanding the internal gas pressure but in some cases be required to withstand external loadings produced by ground movement or inadequate support. A brief review is given here which looks at the ability of the branch connection or tee joint to deal with these loadings.

Mathematical analyses of pressurized outlets under external loadings have been available for some time. Bylaard (1) carried out a theoretical study of outlets in pressure vessels but these were limited to d/D ratios of 0.25. His work has since been developed (2) to cover a wide range of outlets. The stress concentration factors (SCF) associated with tee branches can be assessed using various methods.- Lind's area replacement method (3) provides a simple method for determining the SCF however, the actual stress levels are not known so a high safety margin is used.

Finite element analysis is available for an analytical assessment of the stresses associated with branch connections and has been used successfully as a basis for further experimentations (4).

The widespread use of tubular sections for structural applications has produced various design criteria and strength data for tubular connections. Wardenier (5) outlines the criteria of failure for tubular connections under tension and compression, Figure 1.

Various modes of failure have been exhibited, Figure 2, but generally failure occurs by ductile tearing along the weld toes after considerable plastic deformation. Failure due to plastic instability occurs under compressive loadings. Toprac (6) found that generally all compressive specimens were less rigid than theoretically calculated and plastic flow occurred earlier than predicted.

Hardenbergh and Zamrik (7) noted that external loadings produce larger stresses on the outside surface than the inside. Also transverse bending moments produce higher stress levels than comparable longitudinal moments. This effect is more noticeable with smaller outlets (300 mm) than larger ones (450 mm ).

Joint design has a significant effect upon performance. Hardenbergh andZamrik applied external loadings and internal pressures to tee connections which differed only in their sharpness of contour from the branch pipe to main shell. Their results, reproduced in Table 1, clearly show that contouring the nozzle effectively reduces the critical stresses, particularly for transverse loadings.

The stress distribution around the tee joint groove weld under external loadings has been identified by Riley (8). He tested thin walled branches (508 mm OD x 6.6 mm WT) under various loading modes with and without internal pressure. Some of his results are reproduced in Figures 3,4, and 5.

The diagrams indicate that a transverse moment (Figure 3) produces the most severe stress concentration on the groove weld at the 90° position (with respect to the longitudinal axis). Internal pressure (Figure 4) and longitudinal bending moments (Figure 5) produce a similar stress distribution network with peak stresses occurring at 60 - 70°. This compares well to Decock (9) who found maximum stress concentrations at 45 - 60° depending on the combination of internal pressure and external loading applied. All stress concentrations were found to occur at the inside crotch corner.

Longitudinal bending moments produce a stress reversal on the inside surface of the pipe close to the fillet weld (Figure 6). This effect was noted in this study and is thought to be due to the Poisson effect which raises stresses in the hoop direction as they become lower in the axial direction.

## Reinforcement of Branch Connections

The SCF associated with the tee connection can be reduced by reinforcement of the joint. Hot taps made in the Canadian gas industry are reinforced according to CSA Z184 (10) which requires that the reinforcement in the crotch section of a welded branch connection shall be determined by the rule that the metal area available for reinforcement shall be equal to or greater than d x t, the diameter and thickness of removed header material. Methods available for this include the reinforcing pad or saddle, encirclement sleeve, tee or saddle and the combined encirclement saddle and sleeve. For larger pipe diameters the encirclement saddle is favoured (Figure 7).

This conservative approach provides:
a) Reinforcement to compensate for removal of mainline under branch.
b) Crack arrestor in the event of fracture propagating from the mainline to branch fillet weld.
c) Diversion of stress from mainline/branch junction during external loading.

Previous work has generally been based on unreinforced branch connections with little work on reinforced connections. This work examines the effectiveness of this relatively simple saddle reinforcement.

## PROCEDURES AND RESULTS

### Test #1

Experimental Details. The weldment evaluated was an ex-service 355.6 mm OD x 12.7 mm WT branch connected to a 508 mm OD x 6.6 mm WT mainline. A 3050 mm section of 356 mm diameter pipe was butt welded to the branch allowing a longitudinal bending moment to be applied to the assembly via a hydraulic ram located at the free end of the section. Strain gauges as indicated in Figure 8 were mounted in three general areas. They were:
1) Along a quarter section of the branch, about the saddle/branch fillet weld on the 0°, 45° and 90° axis.
2) Along the longitudinal plane of the branch on the 0° and 180° axis.
3) On the inside surface of the mainline directly underneath the tee groove weld on the 0°, 45° and 90° axis.

The mainline pipe was strapped vertically to a heavy steel framework which opposed the applied bending moment. Loads in the longitudinal plane were applied via the hydraulic ram in increments of 2.45 kN up to a maximum load of 12.25 kN. Strain readings were taken after each load increment to ensure linear consistency, however, the stress distribution is based on the peak loading of 12.25 kN. As an additional measurement a clip gauge was used to determine if any movement occurred between the encirclement sleeve and the mainline.

Results. Graphs of measured strain and calculated strain are shown in Figure 9. Observations of the results indicated a general agreement between theory and the stress distribution along the main body of the branch pipe with the maximum stresses occurring in plane with the hydraulic ram at the 12 and 6 o'clock positions. Approaching the fillet weld connecting the branch line to the encirclement sleeve from the free end, it was noticed that the strain readings became very low at the 6 o'clock region. This disagrees with classic beam theory and indicates that there is a redistribution of stress due to the influence of the saddle. The gauges mounted along a quarter section of the branch pipe give a good indication of this redistribution of stress. A complete review of the strain network can be seen in Table 2 and details in Figure 10 and 11.

The principal strains, and directions for the strain rosettes inside the main pipe around the bore are shown in Figures 12 and 13. Observations indicated that the overall strains inside the main pipe were larger than those on and in the branch. This is to be expected since the wall thickness of the mainline is approximately one half that of the branch pipe.

In accordance with classic beam theory maximum shear stress occurs at the point of minimum bending stress on the neutral axis. This was shown by a high strain reading at gauge #28 (refer to Figure 10, neutral axis). The principal stresses in this region acted at angles of 45° which is in agreement with beam theory.

The stresses at the interface of the tension leg of the joint (rosette #3) indicate that the longitudinal tensile stresses in the branch are transferred into tensile hoop stresses in the mainline. The magnitude of these stresses are of particular interest since they lie in a region of high stress when acted upon by internal pressure (8).

The results of the clip gauge measurements indicated that some combination of lateral and perpendicular outward movement occurred between the encirclement sleeve and the mainline to an extent of 0.28 mm at a loading of 12.25 kN. This tends to suggest that a tight fit was not made between the mainline and the encirclement sleeve, hence lowering the effectiveness of the sleeve. In practice the external sleeve was assembled around the mainline pipe while under internal pressure. Consequently some of the effectiveness of the sleeve is lost after depressurization. The sleeve would actually perform better in the field than results indicate.

## Test #2

Experimental Details. For this test the encirclement saddle was removed and the corresponding strain gauges replaced. The location of the gauges are as shown in Figure 8. The same loading procedures as test #1 were followed and the strain readings measured accordingly.

Results. The gauges mounted inside the main pipe about the bore indicated the same principal stress directions as produced in test #1 with the encirclement sleeve (see Figure 12 and 13). However, the stress magnitudes were much greater in this case.

The gauges mounted on the branch pipe give virtually a mirror image across the neutral axis indicating validity in our results. The gauges mounted inside the pipe agree extremely well with predicted values except for the gauge in the vicinity of the branch/mainline interface. Once again it is felt that the stresses are redistributed towards the 45° axis and that a secondary bending stress is produced due to the geometry of the interface. This is in direct agreement with Riley's (9) findings.

Contrary to beam theory, the strain gauges situated on the tension leg of the branch pipe in the immediate vicinity of the weld interface indicate a compressive strain. This is due to a geometrical bending effect and is beyond the scope of this report.

## EFFECTS OF ENCIRCLEMENT SLEEVE

A direct comparison can be seen in Figures 12 and 13. The major benefits are shown in Figure 12 and indicate that the critical stresses in the vicinity of the tee joint groove weld produced by an external moment are reduced by a factor of 40%. The presence of the sleeve also reduces the overall stress in the branch pipe and redistributes the stress in the pipe segment.

The reinforcement does however produce a SCF at the toe of the fillet weld in the 45° axis and introduces a secondary bending moment in the branch. It must be noted though, that this secondary moment produces stresses lower than would be present without the external sleeve.

## CONCLUDING REMARKS

The main purpose of the encirclement sleeve is its ability to re-distribute external bending stresses away from the key groove weld. The results of the present findings suggest that reductions of 40% are found in the stresses in the vicinity of this weld. In practice this reduction will be somewhat higher due to the tighter fit between the encirclement sleeve and the mainline when internal pressure is maintained.

1. The highest stresses due to bending occur in the region of the tee joint groove weld as predicted.

2. The reinforcement sleeve adequately reduces these stresses at the key girth weld under external bending conditions.

3. The fillet weld at the interface of the branch pipe and encirclement sleeve produces a secondary moment which introduces small increases in hoop and longitudinal stresses. However, these stresses are lower than would be present without the external sleeve.

## ACKNOWLEDGEMENTS

The authors wish to thank their co-workers at TransCanada PipeLines and the Welding Institute of Canada for their assistance and comments on this project.

## REFERENCES

1. Bylaard, P.O., "Stresses from Radial Loads and External Moments in Cylindrical Pressure Vessels", Welding Journal, 1955, 12, p 609s.

2. Wichman, K. R., Hopper, A.G. and Mershon, J. L., "Local Stresses in Spherical and Cylindrical Shells Due to External Loadings", WRC Bulletin 107.

3. Lind, N.C., "Approximate Stress-Concentration Analysis for Pressurized Branch Pipe Connections".

4. Pereira, M.F.S. & Turner, C.E., "The Computation of Stress Intensity Factors in Cracking Thick-Walled Cylinders and T-Junctions Using a Standard Finite Element Program", I Mech. E. Conference C94/78.

5. Wardenier, J., "Design Rules for Predominantly Statically Loaded Welded Joints in Circular Hollow Sections", IIW Doc. XV-436-79.

6. Toprac, A.A., "An Investigation of Welded Steel Pipe Connections", WRC Bulletin 71.

7. Hardenberg, D.E. and Zamrik, S.Y., "Effects of External Loadings on Large Outlets in a Cylindrical Pressure Vessel", WRC Bulletin 96.

8. Riley, W.F., "Experimental Determination of Stress Distributions in Thin-Walled Cylindrical and Spherical Pressure Vessels With Circular Nozzles", WRC Bulletin 108.

9. Decock, J., "External Loadings on Nozzles in Cylindrical Shells", CRIF C78/80.

10. Canadian Standards Association Z184 Gas Pipeline Systems, Pipeline Systems and Materials.

Table 1: Effects of Sharpness of Transition from Branch Pipe to Vessel upon the Magnitude of Critical Stresses (From Riley [7])

| | | P = 1000 psi | | Bending: M = 500 kips in. | | | | P = 30 kips | |
|---|---|---|---|---|---|---|---|---|---|
| | | Pressure | | Longitudinal | | Transverse | | Thrust | |
| Nozzle | Geometry | σ max, ksi | σ/σR | σ max, ksi | σ/σR | σ max, ksi | σ/σR | σ max, ksi | σ/σR |
| R | Direct juncture | 47.5 | 1.00 | 23.7 | 1.00 | 57.1 | 1.00 | 14.4 | 1.00 |
| D | Sharp contoured | 46.8 | 0.99 | 18.1 | 0.76 | 29.3 | 0.51 | 6.8 | 0.47 |
| G | Gradual contoured | 35.1 | 0.74 | 15.4 | 0.65 | 22.2 | 0.39 | 6.2 | 0.43 |

Table 2: Strain Reading Along Branch and Saddle at Maximum Loading (μin/in)

| Location of Gauges | Distance from Datum Point: Reinforcement 55 mm (μin/in) | Branch 75 mm (μin/in) | Branch Reference 406 mm (μin/in) |
|---|---|---|---|
| Hoop / 90° Axis | -3 | 7 | -1 |
| Hoop / 45° Axis | 11 | -1 | -15 |
| Hoop / 0° Axis | 13 | 2 | -46 |
| Axial / 90° Axis | 8 | -21 | -2 |
| Axial / 45° Axis | -68 | 177 | 76 |
| Axial / 0° Axis | 62 | 53 | 125 |

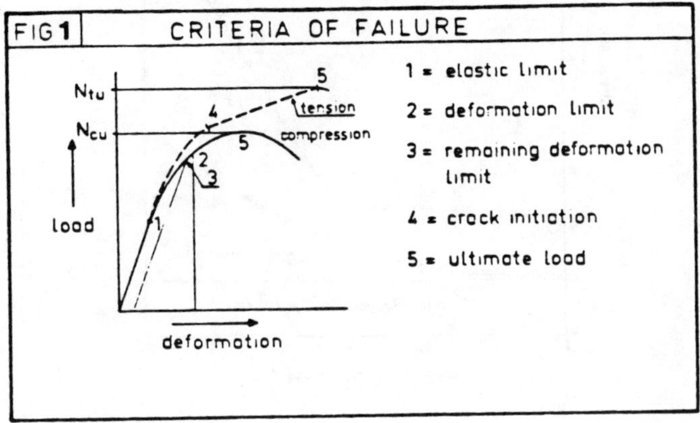


Figure 1. Criteria of Failure (from Wardenier [4])

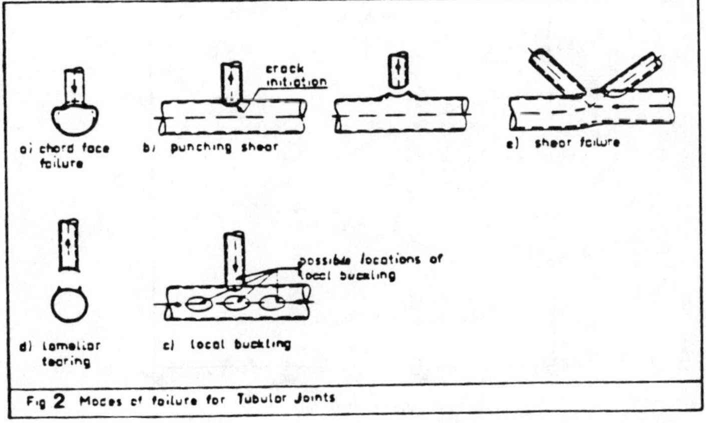


Figure 2. Mode of Failure for Tubular Joints (from Wardenier [4])

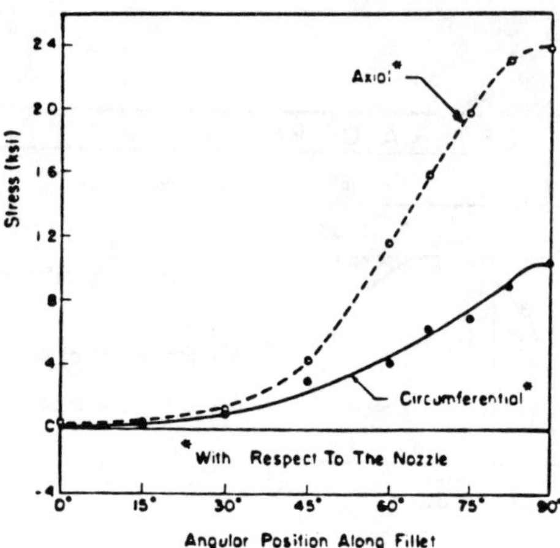


Figure 3. Stress distribution in the fillet of Vessel "C-1" for a transverse moment (clockwise couple) of 3000 in-lb on the nozzle (from Riley [7])

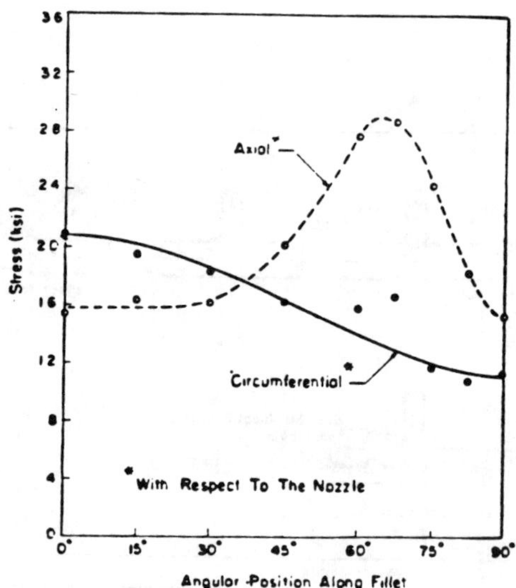


Figure 4. Stress distribution in the fillet of Vessel "C-1" for an internal pressure of 30 psi (from Riley [7])

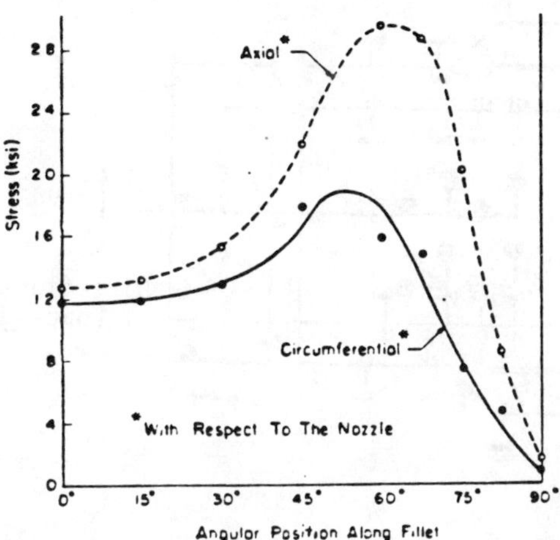


Figure 5. Stress distributions in the fillet of Vessel "C.1" for a longitudinal moment (clockwise couple) of 18,000 in. lb on the nozzle (from Riley [7])

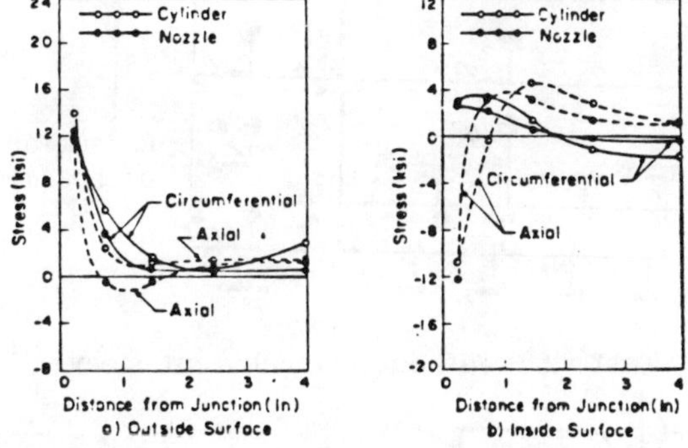


Figure 6. Stress distributions on the longitudinal plane of Vessel "C.1" for a longitudinal moment (clockwise couple) of 18,000 in.lb on the nozzle (from Riley [7])

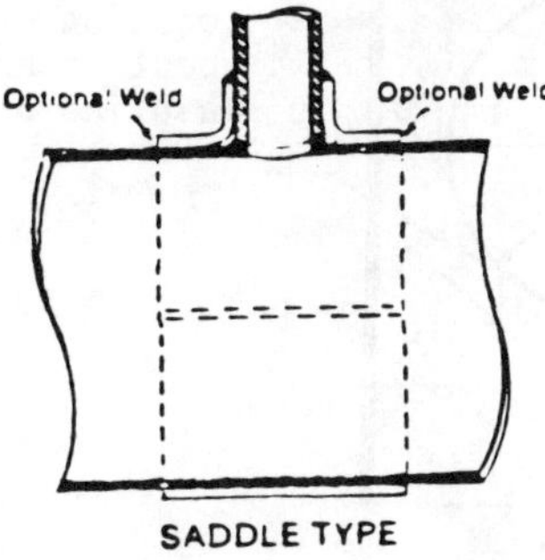


Figure 7. Hot tap connection showing full reinforcement sleeve (from Z184 [10]).

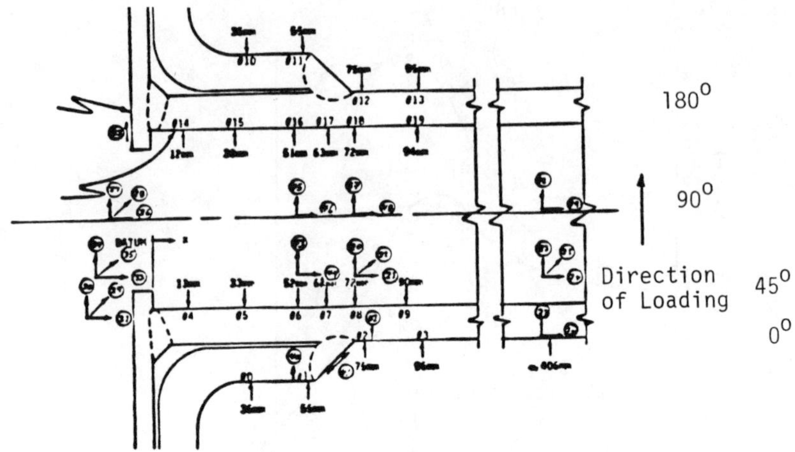


Test #1 - Full encirclement sleeve

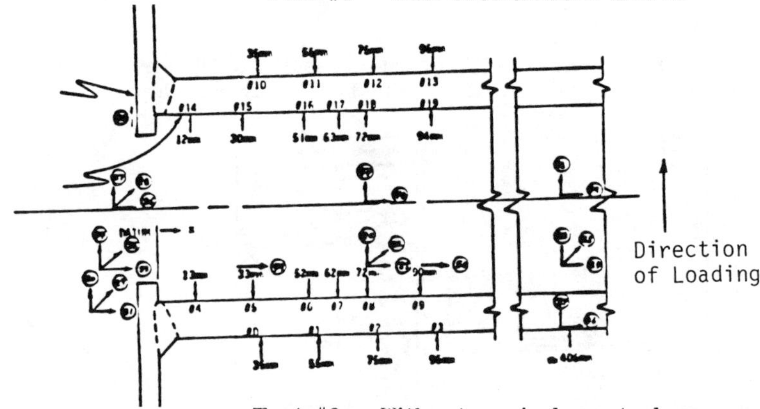


Test #2 - Without encirclement sleeve

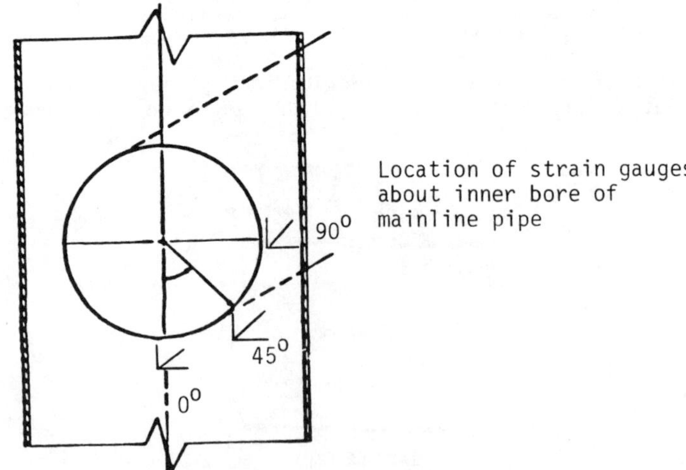


Figure 8. Location of strain gauges

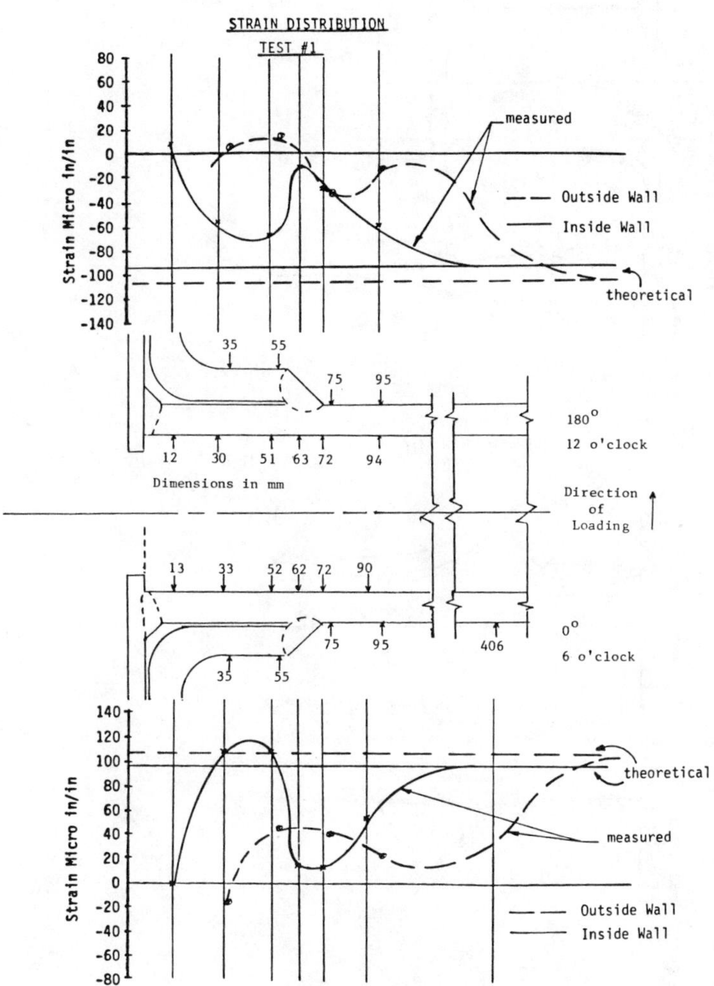


Figure 9. Longitudinal strain distribution on branch with sleeve attached.

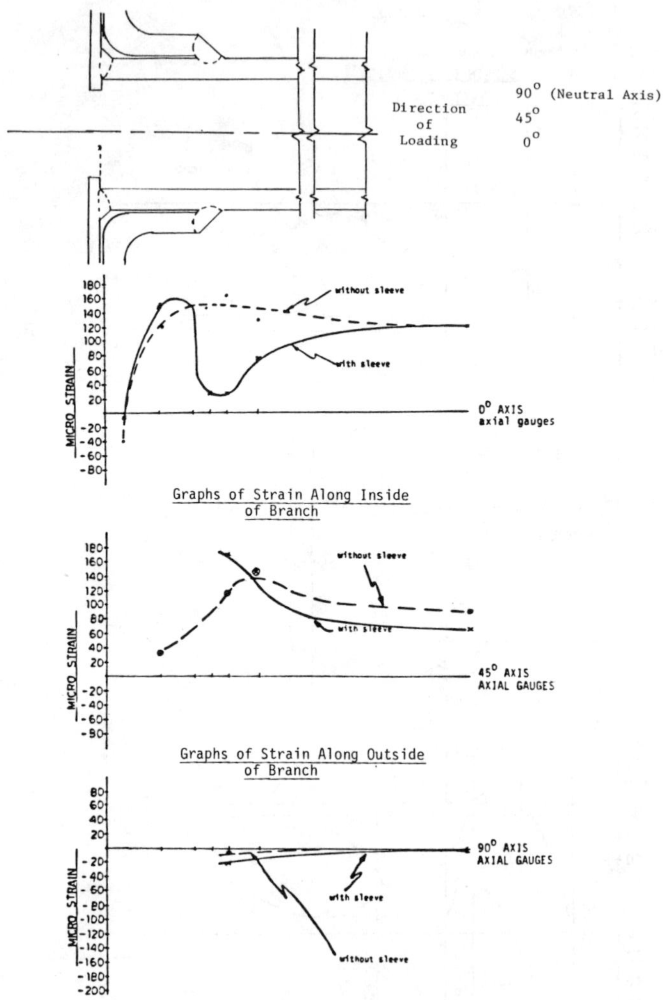


Figure 10. Review of Axial Strain Network on Branch Pipe.

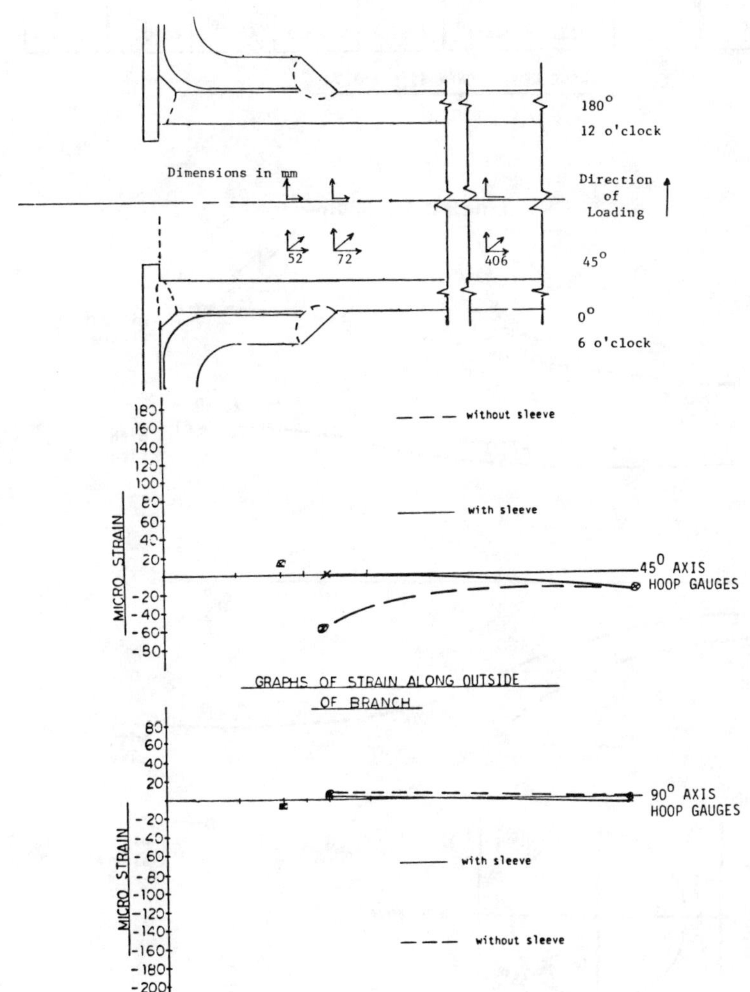


Figure 11. Hoop strain network with and without reinforcing sleeve.

| Angle About Bore | PRINCIPAL STRAINS | | | | | | % REDUCTION | |
|---|---|---|---|---|---|---|---|---|
| | WITH SLEEVE | | | WITHOUT SLEEVE | | | | |
| | $\epsilon_1$ | $\epsilon_2$ | $\theta$ | $\epsilon_1$ | $\epsilon_2$ | $\theta$ | $\epsilon_1$ | $\epsilon_2$ |
| $0^o$ | 240 | -49 | $-0.1^o$ | 378.3 | -78 | $-0.5^o$ | 36.6% | 37.2% |
| $45^o$ | 230.1 | -153.1 | $-77.1^o$ | 419.8 | -273.5 | $-80.2^o$ | 45.2% | 44.0% |
| $90^o$ | 278.1 | -255.1 | $44.1^o$ | 496.6 | -445.3 | $45.2^o$ | 44.0% | 42.7% |

AVERAGE REDUCTION = 41.6 ± 3.74

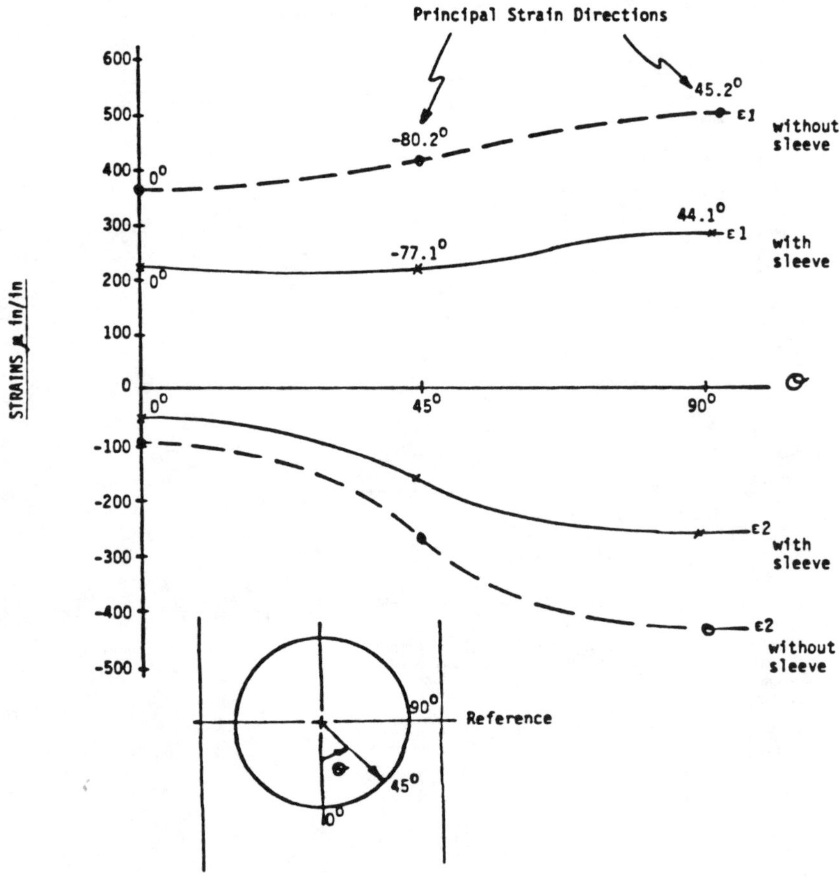


Figure 12. Reduction in strain distribution after sleeve removed.

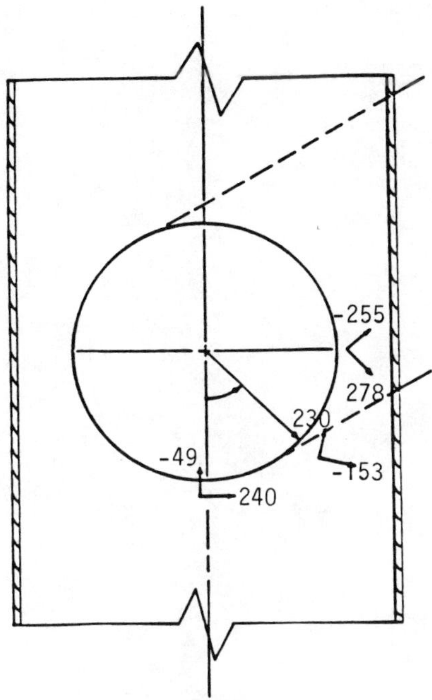


a) Principal strain direction and magnitudes along inside bore of mainline pipe with full encirclement sleeve.

b) Principal strain directions and magnitude along inside bore of mainline pipe with encirclement sleeve removed.

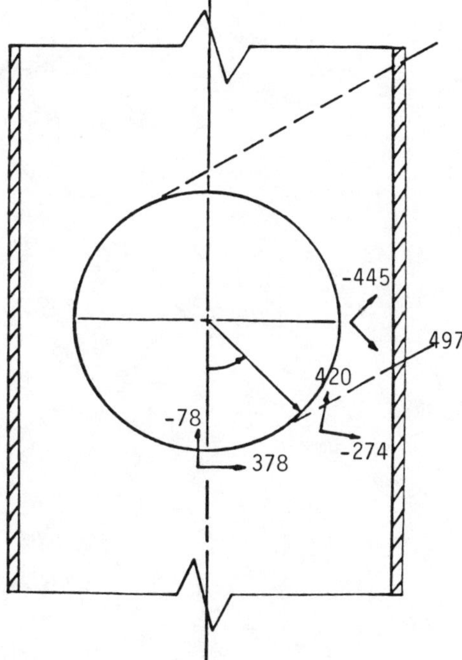


Figure 13. Effect of sleeve removal on principal strain mangitudes and directions.

# COMMON MISCONCEPTIONS IN THE FATIGUE DESIGN OF WELDED STEEL JOINTS

J.G. Wylde and G.S. Booth
*The Welding Institute, Abington*
*Cambridgeshire CB1 6AL England*

**ABSTRACT**

This paper summarises the current philosophy for the fatigue design of welded joints and considers a number of misconceptions which are sometimes held. These include the use of high strength steels and post weld heat treatment. In addition the influence of various other factors such as the location of welded attachments, joint dimensions, environment, heat affected zone hardness and the presence of repair welds on fatigue strength are discussed.

**KEYWORDS**

Fatigue, Welded Joints, Design, Steel.

## INTRODUCTION

Research into the fatigue properties of welded joints has been extensively carried out for many years. The fruit of this work has been the development of fatigue design rules for welded joints (American Welding Society, 1983; British Standards, BS5400, 1980) which provide a rational design method based on an acceptable, high survival probability.

Nevertheless, fatigue failures associated with welded fabrication (International Institute of Welding, 1979) continue to occur with alarming regularity, causing substantial financial losses and, in some instances, tragic loss of life. Further investigation has suggested that in many cases a limited fatigue analysis has been performed but that the analysis has not always been suitable for application to welded joints.

Many of the factors which are well known to influence strongly the fatigue performance of plain components do not exert the same effect on the fatigue behaviour of welded joints. This has led to a number of common misconceptions among designers not particularly familiar with fatigue design of welded structures. This paper considers some of these misconceptions, which in the authors' experience are the most common. In addition, various other aspects of the fatigue design of welded joints are discussed.

## FATIGUE DESIGN PHILOSOPHY FOR WELDED JOINTS

In many respects the fatigue performance of welded joints differs drastically from that of plain materials. This arises for a number of reasons. Firstly, all welded joints contain small slag intrusions at the weld toe as a consequence of the welding operation (Signes, 1967). These intrusions are generally less than 0.4mm deep but are very sharp and act like pre-existing cracks. Fatigue crack growth commences from these intrusions very early in the total fatigue life and it is generally assumed that there is no initiation phase, i.e. the whole life is spent in crack propagation (Gurney, 1979b). This is in direct contrast to plain components which generally exhibit a relatively long initiation period.

Secondly, joints in the as-welded condition contain high tensile residual stresses which have been shown to approach yield magnitude (Gurney, 1979b). Indeed, it is assumed in fatigue design codes that yield tensile residual stresses exist and all applied stresses are superimposed on this level. Thus, the applied stress cycle and the residual stress level combine such that the effective stress cycle at the weld pulsates from yield stress downwards, irrespective of the applied mean stress (Fig. 1). Thus the effective stress ratio (minimum stress/maximum stress) is always high. The whole stress cycle is therefore tensile in the vicinity of the weld even for partly compressive applied stresses and thus the full stress range is damaging and is the major factor controlling fatigue life. In addition, no correction is necessary to account for mean stress as is normally applied to plain materials by using, for example, a Goodman diagram.

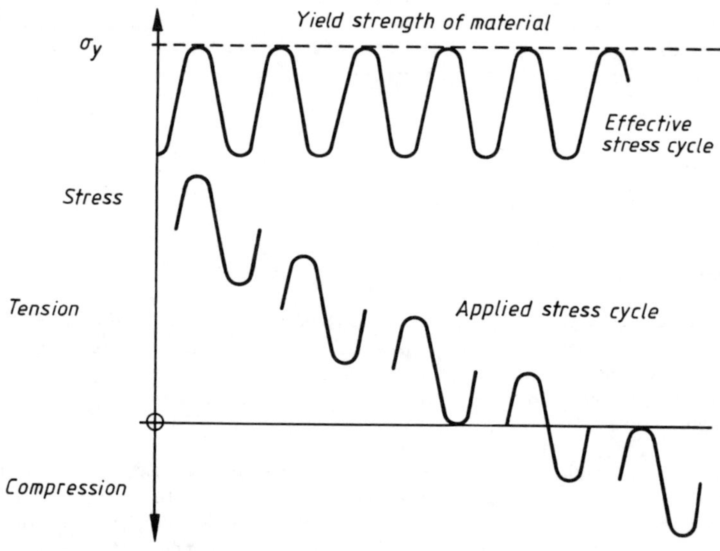


*Fig. 1. The influence of residual stress on applied stress ratio.*

Thirdly, the presence of the joint almost always results in a change of section, leading to a stress concentration factor associated with the presence of the joint itself. In the design rules, welded joints are placed in classes according to their fatigue performance. The joint classifications take into account the influence on the local stress concentrations created by the joints themselves. The design stresses are thus the nominal stresses adjacent to the detail under consideration, i.e. usually in the parent material at the weld toe, without taking into account any increase of stress because of the weld profile. The classes are based on experimental data and, in the British Rules, range from Class B (e.g. a fully machined longitudinal full penetration butt weld free from defects) to Class G (e.g. an attachment on the edge of a stressed member).

In general, the fatigue design codes assume that welding has been carried out to an appropriate fabrication code. This implies certain tolerance levels for welding defects (e.g. undercut, slag, etc.) which are consistent with good welding practice. In many instances welds are made with defects which are outside the acceptance levels stipulated in the fabrication codes. It may, however, be possible to accept the presence of these defects, providing that a fitness for purpose assessment has shown that they will not cause premature failure. The techniques for performing such an analysis have been established (British Standards, PD6493, 1980) but fall outside the scope of this paper.

Finally, the importance of regular routine inspection during service should be emphasised. This is important because there may be uncertainties in the fatigue assessment which will necessitate monitoring critical joints at appropriate intervals.

## HIGH STRENGTH STEELS

The fatigue strength of a welded joint is independent of the tensile or yield strength of the steel (Gurney, 1979b). This phenomenon has been recognised for many years, but because it is contrary to our understanding of the behaviour of plain materials it is still frequently overlooked. Figure 2 compares the fatigue strength at $10^6$ cycles for plain polished specimens, notched specimens and welded joints for various strength levels (Maddox, 1983). This shows that the behaviour of welded joints is quite different from that of plain material. The difference in performance can be explained in terms of crack initiation and propagation behaviour.

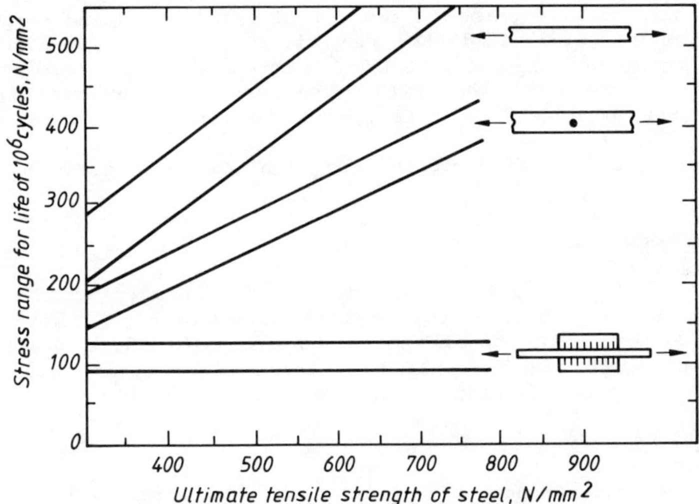


*Fig.2. The effect of material strength on fatigue strength.*

For polished specimens the majority of the total fatigue life is occupied in crack initiation. This is material strength dependent and consequently as the tensile strength of the steel is increased the initiation life and thus the total fatigue life also increases. For a notched specimen, the initiation life is reduced, but still represents a significant proportion of the total life. For welded joints it is widely accepted that there is no initiation period and that the entire life is occupied in crack propagation. Fatigue crack growth rate studies in steels of various strength levels have shown that growth rate is not dependent upon tensile strength (Maddox, 1970).

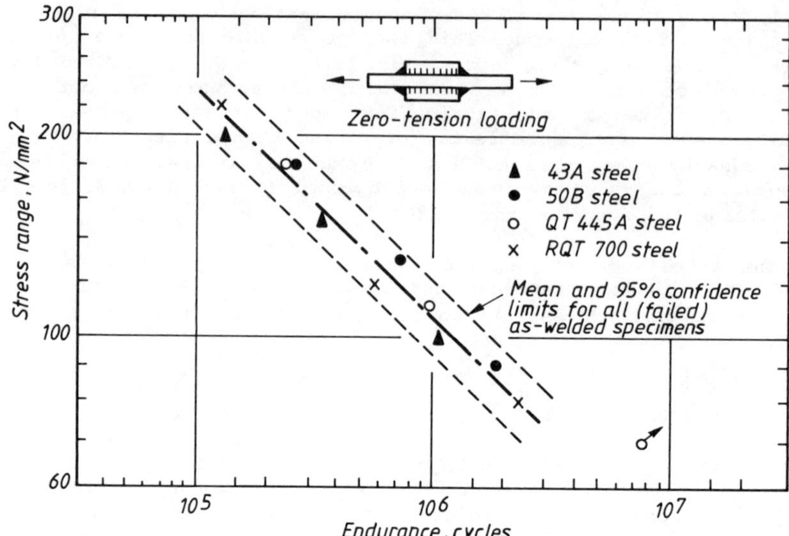


*Fig.3. Fatigue test results for longitudinal joints fabricated from four different steel grades.*

Figure 3 shows typical fatigue test results for specimens with longitudinal attachments tested in the as-welded condition (Maddox, 1980). Four steels were tested, two structural steels and two low alloy quenched and tempered (QT) steels having minimum specified yield strengths within the range 255-700N/mm$^2$ and tensile strengths within the range 430-800N/mm$^2$. Three specimens were tested from each steel and the figure shows that within the range of endurances from $10^5$ to $2 \times 10^6$ cycles there was no difference in fatigue performance.

However, there are some situations where high strength steels may be used to advantage:-

## High Mean Stress

As mentioned previously, the fatigue design of welded structures is now based on stress range irrespective of the applied stress ratio. However, static design criteria (2/3 yield, 1/3 tensile strength are typically quoted) will limit the maximum stress which can be applied in any fatigue cycle. High strength steels will thus be of benefit where the static mean stress is high.

## Low Cycle Fatigue

Although the fatigue strength of steels of various strength levels is identical at lives greater than approximately $5 \times 10^4$ cycles, at lower endurances the S-N curves obtained under load control diverge as shown in Fig. 4 (Frost, 1974). This occurs as the maximum stress in the fatigue cycle approaches the static strength of the material.

## Fatigue Improvement Techniques

Some fatigue strength improvement techniques, for example grinding, remove the small defects inherent at weld toes and thus introduce an initiation period into the fatigue life. As mentioned above, initiation life is dependent on material strength and thus a larger improvement in fatigue strength is seen in higher strength steels (Knight, 1978).

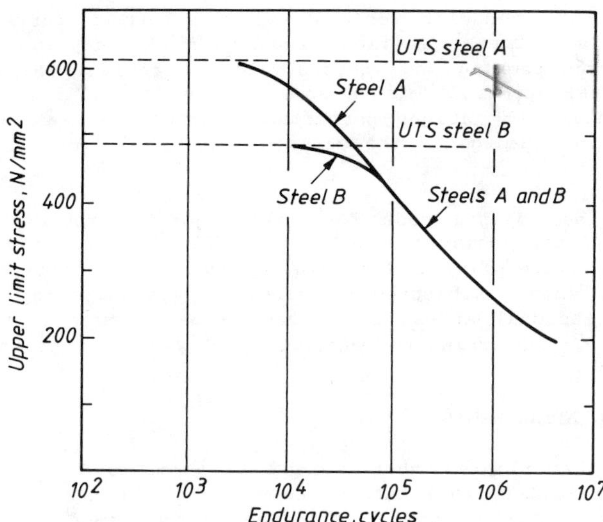


*Fig.4. Influence of tensile strength on the fatigue performance of welded joints in the low cycle regime.*

## THE INFLUENCE OF STRESS RELIEF

Post weld heat treatment (PWHT) of welded fabrications is sometimes carried out in the belief that it will provide an improved fatigue performance. Whilst there are other valid reasons for PWHT, for example, to ensure dimensional stability is retained during service, or to provide a reduced risk of brittle fracture, PWHT will seldom provide any benefit in terms of fatigue strength.

Under tensile loading there is some evidence to suggest that the slope of the S-N curve for stress relieved joints may be slightly higher than for as-welded specimens, and that the endurance limit may be increased (Gurney, 1979b). Recent results (Maddox, 1982) on as-welded and stress relieved joints subjected to various positive stress ratios confirmed the possible increase in the fatigue limit but failed to show any change in the slope or position of the S-N curve.

Different behaviour is observed for joints subjected to a significant compressive component of stress. Tests on as-welded joints subjected to alternating loading (Maddox, 1982) give the same fatigue strength as joints tested under fully tensile loading. This can be explained in terms of the tensile residual stresses which exist in the joint after welding and their effect on the applied stress (Fig. 1). However, tests on stress relieved joints under alternating loading have shown a significant improvement in fatigue strength based on applied stress range. Current research is directed towards explaining this behaviour which is believed to be related to a proportion of stress cycle becoming less or non damaging.

However, it is necessary to offer a cautionary word here in relation to the effectiveness of stress relief, particularly in large fabrications. Measurements in the above tests carried out before and after stress relief have shown that quite high levels of residual stress can still exist in welded joints even after full PWHT. Stress relief of large structures is subject to further doubt since, in addition to residual stresses arising as the result of restraint within a joint, long range residual stresses can occur within the structure as a whole. Consequently, caution should always be exercised when considering the possible benefits of PWHT.

Other techniques for reducing residual stresses either during welding or after fabrication are sometimes specified. These include adopting a balanced welding procedure, interrun peening and vibratory stress relief. Whilst these techniques may reduce residual stress levels in some circumstances, they are unlikely to have any benefit in terms of fatigue performance. Indeed, vibratory stress relief may produce some fatigue damage in the structure and should only be used with extreme caution.

No existing fatigue design code for welded joints provides any allowance for stress relief. On the basis of results available this is certainly justified for joints subjected to tensile loading. There may be justification for relaxing the requirements for joints subjected to partially compressive loading providing an appropriate low residual stress level can be demonstrated. In this respect, the cautionary comments concerning the efficiency of PWHT must be borne in mind.

## WELDED ATTACHMENTS

It is sometimes overlooked that welded attachments represent severe fatigue details. In general, they fall into Classes F or F2 in the UK fatigue design rules (British Standards, BS5400, 1980), although as explained below, in some circumstances, they may be Class G or even lower. Thus, in some structures which are designed to a high fatigue classification, the use of welded attachments may drastically reduce their fatigue strength. A typical example is the use of a temporary attachment such as a lifting lug on a pressure or containment vessel for site erection purposes. Unless this is completely removed after use, and the area beneath it ground to remove the slag intrusions which will have been introduced into the surface of the vessel, the attachment is likely to represent the worst fatigue detail and will thus determine the fatigue strength of the structure.

Careful consideration should also be given to the positioning of permanent attachments, particularly when it is necessary to locate them close to the edge of primary stressed members. Typical examples might be handrails, stairways or secondary stiffeners and such attachments are likely to determine the fatigue strength of the structure. Furthermore, unless care is taken over their precise location, joints with a very low fatigue strength may be produced.

As mentioned above such attachments are generally Class F or F2 depending upon their length in the direction of the applied stress. However, to conform to these classifications they must be located at least 10mm from the plate edge. The reason for this stipulation is that a weld encroaching on the corner of a member may produce a sharp notch with a consequent reduction in fatigue strength. Thus attachments on the edge of a stressed member are downgraded to Class G, the lowest class for joints failing from the weld toe.

However, recent tests (Slater, 1981) have shown that in some cases welds onto a plate edge can be even worse than Class G. This occurs when the weld is carried over the corner of the plate as shown in Fig. 5. In this situation it is inevitable that a severe notch will be produced at the corner as the effective plate width is reduced. The results suggest that a reduction in fatigue strength amounting to approximately two joint classes below Class G is incurred by welding around the plate edge. This may be important when it is necessary to make a sealing weld around the end of an attachment. Whenever possible such attachments should be located on the plate surface 10mm from the edge.

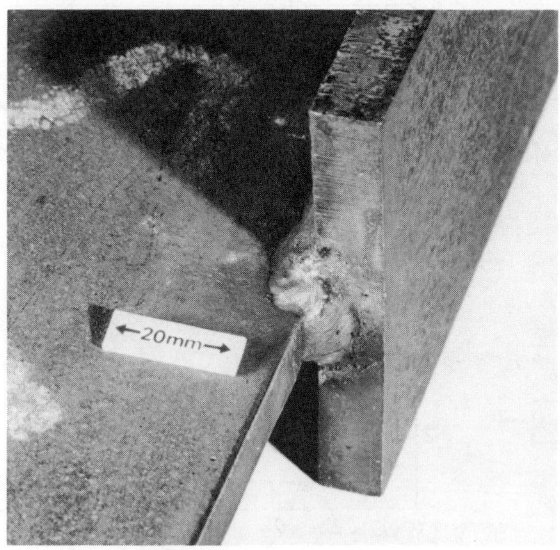


*Fig.5. Weld continued round end of attachment on the edge of a stressed plate.*

## DESIGN OF LOAD CARRYING FILLET/T-BUTT WELDS

A cruciform joint can be made using: i) fillet welds, ii) full penetration butt welds, or iii) partial penetration welds. The full penetration butt weld is often chosen by designers without giving full consideration to the application. However, in many instances a partial penetration or fillet weld will be adequate and the full penetration weld may give rise to additional fabrication problems. For example, it is more expensive to produce and gives a higher risk of lamellar tearing in materials with low through-thickness ductility. Furthermore, there is a risk of introducing lack of penetration defects. Whilst these may be acceptable, if the designer has arbitrarily called for a full penetration joint, it may be necessary to carry out a fracture mechanics assessment of any such defects to demonstrate their acceptability. If this cannot be proven then expensive repairs may be required.

In the fatigue design rules fillet and partial penetration butt welds fall into Class F2 and full penetration welds into Class F. Thus there is only a small difference in fatigue strength between the joint types when considering failure from the weld toe. The danger of using fillet or partial penetration joints is that there is a potential for premature weld throat failure from cracks initiating at the root. However, this type of failure can be avoided by suitable proportioning of the weld. For example, a fracture mechanics analysis of these joint types (Maddox, 1974) and has provided a graphical solution to dimension such details to ensure that throat failures will be avoided. This graph is reproduced here as Fig. 6 and is valid for joints subjected to tensile loading. For a given plate size this graph enables the designer to select a combination of depth of penetration and leg length to ensure that weld failure will not occur. Other work (Ouchida, 1964) has shown that the size of fillet weld necessary to prevent throat failure is smaller under bending than under tension loading.

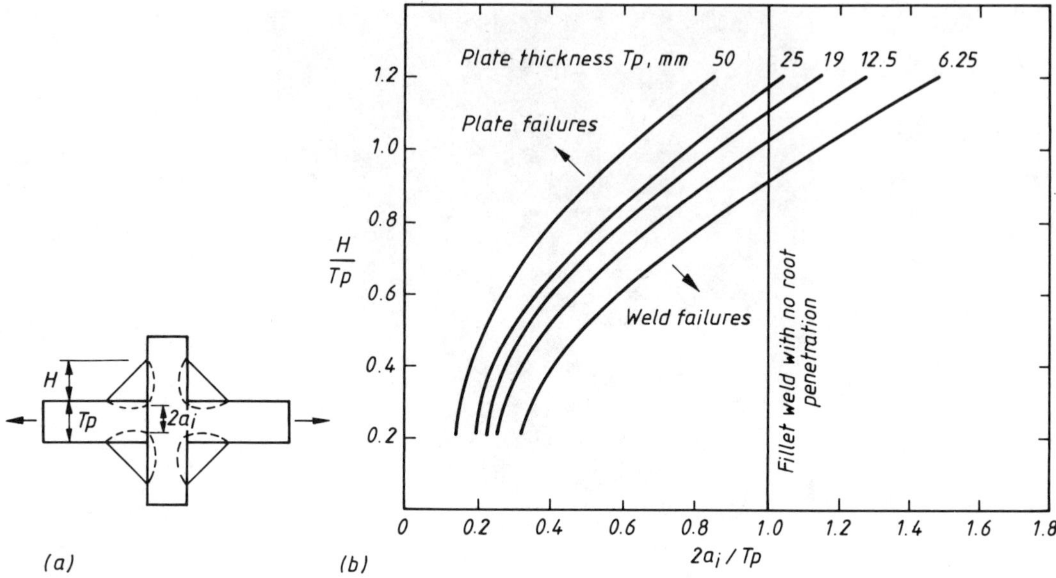


*Fig.6. Design of load carrying joints: a) geometry of cruciform welded joint; b) geometry conditions for optimum design.*

## EFFECT OF JOINT DIMENSIONS

There is now a substantial body of evidence, both theoretical and experimental, that the fatigue life of a welded joint is not determined solely by stress range. For plate and tubular joints subjected to either axial or bending loading, the fatigue strength decreases with increasing joint dimensions (Gurney, 1979a). In addition, it is clear that it is principally the joint thickness which causes this effect although other dimensions such as weld size and attachment size are also important.

This is clearly of some concern since the majority of the current fatigue design rules were based on results from 12.5mm (1/2 in.) thick joints. Use of these rules for thicker joints may overestimate the allowable endurance. This is an area of active research and it is hoped that with the generation of more data it will be possible to derive correction factors for joints of increased thickness.

## ENVIRONMENTAL EFFECTS

The majority of design rules are intended for joints in a non-aggressive environment at room temperature, where frequency of cycling is unimportant. It must be recognised that exposure to other conditions, e.g. seawater in the case of offshore platforms or elevated temperatures in the case of nuclear or petrochemical applications, could lead to a reduction in fatigue strength. The offshore environment has been studied in detail and it is now recommended (Gurney, 1983) that the allowable air endurances are reduced by a factor of 2 for unprotected joints, whereas no correction is required for joints adequately cathodically protected. There is no general rule which can be applied to relate environment to fatigue strength. To create fatigue design rules for one particular application it is necessary to generate fatigue data under appropriate environmental conditions. Furthermore, under corrosion fatigue conditions frequency of loading is an important variable, an increase in frequency being associated with an increase in endurance. One consequence of this is that environmental data should be generated in real time, no accelerated frequency testing is suitable. This can lead to significant difficulties because the low stress/long life regime is usually important but cannot be investigated other than by extrapolation.

## EFFECT OF HEAT AFFECTED ZONE HARDNESS

Many construction codes stipulate that in weld procedural trials the heat affected zone (HAZ) hardness must not exceed a specified value, typically 250Hv. It is not the purpose of this paper to discuss the suitability of this criterion from a general point of view, i.e. avoidance of HAZ cracking, resistance to brittle fracture or susceptibility to stress corrosion cracking. However, strictly from a fatigue standpoint, there is no evidence that hardness level plays any role in determining fatigue strength, when cycling in air.

It is well established (Maddox, 1970) that the rate of fatigue crack propagation is independent of microstructure, similar results being obtained for cracks growing in parent plate, weld metal and simulated HAZ. Furthermore, studies on various steel types (Booth, 1981) have shown that welded joint endurance does not vary when the maximum HAZ hardness is in the range 200 to 350Hv.

One possible area where hardness may be important is in corrosion fatigue applications. Under these conditions, the increased availability of hydrogen may lead to a greater susceptibility to accelerated cracking in hard HAZs. At present, there are no data available to substantiate this concern.

## REPAIR OF FATIGUE CRACKS

In some instances fatigue cracking may occur in a structure during the design life. On such occasions it is desirable to effect a repair and return the structure into service. Whilst this may be quite feasible in many situations, the difficulties in making a satisfactory repair and the limitations of the completed repair should not be overlooked.

Repair welds are invariably made as positional welds, on site, frequently with limited access under unfavourable conditions. Furthermore, they are often made from one side only with the problem of making a suitable root. Thus it is most unlikely that the fatigue life of the repair will match that of the original weld unless the fatigue loading can be reduced. Recent tests (Wylde, 1983) have shown that if the weld repair is not adequately carried out, and a small part of the original fatigue crack is not removed, the repaired joint will have a very low fatigue strength. This shows that the practice of simply welding over a growing fatigue crack is unlikely to prolong the fatigue life by any significant margin.

Furthermore, consideration should be given to the fatigue damage incurred by other parts of the structure. The presence of a fatigue crack is indicative of high cyclic stresses and, unless these have been confined to a small area of stress concentration, then they will have introduced fatigue damage in other areas, even though fatigue cracks may not be visible. Indeed, load shedding from a cracked area may increase the stresses on surrounding areas. Thus such areas may develop fatigue cracking after the structure is returned into service.

In view of the problems associated with repair welds, it is essential that all repairs should be subjected to rigorous non-destructive examination. Consideration should also be given to grinding the surface of the weld repair and the weld toes of surrounding weld details with a view to obtaining the maximum extension in fatigue life.

## CONCLUDING REMARKS

The main objective of this paper has been to explain why some commonly held opinions are misconceptions in the context of contemporary fatigue design of welded joints. In many cases, these misconceptions arise because it is not fully appreciated that welded joint behaviour is significantly different from that of plain components. Nevertheless, using current codes of practice, it is possible to design welded constructions with adequate fatigue lives and it is hoped that this paper has brought the basic elements to a wider audience.

## ACKNOWLEDGEMENTS

The preparation of this paper was funded jointly by the MCRB of the Department of Industry and Research Members of the Welding Institute. The authors wish to thank their colleagues at the Welding Institute for helpful discussions.

## REFERENCES

American Welding Society (1983). Structural Welding Code - Steel, AWS D1.1-83.

Booth, G.S. (1981). Hardness control in respect of fatigue and HAZ toughness. In Hardness control to avoid service failures and fabrication cracking in steel. Welding Institute Seminar.

British Standards Institution (1980). Steel, Concrete and Composite Bridges - Code of Practice for Fatigue. BS5400:part 10.

British Standards Institution (1980). Guidance on Some Methods for the Derivation of Acceptance Levels for Defects in Fusion Welded Joints. PD6493.

Frost, N.E., K.J. Marsh and L.P. Pook. (1974). Metal Fatigue. Oxford University Press.

Gurney, T.R. (1979a). The influence of thickness on the fatigue strength of welded joints. Proc. Int. Conference Behaviour of Offshore Structures, London.

Gurney, T.R. (1979b). Fatigue of Welded Structures, Second Edition, Cambridge University Press.

Gurney, T.R. (1983). Revised fatigue design rules. Metal Construction, 15, (1) 37-44.

International Institute of Welding (1979). Fatigue Fractures in Welded Constructions, Publications de la Soudure Autogene.

Knight, J.W. (1978). Improving the fatigue strength of fillet welded joints by grinding and peening. Welding Research International, 8, (6) 519-539.

Maddox, S.J. (1970). Fatigue crack propgation in weld metal and HAZ. Metal Construction and British Welding Journal, 2, (7) 285-289.

Maddox, S.J. (1974). Assessing the significancce of flaws in welds subject to fatigue. Welding Research Supplement, 53, (9) 401S-409S.

Maddox, S.J. (1982). Influence of tensile residual stress on the fatigue behaviour of welded joints in steel. In Residual Stress Effects in Fatigue. American Society for Testing and Materials, Special Technical Publication 776.

Maddox, S.J. (1983). An introduction to the fatigue of welded joints. In Improving the Fatigue Strength of Welded Joints, Welding Institute.

Maddox, S.J. and J.G. Summers. (1980). Preliminary investigation of controlled shot peening on the fatigue strength of fillet welded joints. Welding Institute Member's Report 127/1980.

Ouchida, H. and A. Nishioka. (1964). A study of fatigue strength of fillet welds. IIW Document XIII-338-64.

Signes, E.E. and co-workers. (1967). Factors affecting the fatigue strength of welded high strength steels. British Welding Journal, 14, (3), 108-116.

Slater, G. (1981). Worse than G! Fatigue test results from plates with welded edge attachments. Welding Institute Research Bulletin, 22, (4) 98-101.

Wylde, J.G. (1983). The fatigue performance of repaired fillet welds. Welding Institute Member's Report 215/1983.

# MODIFIED DESIGN RULES FOR FATIGUE PERFORMANCE OF OFFSHORE STRUCTURES

J.G. Hicks
*Consultant, Cambridge, United Kingdom*

## ABSTRACT

Fatigue design rules for offshore structures in the UK, USA and Norway have been developed from common sources but at any one time exhibit marked differences in the fatigue data. The results of research programmes designed to examine the particular circumstances of offshore structures are being used to revise the rules where this is shown to be necessary. Such revisions must be carefully considered when the rules are used for the re-assessment of existing structures.

Fracture mechanics methods have been introduced into more recent offshore codes for fatigue life prediction and weld defect assessment.

Areas for further development of the rules are to be found in the calculation of hot spot stress in complex joints, fatigue life improvement techniques and weld quality standards.

## KEYWORDS

Offshore structures; design rules; fatigue; tubular joints; hot spot stress; weld defects.

## INTRODUCTION

The development of fatigue design rules specifically for offshore structures has been accomplished over a relatively short period. Initially the rules were just a repeat of those developed for bridges and other land based structures. This paper reviews the progress towards the present rules and shows how they are different from other sets of rules, and, perhaps equally important, where there are similarities. The main aim of the paper is to highlight recent amendments to the UK rules and to indicate where they are still incomplete.

## THE PURPOSE OF DESIGN RULES

The word rule is perhaps a little simple in this context since it usually describes a certain way of behaving or playing a game. Design rules tend to be rather complicated and frequently contain sets of technical data prepared from theoretical or experimental investigations such that when used in accordance with the instructions they will give an adequate engineering design for the relevant circumstances.

In the more complex areas of engineering these rules are embodied in Codes of Practice; such codes cannot anticipate all the requirements and circumstances to which the design must respond and leave some discretion to the designer or the organisation commissioning the design.

It is therefore most important that the users of codes and specifications understand their derivation and use them as an agreed basis of design and not as instructions, a matter which will be referred to in more detail later in this paper.

## THE HISTORY OF FATIGUE DESIGN RULES

This is not the place for a detailed review of the history of design rules which has been described elsewhere, (Gurney and Hicks 1973, Marshall and Toprac 1973) but it is important to appreciate the differences and similarities which are more often a function of their origins than of any fundamental engineering phenomenon. In addition since this paper is entitled "Modified Design Rules ..." it is necessary to establish what rules have been modified.

Historically the majority of design specifications relating to fatigue of welded joints have been concerned with bridges, presumably because the railways tended to use iron and steel bridges and being public utilities had a responsibility for the integrity of their structures.

Bridge structures were generally made of rolled sections or plates and the fatigue rules related to welded joints in such members (AISC 1969 BSI 1962). The offshore industry was the first to make use of large welded tubulars and in the USA developed design rules including fatigue for such structures, which were first published in the early 1970's (API 1973, AWS 1972). It is important to note that until this time most of the United States offshore activity was in the Gulf of Mexico and fatigue cracking was not a pressing problem. That this was so was a function of the sea state history in that part of the world. About 95% of the time the sea conditions were not capable of producing a wave that would cause more than a quarter of the design load (Marshall 1969).

In the early 1970's the development of the northern North Sea fields was commenced and a number of authorities and organisations became aware that the existing rules might be inappropriate to ensure the integrity of the new structures. The wave history in the North Sea would give proportionally more stress fluctuations at the higher stress ranges than in the Gulf of Mexico and the structures were far heavier and more complex. It was recognised even in the context of the AWS Code that there was sparse experimental information on fatigue lives above two million cycles; a platform in the North Sea could expect to experience one hundred million cycles in a twenty five year life.

The UK Government through its Department of Energy recognised that the scale of the subject meritted a formal review of the knowledge and experience relevant to the new operations and funded a study (Hicks 1974) whose aim was "to identify areas of research which could usefully be pursued to reduce the probability of fractures occurring in offshore steel structures in the North Sea."

The scale of the research effort found to be needed required a funding of £3 million and the timescale was such that the results of the research programme would not be available for several years. A field operator had by law to obtain a Certificate of Fitness for his platforms from the Department of Energy which would be based on approval of the design and construction by a certifying authority. It was therefore necessary to establish rules to which approval of the platforms could be given prior to the results of the research programme being available.

For some early fields an agreed set of rules for tubular joints had already been drawn up jointly by Lloyd's Register of Shipping (the certifying authority) and the operator which formed the basis for certification at that time. (Marshall 1974) These were eventually incorporated in a Department of Energy document (DoEn 1977) which was intended to be an interim document pending the publication of a British Standard Code of Practice for Fixed Offshore Structures. The drafting committee for the British Standard first met in 1975 and published a Draft for Development (BSI 1978) in 1978 which became the Code of Practice (BSI 1982) in 1982, hardly reflecting the pace of offshore development. In that document and the Department of Energy's Guidance the fatigue rules for plate and girder construction were basically those under discussion in the mid 1970's for a new British Standard bridge code whilst a development of the Lloyd's curve now labelled the 'Q' curve was used in both documents for tubular joints. The K & T punching shear curves from AWS D1.1 were also included.

During this period the Norwegian classification society Det Norske Veritas had published its own rules in which the fatigue section was very similar to the UK Guidance and BSI documents.

## THE BASIS OF RECENT MODIFICATIONS TO FATIGUE RULES

The research programme which was set up in the UK as a result of the study referred to above was called the United Kingdom Offshore Research Project (UKOSRP) funded by UK Government and eventually augmented by contributions from the European Coal and Steel Community to a total of ∅7 million. Investigations were pursued into the effects on fatigue life of the following major influences together with some associated smaller investigations:

Sea water
Cathodic protection levels
Material thickness
Stress history
Fatigue life improvement techniques

The effects were examined both on a basis of the stress range against life relationship (the SN curve) and on the crack growth rate with particular attention to tubular joints. Associated with this work was an experimental and theoretical study of stress distributions in tubular joints. The whole programme was of considerable complexity and has been described in detail elsewhere (DoEn 1977) and the results have so far been reported at two seminars (The Welding Institute 1978, Institute de Recherches de la Siderurgie Francaise 1981).

Research into the fatigue performance of tubular joints has been conducted in areas other than Europe for example the USA, Canada, Japan, Australia and Norway, ( Wardenier 1981, 1982) but none of these programmes appears to have had the scale, co-ordination and direction of the European and in particular the UK programme. A number of these programmes are directed towards uses of tubulars in structures other than offshore and therefore do not examine some of the environmental effects of the sea.

The UKOSRP programme has been extended to become UKOSRP II (Long 1981) a combined Government and Industry funded programme aimed at resolving some of the uncertainties revealed in the first part.

## MODIFICATIONS TO THE RULES

In the UK the results of the first part of UKOSRP have been digested and used to prepare amendments to the design rules in the Department of Energy Guidance (Gurney 1982). Perhaps the greatest change in these rules has been the shift in the constant amplitude SN curves reflecting the results of the research which demonstrated the effect of thickness. This applies to both tubular and plate structures. The K and T punching shear curves have been deleted in favour of the hot spot stress approach.

A comparison of the various curves for tubular joints is shown in Fig. 1 in which 'UK' refers to the Department of Energy Guidance.

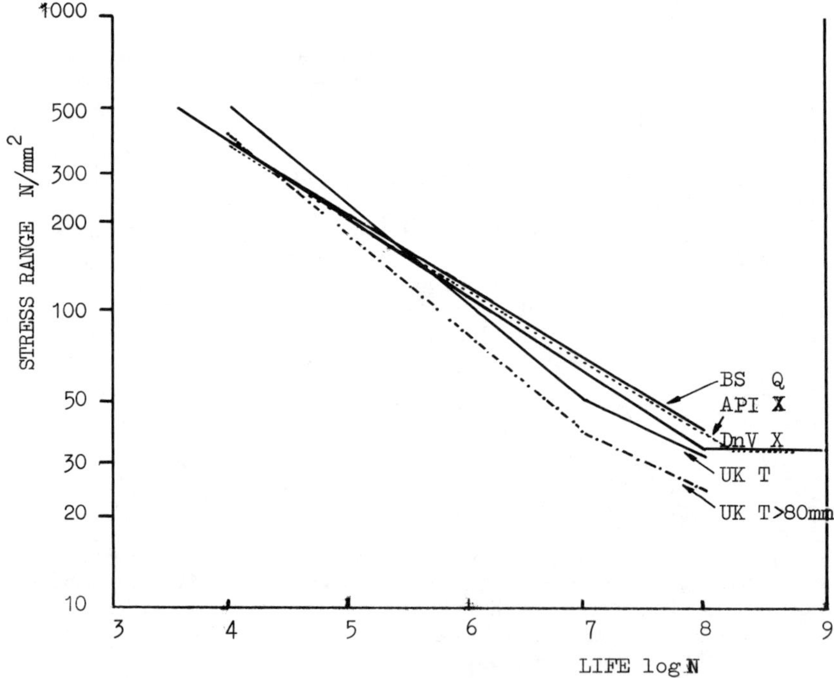


Fig. 1. SN CURVES FOR TUBULAR JOINTS

The basic SN curve for tubular joints - the T curve has replaced the Q curve and it can be seen that in contrast to the constant slope of the Q curve the T curve has a change of slope at $10^7$ cycles; this has been put in to give more realistic results to cumulative damage calculations. The basic T curve is for thicknesses from 22mm to 32mm and for comparison the curve for 80 to 100mm is shown.

A significant change whose effects cannot be easily assessed is in the definition of the peak or hot spot stress which is the variable appearing as the stress in the SN curves for tubular joints. This can be described by parametric equations or obtained by extrapolation from finite element calculations or experimental measurements of stress.

Grinding of weld profiles has been found to be an effective method of improving fatigue life in air and the modified UK rules allow an improvement of up to two on life in sea water provided that an effective corrosion protection system is in operation.

No allowance is given for increased fatigue life as a result of stress relieving because of the difficulties of verification and the effect of unknown long range residual stresses.

In structures which have no cathodic protection the SN curves are reduced in life by a factor of 2 and the change in slope for the higher life end of the curve is deleted.

The DnV Rules (DnV 1982) have SN curves and associated joint categories similar to those in BS 6235. Both BS 6235 and the DnV rules allow the use of fracture mechanics for life assessment of sound and defective welds but the DnV rules go into detail of methods and parameters.

API RP2A Thirteenth Edition 1982 gives for tubular joints subject to wave loading an SN curve X which follows closely the line of the BS 6235 Q curve but has a horizontal cut-off at $2 \times 10^8$ cycles.

For welded joints in plate structures there are various SN curves associated with different joint types. Fig. 2 shows the curves for one particular type of detail and illustrates the differences, as in Fig. 1 'UK' refers to the Department of Energy Guidance.

API RP2A calls for AWS D1.1 but has no special data for a sea water environment. There is far less difference between the BS 6235 curve and the new UK Department of Energy curve than in the case of tubular joints, it consists of a change of slope at $2 \times 10^7$ in place of cut offs at $2 \times 10^7$ and $2 \times 10^8$ for air and sea water environments respectively and a thickness correction.

The largest differences between the various SN curves occur at the high cycle end of the life scale and are naturally a matter of some concern. That totally disparate lives can be calculated from basically the same data is obviously unsatisfactory and emphasises the importance of resolving the situation which, to a large extent, is a function of the time of publication of the various codes in relation to the availability of research results.

## USE OF MODIFIED RULES

When modified rules are introduced a decision must be made as to whether they should be applied retrospectively. Such a decision needs to be taken carefully when the modified rules are more onerous than the original ones. It will be seen that under certain combinations of load history and thickness the revised UK rules will predict shorter lives than those they replace. It would place a tremendous burden on operators if they found that parts of their platforms were no longer considered to have sufficient fatigue life and that they would have to introduce costly structural modifications to retain the Certificate of Fitness.

In considering this matter the UK certifying authorities draw a distinction between design of new structures and the re-assessment of existing ones. In a re-assessment situation one usually has available the as-built dimensions, records of marine growth, wind, sea states and possibly structural response. Such knowledge permits higher confidence in the results of a fatigue analysis than for a new structure. As a result for a structure which has no history of fatigue cracking, it may be feasible to reduce some of the margins on calculated life which may be called for in a new structure. Where calculated lives are found to be lower than would be acceptable in a new structure it is possible to increase the frequency of inspection so as to maintain confidence in the integrity of the structure. Where fatigue cracking is found in service then specific policies have to be established and such a policy has been publicly stated by a UK certifying authority (Fisher 1982); here the area will be assessed on the new rules and any repairs or strengthening will be designed to the new rules.

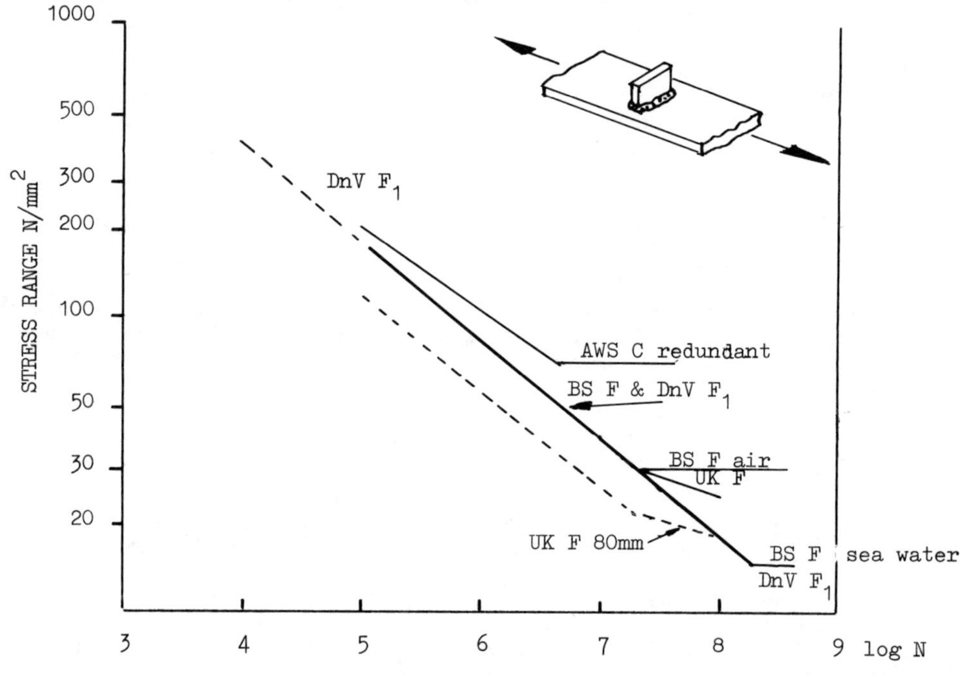


Fig. 2. SN CURVES FOR WELDED ATTACHMENTS

In the general usage of codes there has to be a basic understanding of the technology. That rules were made "for the obedience of fools and the guidance of wise men" may still be very true but in a complex subject such as fatigue the rules cannot say everything so that blind obedience cannot be invoked and all users have to be 'wise' men. On the other hand perhaps compilers of codes ought to go further than they often do in explaining the use of the material. The commentaries accompanying documents such as the AWS Structural Welding Code are most valuable in this respect.

## FUTURE OUTLOOK

Even with all the research undertaken there are still areas which are inadequately covered by the rules themselves or by the associated engineering knowledge and capability. That this is so is a reflection of the scale and pace of development of offshore construction. Some of the more pressing matters will be reviewed here.

The published hot spot stress parametric equations (Kuang 1977, Gibstein 1978, Wordsworth and Smedley 1978) cover a limited range of configurations and load patterns. For the multibrace joints which are frequently used offshore there is no straightforward reliable method of hot spot stress calculation and more development is needed with attention given to load ratio and phase.

When the design of substantial parts of a platform is controlled by fatigue benefit might be obtained by the use of fatigue life improvement techniques. The grinding of weld profiles has already been referred to as a valid method of life improvement provided that an effective corrosion protection system is in operation all the time including the tow out and installation phases of the structure's life; any failure of such a system even for limited periods may result in corrosion pitting as possible sites for premature fatigue cracking and loss of the benefit. Stress relief is considered by some authorities to be an unverifiable technique. Methods relying on the induction of compressive surface stresses such as peening have been shown to be effective in air (Booth 1978) but require close control. This principle is worth examining to establish whether there are other more controllable methods which will be effective in sea water with or without corrosion protection.

Fatigue life improvement over the as-welded condition cannot be safely exploited without attention to weld defect levels. When the surface of a weld is treated to defer fatigue cracking the relative significance of internal defects increases. At present the UK rules do not state any weld quality requirements specifically prepared for offshore structures; BS 6235 refers to the pressure vessel specification BS 5500 (BSI 1982) which is itself similar to ASME VIII (ASME 1982) in this respect. API RP2A quotes the defect acceptance levels from the buildings or tubular structures sections of AWS D1.1 with a dispensation for the root areas of nodal joints in tubulars. The use of the buildings section may be reasonable for statically loaded structures, e.g. modules isolated from platform deflections and other sources of fluctuating load, but where there is a source of fluctuating load it would be prudent to consider the use of the quality levels set for bridges.

Far better of course would be the preparation of quality standards which reflected the needs of offshore structures. Bases for a fitness for purpose approach are given in BS PD6493, (BSI 1980) the DnV Rules, AWS D1.1 and API RP 2X; these methods combined with an appropriate philosophy relating workmanship and fitness for purpose would enable workable and relevant standards to be set up to the benefit of clients and fabricators.

The true test of the rules is whether structures designed in accordance with them perform as required. None of the deep water structures has yet lived out its design life and so the rules cannot be judged. There have been a number of cases of fatigue cracking in various platforms but generally these have been explained by poor detailing, lack of attention to the integrated effects of appurtenances, unpredicted loads or loads being higher than calculated and not because the fatigue data in the rules was substantially in error. However there is a long way to go before it can be said that the rules are adequate and there must be continuous review and refinement of the rules supported by research and feedback of service experience. At the same time designers would be wise to bear in mind the sensitivity of fatigue life to local stress, the difficulty of inspection and the cost of repair and not design for lives just above the minimum required.

## REFERENCES

AISC (1969). Specifications for the Design, Fabrication and Erection of Structural Steel for Buildings.

American Petroleum Institute. (1973). API RP2A Recommended Practice for Planning, Designing and Constructing Fixed Offshore Platforms. Fourth Edition. Washington DC.

American Petroleum Institute. (1982a). API RP2A Recommended Practice for Planning, Designing and Constructing Fixed Offshore Platforms. Thirteenth Edition. Washington DC.

American Petroleum Institute. (1982b). API RP2X Recommended Practice for Ultrasonic Examination of Offshore Structural Fabrication and Guidelines for Qualification of Ultrasonic Technicians. Dallas.

American Society of Mechanical Engineers. (1982). ASME Boiler and pressure vessel code. Section VIII. New York.

American Welding Society. (1972). AWS D1.1 - 72 Structural Welding Code. Miami.

Booth, G. S. (1978). Constant amplitude fatigue tests performed on welded steel joints in air. European Offshore Steels Research Seminar, The Welding Institute, Abington, Cambridge, U.K. 1978.

British Standards Institution. (1962). BS 153 : Parts 3b and 4 : 1962 Steel Girder Bridges. London.

British Standards Institution. (1978). DD55 : 1978 Fixed Offshore Structures. London.

British Standards Institution. (1980). PD 6493 : 1980 Guidance on some methods for the derivation of acceptance levels for defects in fusion welded joints.

British Standards Institution. (1982a). BS 5500 : 1982 Unfired fusion welded pressure vessels.

British Standards Institution. (1982b). BS 6235 : 1982 Code of Practice for Fixed Offshore Structures.

Department of Energy. (1977a). Offshore Installations : Guidance on Design and Construction. Her Majesty's Stationery Office, London.

Department of Energy. (1977b). A description of the UK Offshore Steels Research Project. Ref. MA.P.015G(1)/20-2. London. April 1977.

Det Norske Veritas. (1982). Rules for the Design Construction and Inspection of Offshore Structures. 1977 Appendix C Steel Structures. Oslo 1982.

Fisher, P. J. (1982). Revised fatigue guidance and the certifying authority. Second International Conference on Offshore Welded Structures. The Welding Institute, Abington, Cambridge, U.K.

Gibstein, M. B. (1978). Parametric stress analysis of T joints. European Offshore Steels Research Seminar, The Welding Institute, Abington, Cambridge, U.K.

Gurney, T. R. and Hicks, J. G. (1973). On fatigue design rules for welded structures. Paper OTC 1907, Fifth Annual Offshore Technology Conference, Houston, Texas.

Gurney, T. R. (1982). The basis of the revised fatigue design rules in the Department of Energy Offshore Guidance Notes. Second International Conference on Offshore Welded Structures. The Welding Institute, Abington, Cambridge, U.K.

Hicks, J. G. (1974). A Study of material and structural problems in offshore installations. Welding in Offshore Constructions Conference Proceedings. The Welding Institute, Abington, Cambridge, U.K.

Institute de Recherches de la Siderurgie Francaise, Paris. (1981). Steel in Marine Structures. International Conference.

Kuang, A. B. et.al. (1977). Stress concentration in tubular joints - Soc. Pet. Eng. J.

Long, D. (1981). Description of the programme of work for phase II of UKOSRP. United Kingdom Atomic Energy Authority, Warrington, U.K.

Marshall, P. W. et.al. (1969). Materials Problems in Offshore Platforms. Paper No. OTC 1043. First Annual Offshore Technology Conference. Houston, Texas.

Marshall, P. W. (1974). General considerations for tubular joint design. Welding in Offshore Constructions Conference Proceedings. The Welding Institute, Abington, Cambridge, U.K. 1974.

Marshall, P. W. and Toprac, A. A. (1973). Basis for tubular joint design codes. ASCE National Structural Engineering Meeting, San Francisco, California.

The Welding Institute. (1978). European Offshore Steels Research. Seminar Proceedings. Abington, Cambridge, U.K.

Wardenier, J. (1981). Questionnaire on work in progress - Welded Tubular Joints. Doc. XV-489-81 International Institute of Welding. (Unpublished).

Wardenier, J. (1982). Questionnaire on work in progress. Doc. XV-516-82. International Institute of Welding. (Unpublished).

Wordsworth, A. C. and Smedley, G. P. (1978). Stress concentration of unstiffened tubular joints. European Offshore Steels Research Seminar, The Welding Institute, Abington, Cambridge, U.K. 1978.

## DISCUSSION

Question:

The author commented in his presentation that the worst HAZ toughness in microalloyed steels was found in those containing Nb and V together. The other authors T. Wada and I. Hoshi have been working with steels containing both these elements. Have they encountered any problems with HAZ toughness?

Submitted by: R. J. Pargeter

Response:

Our paper concerned grades X75 - X80. The steels were micro-alloyed Nb-V. For these grades we have to use these elements. However it is well known that an exaggerated addition of them can be injurious.

It is the reason why the Nb and V content has to be very well controlled and maintained to a weldable level.

Speaker: Y. Provou

Paper title: Steels for Pipes Over Grade X70 and Their Weldability

..........................................................................

Question:

The author commented in his presentation that the worst HAZ toughness in micro-alloyed steels was found in those containing Nb and V together. The other authors I. Hoshi and Y. Provou have been working with steels containing both these elements. Have they encountered any problems with HAZ toughness?

Submitted by: R. J. Pargeter

Response:

The normalized and aged Mn-Mo-V-Nb steel which I presented exhibits fairly good HAZ toughness as described in our previous paper (Reference 3 in the present paper). The HAZ toughness is generally influenced not only by the content of V plus Nb but also by the contents of other elements and the HAZ microstructure; the carbon content is one of the most significant factors (see Reference D-1 below).

D-1. J.M. Sawhill, Jr., T. Wada and D.E. Diesburg, "Heat-Affected Zone Toughness of Molybdenum Pipeline Steels" in "Welding of Line Pipe Steels", p. 86, Welding Research Council, (1977).

Speaker: T. Wada

Paper title: Development of Normalised and Aged Mn-Mo-V-Nb Steel for Fittings.

..........................................................................

Question:
Please explain why the vertical down welding produced higher mechanical properties than vertical up.
Submitted by: S. Skarborn

Response:
The mechanical properties of the welded joints depend on the analysis and the thermal cycles.

The thermal cycle is different depending on the vertical up or down position and the structure obtained in the vertical up is less ductile. However, in the vertical down, the penetration is better.
The vertical down technique has only been used for the root pass in field applications in order to obtain the best quality.
Speaker: Y. Provou
Paper title: Steels for Pipes Over Grade X70 and Their Weldability.

.........................................................................

Question:
In the assessment of field welding performance, what observations led to the conclusion that cellulosic coated electrodes could not be used for the filler passes? Were mechanical properties or radiographic quality the major limitations? What types of cellulosic consumbables were evaluated?
Submitted by: B. Rothwell

Response:
In our study, for the performance tests we used a gas pipe-line working at 100 bars and a very low temperature. For the charpy we heat in minus 65 Joules at -20°C. Under those conditions the tests carried out in the laboratories and the field with the low hydrogen electrodes (with a possibility to use the cellulosic one for the root pass only) gave much better results than the cellulosic.
The major limitations are the mechanical properties, both tensile and charpy tests.
We have used, Lincoln 70+, Boehler Foxcel 90, Thyssen cel 80 and Oerlikon PLS2.
Speaker: Y. Provou
Paper title: Steels for Pipes Over X70.

.........................................................................

Question:
1) You mentioned the use of low hydrogen electrodes for the vertical down welding of pipe. Is this being successfully done in the field or is it limited to lab work? Is the root also deposited in the same manner?
2) Where can we obtain such electrodes and what is their AWS classification?
Submitted by: A.A. Omar

Response:
The vertical down low hydrogen electrodes (Boehler BVD90, Philips 207, SAF) are being used in Europe, for the high strength grades (X65, X70), quite frequently.
The root pass can be realized with a cellulosic or a vertical up low hydrogen one, in order to get the best penetration. A good welder can also

realize a vertical down one (Philip's welders were able to do that).
AWS Classification: They are different for the grade chosen - For example: E80 18 9 for the grades X65/X70.

Speaker: Y. Provou
Paper title: Steels for Pipes Over Grade X70 and Their Weldability.

.................................................................

Question:
Regarding your work with tellurium treated steels, besides an observed increase in hot cracking susceptibility, did you see a porosity problem either with GMAW or SMAW?
Submitted by: N.L. Vaughn

Response:
No porosity was observed in our welding studies made on tellurium-treated steels within the 0.004/0.006 percent range.
Porosity could possibly be a problem at relatively higher tellurium levels.
Speaker: G.A. Ratz
Paper title: The Weldability of Rare-earth Treated Linepipe Steel.

.................................................................

Question:
You indicated that preheat is not required for the welding of these steels. What is the maximum wall thickness of pipe that can be welded without preheat?
Submitted by: A.A. Omar

Response:
Steel of maximum PCM value of 0.13% can be applicable up to wall thickness of 35mm for SMAW or MIG welding without preheat. The maximum PCM value of 0.15% can be welded without preheating up to wall thickness of about 30mm in SMAW.
Speaker: T. Hashimoto
Paper title: Development of Low PCM X70 Grade Line Pipe for Prevention of Cracking at Girth Welding.

.................................................................

Question:
Which is the cooling velocity of the plate in the DAC system and what are the differences between DAC I and DAC II.
Submitted by: M. Lafrance

Response:
DAC I is applied for interrupted cooling down to about 550°C after finishing rolling. The cooling rate can be varied between 8 and 15°C/sec. at plate of wall thickness 25mm. DAC II is applied for continuous cooling down to room temperature. The cooling rate can be varied between 10 to 25°C/sec. at 25mm in wall thickness.
Speaker: T. Hashimoto
Paper title: Development of Low PCM X70 Grade Line Pipe for Prevention of Cracking at Girth Weld.

.................................................................

Question:
What was the range of diameters of pipe tested with Cu backing? How would you use Cu backing on small dia. 12" or 300 mm when welding long sections? Did your work relate to only large dia. pipe?
Submitted by: L.E. Baxter

Response:
Steel pipe with 16 inch outer diameter only was used for testing because the oil pressure internal clamp which we have now is applied to one kind of diameter size. The copper backing band is wound around the outer periphery of the clamp and the clamp with copper backing is inserted into the pipe before circumferential welding.
Initially only medium diameter pipes will be welded for some time, not large diameter pipes.
Speaker: T. Fujimoto
Paper title: A New High Speed Circumferential Root Pass Welding for Pipeline Construction.

..........................................................................

Question:
1) Do temper bead procedures developed at Ontario Hydro give 100% refinement of the coarse grained HAZ of the first layer of beads?
2) Do you anticipate any problems in producing defect free repair welds using automated equipment, particularly at the ends of the repair cavity?
Submitted by: R.H. Leggatt

Response:
1) Considerable effort is made to provide effective procedural control in the actual field implementation of the temper-bead procedure which, when properly maintained, will result in efficient tempering and close to 100% refinement. Precautions include proper welder education, limits on weld cavity geometry, stringent preheat control, and strictly defined rules for welding technique (e.g. heat input, stringer beads, overlap, deposit sequence). Recently examined qualification coupons, prepared in-situ for a heavy water plant vessel repair, were found to contain a few small ($\simeq$ 0.5 mm x 0.5mm) isolated patches of incompletely refined microstructure, but the very conservative nature of the temper-bead procedure developed for heavy water plant service (230°C preheat) guaranteed an acceptable hardness level (i.e. VHN 250 for $H_2S$ service) even in these regions.
2) Work to date has concentrated on the development of computer software and interfacing electronics to achieve a remote controlled, automated welding capability. Naturally, problems in producing defect-free welds might be anticipated initially in the welding procedure development phase of the research program, particularly as pulsed-gas metal arc welding is being applied to the temper-bead technique for the first time in Ontario Hydro. If the possibility of weld stop/start defects at the ends of the cavity is of concern, it should be noted that this was dealt with previously in the heavy water plant tower repair, by depositing a series of vertical-up beads at either end of the otherwise horizontally-filled weld repair areas.
Speaker: M.J. Tinkler
Paper title: Automated Welding Development for Electrical Power Plant Maintenance and Construction.

..........................................................................

Question:
How well does the process handle misalignment of ovality of the pipe ends? What tolerances exist?
Submitted by: J. O'Beirne

Response:
The oil-pressure internal clamp is used to correct the misalignment. The groove is constrained right as the clamp-pieces which bridge both pipe ends shift radially out through the moving of a cone linked to a piston. Within a 1.5 mm high-low, sound beads are obtained.
Speaker: T. Fujimoto
Paper title: A New High Speed Circumferential Root Pass Welding for Pipeline Construction.

..........................................................................

Question:
The speaker has outlined the features of four "one-shot" processes, at least three of which are already at the prototype or early application stage for welding of offshore pipeline systems. Can he give any guidance as to the appropriate fields of application for these processes; despite their quite widely differing characteristics and performance in the fields of, for example, energy efficiency, all appear to be targetted at similar areas.
Submitted by: B. Rothwell

Response:
The three processes are in different stages of development with respect to pipe laying. The EB process is the most advanced and an industrial machine is available for use on pipes of 20" diameter. Modifications can be made to suit other sizes. It has been reported that this machine has already been used for riser pipe welding. EBW at present needs to be used in the HV position due to metal flow control requirements and is therefore suited to vertical pipe laying i.e. J configuration and risers. Friction Welding is being developed in both the conventional, (one side rotating) and radial modes. Our licences will have available a production machine for applying the radial process to 6" diameter feeder lines next year. This will be extended to large diameter pipes later.

The laser process is being adapted for pipe circumferential welds by at least one company but the stage of development is confidential at the present.

All these processes are energy efficient compared to arc welding, the main difference between them being the rate at which they apply energy. The friction processes must input all the required energy for given joint area in a single shot over say 30-60 secs whereas the EB and laser processes apply the available power over a large period depending on travel speed used. Thus a friction welder could be rated at 1000kW whereas the EB and laser processes would need 10-25kW depending on thickness, EB is of course much more efficient in energy conversion than the laser.
Speaker: J.D. Russell
Paper title: Potential Use of Non-Arc Welding Processes in Energy Related Fabrications.

..........................................................................

Question:
1) What was pipe wall thickness?
2) What microstructure resulted from cooling?
3) Was any post weld heat treatment used?
Submitted by: J.O'Beirne

Response:
1) Existing equipment and technology make it possible to weld pipelines of certain dimensions. Self propelled system "Styk" allows welding of pipeline joints of 1220 and 1420mm diameter with wall thickness from 10-24mm. System "Styk 2" allows the welding of pipeline joints of 325-1020mm diameter with a wall thickness from 10-30mm.
2) Structure of weld metal is a mixture of bainite, pro eutectoid-ferrite, and ferrite-pearlite. The HAZ has a structure of bainite and ferrite-pearlite. The hardness of the HAZ is 160-170 Hv. The hardness of the weld metal and base metal is 140-160H.
3) We do not perform a thermal treatment after welding. We do not perform thermal treatment on pipeline joints.
Speaker: V.N. Shlepakov
Paper title: 1) Position Butt Welding of Technological Pipelines of Power Engineering Structures.
2) Resources Saving Technology of Position Automatic Flux-Cored Wire Welding of Gas Pipelines with a Forced Weld Formation.

..............................................................................

Question:
1) Could you explain the differences in the behaviour of the three different steels? Is it due solely to Nb?
2) 5 minutes time for normalizing is too small for heat to flow to and normalize the center of the full thickness weld? I think the 1 hr./inch soaking time as stipulated and coded by ASME is for the purpose of homogenizing the temperature. Your results look very promising but in view of the above do you think that the soaking time of 5 minutes is practical?
Submitted by: R.S. Chandel

Response:
1) The impact energy transition curves for the three different A51b Gr 70 steels used are not the same as shown in the figures. This is because, even though the steels are produced to the ASTM specification, their production methods are different and they meet only the tensile requirements of the specification. Thus the toughness evaluation of the base material and the HAZ after normalizing is an important factor. Nb may have some effect on the properties, but there are other factors that also have the influence on the mechanical properties.
2) In our study the normalizing temperature was measured at the geometic centre of the weld. In industries, pressure vessels are heated in a furnace and when the surface temperature is the normalizing temperature, they were held at that temperature for up to 1 hour/25mm thickness. However, the heat transfer calculations show that the centre of the thickness would reach the normalizing temperature in less than ten minutes after the surface has reached this temperture. Again time required for the centre of the thickness to reach the normalizing temperature depends on the

heating rate, and it is a simple heat transfer calculation. Thus, 5 minutes of normalizing after the centre of the thickness has attained the normalizing temperature seems reasonable and practical.

Speaker: S.R. Bala

Paper title: Minimum Holding Time for Normalizing of Electroslag Weldments.

..........................................................................

Question:

Is the operator of your welding system of a welding, electronic or mechanical background?

Submitted by: W. Boicey

Response:

The welding machine has been in production for about 9 months, and as far as I know, our customer uses the same people who used to do the manual welding. So the specific answer to your question is that the operator has a welding background.

On a more general note, I'd like to take this opportunity to stress that the "digital image" approach, when combined with a "direct-entry" method for data input, eliminates the need for any operator programming as such. This makes the system more user-friendly, and would make it more readily acceptable in a production environment.

As implied in the discussion part of my paper, the "digital image" approach could be taken to a much greater level of sophistication than was possible on the Automatic Engine Welding Machine. The digital image could have a number of "layers" of information, each one pertaining to one aspect of the complete welding task. To be sure, the digital image would require a lot of memory, but the technology to handle images in layers does exist: It has been developed for CAD/CMA systems.

This opens the door for welding information to be conveyed to the system by literally "showing it what to do". The different layers of the digital imge could be built up (and edited) through a series of teaching runs. And it is likely that the best person to do this is the skilled welder himself, who knows about the job in hand.

Therefore, in general terms, the more sophisticated the control system is, the more important it is that the operator is a person who knows how the job should be done.

Speaker: C. Wilson

Paper title: Microprocessor Technology in Automatic Integrated Welding Systems.

..........................................................................

Question:

You mentioned the use of narrow gap TIG welding to fabricate the liner welds of the AGR penetrations. Can you describe how this is accomplished? Do you use a conventional TIG welding torch and simply increase the tungsten electrode stick out or is it a miniature welding head that is inserted into the narrow gap? If so, how easy do you start the arc and do you oscillate the arc and wire to insure side wall fusion?

Submitted by: A. A. Omar

Response:

A purpose built welding torch is utilized for this application. This is a Babcock-Hitachi water cooled torch with twin gas ports and an additional gas shield. Normal stick-out is employed. Arc initiation is by a high frequency voltage and presents no difficulties. Side wall fusion is very good and there is no necessity to oscillate either the electrode or the wire.

Speaker: M. Crowther

Paper title: Five Energy-Related Case Stories of Using Stainless Welding Consumables.

.........................................................................

Question:

Did your investigation include a study of the electrical and magnetic fields around the arc, and if so, what were your conclusions?

Submitted by: C. Wilson

Response:

We made no direct observations on possible electromagnetic effects around the arc. It is of course well know that electromagnetic forces can cause motion in the weld pool as pointed out in the discussion. However, at the relatively low currents ( 50A) we were using, we feel that electromagnetic effects have no major influence. The sensitivity of the heat sinks to contact with the tube surface also indicates that their effect is a thermal, rather than electromagnetic one.

Speaker: I. Grant

Paper title: Welding 304L Stainless Steel Tubing Having Variable Penetration Characteristics.

.........................................................................

Question:

1) You indicated the necessity of taking hardness reading at the fusion line within 1/2mm from it in the HAZ. While this is easily achieved on test specimens in the lab, how could it be accomplished on field or yard fabricated weld joints of the structure? Do we rely only on weld specimens taken from the weld procedure qualification test coupons or should we do field testing? If so, how could we do this without cutting out a weld joint?

2) What is the best portable field hardness tester you recommend? Please indicate name and source.

Submitted by: A.A. Omar

Response:

1) Field testing may be done in two ways:

A. Tacking a test coupon on to the end of production weldjoints, continuing production weld on to coupon using same weld procedure. Cutting and lab. testing (Obviously not possible for pipe joints.).

B. Grinding a small area of weld reinforcement and HAZ, polishing the flat area and etching to find fusion line clearly. Setting a number of indentations with portable hardness tester (see 2) closely spaced across HAZ into weld and base metal either side. Thus detecting peak hardness, which seems to correlate rather well with hardness measured 1mm below surface, in a transverse macro.

2) There are a number of portable hardness testers on the market, I do not know them all, but have heard that fabricators have good experience (correlation with lab. Vickers testing) with:

EQUOTIP, By Proceq. in Zurich

DHV 10, by Krautkramer

The first one is probably best for field work, the other being more sensitive to operator skill.

Speaker: S. Tandberg

Paper title: Offshore Structures for North Sea HAZ Hardness Requirements and Practical Implications.

........................................................................................

Question:

In the fracture mechanics analysis of repair welds, in how many cases has LEFM been applied? When LEFM is used what levels of residual stress are used in the analysis?

Submitted by: J.T. McGrath

Response:

To date there have been no discontinuities found in the region of a temper-bead repair weld that require acceptance by analysis. Therefore, LEFM has not been applied to repairs.

If an analysis were required and specific information regarding residual stresses for the geometry in question was not available, yield level tensile residual stresses would be used.

Speaker: J.W. Prince

Paper title: Integrity of Non PWHT Heavy Section Weld Repairs.

........................................................................................

Question:

1) You have indicated that welds fabricated with low hydrogen electrodes in the vertical down direction exhibited superior notch toughness compared with welds fabricated with the same electrodes in the vertical up direction. What do you attribute this to be caused by?

2) Is weaving practiced in the vertical down welding?

3) You have indicated that in order to produce good welds using low hydrogen electrodes in the vertical down direction, control of the weld joint is essential. Can you elaborate on this point? What is the optimum joint design for the vertical down welding?

4) Has the British Gas constructed any lines using low hydrogen electrodes in the vertical down welding direction?

Submitted by: A. A. Omar

Response:

1) We would not normally consider using the LHVD electrodes in the vertical up direction, and therefore, have not made the comparison referred to. However, we have found that girth welds produced using LHVD electrodes can have superior notch toughness to those produced using conventional vertical up low hydrogen electrodes. This behaviour was attributed to a higher proportion of refined microstructure in the vertical down welds which was a consequence of the necessity to use smaller bead sizes with this technique.

2) The answer to this question is Yes and No. Some LHVD electrodes can be used with a normal weave technique, while others must be used with a stringer bead technique in order to maintain slag control.

3) Although some electrode manufacturers recommend a 'feather edge' root face preparation, it is normally possible to use pipe with a standard API bevel ($30^{o}$ bevel angle, 1.5mm root face). However, because the LHVD electrode does not have the same arc 'punch' as the cellulosic electrode it

is not as tolerant to situations where there is a tight root gap. A root gap of 2.5 +1 -0mm must, therefore, be maintained when using the LHVD electrodes with API bevelled pipe.

4) British Gas has used LHVD electrodes for hot tap welding operations and for the repair welding of mechanised girth welds. We have not yet used LHVD electrodes for initial pipeline construction.

Speaker: K. Prosser

Paper title: The Properties of Pipeline Girth Welds Produced by Arc Welding Processes.

.....................................................................

Question:

Virtually all the design guidelines for fatigue which have been discussed are expressed in terms of S-N waves; since it is generally assumed that crack growth accounts for the entire fatigue life of welded structures, is there any role for fracture mechanics based crack-growth calculations in design, or is the calculation of $\Delta K$ values for complex details simply too difficult?

Submitted by: Brian Rothwell

Response:

The SN curve approach has the benefits of simplicity and the data can readily be incorporated in a computer data base for use in structural analysis. Levels of probability can be attributed to the data giving flexibility in usage for life assessment. The methods of fracture mechanics are much more specific and require detailed knowledge of the local stress field and stress history. Such methods are of value in predicting the residual life of cracked members and in failure analysis. Further developments may permit the use of fracture mechanics for design predictions but at present such an approach can be used only for the simplest details.

Speaker: J.G. Hicks/ R. Leggatt

Paper title: Fatigue Design of Welded Joints.

.....................................................................

Additional Contribution:

My contribution relates to the paper by Derek Russell on the Potential Use of Non-Arc Processes and concerns a friction welding machine designed for attaching studs underwater.

It ws built and commissioned by the Welding Institute, U.K., Ian Thomson Welding and Inspection Ltd., a welding services company in Inverness, Scotland and as consultant to them I was present at some recent sea trials when the equipment was tested down to 500 feet at Fort William, Scotland. I have a few slides to show you of this occasion.

In their regular welding work in the North Sea, Thomson's recognized the need for a reliable welding process for various applications in deep sea water and particularly for the retrofitting of sacrificial anodes and their electrical connectors to pipelines or jacket legs etc.

The first slide shows a typical anode mounted on a length of pipe with its connector tail welded on by a friction welded stud. The macrograph shows the quality of weld produced at 500 feet. It is extremely strong and tough being a fine-grained forge weld and in 50D grade materials maximum hardness values around 350 $HV_{10}$ are usual. This can be controlled by adjustment of the welding parameters. In the absence of any fused zone, common weld metal defects such as cracks and porosity do not occur and reliably consistent results are achieved.

The friction welding bead is mounted firmly onto the work by a diver or remote vehicle and includes a hydraulic motor to provide the necessary stud rotation. This is fed via hoses from the sled mounted power source which comprises an oil reservoir primary hydraulic pump, accumulator bottles, pressure compensation and control gear.

The power source is powered from the surface by a 415 volt, 3 phase umbilical but triggering of the weld cycle is done locally by a diver. A skilled welder is not required as the cycle is preprogrammed and diver handling and familiarisation work has just been completed by Subsea Offshore Ltd., Aberdeen, who will operate the equipment in its various applications. The machine welds up to 20mm diameter steel studs and is designed to operate at depths to 1000 feet. It commences work offshore this week.

Submitted by: K.P. Bentley

Speaker: J.D. Russell
Paper title: Potential Use of Non-Arc Processes in Energy Related Fabrications.

# AUTHOR INDEX

# KEYWORD INDEX